Battery Materials Characterization

Advanced Techniques and Applications

Ning Li
Liang Zhang

WILEY VCH

Authors

Ning Li
Beijing Institute of Technology, and Chongqing Innovation Center Beijing Institute of Technology
China

Liang Zhang
Institute of Functional Nano & Soft Materials (FUNSOM) of Soochow University
China

Cover Design: Wiley
Cover Image: © Ning Li

Typesetting: Lumina Datamatics, Madurai, India

Library of Congress Card No.: applied for

British Library Cataloguing-in-Publication Data A catalogue record for this book is available from the British Library.

Bibliographic information published by the Deutsche Nationalbibliothek The Deutsche Nationalbibliothek lists this publication in the Deutsche Nationalbibliografie; detailed bibliographic data are available on the Internet at http://dnb.d-nb.de.

Print ISBN: 9783527355785
ePDF ISBN: 9783527854349
ePub ISBN: 9783527854332
oBook ISBN: 9783527854356

Boschstraße 12, 69469 Weinheim, Germany

Printed and bound by CPI Group (UK) Ltd, Croydon, CR0 4YY

C9783527355785_130726

Contents

About the Author

Ning Li is a Researcher and Tenured Associate Professor at Beijing Institute of Technology and Chongqing Innovation Center, Beijing Institute of Technology, China.

Liang Zhang is a Full Professor at the Institute of Functional Nano & Soft Materials (FUNSOM) of Soochow University, China.

Preface

The global transition toward sustainable and low-carbon energy systems has significantly accelerated the development of renewable energy technologies, including wind power, solar energy, and electrified transportation. In this context, electrochemical energy storage systems—particularly rechargeable batteries—have become indispensable for enabling large-scale energy storage, grid stabilization, and the widespread adoption of electric vehicles. Driven by these societal and technological demands, battery research has experienced unprecedented growth worldwide, with continuous innovations in materials design, cell architecture, and system integration.

However, the pursuit of higher energy density, longer cycle life, faster charging capability, and improved safety has led to battery materials with increasingly complex physicochemical properties. Modern battery materials and interfaces often exhibit complex phase behavior, intricate electronic and chemical states, and pronounced structural and compositional evolution across multiple length and time scales under realistic operating conditions. Despite remarkable progress in material synthesis and performance optimization, many fundamental questions related to degradation mechanisms, interfacial reactions, and dynamic structural evolution remain unresolved. These challenges highlight a growing gap between rapid materials development and a comprehensive understanding of their behavior during battery operation. Addressing these challenges requires not only advanced characterization techniques, but also a systematic and collaborative effort to integrate diverse methodologies and expertise across the battery research community.

This book is organized according to the physical and chemical dimensions probed by advanced characterization techniques, rather than by individual materials or isolated methods. Following an introductory chapter that outlines the fundamentals of battery materials and the evolution of characterization strategies, the subsequent chapters are broadly grouped into several categories, including structural characterization across multiple scales, advanced imaging and multiphysics visualization, electronic structure and chemical state analysis, and interfacial reactions and gas evolution during battery operation.

Each chapter provides a comprehensive discussion of the underlying physical principles, experimental instrumentation, and representative measurement modes, with particular emphasis on in situ and operando configurations. More importantly, the characterization techniques are presented in close connection with practical battery research scenarios. Typical applications of each characterization are discussed for different battery materials and systems, illustrating how specific techniques can be selected to address key scientific and engineering questions related to structure evolution, electronic states, interfacial chemistry, and performance degradation.

This book is edited by Prof. Ning Li of Beijing Institute of Technology from Chapter 1 to Chapter 5, and from Chapter 9 to Chapter 12, and Prof. Liang Zhang of Soochow University from Chapter 6 to Chapter 8, who are responsible for the overall conceptual design of the book, the organization of the chapter framework, the coordination of content development, and the academic quality control. During the preparation of this volume, graduate students from the editors' research groups made substantial contributions by assisting with literature surveys, data compilation, and manuscript proofreading. More specifically, the editors would like to acknowledge the great contribution of Yongzhi Shi and Mengyao Yan in Chapter 1, Yongjian Li in Chapter 2, Yihong Wang and Xinyu Zhu in Chapter 3, Siyuan Ma and Wenlong Kang in Chapter 4, Wengang Yan and Yizhi Zhai in Chapter 5, Qianjie Niu in Chapter 6, Zheng Zhou in Chapter 7, Lei Wang in Chapter 8, Tao Liu and Feifei Hong in Chapter 9, Wengang Yan and Yu Dong in Chapter 10, Jizhuang He and Jie Liu in Chapter 11, and Chenxing Yang and Zian Wang in Chapter 12.

The editors would also like to express their sincere gratitude to the editors and reviewers at Wiley for their valuable comments and constructive suggestions, which greatly improved the quality of this book. In the preparation of this volume, numerous books, journal articles, and illustrative materials published by researchers in the field were consulted and cited, and the editors gratefully acknowledge these contributions. We also acknowledge the public technical resources and documentation provided by major synchrotron radiation facilities, including APS, ESRF, PETRA III, Diamond Light Source, and SPring-8. More importantly, we sincerely thank BSRF, HEPS, SSRF, and so on, in China, for providing valuable reference materials and sharing experimental experience.

1

Fundamentals of Battery Materials and the Evolution of Advanced Characterization Techniques

1.1 Global Energy Transition

1.1.1 Core Demand for Energy Transformation

Guided by the china's dual-carbon goals of "carbon neutrality by 2050 and net-zero emissions by 2060," the proportion of renewable energy sources such as wind energy, solar energy, and tidal energy in the global energy market has been continuously increasing [1]. Promoting the transformation of the energy system from "fossil energy-dominated" to "renewable energy-centered" has long become a universal consensus of all mankind. However, new renewable energy sources generally have inherent defects of "intermittency, volatility, and regionality," which greatly restrict the progress of energy transformation. For instance, the power generation intensity of solar energy fluctuates by up to 80% during the day, while the output of wind energy drops by more than 50% at night. Therefore, developing supporting high efficiency energy storage technologies to achieve the dynamic balance of "development-storage-utilization" of new energy will become a key driving force for the implementation of new energy reform. According to the prediction of the International Energy Agency, the global energy storage demand will exceed 15 TWh by 2030. Among various energy storage technologies, battery energy storage has become the core solution for grid peak regulation and distributed energy storage scenarios, owing to its millisecond-level fast response speed and high energy density exceeding 100 Wh/kg [2].

1.1.2 Performance Driven by Terminal Application Scenarios

In terminal application fields, popular sectors such as electric vehicles (EVs), portable consumer electronics (including smartphones, drones, etc.), and large-scale energy storage are imposing increasingly higher requirements on the technological advancement of battery energy storage [3, 4]. For example, in the transportation field, electric vehicles are developing toward the goals of "ultra-fast charging (charging to 80% capacity in 10 minutes), long driving range (exceeding 1000 km), and high safety (no thermal runaway)," while also requiring the battery energy density to increase from the current 300 Wh/kg to more than 500 Wh/kg. For 3C consumer electronics, the pursuit of "slimmer and lighter design, long standby time (exceeding 72 hours), and fast charging (full charge in 30 minutes)" demands the battery's volumetric energy density to exceed 700 Wh/L. In the energy storage field, large-scale energy storage power stations require batteries to have a cycle life of over 10,000 cycles and a cost lower than 0.5 yuan/Wh. In addition, a few special fields such as polar scientific research and high-altitude power grids have put forward more stringent requirements, i.e., batteries must maintain a capacity retention rate of over 70% at −40 °C. These high standards and requirements for the performance of energy storage batteries in diverse application scenarios are driving battery technology to continuously make breakthroughs toward more diverse systems and better performance.

1.1.3 Industrialization and Scientific Research Pattern of Battery Technology

At present, the global battery industry has formed a development pattern of "lithium-ion batteries as the dominant type, with multiple new battery systems under parallel R&D" [5, 6]. In recent years, the global annual production capacity of lithium-ion batteries has exceeded 2 TWh, accounting for more than 85% of the energy storage market share, and has become the most mature commercialized energy storage technology. Meanwhile, other new battery systems are also continuously achieving technological breakthroughs and accelerating their commercialization process. For example, sodium-ion batteries have entered the pilot-scale production stage, and mass production of 1 GWh level is expected to be realized by 2025; all-solid-state batteries have achieved breakthrough results at the laboratory level and are gradually transitioning to the pilot-scale stage, with enterprises such as Toyota and QuantumScape planning to commercialize solid-state batteries by 2030. In the global scientific research field, the annual number of published Science Citation Index Expanded papers related to batteries has exceeded 80,000 in recent years, and the annual growth rate of patent applications has reached 15%, indicating an extremely high enthusiasm for R&D in the industry. However, it should be noted that there is a significant gap between the laboratory performance of many battery technologies and their actual industrial applications. For example, the cycle life of silicon-based anodes can exceed 1000 cycles under laboratory conditions, but after industrialization, their cycle life is far less than half of the laboratory level. The core reason lies in the insufficient in-depth understanding of the dynamic failure mechanisms of materials under actual working conditions. Therefore, it is urgent to use advanced characterization technologies to build a bridge between "scientific research achievements and industrial applications" and promote the rapid implementation of technologies.

The diagram depicted in Figure 1.1 includes the dominant position of lithium-ion batteries and R&D progress of new battery technologies. It illustrates the global battery industry pattern: lithium-ion batteries (annual capacity over 2 TWh, >85% market share) dominate, while sodium-ion and all-solid-state batteries advance toward commercialization, reflecting industry-science synergy.

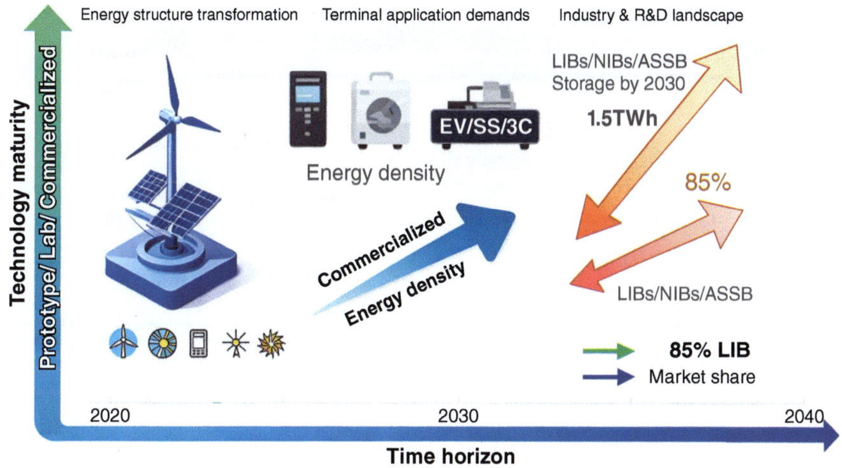

Figure 1.1 Schematic diagram of the global battery technology industry and scientific research pattern.

1.2 Basic Principles and Core Composition of Lithium-Ion Batteries

1.2.1 Working Principle of Lithium-Ion Batteries

In essence, lithium-ion batteries realize the storage and release of electrical energy through the reversible intercalation, deintercalation, and migration of lithium ions between the cathode and anodes. Precisely due to this "back-and-forth shuttling" characteristic of lithium ions, they are vividly called "rocking chair batteries" [7]. In Figure 1.2, a schematic diagram of the "rocking chair"-type working principle of lithium-ion batteries is shown. Taking $LiCoO_2$ as the cathode and graphite as the anode for specific illustration, its core electrochemical process is as follows: during charging, driven by an external electric field, the active material of the cathode undergoes an oxidation reaction, and lithium ions are deintercalated from the lattice of the cathode, then migrate to the anode through the electrolyte and separator. At the same time, the anode undergoes a reduction reaction, where lithium ions are intercalated between the layers of graphite to form the intercalation compound LiC_6, and electrons flow from the cathode to the anode through the external circuit.

Cathode oxidation reaction: $LiCoO_2 \rightarrow Li_{1-x}CoO_2 + xLi^+ + xe^-$
Anode reduction reaction: $xLi^+ + xe^- + 6C \rightarrow LiC_6$

The discharge process is completely opposite to charging. Lithium ions are deintercalated from the interlayers of graphite in the anode, migrate back to the cathode through the electrolyte, and re-intercalate into the lattice. Electrons also flow to the cathode through the external circuit, thereby realizing the discharge to external electrical devices. In the entire charge-discharge cycle system, the structure of the active materials in the cathode and anodes must have good reversibility

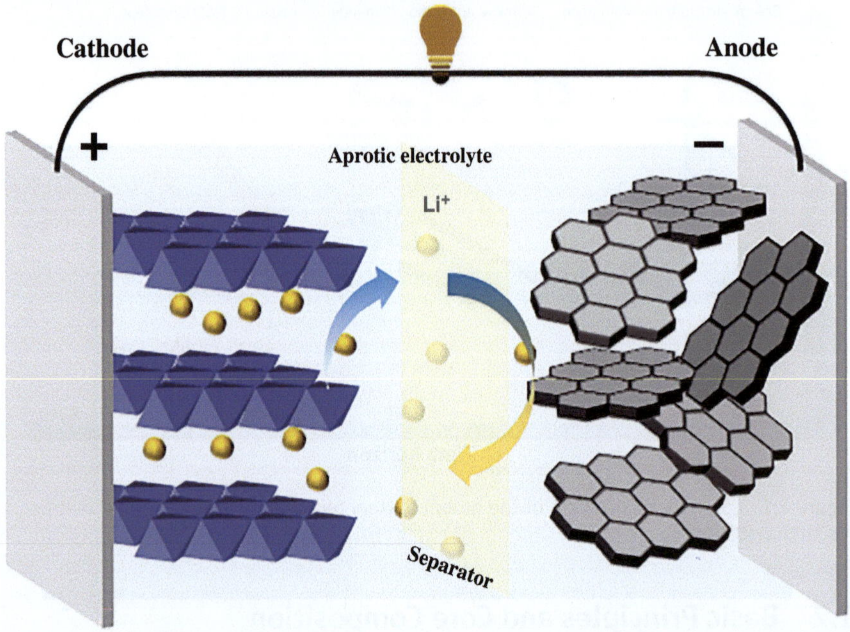

Figure 1.2 Schematic diagram of the "rocking chair"-type working principle of lithium-ion batteries.

to ensure the cycle stability of the battery; the electrolyte needs to have a high ionic conductivity of no less than 10^{-2} S/cm to ensure the rapid response of lithium ions during charge and discharge; the separator must block the passage of electrons while allowing the smooth migration of lithium ions. These components work together to ensure the efficient and stable cyclic operation of the battery.

1.2.2 Core Composition of Lithium-Ion Batteries

Figure 1.3 indicates a schematic diagram of the functional positioning of the four core components of lithium-Ion batteries (marking the core functions of the cathode, anode, electrolyte, and separator respectively) [7]. It defines each component's role: cathode (energy density determinant), anode (Li^+ storage), electrolyte (Li^+ transport), separator (electron barrier), highlighting their synergistic support for battery stability.

1. Cathode
   The cathode of a battery is mainly composed of active material, conductive agent, binder, and aluminum foil current collector. Among them, the active material accounts for 80–90% of the total mass of the cathode and is the core component that directly determines the operating voltage and capacity of the battery. At present, mainstream cathode materials are mainly divided into three categories: Layered oxides: Common types include $LiCoO_2$ and $LiNi_xCo_\gamma Mn_{1-x-\gamma}O_2$ (abbreviated as NCM). $LiCoO_2$ has an operating voltage of 3.9 V

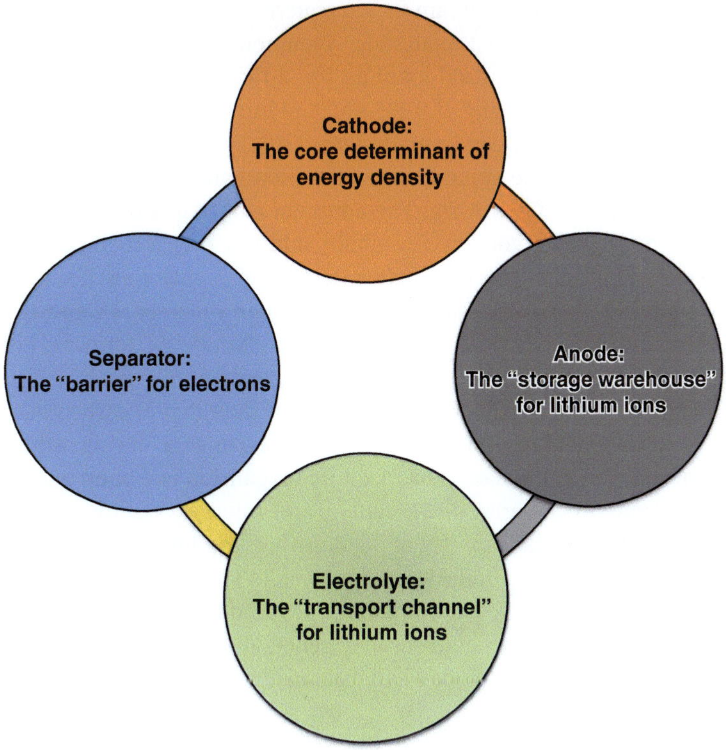

Figure 1.3 Schematic diagram of the functional positioning of the four core components of lithium-Ion batteries.

and a theoretical capacity of up to 274 mAh/g. For the NCM series, the higher the Ni content, the higher the capacity—for example, the actual capacity of NCM811 can reach 200 mAh/g. The core advantage of this type of material is its high energy density, but it is prone to oxygen release and phase transition under high-voltage conditions, which affects cycle stability. Polyanionic materials: A typical representative is $LiFePO_4$, which has an operating voltage of 3.4 V and a theoretical capacity of 170 mAh/g. Its greatest advantages are stable crystal structure and low preparation cost, making it suitable for scenarios sensitive to safety and cost, but its obvious shortcoming is relatively low energy density. Spinel-type materials: An example is $LiMn_2O_4$, which has an operating voltage of 4.1 V and a theoretical capacity of 148 mAh/g. Benefiting from its unique three-dimensional ion channel structure, it exhibits excellent rate performance, but Mn dissolution occurs under high-temperature conditions, which directly leads to rapid capacity fading of the battery.

2. Anode
The anode of a battery is composed of active material, conductive agent, binder, and copper foil current collector. Its core function is to realize the reversible storage of lithium ions, which is the key to the repeated charge and discharge

of the battery. Currently, mainstream anode materials are mainly divided into three categories: Carbon-based materials: Common types include natural graphite and hard carbon. Natural graphite has a theoretical capacity of 372 mAh/g and an interlayer spacing of 0.335 nm, and lithium ions realize lithium storage by intercalating between layers to form $LiC_{6.}$ Hard carbon has a disordered structure and is more suitable for sodium-ion batteries, with a sodium storage capacity of 300–350 mAh/g. The common advantages of this type of carbon-based material are good cycle stability and low preparation cost, and they are widely used in commercial batteries. Alloy-type materials: Examples include silicon and tin. Silicon has an extremely high theoretical capacity of 4200 mAh/g and achieves high-capacity lithium storage by forming Li-Si alloys with lithium. Tin also has a theoretical capacity of 994 mAh/g. Their core advantage is that their capacity is much higher than that of carbon-based materials, but they undergo significant volume expansion during lithium intercalation—silicon, for instance, has a volume expansion rate of up to 300%, which easily leads to material pulverization and interface failure. Metallic lithium: It has a theoretical capacity of 3860 mAh/g and high energy density potential, making it an ideal anode choice for all-solid-state batteries. However, in traditional liquid electrolytes, lithium dendrites tend to grow on the surface of metallic lithium, and these dendrites may penetrate the electrolyte to cause battery short circuits, posing serious safety hazards.

3. Electrolyte
The electrolyte is the transport medium for lithium ions inside the battery. To ensure the normal operation of the battery, it must meet three core requirements: "high ionic conductivity, wide electrochemical window, and high chemical stability." Its composition usually includes lithium salt, organic solvent, and various functional additives. Lithium salt: The currently mainstream lithium salt is $LiPF_6$ with a concentration of 1 mol/L. It has a high ionic dissociation degree and can provide sufficient lithium ions for the electrolyte, but its disadvantage is that it is prone to hydrolysis to generate hydrogen fluoride (HF), which corrodes electrode materials. Auxiliary lithium salts such as LiFSI ($LiN(SO_2F)_2$) have good hydrolysis resistance and high-temperature stability, and are often used in high-voltage battery systems to improve reliability. Organic solvent: Mainly carbonate-based, including ethylene carbonate (EC), dimethyl carbonate (DMC), and ethyl methyl carbonate (EMC). EC has a high dielectric constant and can effectively promote lithium salt dissociation; DMC has low viscosity and can improve the migration rate of lithium ions; EMC can balance dielectric constant and viscosity. These three solvents are usually mixed in a volume ratio of 1:1:1 to balance dissociation efficiency and ion transport speed. Additives: Although the addition amount is only 1–5%, they play a crucial role. For example, vinylene carbonate (VC) can help form a stable solid electrolyte interphase (SEI) layer on the electrode surface; $LiPO_2F_2$ can improve the stability of the battery under high voltage; and phosphate-based additives, as flame retardants, can significantly reduce the risk of electrolyte combustion and explosion.

4. Separator
The separator is a porous insulating material in the battery. Its core functions are, on the one hand, to physically separate the cathode and anode to prevent electronic short circuits caused by direct contact between them, and on the other hand, to allow the smooth passage of lithium ions to complete the charge-discharge cycle. Therefore, it needs to meet the key characteristics of "high porosity (30–50%), high mechanical strength, and resistance to electrolyte corrosion." Currently, mainstream separators are mainly divided into three categories: Polyolefin separators: Common types include polypropylene (PP) membranes, polyethylene (PE) membranes, or PP/PE/PP composite membranes. Their advantages are low production cost and good air permeability, and they are widely used in commercial batteries. However, their disadvantage is poor high-temperature stability and easy shrinkage—PE has a melting point of 135 °C and PP has a melting point of 165 °C, and the shrinkage risk increases significantly when the temperature exceeds the corresponding melting point. Ceramic-coated separators: They are made by coating ceramic particles such as Al_2O_3 and SiO_2 on the surface of polyolefin base membranes. They can greatly improve high-temperature anti-shrinkage performance, with a shrinkage rate of less than 5% at 150 °C, making them particularly suitable for battery products with high safety requirements. Solid electrolyte separators: Examples include lithium lanthanum zirconium oxide ceramic sheets and lithium germanium phosphorus sulfide sheets. They have both the ion-conducting function of electrolytes and the isolation function of separators, and are indispensable core components in all-solid-state batteries.

1.3 Development Trend and Core Characteristics of Secondary Batteries

1.3.1 Lithium-Ion Batteries

Since lithium-ion batteries were commercialized by Sony in 1991, as shown in Figure 1.4, they have developed a mature technical route of "layered oxide or polyanionic cathode + carbon-based or alloy-type anode + liquid electrolyte" and are widely used in various electronic devices and new energy vehicles. At present, the industry's R&D mainly focuses on three core directions: High energy density improvement: The core combination is "high-nickel ternary cathode (e.g., NCM811, NCM955) + silicon-based anode." By gradually increasing the Ni content in the cathode from 60% to 95%, the actual capacity of the cathode is increased from 160 mAh/g to approximately 220 mAh/g. For the silicon-based anode, nanostructuring (e.g., preparing Si nanoparticles) or compounding with carbon materials (e.g., constructing Si/C core-shell structures) is used to alleviate the volume expansion of up to 300% during lithium intercalation. High safety optimization: The focus is on developing cobalt-free cathodes (e.g., $LiFePO_4$/C composite cathodes, $LiNi_{0.9}Mn_{0.1}O_2$), flame-retardant electrolytes (added with

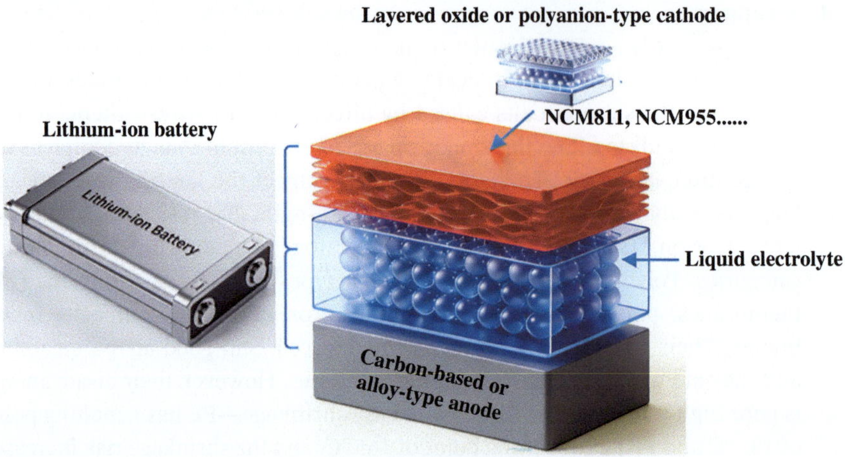

Figure 1.4 Schematic diagram of lithium-ion battery.

phosphate or fluorinated solvents), and separators with ceramic coatings on the surface, so as to reduce the risk of battery thermal runaway. Fast charging performance breakthrough: Single-crystal high-nickel cathodes are developed to reduce grain boundary impedance, and graphite/hard carbon composite anodes are designed to improve the migration rate of lithium ions, ultimately achieving a 10C fast charging effect (i.e., charging the battery to 80% capacity in 10 minutes). However, the lithium-ion battery system still faces many bottlenecks at present. For example, high-nickel cathodes are prone to lattice oxygen release during deep deintercalation, silicon-based anodes have interface failure problems, and electrolytes are prone to oxidative decomposition under high-voltage scenarios. All these factors restrict the further performance upgrade of lithium-ion batteries.

1.3.2 Sodium-Ion Batteries

Sodium-ion batteries have become a core candidate technology for large-scale energy storage scenarios due to their significant advantages: the abundance of sodium in the Earth's crust is as high as 2.36%, while that of lithium is only 0.002%, showing a prominent resource advantage; at the same time, the price of sodium salts is only 1/20 of that of lithium salts, with a distinct cost advantage; and they can maintain a capacity retention rate of over 80% at −20 °C, whose low-temperature adaptability is far superior to that of many battery systems. At present, the technical characteristics of sodium-ion batteries have gradually become clear, as shown in Figure 1.5: Cathode materials: Layered oxides (e.g., $NaNi_{0.3}Co_{0.3}Mn_{0.4}O_2$, with an actual capacity of 150–180 mAh/g) and polyanionic materials (e.g., $Na_3V_2(PO_4)_3$, with a cycle life of over 10,000 cycles) are the current mainstream choices. However, they generally face the problem of low Na^+ diffusion rate (the diffusion coefficient is only 10^{-11} cm^2/s), which restricts rate performance. Anode materials: Mainly hard carbon, which achieves a sodium storage capacity of 300–350 mAh/g through the dual mechanism

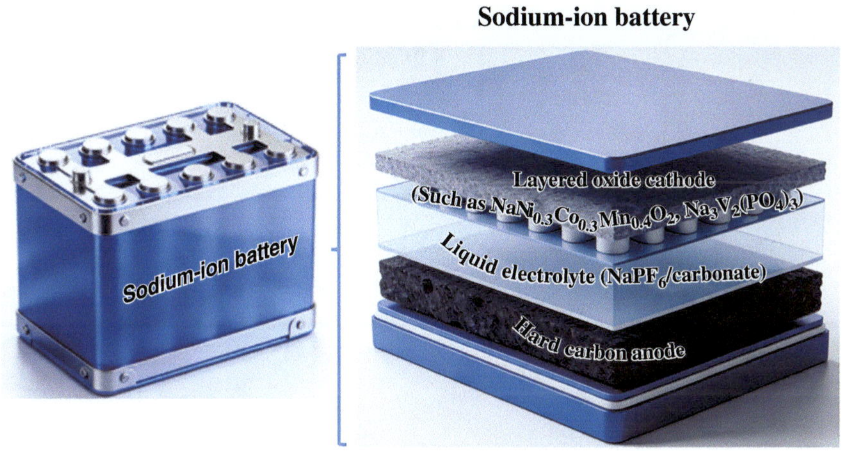

Figure 1.5 Schematic diagram of sodium-ion batteries.

of "interlayer intercalation-micropore filling." However, its first-cycle Coulombic efficiency is less than 85%, which usually needs to be improved through surface modification (e.g., coating with amorphous carbon). Electrolytes: Mainly adopt the $NaPF_6$/carbonate system, which has obvious defects. On the one hand, the solubility of $NaPF_6$ in carbonate solvents is only 60% of that of $LiPF_6$; on the other hand, the interface stability between this system and the electrode is poor, which affects the battery cycle life. At present, the energy density of sodium-ion batteries is approximately 120–160 Wh/kg, only half of that of lithium-ion batteries. To achieve wider applications, it is necessary to break through the two core bottlenecks of slow electrode kinetics and poor interface compatibility [5].

1.3.3 All-Solid-State Batteries

All-solid-state batteries replace traditional liquid electrolytes with solid electrolytes, which can not only completely solve the safety hazards caused by lithium dendrite growth but also eliminate the risk of electrolyte combustion and explosion. Their theoretical energy density can reach 600–800 Wh/kg, far exceeding the level of existing lithium-ion batteries [8, 9]. As shown in Figure 1.6, a schematic diagram of the all-solid-state battery structure (including the cathode, anode, solid-state electrolyte, and MIEC/HEO material composition) is depicted. It presents the structure: solid electrolytes replace liquid ones (solving Li dendrite/flammability issues), with MIEC/HEO optimizing electrode-electrolyte compatibility, enabling high energy density and safety.

At present, the industry has mainly formed three technical routes: Oxide system: Represented by LLZO ($Li_7La_3Zr_2O_{12}$), it has an ionic conductivity of 10^{-4}–10^{-3} S/cm at room temperature and excellent chemical stability. However, this type of material has high brittleness (with a fracture toughness of only 1–2 $MPa \cdot m^{1/2}$) and is prone to poor interface contact with the electrode [9]. Sulfide system: Typified by LGPS

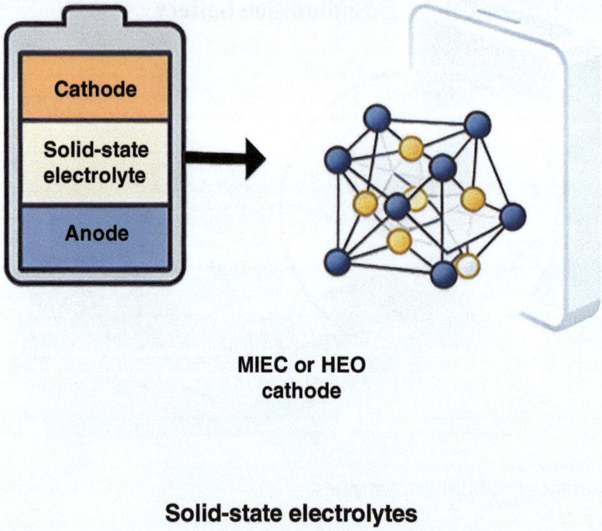

Figure 1.6 Schematic diagram of the all-solid-state battery structure.

($Li_{10}GeP_2S_{12}$), it has better room-temperature ionic conductivity (10^{-3}–10^{-2} S/cm) and good mechanical flexibility, which is convenient for processing and forming. However, it is prone to hydrolysis with moisture in the air (generating H_2S) and reacts violently with metallic lithium (generating Li_3P), so its stability needs to be improved [9]. Polymer system: Such as the PEO-LiTFSI system, which has the characteristics of bendability and stretchability and is suitable for flexible battery scenarios. However, its low-temperature performance is poor—the ionic conductivity drops to less than 10^{-6} S/cm below 0 °C, and its oxidation stability is poor (decomposing at voltages above 4.0 V) [10]. Overall, the core bottlenecks of all-solid-state batteries are concentrated on three aspects: "high solid-solid interface impedance, severe interface reactions, and poor mechanical compatibility." To achieve technological breakthroughs, it is necessary to make simultaneous efforts and conduct collaborative optimization from two aspects: interface regulation and electrolyte design [8].

1.3.4 Other New Battery Systems

Lithium-sulfur batteries: Their theoretical energy density is as high as 2600 Wh/kg, which is achieved through the multielectron transfer reaction of "$S_8 \rightarrow Li_2S$." However, they have obvious shortcomings: sulfur itself has extremely low conductivity (only 10^{-30} S/cm), and polysulfide shuttling occurs during the discharge process, which directly leads to rapid capacity fading of the battery. Zinc-ion batteries: They use aqueous electrolytes, which are safe without combustion and explosion risks, and their cost is only one-third of that of lithium-ion batteries, making them particularly suitable for scenarios such as low-speed electric vehicles and backup power supplies. However, they also have prominent problems: the cycle stability of cathode materials (e.g., MnO_2) is poor (cycle life less than 500 cycles), and zinc dendrites tend to grow on the surface of the zinc anode, affecting battery life and safety. Potassium-ion batteries: They stand out due to the advantage of potassium resources—the abundance of potassium in the Earth's crust is 2.09%, and it can be directly compatible with graphite anodes for potassium intercalation, without the need to develop new anode materials specifically. However, due to the large radius of K^+ (1.38 Å), the structure of the cathode material is prone to collapse during intercalation and deintercalation. At present, their energy density can only reach 80–120 Wh/kg, which is difficult to meet high-capacity requirements. These emerging battery systems have shown unique potential in specific application scenarios, but to achieve large-scale commercialization, they all need to break through the two core bottlenecks of insufficient intrinsic material performance and difficult interface regulation [11].

1.4 Research Trend and Core Bottlenecks of Key Battery Components

1.4.1 Electrode Materials

1.4.1.1 Cathode Materials

High-nickel ternary cathode materials of lithium-ion batteries (e.g., NCM811) face a series of performance degradation problems during deep deintercalation: Ni^{3+}/Ni^{4+} undergoes strong hybridization with O 2p orbitals, which triggers lattice oxygen release—especially at voltages above 4.5 V, the amount of O_2 evolution increases sharply; at the same time, the H1−3→O1 phase transition occurs, leading to a 5–7% shrinkage of the material's lattice c-axis. The internal stress of the particles accumulates continuously, eventually generating microcracks. Observations via FIB-SEM show that these cracks will penetrate the entire particle after 100 cycles. Although lithium-rich manganese-based cathodes (e.g., $xLi_2MnO_3 \cdot (1-x)NCM$) can achieve a high capacity of 250–300 mAh/g, the Jahn-Teller effect caused by Mn^{3+} will lead to significant battery voltage fading (dropping from the initial 4.2 V to approximately 3.8 V), which affects the actual user experience. The cathode materials of sodium-ion batteries also have their own shortcomings: due to the large radius of Na^+ (1.02 Å), the diffusion coefficient of layered oxide $NaNi_{0.3}Co_{0.3}Mn_{0.4}O_2$ in the lattice is only 10^{-11} cm^2/s, resulting in poor rate performance (the capacity at 1C rate is only 70% of that at 0.1C rate); polyanionic $Na_3V_2(PO_4)_3$ has poor conductivity (only 10^{-6} S/cm), and carbon coating treatment is usually required to improve electron transport efficiency. At present, research hotspots in related fields focus on targeted performance optimization. For example, designing a "core-shell-gradient" structure for high-nickel cathodes—using high Ni content in the core to ensure capacity and high Mn/Al content in the outer layer to improve structural stability; adopting Ti/Mg doping for high-lithium cathodes to inhibit Mn migration and alleviate voltage fading. However, most of these studies are limited to macro-performance improvement, and there is a lack of clear and in-depth understanding of atomic-level details such as the specific occupancy of doping elements in the lattice and the lattice matching degree between the coating layer and the matrix material [7].

1.4.1.2 Anode Materials

Although silicon-based anodes of lithium-ion batteries have outstanding theoretical capacity, they undergo a volume expansion of up to 300% during lithium intercalation. This severe volume change will lead to the pulverization of anode particles and repeated rupture and regeneration of the surface SEI layer, eventually resulting in a capacity retention rate of less than 50% after 50 cycles. Lithium metal anodes also face tricky problems: lithium dendrites tend to grow in liquid electrolytes (in-situ optical microscopy observations show that the length of these dendrites can reach 5–10 μm), and stripping pores appear in solid-state batteries (in-situ transmission electron microscopy (TEM) detection shows that the volume ratio of these pores exceeds 20%), which seriously affects battery safety and cycle stability. Hard carbon, a commonly used anode material for sodium-ion batteries, has a sodium storage capacity of 300–350 mAh/g, but its first-cycle Coulombic efficiency is less than 85%. Moreover, there are two controversial views in the industry regarding its sodium storage mechanism: "intercalation-adsorption" and "filling-adsorption," and no unified understanding has been formed yet. Current research hotspots in related fields focus on targeted modification and optimization. For example, nanostructuring (e.g., Si nanowires) or compounding with carbon materials (e.g., Si/graphene composite systems) is used for silicon-based anodes; and artificial SEI layers (e.g., Li_3N, LiF) are constructed for lithium metal anodes to improve performance. However, most of these studies focus on macro-effect improvement, and there is a lack of in-depth and systematic understanding of the dynamic correlation between "volume expansion—interface evolution—electrochemical performance" [12].

1.4.2 Electrolyte

1.4.2.1 Liquid Electrolyte

The $LiPF_6$/carbonate electrolyte system commonly used in lithium-ion batteries (mostly a mixed system of EC, EMC, and DMC) will undergo obvious oxidative decomposition when the charging voltage exceeds 4.5 V: the C-O bond in EC molecules is prone to cleavage (density functional theory (DFT) calculations show that its oxidation barrier is only 2.8 eV), thereby generating by-products such as CO_2 and Li_2CO_3. DEMS detection shows that the CO_2 gas production rate increases by 10 times at voltages above 4.5 V. At the same time, $LiPF_6$ in the electrolyte is prone to hydrolysis to generate HF—when the water content in the system exceeds 50 ppm, the HF concentration will exceed 10 ppm, which corrodes the cathode material and current collector. The electrolytes of sodium-ion batteries face another type of problem: the solubility of commonly used $NaPF_6$ in carbonate solvents is only 60% of that of $LiPF_6$, and there is a strong solvation interaction between Na^+ and solvent molecules (the solvation energy is 15–20 kJ/mol higher than that of Li^+), which significantly increases the desolvation barrier of Na^+ and affects ion transport efficiency. At present, research hotspots in related fields focus on fluorinated electrolytes (e.g., fluorinated EC) and flame-retardant electrolytes (added with flame retardants such as triethyl phosphate). However, these solutions also have obvious shortcomings: the cost of fluorinated solvents is three to five times that of traditional solvents, and the addition of flame retardants will reduce the ionic conductivity of the electrolyte from approximately 10^{-2} S/cm to 5×10^{-3} S/cm [13].

1.4.2.2 Solid Electrolyte

Among the oxide electrolytes commonly used in solid-state batteries, LLZO ($Li_7La_3Zr_2O_{12}$) has an ionic conductivity of 10^{-4}–10^{-3} S/cm at room temperature, but it needs to be sintered at a high temperature above 1000 °C for forming. During this process, Li element volatilization is prone to occur, which further leads to an interface porosity of over 15% between the electrolyte and the electrode, affecting the interface contact effect. LGPS ($Li_{10}GeP_2S_{12}$) in sulfide electrolytes has more outstanding performance, with a room-temperature ionic conductivity of up to 10^{-3}–10^{-2} S/cm. However, it has obvious stability defects: it not only easily undergoes hydrolysis with moisture in the air but also reacts violently with metallic lithium to generate Li_3P, leading to a sharp increase in interface impedance to over 1000 $\Omega \cdot cm^2$. Polymer electrolytes such as the PEO-LiTFSI system have the advantages of good mechanical properties and bendability, which can be adapted to flexible battery scenarios. However, their low-temperature performance is extremely poor—the ionic conductivity drops to less than 10^{-6} S/cm below 0 °C, which is difficult to meet low-temperature application requirements. At present, research hotspots in the field of solid electrolytes focus on composite electrolytes, such as the LLZO-PEO composite system and the LGPS-carbon nanotube composite system. However, there are still cognitive blind spots in these composite systems: there is no clear understanding of the specific ion transport paths in them, the products generated after interface reactions, and their dynamic evolution processes [14].

1.4.3 Interface

1. Solid-Liquid Interface of Liquid Batteries (SEI/Cathode Electrolyte Interphase (CEI) Layers)
   At the solid-liquid interface of liquid lithium-ion batteries, an ideal SEI layer needs to meet the core requirements of "thickness less than 10 nm, dense structure, and high lithium-ion conductivity." However, the silicon-based anode undergoes a volume expansion of up to 300% during lithium intercalation. This severe expansion will lead to repeated rupture and regeneration of the surface SEI layer, which not only causes the electrolyte consumption to exceed 50% of the initial total amount but also significantly deteriorates

the battery cycle performance. On the cathode side, the CEI layer is affected by oxygen free radicals (e.g., O_2^-) released by the cathode. X-ray photoelectron spectroscopy (XPS) detection shows that the proportion of organic phase in the CEI layer increases from the initial 40% to approximately 70%. An excessively high proportion of organic phase will lead to a sharp increase in interface impedance, which impairs the kinetic performance of the battery. At present, there are still obvious cognitive gaps in the research on SEI and CEI layers: key issues such as the formation order of each component of the two layers and whether a self-healing mechanism exists during cycling have not been clarified. At the same time, there is a lack of effective methods to quantify the specific role of inorganic phases such as LiF in the interface layer.

2. Solid-Solid Interface of Solid-State Batteries
   In the solid-solid interface problem of all-solid-state batteries, high contact impedance is the primary challenge—in-situ atomic force microscopy (AFM) observations show that the actual contact area between metallic lithium and solid electrolyte is less than 30%, which makes the interface impedance account for over 60% of the total battery impedance and seriously affects ion transport efficiency. At the same time, interface reactions will exacerbate performance degradation: the reaction between metallic lithium and sulfide electrolyte LGPS generates Li_3P (the calculated reaction barrier is only 0.8 eV), and the contact between metallic lithium and oxide electrolyte LLZO generates $LiAlO_2$. TEM observations show that the thickness of such interface reaction layers is between 5 and 10 nm. In addition, the impact of mechanical incompatibility cannot be ignored. Metallic lithium undergoes volume changes during deposition and stripping, which leads to interface separation. The volume ratio of the formed pores can exceed 20%, further worsening the interface contact. At present, there are still obvious cognitive gaps in the research on the interface of solid-state batteries: the complete process from nucleation to growth of interface pores has not been clarified, and the coupling mechanism between electrochemical processes and mechanical effects has not been fully understood. These factors restrict the performance breakthrough of all-solid-state batteries [14, 15].

1.5 Evolution of Characterization Technology

As shown in Figure 1.7, a schematic diagram of the correlation between advanced characterization technologies and theoretical calculation methods (including in-situ characterization technologies, X-ray related technologies, and theoretical calculation methods such as DFT and molecular dynamics) is depicted. It links in-situ characterization (e.g., in-situ x-ray diffraction (XRD)/XPS) with DFT/MD, overcoming static offline method limits to enable multiscale analysis of battery material failure mechanisms.

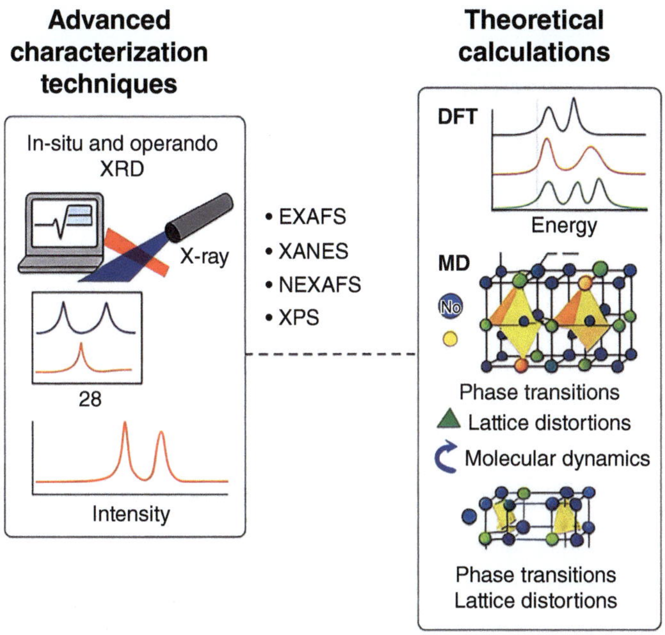

Figure 1.7 Schematic diagram of the correlation between advanced characterization technologies and theoretical calculation methods.

1.5.1 Limitations of Traditional Offline Characterization

Traditional offline characterization methods, such as ex-situ XRD, TEM, and XPS, can only capture "instant snapshots" of the battery in a static state. Specifically, ex-situ XRD can only compare the phase transition differences before and after battery cycling, and cannot track the onset voltage and evolution rate of phase transitions such as H1−3→O1. Ex-situ TEM cannot observe the complete real-time growth process of lithium dendrites, and can only infer their formation mechanism based on the morphology after cycling. Ex-situ XPS is also unable to clarify the formation order of each component of the SEI and CEI layers, and ultimately it is difficult to establish a direct correlation between "structure-performance-failure." For example, it cannot clearly explain the dynamic correlation between the capacity fading of silicon-based anodes and the rupture of the SEI layer.

1.5.2 Technical Requirements for Advanced Characterization

With the continuous development of battery materials toward high capacity, high voltage, and compounding, as shown in Figure 1.8, the industry has put forward three clear requirements for characterization technology: Spatial resolution: It needs to be further upgraded from the micron level (capable of observing particle morphology) to the atomic level (capable of identifying the occupancy of doping elements and the composition of interface phases). Temporal resolution: It needs

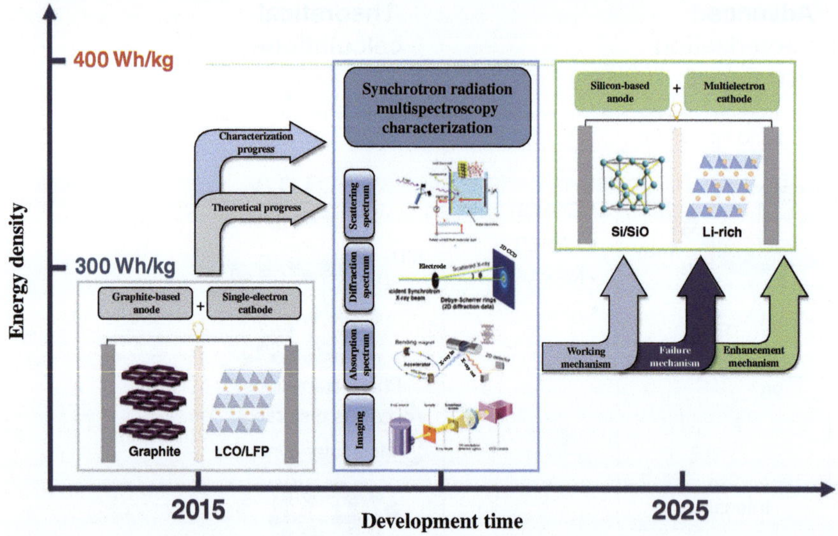

Figure 1.8 The urgent demand of the industry for advanced characterization technologies.

to be accelerated from the minute level (corresponding to charge-discharge curves) to the millisecond level (capable of capturing ion diffusion and interface reactions). Dimensional integrity: It needs to break through the limitation of single structural characterization (e.g., XRD) and realize multidimensional coupling analysis of structure, composition, electrochemistry, and mechanics [16].

1.5.3 Core Value of Advanced Characterization

Advanced characterization technology realizes the comprehensive analysis of the dynamic behavior of battery materials through in-situ testing modes and multitechnology coupling [17]. For example: In terms of dynamic structure tracking: In-situ synchrotron radiation XRD (with a spatial resolution of less than 1 μm and a temporal resolution of less than 10 s) can accurately capture the phase transition process of the cathode; in-situ aberration-corrected TEM (with atomic-level resolution) can clearly observe the microscopic arrangement of lithium deposition. In terms of composition and interface analysis: in-situ XPS (with a detection depth of less than 1 nm) can track the formation law of LiF in the CEI layer in real time; in-situ TOF-SIMS (with a spatial resolution of less than 10 nm) can reconstruct the elemental gradient distribution of the SEI layer. In terms of electrochemical-mechanical coupling characterization: in-situ AFM (with a mechanical resolution of less than 1 nN) can quantify the volume change of electrode particles; in-situ tensile testing combined with voltage measurement can establish the correlation between Co-O bond vibration and Co valence state. In terms of material design guidance, these in-situ characterization technologies can optimize the coating thickness of the cathode (to ensure that the proportion of LiF in the CEI layer exceeds 50%) and the volume expansion rate of the silicon-based anode (to control it within 150%), and ultimately build a complete research closed loop of "characterization-mechanism-design."

References

1 H. Du, H. Zhang, Climate neutrality in the EU and China: An analysis of the stringency of targets and the adaptiveness of the relevant legal frameworks, Review of European, Comparative & International Environmental Law 31(3) (2022) 495–509. https://doi.org/10.1111/reel.12453.

2 N. Azizi, H. Moradi, K. Rouzbehi, F. Mohammadi, HVDC grids stability enhancement through the integration of battery energy storage systems, IET Renewable Power Generation 18(15) (2024) 2838–2853. https://doi.org/10.1049/rpg2.12788.

3 X. Hu, Z. Wang, H. Zhang, Y. Song, J. Cui, J. Guo, M. Cao, Z. Wang, Y. Yao, Y. Huang, High-entropy design in battery materials for high performance electrochemical energy storage, Interdisciplinary Materials 4(6) (2025) 795–811. https://doi.org/10.1002/idm2.70013.

4 T.-H. Kim, J.-S. Park, S.K. Chang, S. Choi, J.H. Ryu, H.-K. Song, The current move of lithium ion batteries towards the next phase, Advanced Energy Materials 2(7) (2012) 860–872. https://doi.org/10.1002/aenm.201200028.

5 C. Li, L. Du, J. Ju, J. Niu, G. Xu, J. Zhao, G. Cui, Assessment of safety concerns in sodium-ion batteries, Advanced Energy Materials 15(40) (2025) e03209. https://doi.org/10.1002/aenm.202503209.

6 K. Kubota, M. Dahbi, T. Hosaka, S. Kumakura, S. Komaba, Towards K-ion and Na-ion batteries as "beyond Li-ion," The Chemical Record 18(4) (2018) 459–479. https://doi.org/10.1002/tcr.201700057.

7 L. Chang, A. Wei, S. Luo, S. Cao, X. Bi, W. Yang, R. Yang, J. Liu, Lithium-ion battery: A comprehensive research progress of high nickel ternary cathode material, International Journal of Energy Research 46(15) (2022) 23145–23172. https://doi.org/10.1002/er.8618.

8 S. Chai, Q. He, J. Zhou, Z. Chang, A. Pan, H. Zhou, Solid-state electrolytes and electrode/electrolyte interfaces in rechargeable batteries, ChemSusChem 17(3) (2024) e202301268. https://doi.org/10.1002/cssc.202301268.

9 K.J. Kim, M. Balaish, M. Wadaguchi, L. Kong, J.L.M. Rupp, Solid-state Li–metal batteries: Challenges and horizons of oxide and sulfide solid electrolytes and their interfaces, Advanced Energy Materials 11(1) (2021) 2002689. https://doi.org/10.1002/aenm.202002689.

10 J. Hou, J. Zhang, S. Wu, X. Liu, Y. Li, Z. Yan, K. Zhang, Y. Lu, J. Chen, Advanced design and characterization of polyether-based solid-state electrolytes for high-energy-density lithium batteries, Advanced Materials n/a(n/a) (2025) e15430. https://doi.org/10.1002/adma.202515430.

11 S. Kaskel, Lithium–Sulfur Batteries, Energy Technology 7(12) (2019) 1900940. https://doi.org/10.1002/ente.201900940.

12 M. Ezzedine, F. Jardali, I. Florea, M.-R. Zamfir, C.-S. Cojocaru, Nanostructuring strategies for silicon-based anodes in lithium-ion batteries: Tuning areal silicon loading, SEI formation/irreversible capacity loss, rate capability retention and electrode durability, Batteries & Supercaps 6(3) (2023) e202200451. https://doi.org/10.1002/batt.202200451.

13 W. Teng, J. Wu, Q. Liang, J. Deng, Y. Xu, Q. Liu, B. Wang, T. Ma, D. Nan, J. Liu, B. Li, Q. Weng, X. Yu, Designing advanced liquid electrolytes for alkali metal batteries: Principles, progress, and perspectives, Energy & Environmental Materials 6(2) (2023) e12355. https://doi.org/10.1002/eem2.12355.

14 Q. Zhou, X. Xiong, J. Peng, W. Wu, W. Fan, H. Yang, T. Wang, Y. Ma, F. Wang, Y. Wu, Tailored engineering on the interface between lithium metal anode and solid-state electrolytes, Energy & Environmental Materials 8(1) (2025) e12831. https://doi.org/10.1002/eem2.12831.

15 S.A. Pervez, G. Kim, B.P. Vinayan, M.A. Cambaz, M. Kuenzel, M. Hekmatfar, M. Fichtner, S. Passerini, Overcoming the interfacial limitations imposed by the solid–solid interface in solid-state batteries using ionic liquid-based interlayers, Small 16(14) (2020) 2000279. https://doi.org/10.1002/smll.202000279.

16 J. Wu, F. Ma, X. Liu, X. Fan, L. Shen, Z. Wu, X. Ding, X. Han, Y. Deng, W. Hu, C. Zhong, Recent progress in advanced characterization methods for silicon-based lithium-ion batteries, Small Methods 3(10) (2019) 1900158. https://doi.org/10.1002/smtd.201900158.

17 Z. Deng, X. Lin, Z. Huang, J. Meng, Y. Zhong, G. Ma, Y. Zhou, Y. Shen, H. Ding, Y. Huang, Recent progress on advanced imaging techniques for lithium-ion batteries, Advanced Energy Materials 11(2) (2021) 2000806. https://doi.org/10.1002/aenm.202000806.

2

Neutron Scattering and Pair Distribution Function Analysis for Battery Materials

2.1 Introduction

The discovery of the neutron in 1932 marked a fundamental breakthrough in modern nuclear science. When Chadwick bombarded beryllium with energetic α particles as shown in Figure 2.1, the reaction (1) produced a new and electrically neutral particle with remarkable penetrating power [1]. For this discovery, James Chadwick was awarded the 1935 Nobel Prize in Physics, recognizing the neutron as a foundational constituent of matter and establishing the basis for neutron scattering science.

$$ {}^{4}_{2}\alpha + {}^{9}_{4}\beta \rightarrow {}^{1}_{0}n + {}^{12}_{6}C \tag{2.1} $$

Unlike charged particles, neutrons interact weakly with matter, enabling them to probe atomic nuclei and light elements with exceptional sensitivity. Nowadays, neutron-based characterization has become an indispensable component of modern materials research, particularly in the study of electrochemical energy storage systems such as lithium-ion and sodium-ion batteries [2]. The advancement of high-flux research reactors, accelerator-driven spallation neutron sources, and operando instrumentation has significantly broadened the scientific reach of neutron scattering techniques, enabling multiscale insights into the origin of battery performance degradation [3, 4].

For lithium-ion batteries (LIBs), common electrons and X-ray interact primarily with electron clouds and therefore strongly favor heavy elements, while neutrons interact with atomic nuclei through the strong nuclear force. This feature endows neutron scattering with two remarkable advantages. First, neutrons exhibit exceptional sensitivity to light elements—especially Li, Na, H elements and their isotopes which play critical roles in ionic conduction, interfacial chemistry, and structural evolution in rechargeable batteries. Measurements of lithium occupancy, diffusion pathways, local distortions and hydrogen-containing electrolytes and solid/electrolyte interphases are therefore much more accessible with neutrons than with conventional X-ray techniques. Second, neutron scattering lengths vary irregularly across the periodic table and are often highly sensitive to isotopic substitution. In widely used Ni-Co-Mn layered oxide cathodes, these transition metals possess closely related electronic structures and adjacent atomic numbers, making their site occupancies

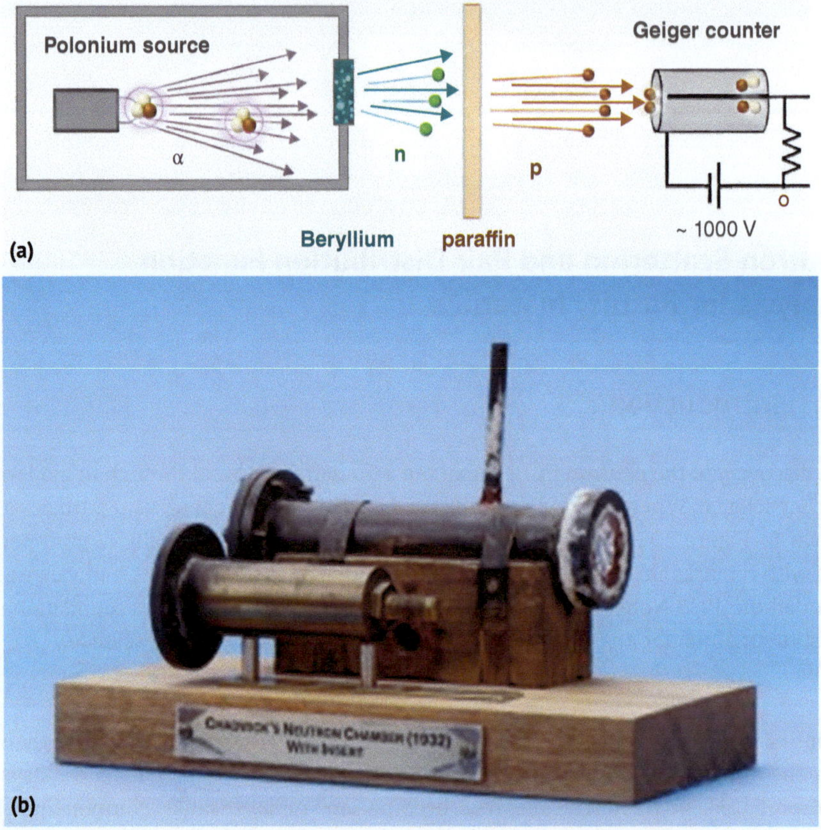

Figure 2.1 Chadwick's discovery of the neutron in 1932. (a) Schematic of the α-Be experiment. When this beam impinges on paraffin, energetic recoil protons are generated and detected using a Geiger counter operated at ~1000 V, providing indirect evidence for the existence of the neutron. (b) Reconstructed model of Chadwick's original neutron chamber. *Source*: Triestino Minniti et al. (2018)/with permission of Elsevier

and cation disorder difficult to distinguish by conventional X-ray diffraction. In contrast, the irregular and element-specific neutron scattering lengths of Ni, Co, and Mn provide sufficient contrast to resolve their crystallographic distribution and mixing behavior. This capability is essential for understanding the intrinsic structural evolution mechanisms of cathode materials during electrochemical cycling, as Ni, Co, and Mn exhibit distinct migration features, valence-state evolution and local coordination structural preferences during repeated Li insertion and extraction. By enabling direct visualization of their individual behaviors within the lattice, neutron diffraction offers critical insight into how transition-metal ordering, charge compensation, and defect formation govern the performance and long-term stability of layered oxide cathodes.

Another defining characteristic of neutrons is their high penetration depth. Even at thermal or cold energies, neutrons readily traverse millimeter- to centimeter-thick electrodes and current collectors, making them ideally suited for nondestructive operando measurements. This enables researchers to follow structural transitions in cathode and anode materials, monitor ion transport across interfaces, visualize lithium distribution, and quantify the formation and evolution of solid-electrolyte interphases (SEI/CEI) without dismantling the cell [5]. The development of neutron-transparent coin cells, pouch cells, and Ti-Zr alloy operando cells has facilitated true device-level studies that capture realistic electrochemical conditions, including temperature gradients, compositional inhomogeneity and mechanical deformation.

More importantly, neutron science is a vital complement to X-ray and electron techniques, offering deep penetration and strong sensitivity to light elements and buried structures. For energy storage systems, neutrons enable nondestructive tracking of structural changes, ion transport, and degradation under realistic conditions. Combined with X-ray, electron and computational methods, neutron tools provide a multiscale view of how batteries evolve, supporting the design of more durable next-generation materials.

2.2 Neutron Facilities: Global Distribution and Specialized Capabilities

Neutron facilities constitute indispensable large-scale infrastructures for contemporary scattering and diffraction research. Their global deployment reflects strategic investments in fundamental science, advanced materials, and energy technologies. Distinct neutron-generation principles give rise to complementary facility capabilities and experimental modalities. Reactor-based and spallation-based sources define the core technological landscape. Their differences shape beam characteristics, instrument design, and application domains. An overview of global neutron facilities therefore provides essential context for understanding specialized capabilities, national strategies, and emerging trends in neutron science.

2.2.1 Neutron Production Mechanisms: ^{235}U Fission and Spallation Reactions

Neutron sources for modern scattering and diffraction experiments are primarily based on two fundamental nuclear processes: thermal neutron generation from ^{235}U nuclear fission and high energy neutron production through spallation reactions. Although both approaches ultimately provide intense neutron beams suitable for scientific research, their production physics, energy spectra, and facility characteristics differ significantly.

Nuclear reactors traditionally serve as steady-state neutron sources by exploiting the fission of uranium-235 (^{235}U). As shown in Figure 2.2, when a thermal neutron is absorbed by a ^{235}U nucleus, it becomes unstable and undergoes fission, splitting into two fission fragments and releasing approximately two to three neutrons. These neutrons have typical energies of ~1–2 MeV, with a broad Maxwellian-like distribution. A small fraction of delayed neutrons emitted from β-decay of fission fragments have much lower intensities but are crucial for reactor control.

The high fission cross-section of ^{235}U for thermal neutrons enables a self-sustaining chain reaction. A reactor core maintains a continuous neutron flux of 10^{13}–10^{15} n/(cm^2·s), offering exceptional stability and long-term operation for neutron scattering

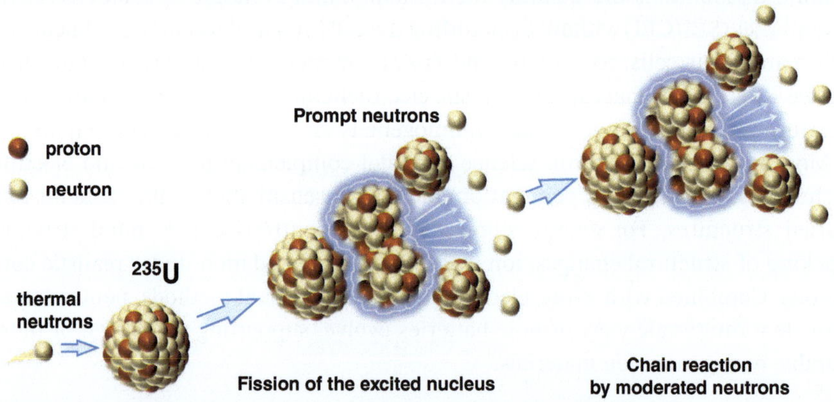

Figure 2.2 Schematic diagram of neutrons-fission process of ^{235}U.

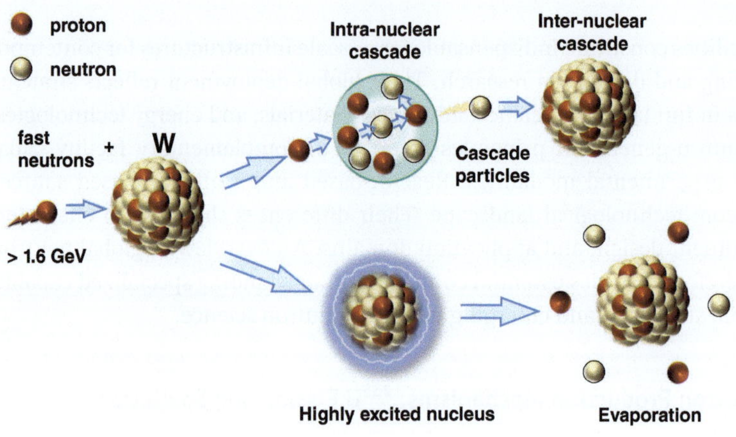

Figure 2.3 Schematic diagram of neutrons spallation.

instruments. The prompt neutrons emitted from fission are rapidly moderated using light water and heavy water, or graphite to produce thermal or cold neutron beams suitable for spectroscopy and imaging. The key advantage of reactor sources is their continuous and high-flux neutron output, which simplifies beamline design and enables high-throughput experiments. However, safety constraints and aging infrastructure impose limitations on global deployment.

Spallation sources represent the modern alternative to reactor-based neutron facilities. In spallation, high energy protons (typically 0.8–1.5 GeV) are accelerated and directed onto a heavy metal target such as tungsten or lead, with the neutron production process illustrated schematically in Figure 2.3. Upon impact, the proton induces an intranuclear cascade that ejects 20–40 neutrons per incident proton—an order of magnitude more neutrons per primary reaction than fission. The emitted neutrons have a broad high energy spectrum up to several hundred MeV, and thus requiring moderation to produce thermal or cold neutrons.

Unlike reactors, spallation sources operate in pulsed mode, inherently providing a time structure that is particularly advantageous for time-of-flight (ToF) techniques. This enables precise energy definition, making spallation sources ideal for inelastic scattering, quasi-elastic measurements and high-resolution diffraction. Furthermore, spallation facilities avoid the criticality issues associated with fission reactors, offering potential advantages in safety and scalability. Their main challenges include the technological complexity of megawatt-class proton accelerators and the thermal/mechanical demands on the target systems. The detailed contrasts are summarized in Table 2.1, which highlights their individual advantages and limitations of the two neutron production methods.

Table 2.1 The comparison between the two neutron production methods.

	^{235}U Nuclear Fission	Spallation Reactions
Generation principle	Neutrons released from thermal neutron-induced fission of ^{235}U, producing 2–3 prompt neutrons per event	High energy protons striking a heavy metal target, inducing an intranuclear cascade that ejects 20–40 neutrons
Chain reaction	Yes (self-sustaining chain fission reaction)	No (single-event neutron production)
Initial neutron energy	~1–2 MeV (prompt), with tail up to ~10 MeV	Very broad spectrum from ~1 MeV to several hundred MeV (up to ~500 MeV)
Need for moderation	Required to obtain thermal/cold neutrons (H_2O, D_2O, graphite, cold moderators)	Required to obtain thermal/cold neutrons (H_2O, D_2O, cold hydrogen, cold deuterium, cryogenic moderators)
Output mode	Continuous, steady state neutron emission	Pulsed neutron emission inherent to accelerator cycle
Neutron flux	10^{13}–10^{15} n/(cm* s) (high and stable)	High peak flux; average flux typically lower than high-power reactors
Beam stability	Excellent long-term stability	High stability but depends on accelerator operation cycles
Safety considerations	Criticality safety, nuclear fuel handling, long-lived radioactive waste	No criticality risk; lower long-lived waste, but high-power target hazards
Infrastructure complexity	High but technologically mature	Very high: GeV linear accelerators/synchrotrons + MW target stations
Key advantages	Continuous high flux, operational stability, well-established instruments	Pulsed beams, superior ToF capabilities, high energy resolution, improved safety
Best-fit applications	Steady-state diffraction, imaging, SANS, classical spectroscopy	ToF diffraction, INS, QENS, pulsed imaging, high-resolution spectroscopy
Typical facilities	ILL (France), HFIR (USA), OPAL (Australia), JRR-3 (Japan), CARR (China)	SNS (USA), J-PARC (Japan), ISIS (UK), CSNS (China)

2.2.2 Representative Neutron Facilities Worldwide

Building on these two fundamental neutron production approaches, many countries have strategically invested in large-scale neutron facilities to advance materials science, energy research, and high-technology manufacturing. Their global distribution including synchrotron radiation facilities and X-ray free-electron laser sources is shown in Figure 2.4.

Although the technical routes differ between nations, their motivations includes enabling world-leading research infrastructure, supporting national industrial competitiveness and securing long-term access to high-flux neutron beams as older research reactors approach end of life.

Historically, Europe and the United States prioritized high-performance reactor-based neutron sources—exemplified by the ILL (France) and HFIR (United States). In parallel, Germany established a complementary flagship at Garching—the Heinz Maier-Leibnitz Research Neutron Source (FRM II), a reactor-based high-flux beam-tube facility integrated into the MLZ user ecosystem, as shown in Figure 2.5. FRM II operates as a dedicated research neutron source (20 MW class) and employs a pool-type configuration with a large volume of fully demineralized water for shielding and in-pool operation. Neutrons are delivered to instruments through 11 beam tubes (10 horizontal and one inclined) and conditioned by neutron guides/monochromating optics, with first optical elements typically positioned a few meters from the fuel element to mitigate irradiation damage. Thermal management relies on three hydraulically separated but thermally coupled cooling circuits, with a high-flow

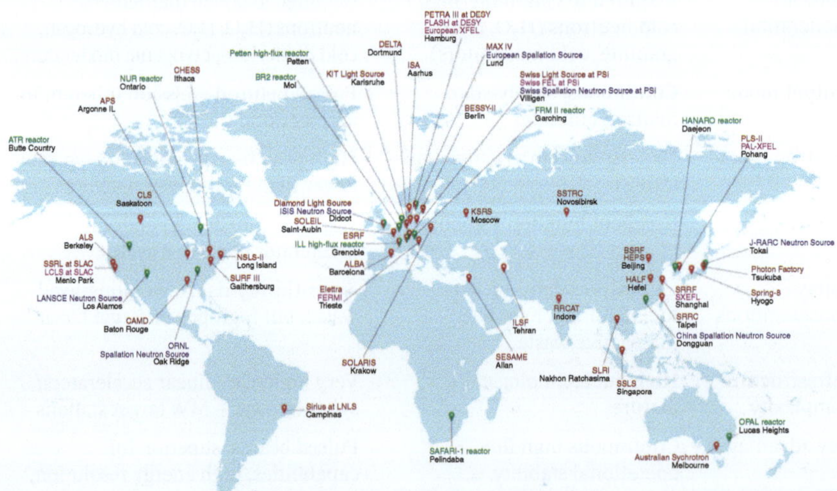

Figure 2.4 Distribution of major light sources (including synchrotron radiation X-ray sources and X-ray free-electron lasers) and neutron sources (including neutron reactors and spallation neutron sources) around the world. (Map by Free Vector Maps http://freevectormaps.com). The locations of cities/areas marked on the map are estimates only and do not represent their precise geographic locations [6] / with permission of John Wiley & Sons.

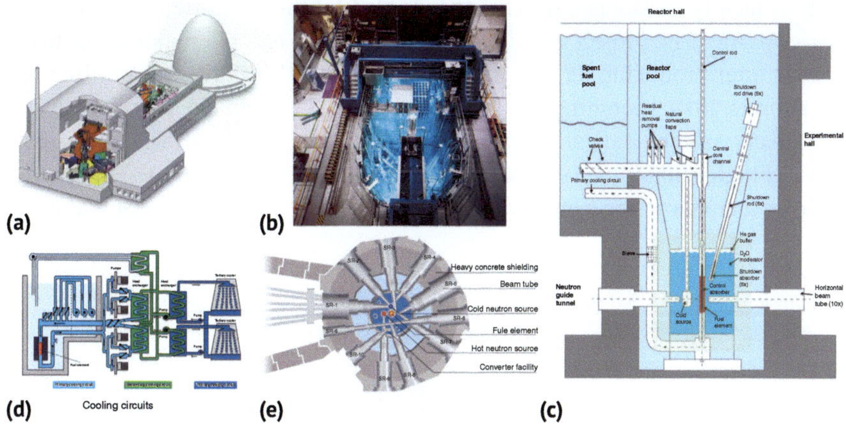

Figure 2.5 Overview of the FRM II research reactor (Heinz Maier-Leibnitz Research Neutron Source) and its neutron beam infrastructure [7]/F. Tralmer, FRM II/TUM. (a) Schematic layout of the reactor core and user areas (reactor building and experimental halls). (b) Photograph of the reactor pool region showing the "swimming-pool" configuration; the pool contains ~700 m^3 of fully demineralized water that provides shielding and operational access. *Source*: Wenzel Schürmann/TUM (c) Simplified facility cross-section corresponding to (b). (d) Conceptual diagram of the three hydraulically separated but thermally coupled cooling circuits; the primary circuit runs through the fuel element and is driven by four main pumps (~300 L/s), heating from ~35 °C to ~50 °C across the core before transferring heat via heat exchangers to the secondary/tertiary loops. (e) Beam extraction and transport. Neutrons are delivered to instruments via beam tubes and neutron guides.

primary loop traversing the fuel element and transferring heat via heat exchangers to secondary/tertiary loops and ultimately to cooling towers. In today's mixed global landscape where many countries have expanded spallation capacity, FRM II remains a strategically important continuous-source platform, supporting high-precision and low-background beam experiments and specialized capabilities within Europe.

In recent decades, next-generation spallation projects have been launched to achieve unprecedented brightness and efficiency. European Spallation Source, funded jointly by multiple nations, aims to become the world's brightest neutron facility. Its construction is explicitly tied to European Union strategic initiatives in sustainable materials and digital manufacturing. Similarly, the United States operates the Spallation Neutron Source (SNS) at Oak Ridge National Laboratory, developed under Department of Energy programs to advance condensed matter physics, nuclear engineering, and applied materials science. J-PARC program integrates neutron scattering with particle physics and accelerator innovation, reflecting a coordinated national strategy for cross-disciplinary science.

China has pursued rapid development of neutron capabilities through both reactor-based and accelerator-based initiatives. The China Advanced Research Reactor (CARR) provides a stable high-flux platform [8], while the China SNS (CSNS) represents a major commitment to modern pulsed neutron science, aligning with national priorities in high-performance materials and next-generation manufacturing [1]. The CSNS is China's flagship accelerator-driven pulsed neutron facility in Dongguan, Guangdong, led by the Institute of High Energy Physics (CAS). The project was approved in 2005 and entered major construction in 2011; after producing its first neutron beam on 2017, CSNS was put into user operation in 2018, becoming the fourth pulsed SNS worldwide after the UK, US, and Japan. As schematically summarized in Figure 2.6a, CSNS adopts a compact and high-reliability accelerator chain consisting of an 80 MeV H^- linac followed by a rapid cycling synchrotron that accelerates protons to 1.6 GeV at 25 Hz,

(a)

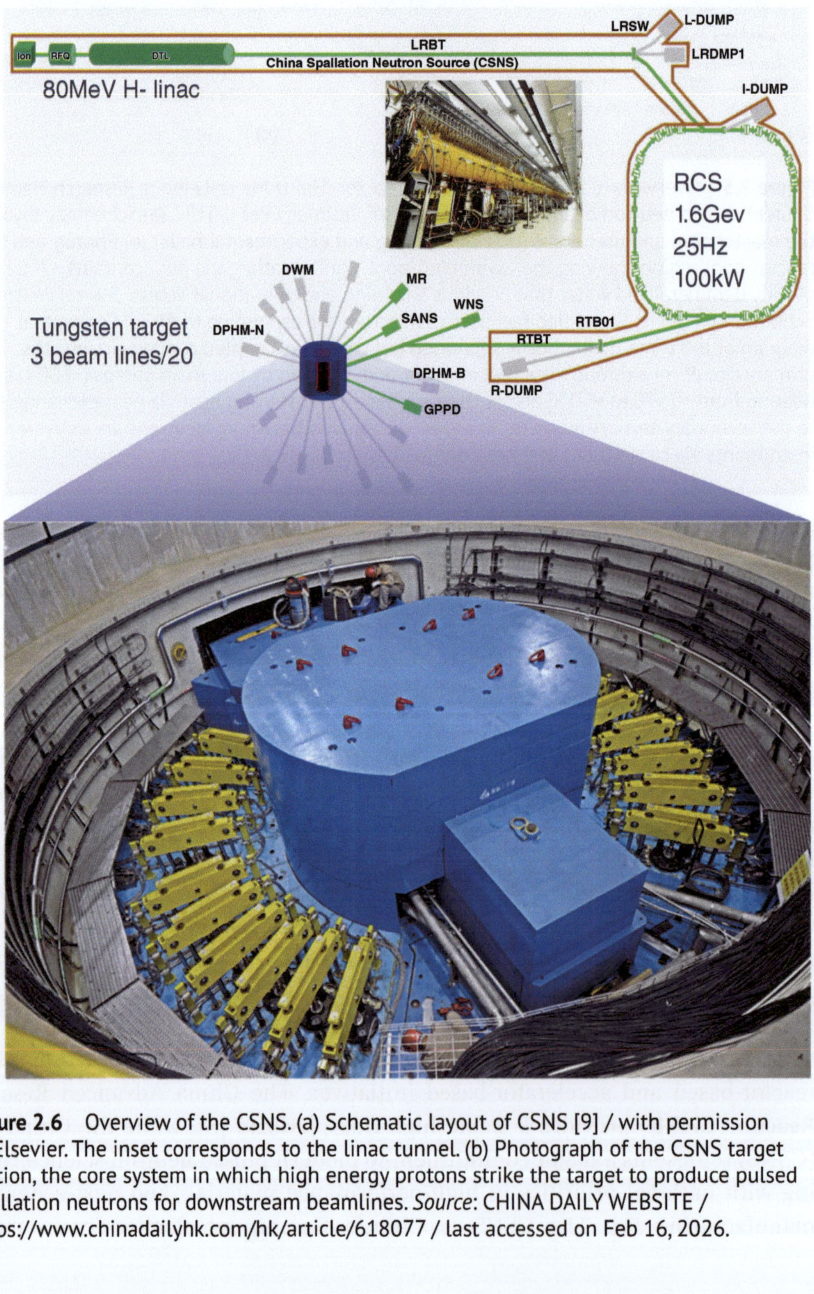

(b)

Figure 2.6 Overview of the CSNS. (a) Schematic layout of CSNS [9] / with permission of Elsevier. The inset corresponds to the linac tunnel. (b) Photograph of the CSNS target station, the core system in which high energy protons strike the target to produce pulsed spallation neutrons for downstream beamlines. *Source*: CHINA DAILY WEBSITE / https://www.chinadailyhk.com/hk/article/618077 / last accessed on Feb 16, 2026.

delivering a ~100 kW proton beam onto the target in the first-phase configuration. The high energy protons strike a tungsten target station (Figure 2.6b), where spallation reactions generate intense short-pulse neutron bursts that are moderated and distributed to surrounding instruments.

Within the CSNS instrument suite, neutron scattering and diffraction constitute important techniques for resolving atomic and magnetic structures, and the General Purpose Powder Diffractometer (GPPD) serves as a representative beamline dedicated to high-quality time-of-flight neutron powder diffraction. The GPPD is the first dedicated powder diffractometer at the CSNS (Figure 2.7d). As illustrated in Figure 2.7a, b, the instrument adopts a 30 m moderator-to-sample flight path with a pulsed-beam optical train in which a T0 chopper placed near the

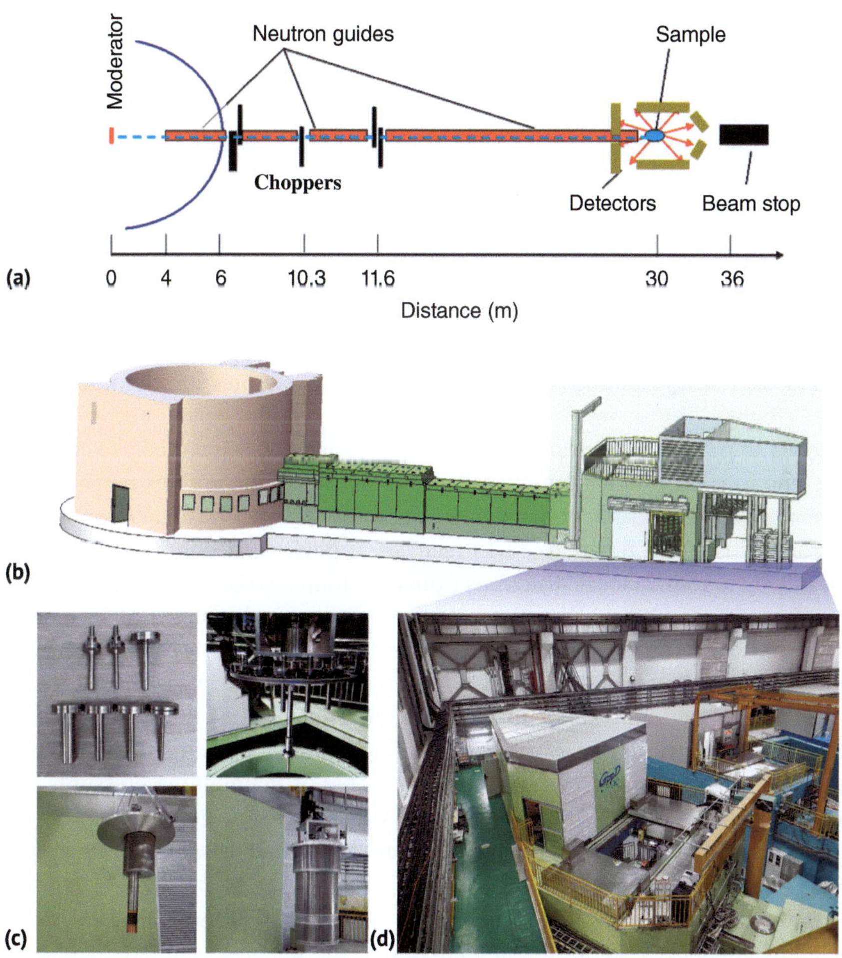

Figure 2.7 (a) Schematic layout and (b) 3D representation of the GPPD [9] / with permission of Elsevier. (c) GPPD ancillary equipments: custom cans, auto changer, one of the cryostats and superconducting magnet [10] / with permission of Elsevier. (d) Photograph of GPPD in CSNS. *Source*: Beijing News / https://m.bjnews.com.cn/detail/166174963514251.html / last accessed on Feb 16, 2026.

source suppresses prompt-pulse background, while subsequent bandwidth choppers define the usable wavelength band and further reduce fast-neutron/γ-related backgrounds [9, 10]. Beam transport is realized by a guide system beginning at the shutter exit (~2.55 m, 40 mm × 80 mm), followed by an 18 m tapered focusing guide that compresses the beam to ~20 mm × 40 mm at the end; this focusing section substantially enhances neutron transmission (reported >50% gain relative to a non-focusing straight guide), enabling "reasonable intensity" without sacrificing resolution. A downstream interchangeable guide/collimator module provides a practical switch between high-resolution and high-intensity operation, supporting both precision refinement and throughput-oriented measurements. The sample is hosted in a large vacuum chamber (~1.2 m diameter) (Figure 2.7c), while the detector system comprises scintillator-based modules arranged into three detector banks (high-/medium-/low-angle), delivering an exceptionally wide d-spacing coverage of ~ 0.05–50.48 Å that benefits both small-unit-cell crystallography and large-period magnetic/complex structures.

Commissioning and subsequent routine operation demonstrate low background, symmetric peak shapes, and sub-0.2% resolution (with reports of ~0.15% at the high-angle bank), providing reliable *Rietveld* refinement and enabling applications spanning advanced steels, energy materials (Li-ion, Zn-ion, Na-ion systems) and magnetic-structure studies [11–13].

In addition, CSNS also deploys dedicated neutron scattering instruments to probe nanoscale morphology and mesoscopic correlations. The Multipurpose Instrument (MPI) provides a versatile scattering platform to access real space length scales beyond the reach of powder diffraction [9]. The MPI situated at beamline #16 of CSNS (Figure 2.8), serves as a high-performance total scattering diffractometer designed to probe the intricate structural characteristics of materials across multiple length scales, ranging from long-range crystalline order to short-range atomic correlations. The instrument operates as a time-of-flight diffractometer optimized for pair distribution function (PDF) analysis, a powerful technique that allows researchers to resolve local structures from the atomic to the nanoscale with high precision. At its technical core, the instrument utilizes a decoupled water moderator at 300 K and features a sophisticated detector system initially comprised of 240 high-pressure helium-3 tubes arranged in seven banks. This detector array, which covers a wide scattering angle from 2.9° to 170°, effectively doubled the detector coverage and instrument performance. The hardware architecture is further supported by a massive 12 m^3 vacuum vessel capable of maintaining pressures between 100 and $1*10^{-6}$ mbar, ensuring minimal background interference during experiments. With a high neutron flux of approximately 10^7 $n/s/cm^2$ and an expansive momentum transfer (Q) range reaching up to 50 $Å^{-1}$ at a resolution of 0.3%, MPI provides an unparalleled experimental platform for diverse scientific fields including condensed matter physics, chemistry, life sciences, and nanomaterials. The instrument is particularly adept at conducting *in-situ* structural analysis of new metallic alloys, measuring environmental effects on nanostructures, and determining the atomic local distribution and short-range chemical bonds of disordered

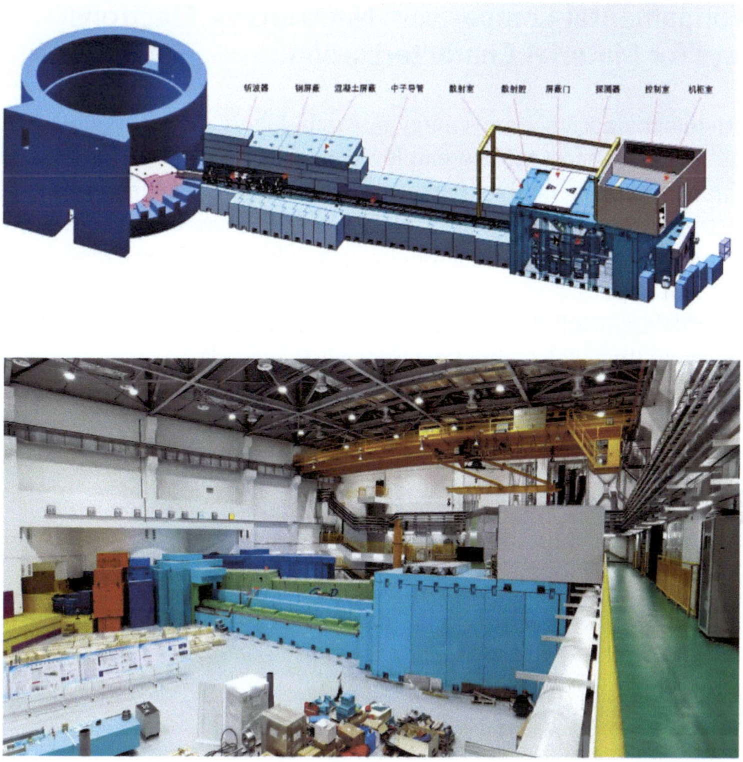

(a)

(b)

Figure 2.8 Multiphysics Instrument (MPI). (a) 3D representation of MPI. (b) Overview of MPI in CSNS [17] / City University of Hong Kong.

materials. By facilitating the study of material transition states under extreme conditions and bridging the gap between traditional diffraction and local structural probes, MPI serves as a versatile gateway for international researchers to explore the structural complexities of modern materials science and engineering [14–16]. This capability is essential for investigating two-body correlation information in crystals and understanding how local fluctuations drive macroscopic properties in functional materials.

Collectively, recent progress in neutron facilities and instruments underscores the close coupling between frontier science and national research strategy. Reactor and spallation sources play complementary roles in which continuous reactors provide highly stable flux for precision diffraction and fundamental studies, and pulsed spallation platforms enable time-of-flight methods and broad spectral coverage. Beyond source choice, modern competitiveness increasingly depends on instrument design-advanced choppers, low-background shielding, and versatile sample environments. Together, these infrastructures are reshaping capabilities in crystallography, magnetism, soft matter, and energy storage materials.

2.3 Fundamental Comparison: Neutrons vs. Electrons vs. X-rays for Material Characterization

In modern materials science and energy research, X-rays, electrons, and neutrons constitute the three most widely employed probes because they interact with matter through fundamentally different mechanisms and therefore access different scale structural and dynamic information. As summarized in Table 2.2, X-rays, electrons, and neutrons differ substantially in their interaction mechanisms, probing depths and structural sensitivities. Their combined use forms a comprehensive multiscale approach that spans atomic to mesoscale structures and probes dynamics across time scales from femtoseconds to seconds.

Among these probes, neutrons occupy a particularly important and irreplaceable role due to their unique nuclear and magnetic interactions, their ability to penetrate

Table 2.2 The comparisons between X-ray, electron, and neutron.

	X-ray	Electron	Neutron
Is it an elementary particle?	No—electromagnetic wave (but photon is an elementary particle, meanwhile X-ray refers to radiation, not a particle in the same sense)	Yes—electron is a fundamental lepton	Yes—neutron is a fundamental baryon
Nature of probe	Electromagnetic waves (keV range)	Charged particles (highly accelerated electrons)	Neutral particles (thermal/cold neutrons)
Primary interaction with matter	Scattering by electron clouds (Z-dependent)	Strong Coulomb interaction with electrons and nuclei	Nuclear scattering + magnetic interaction with unpaired spins
Interaction strength	Moderate	Very strong	Weak → high penetration
Penetration depth	Tens of microns (depends on energy)	Tens to hundreds of nanometers	Millimeters to centimeters
Isotope sensitivity	No	No	Yes (key neutron advantage)
Magnetic sensitivity	Weak (needs resonant techniques)	Moderate (Lorentz TEM, EMCD)	Strong (direct spin coupling)
Representative technique	XRD, SAXS, XAS, tomography	TEM/STEM, EELS, 4D-STEM	Neutron diffraction, SANS, INS, imaging
Best-suited scientific questions	Crystal & electronic structures	Defects, interfaces, local chemistry	Li/H location, cation mixing, magnetism, diffusion, phonons, operando bulk processes

deeply into bulk materials, and their sensitivity to light elements that are essential in functional systems such as batteries, hydrogen-storage materials, and functional oxides.

X-rays are electromagnetic waves with wavelengths on the order of angstrom and interact primarily with the electron clouds surrounding atoms. Their scattering power increases with atomic number, making X-ray based methods particularly effective for determining average crystal structures and long-range ordering. Similarly, electrons—charged particles with extremely short de Broglie wavelengths—interact strongly with matter via Coulomb forces, resulting in large scattering cross-sections and inherently surface- and interface-sensitive signals. Electron microscopy techniques such as TEM, STEM, EELS, and emerging 4D-STEM provide sub-angstrom spatial resolution and allow direct visualization of atomic-scale defects and grain boundaries. These techniques are indispensable for characterizing nanoscale heterogeneity, but their strong interactions limit penetration depths and make *in-situ* or bulk measurements challenging.

Specifically, neutrons offer a unique set of intrinsic physical advantages arising from their nuclear and magnetic interactions with matter. As electrically neutral particles, neutrons traverse matter with relatively weak attenuation compared with X-rays and electrons, enabling measurements deep within bulk samples, thick electrodes, and integrated devices. Their de Broglie wavelengths in the thermal and cold regimes coincide naturally with interatomic distances and lattice vibrations, making them ideally suited for structural and dynamical characterization. Most critically, neutrons interact with atomic nuclei via strong nuclear forces rather than with electron clouds. As a result, neutron scattering lengths vary irregularly across the periodic table and even between isotopes of the same element. This property makes neutrons uniquely capable of distinguishing neighboring elements that X-rays cannot differentiate, such as Ni, Co, and Mn in layered oxides. More importantly, neutrons also enable sensitive detection of light elements such as Li, H, D, B, and O, which are nearly invisible to conventional X-ray methods.

Neutrons also carry a magnetic moment and therefore couple directly to unpaired electron spins. This interaction enables direct probing of magnetic structures, spin correlations, and magnetically driven phase transitions, which are inaccessible to X-rays unless specialized resonant techniques are employed. For functional materials that rely on electron-phonon coupling, spin-charge interactions or orbital ordering, neutrons provide a unique perspective on structure, magnetism, and dynamics.

These distinct interactions give rise to a rich suite of neutron techniques that span wide spatial and temporal scales. Neutron diffraction allows precise determination of crystal structures, cation distribution, occupancy changes, anti-site defects, and subtle lattice distortions even in multicomponent compounds. Small-angle neutron scattering (SANS) is powerful for probing mesoscale structural features such as nanopores, precipitates, phase separation, and polymer self-assembly with excellent contrast from isotopic substitution. Neutron imaging, enabled by deep penetration and strong sensitivity to hydrogen-containing phases, offers real time visualization of bulk processes such as electrolyte flow, electrode wetting, gas evolution, and structural deformation during battery operation.

When contrasted with these capabilities, the limitations of X-rays and electrons highlight the unique position of neutrons in materials characterization. X-ray diffraction offers excellent resolution for average structures but struggles to detect light elements and to differentiate transition metals with similar atomic numbers. Spectroscopic X-ray methods provide valuable electronic information but are often limited to surface or local regions depending on sample geometry and photon energy. Electron-based methods provide unmatched spatial resolution, yet their strong interaction with matter restricts sample thickness and complicates bulk or device-level studies. Electrons are also insensitive to isotopic variation and cannot probe magnetic excitations with the same clarity as neutrons.

Thus, while every probe has its strengths, neutrons fill a critical gap in the characterization landscape by offering broad penetration, isotope sensitivity, magnetic sensitivity, and direct dynamical information. These advantages make neutron scattering indispensable for understanding energy-related materials, particularly systems containing light elements, disordered or partially crystalline structures, and dynamic processes essential for material function. In energy storage research, neutron methods have become essential for resolving lithium distribution, quantifying transition-metal migration, studying hydrogen-bearing electrolytes, and probing phonon-mediated transport, thereby providing insights that are fundamentally inaccessible by X-rays or electrons alone.

Together, electrons, X-rays, and neutrons form a complementary set of probes that enable multiscale and multimodal characterization of modern materials. Among them, neutrons offer distinct capabilities for resolving the structural, chemical, magnetic, and dynamic features that govern the behavior of functional and electrochemical materials.

2.4 Neutron Scattering Techniques: Principles and Variants

Broadly, neutron-based characterization technologies can be categorized into neutron scattering, neutron imaging, and nuclear-reaction-based methods, shown in Figure 2.9. Neutron scattering encompasses diffraction, total scattering, small-angle scattering, reflectometry, and inelastic or quasi-elastic measurements, collectively enabling the determination of atomic arrangements, site occupancies, short-range order (SRO), defects, and ionic or molecular dynamics. Neutron imaging, including radiography, tomography, and energy-resolved Bragg-edge imaging, provides nondestructive visualization of internal morphology, phase distribution, electrolyte transport, gas evolution, and mechanical degradation in two and three dimensions. In contrast, nuclear-reaction-based techniques—such as neutron activation analysis, prompt gamma neutron activation analysis, and neutron depth profiling (NDP)—offer highly selective quantification and depth-resolved mapping of light

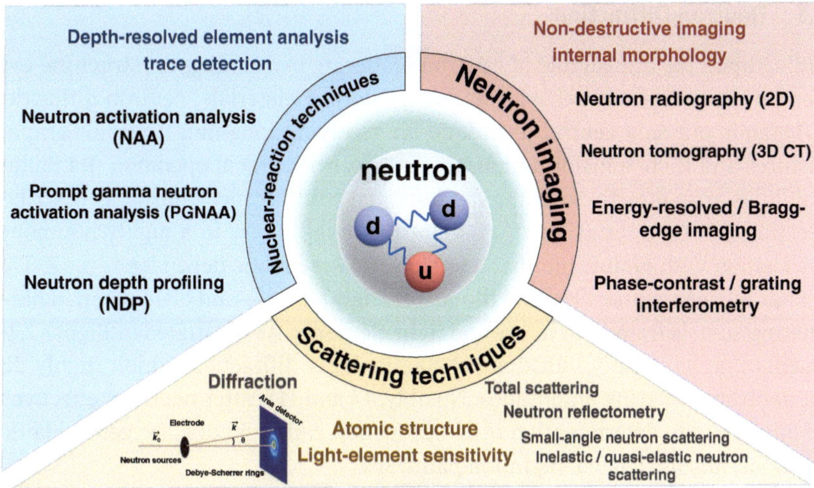

Figure 2.9 Neutron-based characterization techniques.

and trace elements, especially Li, B, F, and other species that are difficult to detect using X-ray and electrons-based tools.

While these three categories collectively establish a comprehensive neutron framework, their relative importance varies with the scientific question and materials system under investigation. In the field of electrochemical energy storage—particularly LIBs and related emerging chemistries—understanding the evolution of crystal structure and ion transport pathways under continuous (de)lithiation is of central importance. The performance and degradation of layered oxides, spinel phases, solid electrolytes, and conversion-type electrodes are intimately associated with subtle changes in cation ordering, anion redox, lattice strain, oxygen activity, and Li-ion mobility across a wide range of states of charge.

Because neutron scattering provides nuclear—not Z-dependent—contrast, neutron scattering provides a set of advantages that are particularly well-suited to probing these structural and transport phenomena. The comparable scattering lengths of neighboring 3D transition metals (especially Ni, Co, Mn) and the strong sensitivity to light elements such as Li and O allow neutron diffraction and PDF analysis to quantify cation mixing, detect minor phase transformations, resolve short-range distortions, and track the evolution of both long-range and local structures with high fidelity. Meanwhile, inelastic and quasi-elastic neutron scattering directly access ionic hopping processes, rotational dynamics, and vibrational modes that dictate charge-carrier transport and interfacial kinetics.

For these reasons, neutron scattering has become the most widely employed neutron-based approach for investigating the intrinsic structural evolution and transport mechanisms of battery materials, forming an indispensable foundation for elucidating degradation pathways, optimizing electrode architectures, and guiding the design of next-generation high-performance energy storage systems.

2.4.1 Neutron Diffraction

Building upon the critical role of neutron scattering in decoding the structural evolution and ion transport behavior of energy storage materials, neutron diffraction (ND) stands out as a central technique for resolving long-range crystallographic structures and their transformations during electrochemical operation. Its unique capabilities arise directly from the fundamental contrast mechanism, as described in Section 2.3. As a result, neutron scattering lengths vary in a highly nonmonotonic fashion with atomic number and show no systematic dependence on Z. This irregular nuclear contrast leads to the surprisingly strong—and often comparable—scattering signals from both heavy and light elements. As illustrated in Figure 2.10, elements that are nearly indistinguishable in X-ray diffraction exhibit highly distinct neutron scattering amplitudes. Notably, Li and O scatter neutrons effectively, enabling accurate determination of Li occupancy, cation mixing, oxygen sublattice distortions, and associated migration pathways.

The distinct characteristics of X-ray diffraction, electron diffraction, and neutron diffraction are compared in Table 2.3. Among them, ND is particularly advantageous for bulk-sensitive structural characterization. Another intrinsic advantage of ND is its exceptional penetration depth, allowing truly bulk-sensitive measurements on thick electrodes, composite laminates, multilayer architectures, and even fully assembled battery cells. This feature makes ND uniquely suited for operando and *in-situ* tracking of crystallographic evolution under realistic cycling conditions. Phase transitions, solid solution behaviors, domain growth, and interfacial reaction fronts can be quantitatively monitored as a function of state of charge. Combined with *Rietveld* refinement, ND provides high-precision information on lattice parameters, site occupancies, anisotropic strain, and the evolution of oxygen positional parameters. When further integrated with total scattering or PDF analysis, neutron measurements simultaneously capture both long- and short-range ordering,

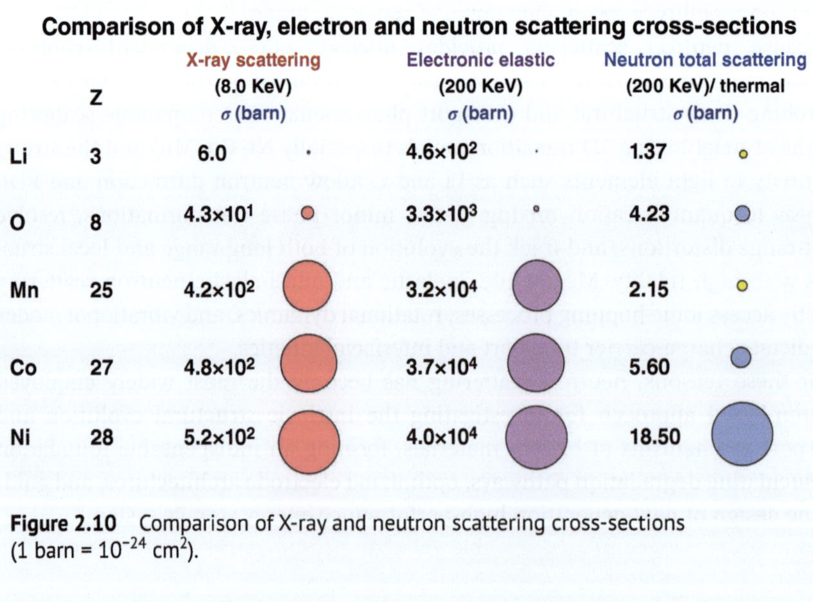

Figure 2.10 Comparison of X-ray and neutron scattering cross-sections (1 barn = 10^{-24} cm^2).

Table 2.3 The comparisons between X-ray diffraction, electron diffraction, and ND.

	X-ray Diffraction	Electron Diffraction	Neutron Diffraction
Typical sample form	Powders, bulk solids, thin films	Ultrathin samples (<100 nm), nanoparticles	Powders, bulk materials, electrode
Ability to detect light elements (Li, H, O)	Poor (Li nearly invisible)	Poor-moderate	Excellent (isotope-sensitive)
Sensitivity to heavy elements	Excellent	Excellent but with strong dynamical effects	Irregular but often sufficient
Multiple scattering effects	Generally weak → kinematical approximation valid	Strong → dynamical diffraction dominates	Weak-moderate; kinematical approximation often usable
Spatial resolution	Micron-scale probing volume	Sub-Å structural resolution (local)	Larger probing volume (bulk averaging)
Chemical state sensitivity	Accessible via XAS/ resonant techniques	Accessible via EELS	Limited or indirect
Operando capability	Good (synchrotron XRD widely used for operando cells)	Limited (sample geometry constraints)	Excellent for bulk electrodes and working devices
Representative applications	Phase identification, Rietveld refinement, thin-film epitaxy, strain analysis	Defect mapping, nanocrystal structure, zone-axis diffraction, interface structure	Cation ordering, Li/H location, magnetic structure, quantifying TM migration, bulk operando battery analysis
Main advantages	Mature, high-throughput, broad accessibility, strong for crystalline materials	Highest spatial resolution; excellent for nanoscale and defect-sensitive studies	Nondestructive bulk probing; sensitive to light elements & magnetism; ideal for operando battery studies
Main limitations	Weak for light elements; poor differentiation of neighboring 3D metals (Ni/Co/Mn)	Requires ultrathin samples; strong beam damage; complex dynamical effects	Requires large facilities; lower instrumental flux; coarser spatial resolution

enabling the study of local distortions and structural heterogeneity that are invisible to conventional diffraction.

Despite these substantial strengths, ND also faces several intrinsic limitations. First, ND experiments require access to high-flux research reactors or spallation neutron sources, which are available only at a limited number of large-scale facilities worldwide. The relatively low neutron flux compared with synchrotron X-rays leads to longer acquisition times, making rapid operando measurements challenging. Second, neutrons are strongly scattered or absorbed by hydrogen, necessitating careful selection of cell components, electrolytes, and sample environments to reduce background noise. Third, the spatial resolution is generally lower than that achievable with synchrotron micro-diffraction, limiting its ability to probe highly heterogeneous microstructures with fine spatial precision.

Nevertheless, the unique light-element sensitivity, nonsystematic elemental contrast, and true bulk structural fidelity of ND make it indispensable for studying electrochemical materials. In the context of lithium-ion and solid-state batteries, where Li transport pathways, cation ordering, oxygen activity, and lattice distortions are critically correlated with electrochemical performance, ND provides access to crystallographic descriptors that no other probe can resolve with comparable clarity. As such, ND continues to serve as a cornerstone technique for unraveling the complex structural chemistries underlying next-generation energy storage systems.

2.4.2 Total Scattering

While ND provides indispensable information on the long-range crystallographic evolution of battery materials, it remains intrinsically limited by its reliance on periodic and average structural models. *Rietveld* refinement effectively quantifies lattice parameters, site occupancies and phase fractions, yet it necessarily filters out structural features that deviate from ideal periodicity. As a result, many phenomena critical to electrochemical performance—such as short-range transition-metal migration, lattice oxygen distortions, defect clustering, and electrochemically induced spinel-like or rocksalt nanodomains—remain invisible or only weakly expressed in conventional diffraction patterns. These challenges are especially obvious in oxide cathodes, where structural complexity grows continuously during cycling. Consequently, capturing both long-range order and local, nonperiodic atomic correlations is essential, motivating the use of neutron total scattering and related PDF analysis.

Neutron total scattering addresses these limitations by simultaneously capturing the sharp Bragg reflections associated with long-range periodicity and the diffuse scattering contributions arising from disorder and short-range structural heterogeneity (as shown in Figure 2.11).

Rather than discarding the diffuse component as background, total scattering treats it as an equally important source of structural information. To fully exploit this information, the measured total scattering structure function $S(\vec{Q})$ is Fourier-transformed into the real space PDF G(r), which represents the probability distribution of atomic pair distances (the comparison between ND and total scattering-based neutron pair distribution function (NPDF) workflows are shown in Figure 2.12). Because PDF analysis does not rely on translational symmetry, it can resolve structural features on nanometer scales—whether the material is crystalline, nanocrystalline, partially amorphous, or undergoing dynamic reconstruction during

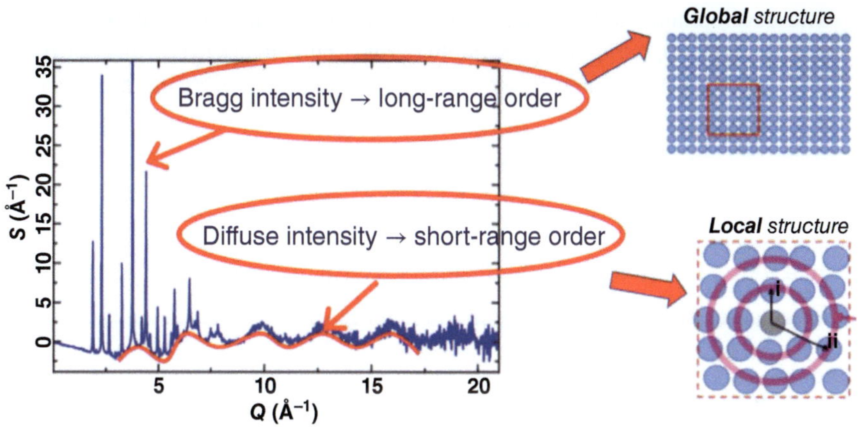

Figure 2.11 Total scattering data of the structure function $S(Q)$ as a function of the scattering vector $\vec{Q}$(Å^{-1}). The data (blue line) is composed of sharp Bragg peaks, which correspond to the long-range order of the material's average crystal structure. The underlying smooth red curve represents the diffuse scattering intensity, which arises from local structural disorder or short-range order [18] / with permission of Elsevier.

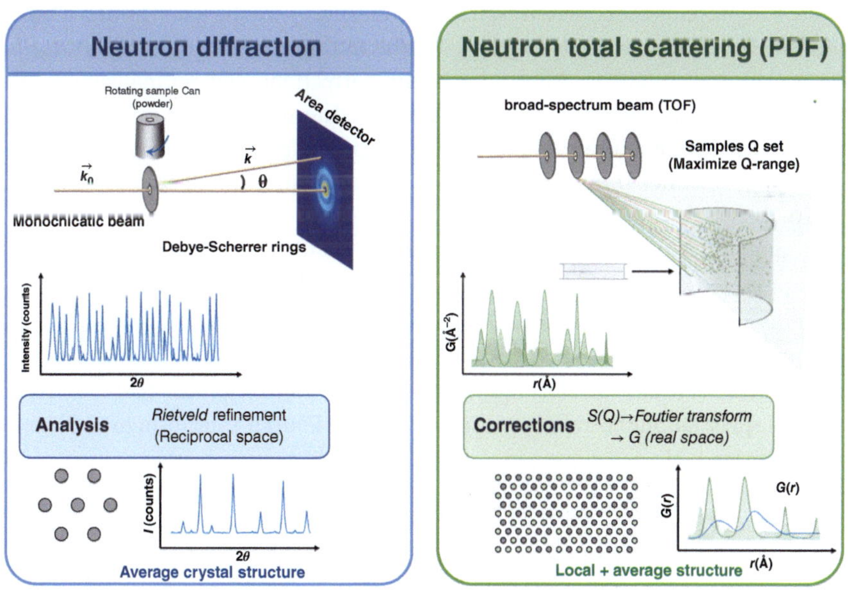

Figure 2.12 The experimental workflow differences between ND vs. neutron total scattering (PDF).

electrochemical cycling. This ability to bridge reciprocal-space and real space correlations makes total scattering and PDF particularly powerful for uncovering local distortions, SRO, and defect-driven transformations that accompany long-range structural evolution in battery electrodes.

The development of PDF method can be traced back to the Debye scattering equation (Eq. 2.2), which was originally derived by Peter Debye in 1915 [19]. The Debye scattering equation expresses the total coherent scattering intensity as a sum over all atomic pairs,

$$I(\vec{Q}) = \sum_{i}\sum_{j\neq i}\frac{\sin(Qr_{ij})}{Qr_{ij}} \tag{2.2}$$

where $I(\vec{Q})$ represents the total coherent scattering intensity as a function of the scattering vector $\vec{Q}$; r_{ij} represents the real space distance between atoms i and j, and $\frac{\sin(Qr_{ij})}{Qr_{ij}}$ is the spherical Bessel function that describes the interference contribution from each atomic pair to the scattering intensity. Because this formulation does not assume translational symmetry, the Debye scattering equation applies equally well to crystalline, nanocrystalline, and amorphous materials. This is also a key reason to support modern total scattering approaches.

In practice, the measured scattering $I_T(Q)$ intensity contains several contributions:

$$I_T = I_C + I_{IC} + I_{MC} + I_{BG} \tag{2.3}$$

where I_C represents the coherent scattering intensity, I_{IC} for the incoherent scattering intensity (nuclear spin scattering), I_{MC} for the multiple scattering intensity, I_{BG} for the background intensity. After appropriate corrections and normalizations, the coherent signal is converted into the total scattering structure function $S(Q)$. This quantity carries signatures of both long-range periodicity (sharp Bragg peaks) and short-range disorder (broad diffuse scattering), making it more structurally informative than diffraction intensity alone.

To amplify high-Q oscillations which contain critical information about short-range structure, $S(Q)$ is transformed into the reduced total scattering function:

$$F(Q) = Q\left[S(Q) - 1\right] \tag{2.4}$$

This form serves as the starting point for converting reciprocal-space scattering into real space atomic correlations. Applying a sine Fourier transform to $F(Q)$ yields the real space PDF:

$$G(r) = 4\pi r\left[\rho(r) - \rho_0\right] = 4\pi r\rho_0\left[g(r) - 1\right] \tag{2.5}$$

In this expression, $\rho(r)$ is the microscopic atomic pair-density distribution, ρ_0 for the average atomic number density, and $g(r)$ is the dimensionless PDF representing the relative probability of finding an atom at distance r. In detail,

$$g(r) = \frac{\rho(r)}{\rho_0} = \frac{1}{4\pi r^2 N\rho_0}\sum_{i}\sum_{j\neq i}\delta(r - r_{ij}) \tag{2.6}$$

When $r \to 0$, $g(r) \to 0$; $r \to \infty$, $g(r) \to 1$.

The key microstructural information such as atomic distances, coordination number, and structural disorder are extracted by analyzing the peak position, peak area, and peak width of PDF curves derived from total scattering data (Figure 2.13).

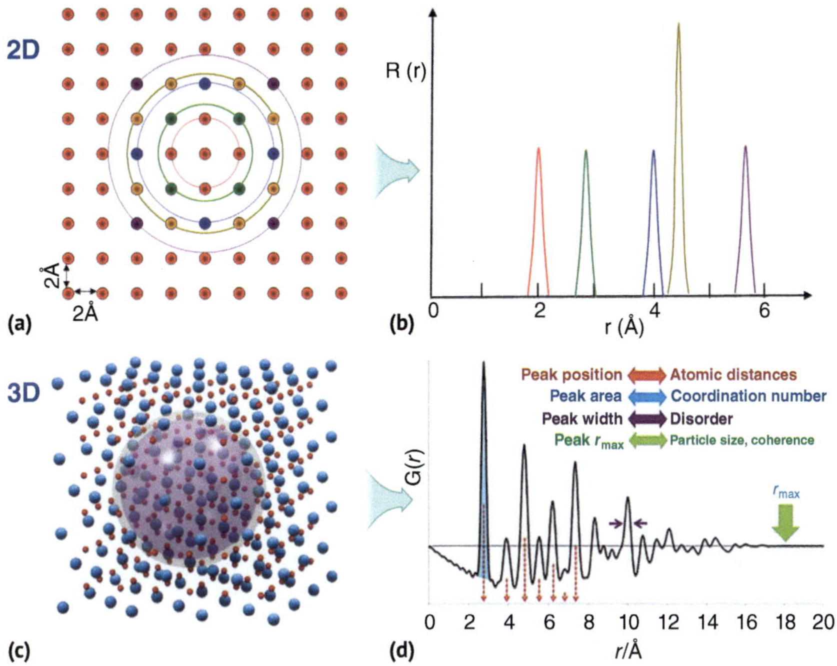

Figure 2.13 Schematic illustrations of (a) generation of PDF from a 2D crystal (the colored circles indicate distances of neighbors to the origin atom), and (b) corresponding peaks in the radial distribution function in the same color [20] / Elsevier / CC BY 4.0. (c) Atom distribution of CeO_2 in real space, and a sphere centered at an atom with radius *r* and annulus thickness *dr* [21] / with permission of John Wiley & Sons. (d) total scattering derived PDF in real space, *G*(*r*), provides real space structure information which is independent of structure modeling [22] / with permission of Cambridge University Press.

In a perfectly random distribution, $g(r) = 1$, whereas near-neighbor atomic correlations lead to pronounced peaks at characteristic bond lengths. Thus, $G(r)$ provides a direct, model-independent depiction of local atomic arrangements, capturing both ordered and disordered components.

In parallel with $G(r)$, the radial distribution function $R(r)$ is defined as:

$$R(r) = 4\pi r^2 \rho_0 g(r) \tag{2.7}$$

which gives the number of atoms within a spherical shell of thickness *dr* at distance *r*. $R(r)dr$ gives the number of atoms in an annulus of thickness *dr* at distance from another atom, which reflects the precise shape of the pair-probability distribution.

Neutrons provide several advantages for total scattering studies of energy storage materials. Their sensitivity to light elements, particularly Li and O, allows accurate extraction of Li-O, TM-O, and TM-TM correlations, which are intimately linked to ion transport pathways and anion-redox chemistry. The nonmonotonic variation of neutron scattering lengths with atomic number creates strong contrast among Ni, Co, Mn, and Fe—elements that are indistinguishable to X-rays at the short-range structural level. Combined with their high penetration depth, neutrons enable bulk-representative PDF data to be collected from thick electrodes and even sealed cells.

Beyond crystalline electrodes, total scattering is uniquely suited for characterizing partially amorphous phases such as CEI/SEI layers, surface reconstruction products, disordered rock salt domains, and glassy conversion phases. These components often dominate cycling behavior but produce only diffuse scattering signals, making them ideally analyzed by PDF. Operando neutron total scattering further enables tracking of bond-length fluctuations, disorder evolution, and phase competition in real time during cycling, heating, or external perturbation.

Despite its strengths, neutron total scattering still faces challenges: high-quality measurements require broad Q-range and long counting times, modeling may involve non-unique solutions, and hydrogen-rich sample environments must be carefully controlled due to background scattering. Nevertheless, total scattering serves as a critical complement to ND, enabling multiscale structural insight that fully captures the complexity of modern battery materials.

2.5 Applications of Neutron Techniques in Energy Storage Materials

ND is widely used in LIBs and SIBs to reveal long-range crystal structures, quantify cation distributions, and investigate structural evolution during electrochemical cycling. Because neutrons interact with atomic nuclei rather than electron density, ND provides strong sensitivity to light elements such as Li and Na and can distinguish adjacent 3d transition metals that are difficult to resolve by X-ray diffraction alone. This makes ND particularly valuable for identifying Li/Na positions, measuring cation disorder, locating oxygen atoms, and detecting subtle phase transitions associated with anionic redox or migration-driven degradation.

2.5.1 Cathode Materials

Cathode materials play a central role in determining the energy density, cost, and long-term cycling stability of LIBs and SIBs. Because these cathodes are crystalline solids whose performance is closely coupled to structural evolution, migration tendency, cation ordering, and the accessibility of Li/Na storage, ND has been widely applied to study their long-range structural behavior and its impact on electrochemical performance.

In commercial layered NCM cathodes, even small degrees of Li/Ni cation mixing strongly influence Li-ion transport, phase stability and long-term performance, making its accurate quantification essential [13, 23]. By using the high sensitivity of neutrons to light elements and their unique elemental contrast, our team quantitatively resolved Li/Ni anti-site defects in high-nickel oxide cathode with unprecedented precision (shown in Figure 2.14), even at low concentrations—a task extremely challenging for XRD or electron microscopy [24]. Using ND in combination with first-principles calculations, we further uncovered the halogen-induced favorable anti-site mechanism, demonstrating that F substitution promotes the formation of a more stable local octahedral structure (LOSH), as evidenced by the linear increase

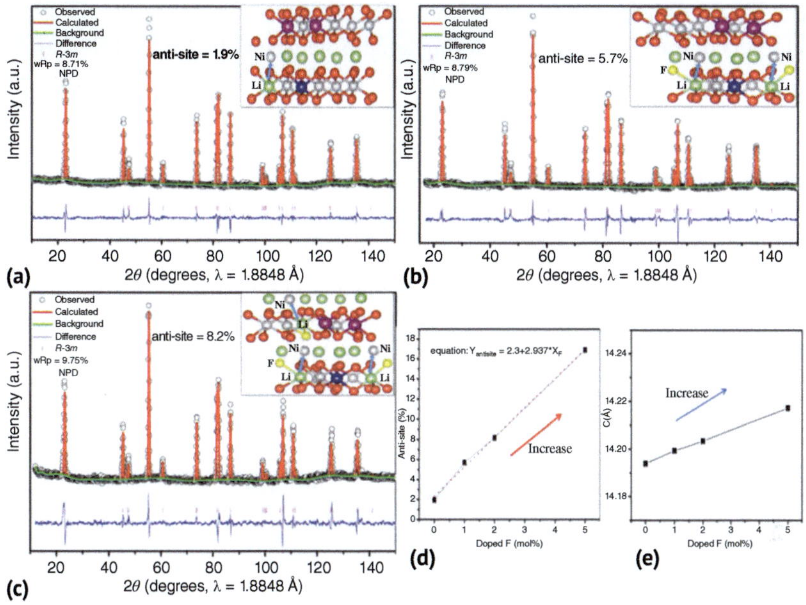

Figure 2.14 Neutron *Rietveld* refinement and the corresponding ground state crystal structures for (a) LNCM, (b) 1% F-LNCM, and (c) 2% F-LNCM, respectively. (d) The anti-site concentration and (e) changes in lattice parameter c with doped F content in the LNCM [25] / with permission of John Wiley & Sons / CC BY NC 4.0.

in anti-site concentration with F content and the corresponding refinement results. This ND-enabled insight allowed us to correlate anti-site tuning with enhanced lattice stability, increased c-parameter, improved Li-ion mobility, and markedly superior cycling and thermal performance.

In addition, in $Li[Ni_{0.5}Mn_{0.5}]O_2$ and Li-rich oxides, Ma et al. used time-of-flight ND and *Rietveld* refinement to obtain accurate site occupancies for Li, Ni and Mn and to show that excess Li substitutes into transition-metal 3b sites rather than forming a $Li[Ni_{0.5}Mn_{0.5}]O_2$-Li_2MnO_3 solid solution [26]. This work illustrates ND for quantifying small amounts of Li/Ni exchange between 3a and 3b positions and for relating suppressed structural disorder to improved electrochemical behaviors.

Cahill et al. combined high-resolution ^{6}Li NMR with X-ray and ND on $Li[Ni_{1/3}Mn_{1/3}Co_{1/3}]O_2$. ND provided an independent measure of Li/Ni anti-site disorder that agreed with *Rietveld* refinement and served as a benchmark for interpreting the local NMR environments [27]. ND is used not only to refine lattice parameters but also to extract precise occupancies and thereby quantify cation mixing in layered NMC cathodes.

For more complex Li-rich layered oxides, ND is applied to resolve phase symmetry and transition-metal ordering that are difficult to see by X-rays. In the Li-rich NMC ($Li_{1.2}Mn_{0.55}Ni_{0.15}Co_{0.1}O_2$), Ma et al. show that the ND pattern contains superstructure reflections that cannot be indexed by a simple trigonal $R\bar{3}m$ model [27]. Instead, a monoclinic C2/m Li_2MnO_3-type framework with mixed occupancy of

Mn, Co, and Ni on 2b/4g sites and Li on 2c/4h sites is required to fit the data. The much larger contrast in neutron scattering lengths between Mn, Co, and Ni than for X-rays is essential, enabling to refine multication ordering and to address questions about cation site interchange in Li- and TM-layers. In this way, ND directly supports structural models that underlie explanations of voltage fade, first-cycle irreversibility in Li-rich NMC cathodes.

In recent years, oxygen anionic redox has emerged as a crucial direction for boosting the capacity of layered oxides beyond the limits of traditional transition-metal redox, offering additional charge-compensation pairs $\left(O^{2-}/O_2^{n-}, 1 < n < 3\right)$ to higher energy density using abundant and low-cost elements. However, the low atomic number of oxygen makes its electronic and structural responses extremely difficult to probe, and most spectroscopic signatures remain indirect or ambiguous.

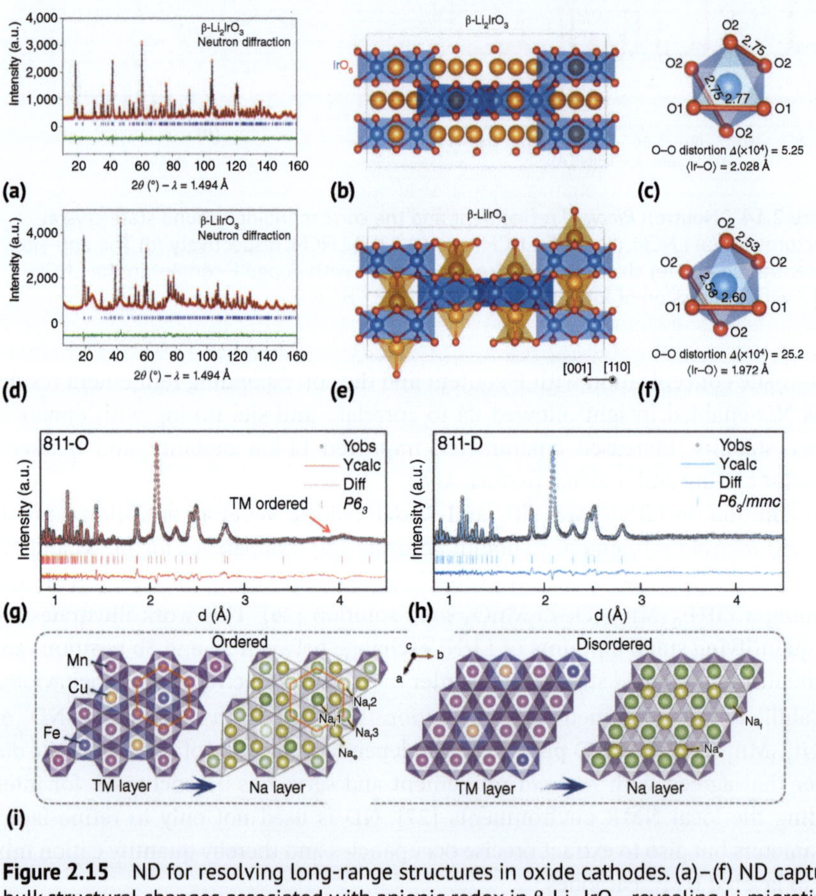

Figure 2.15 ND for resolving long-range structures in oxide cathodes. (a)–(f) ND captures bulk structural changes associated with anionic redox in β-Li_2IrO_3, revealing Li migration from octahedral to tetrahedral coordination and contraction of Ir-O and O-O distances upon charging [28] / with permission of Springer Nature. (g)–(h) ND identifies long-range cation ordering in layered Na transition-metal oxides, distinguishing ordered (811-O) and disordered (811-D) arrangements through *Rietveld* refinement and corresponding structural models. (i) Crystal structure schematics with TM ordered and disordered arrangement along [001] zone axis [16] / with permission of John Wiley & Sons.

ND (Figure 2.15a, d) provides a powerful and element-sensitive route to track oxygen lattice displacements, O-O distance changes, and redox-driven structural rearrangements that are inaccessible to conventional X-ray or electron-based techniques. Yet, many key questions remain unresolved, including how the oxygen sublattice distorts during oxidation, whether O-O dimers form, and how these processes couple to transition-metal environments. These challenges persist largely because oxygen with low atomic number is difficult to probe directly using conventional X-ray or electron-based methods. ND, however, is uniquely sensitive to oxygen positions and O-O distance changes, making it an indispensable tool for clarifying the structural rearrangements of anionic redox. Thus, neutron studies on Ir-based model systems have provided some of the crystallographic insights to date. For Li_2IrO_3, ND refinements of pristine and charged samples capture collective displacements of the oxygen sublattice that correspond to the formation of O-O dimers [28]. These bulk structural changes are consistent with local ABF-STEM images and DFT calculations. Upon Li removal, Li migrates to new tetrahedral sites, Ir-O distances contract, and the shortest O-O dimerization within IrO_6 octahedra decrease from 2.75 to 2.53 Å, providing direct structural evidence for peroxo-like species and simultaneous cationic-anionic redox (Figure 2.15b, c, e, f).

SIBs have gained increasing attention as a complementary technology to LIBs because sodium is far more abundant and widely distributed. The main advantages, especially the lower cost of Na elements, make SIBs particularly attractive for large-scale energy storage. Layered transition-metal oxides remain one of the most promising cathode families for SIBs, yet they face structural challenges that differ in important ways from their lithium analogues. Compared with LIBs layered oxides, Na-based compounds exhibit stronger Na^+/vacancy ordering, larger ionic radii, and more obvious phase heterogeneity during cycling, as shown in Figure 2.15g, h, i. Those features make their long-range structures particularly sensitive to cation arrangement and oxygen coordination. Enyue Zhao et al. adopted ND to provide the key crystallographic evidence for multiple superstructure units (Li@Mn_6, Mg@Mn_6 and Ni@Mn_6) that enable the gradient and de-clustered anionic-redox design [29]. As shown in Figure 2.16, joint refinement of XRD and ND data shows that models lacking these units fail to reproduce the superstructure reflections, confirming their structural reality. Moreover, ND verifies that these superstructures still persist even after charging to 4.5 V, indicating that high reversibility of oxygen redox proceeds without triggering extensive irreversible TM migration or O_2 evolution. Through these results, ND directly substantiates both the structural foundation and the electrochemical stability mechanism proposed for this Mn-based SIBs cathode.

While ND reveals the long-range evolution of the oxygen lattice, many key structural transformations to electrochemical cycling occur locally and are often heterogeneous across different regions of the cathode. Oxygen coordination, short-range lattice distortions, and local TM-O environments can evolve in ways that are not detected by average crystallographic models. Neutron total scattering and PDF analysis therefore become indispensable for resolving these local structural changes, providing real space insight that complements ND and enables a unified understanding of both long-range order and short-range anionic-redox reactions (ARR).

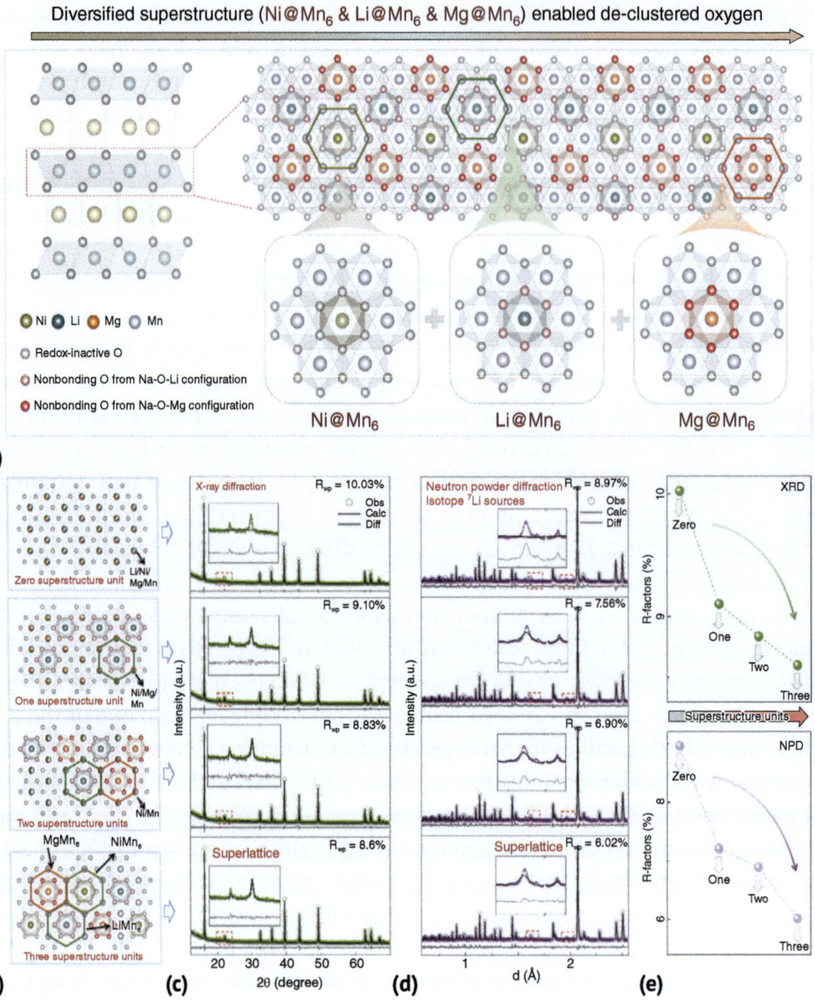

Figure 2.16 Schematic of diversified superstructure units (Ni@Mn_6, Li@Mn_6, and Mg@Mn_6) in the O-Mn-LNM sample. The declustered spatial distribution of oxygen ions kinetically avoids the anionic redox-induced stress concentration. (a) Illustration of cation distributions in the TM layer, including zero (cation disorder), one (Li@Mn_6), two (Li@Mn_6 & Mg@Mn_6), and three (Li@Mn_6 & Mg@Mn_6 & Ni@Mn_6) superstructure units; refinement results of the (b) XRD and (c) NPD data for the O-Mn-LNM sample using structure model including zero, one, two, and three superstructure units, insets show the enlarged superstructure diffraction peaks. The comparison of obtained fit indicators based on the (d) XRD and (e) NPD data [29] / with permission of John Wiley & Sons.

As shown in Figure 2.17, neutron total scattering combined with NPDF analysis is employed to directly probe oxygen-related, short-range structural correlations. Combining with the sensitivity of neutrons to oxygen, NPDF enables quantitative tracking of nearest-neighbor O-O distances and local lattice distortions during electrochemical cycling. This capability has distinguished localized oxygen lattice rearrangement associated with reversible ARR from extended distortions that would

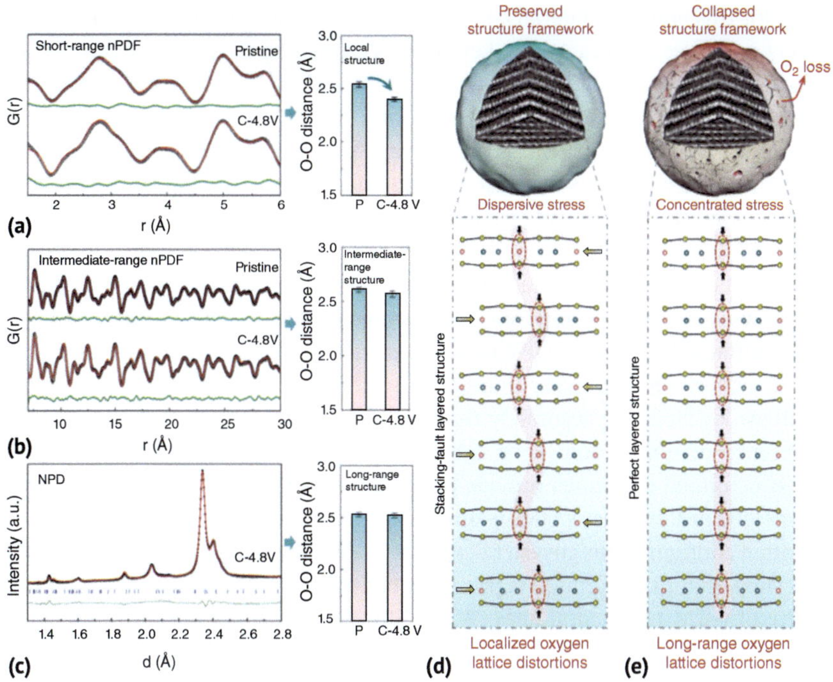

Figure 2.17 NPDF analysis and stable ARR mechanisms. For the pristine and 4.8 V-charged hSF-LMO samples: (a) Local nPDF data (left panel) and the refined nearest-neighbor O-O distance (right panel). (b) Same as with (a) but with intermediate-range nPDF data. (c) NPD data for the 4.8 V-charged hSF-LMO sample (left panel), and the refined nearest-neighbor O-O distance (right panel). (d, e) Schematic illustration of oxygen lattice evolutions during ARR process for the all-Mn-based Li-rich layered samples (d) with and (e) without stacking-fault defects [30] / John Wiley & Sons.

destabilize the layered framework. In addition, NPDF provides direct insight into how stacking-fault defects regulate oxygen lattice evolution at the atomic scale, thereby linking localized structural behavior to suppressed oxygen release and improved voltage stability [30].

Beyond layered oxide cathodes, cation-disordered rock salt (DRX) cathodes represent a new class of lithium-ion battery materials in which cations are randomly distributed on a NaCl-type lattice [31, 32]. From a long-range crystallographic perspective, such complete cation disorder would intuitively be expected to hinder Li^+ transport, as the absence of well-defined diffusion planes and the presence of randomly distributed transition metals should increase migration barriers and disrupt percolation pathways. However, Li-excess DRX materials challenge this intuition. Despite long-range cation disorder, excess lithium enables the formation of low-barrier Li migration channels through tetrahedral sites coordinated by zero transition metals configurations (0-TM) or 1-TM, and once these channels become three-dimensionally connected, an effective Li^+ percolation network emerges, allowing reversible lithiation and delithiation.

Thus, for this class of materials, long-range average structural probes are fundamentally insufficient, because the key physics governing Li transport and electrochemical performance is dictated by short-range cation configurations rather than long-range periodicity. Neutron total scattering and relative NPDF analysis therefore become particularly critical, as they directly probe local atomic correlations and SRO in chemically complex and disordered lattices. NPDF provide direct evidence that DRX cathodes are not truly random at the local scale, but instead exhibit pronounced SRO manifested in specific Li-Li and Li-F pair correlations. These local arrangements strongly influence the population and connectivity of 0-TM diffusion channels, which cannot be inferred from average rock salt diffraction alone [33].

By quantitatively fitting NPDF data obtained from neutron total scattering experiments, structural models incorporating realistic SRO reproduce local peak positions and intensities far more accurately than fully random configurations, confirming the presence of nonrandom local cation arrangements in DRX cathodes. These NPDF-constrained structures provide physically meaningful starting points for ab initio molecular dynamics (AIMD) simulations, enabling direct evaluation of how local cation configurations govern Li^+ transport. As shown in Figure 2.18, the AIMD results show that Li^+ migration is primarily controlled by the distribution and connectivity of 0-TM tetrahedral sites. While SRO has a limited effect on individual migration barriers, it strongly disrupts the long-range percolation of low-barrier

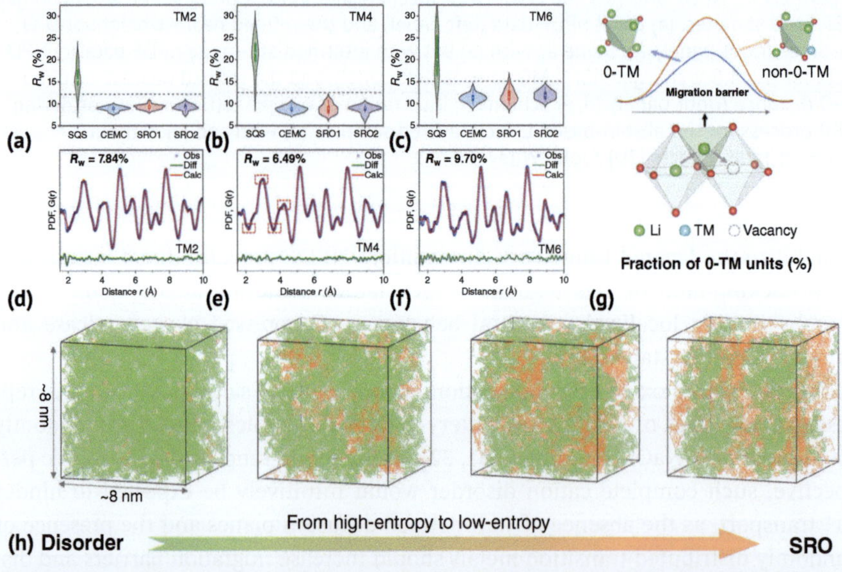

Figure 2.18 NPDF analysis and Li-ion percolation in Li-excess disordered rock salt cathodes, highlighting the role of SRO. (a)–(c) Comparison of PDF fitting quality for different structural models. (d)–(f) Representative NPDF refinements capturing short-range atomic correlations. (g) Schematic of Li^+ migration through 0-TM and non-0-TM tetrahedral sites. (h) Schematic illustrations of percolating (colored in green) and isolated (colored in orange) 0-TM Li when the structure changes from disorder to SRO in a supercell consisting of 64,000 atoms [33] / with permission of John Wiley & Sons.

pathways. As SRO increases, percolating 0-TM networks progressively fragment into isolated clusters, as visualized in large supercell percolation maps. In contrast, high configurational entropy suppresses SRO, preserves a homogeneous Li distribution, and maintains 3D connected Li migration networks.

2.5.2 Interfaces (SEI/CEI)

Fast charging in thick graphite electrodes is essential for achieving high energy LIBs without sacrificing charging time, yet the transport limitations inside such electrodes remain poorly understood. Fast charging in thick graphite electrodes is limited by complex solvent-ion transport, yet existing methods cannot directly resolve how electrolyte redistributes inside porous structures under large current. The mechanistic diagrams in Figure 2.19a illustrate that Li^+ migration can withdraw coordinated solvent molecules from the electrode by electro-osmotic drag. However, this process had never been experimentally visualized because X-ray techniques lack contrast for carbonate solvents, impedance data provide only indirect signatures, and modeling cannot be validated without spatially resolved measurements. The neutron attenuation difference shown in Figure 2.19b enables the key breakthrough. Protium-containing solvents exhibit strong neutron absorption, whereas ^7Li metal, $^7LiPF_6$ salt, graphite, and the separator are nearly transparent. By constructing a cell where only the solvent contributes significantly to neutron

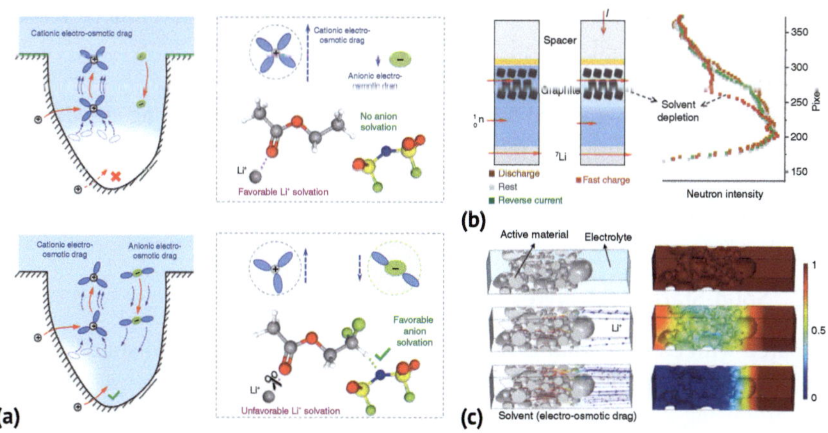

Figure 2.19 (a) Schematic representation of electro-osmotic drag in LIBs with different electrolytes. Upper: Schematic illustration of the mechanism of electro-osmotic drag overpotential. In conventional carbonate-based electrolytes, lithium ions are strongly solvated by solvents, whereas anions remain unsolvated. Lower: The difluorination electrolytes developed in this work weaken lithium-ion solvation and enhance anion solvation, thereby reducing the net electro-osmotic drag flow, preventing solvent depletion within the pore structure, and mitigating electro-osmotic drag overpotential. (b) The cross-sectional gray value of neutron images of the ^7Li/graphite coin cell at different states. (c) Schematic illustration of the finite element model together with the ion and solvent transport behaviors in a thick porous electrode during charging [34] / with permission of AAAS.

contrast, the authors isolate the ^{1}H signal and capture real time changes in solvent concentration during fast charging.

Under high-rate delithiation, neutron transmittance profiles reveal a pronounced increase in intensity within the electrode interior, directly corresponding to a decrease in local ^{1}H content. This provides unambiguous evidence of solvent depletion—an effect that cannot be attributed to electrode breathing or lithium redistribution, because those components contribute negligibly to neutron attenuation. Thus, neutron imaging uniquely verifies that solvent leaves the graphite pores during fast charging, confirming the directional solvent migration proposed in Figure 2.19a. The images also demonstrate that solvent depletion reaches its maximum near the separator side and extends deep into the porous network, establishing the spatial heterogeneity of electrolyte availability.

The simulations in Figure 2.19c integrate the neutron-derived solvent profiles into a transport model and show how Li^{+} flux induces asymmetric solvent flow. The calculated solvent velocity fields and concentration maps quantitatively reproduce the patterns observed experimentally, validating the cation-driven electro-osmotic mechanism. These results also clarify the reasons that fast charging becomes inefficient in thick electrodes. Solvent is physically displaced from the active regions, increasing viscosity, reducing ionic conductivity, and generating concentration overpotentials that cannot be mitigated by simply improving electrode porosity or reducing tortuosity.

By combining targeted neutron imaging with solvation-engineered electrolyte design, the study demonstrates that weakening Li^{+} solvation and strengthening anion solvation suppress electro-osmotic drag, as shown by the reduced solvent depletion under modified electrolyte conditions. This establishes a direct causal link between solvation chemistry, ion-solvent transport, and fast-charging performance. Through these neutron-enabled observations, the work resolves a long-standing hypothesis, provides quantitative mechanistic evidence for electro-osmotic drag, and reframes electrolyte design as a crucial lever for overcoming thick-electrode fast-charging limitations.

2.5.3 Solid-State Electrolytes (SSEs)

Compared with conventional LIBs with liquid electrolytes, all-solid-state batteries offer intrinsic advantages in safety, thermal stability and energy density, making them promising candidates for next-generation energy storage systems. In SSEs, lithium-ion transport is governed by local structural environments, where subtle variations in short-range anion configurations can strongly modulate migration bottlenecks and activation barriers.

Neutron total scattering combined with NPDF analysis is used to directly resolve temperature-dependent short-range structural evolution in Li_3YCl_6-type halide solid electrolytes, which cannot be captured by average-structure diffraction, as shown in Figure 2.20 [35]. Although conventional neutron and synchrotron XRD show that the long-range crystal structure and symmetry remain essentially unchanged across the superionic transition, NPDF reveals pronounced changes in local anion

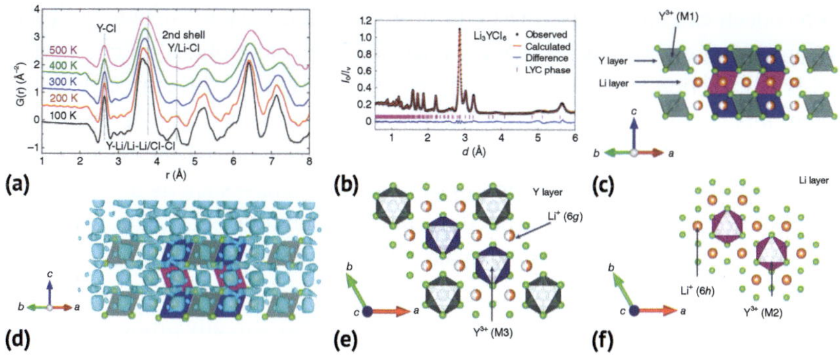

Figure 2.20 Short-range NPDF data of LYC collected at different temperatures. FDMs generated from the ND pattern of Li_3YCl_6 at 500K with a cut-off of $-0.02\,fm\,Å^{-3}$ (f) [35] / with permission of Springer Nature.

correlations that govern lithium-ion transport. The primary Y-Cl coordination environment remains rigid, with negligible variation in Y-Cl bond lengths, ruling out framework collapse or cation rearrangement as the origin of enhanced conductivity. In contrast, distinct weakening and broadening of short-range Cl-Cl and second-shell Y-Cl correlations are observed at elevated temperatures, demonstrating a loss of short-range anion order. These changes directly reflect a transition in collective anion dynamics, in which Cl^- ions evolve from correlated rocking motions to a more delocalized breathing mode of the YCl_6 octahedra.

Crucially, the nPDF-derived short-range anion disorder is shown to expand local diffusion bottlenecks within the Li sublattice, particularly in the 2D ab plane, thereby reducing migration barriers without altering the average lattice frameworks. This mechanism establishes that the superionic behavior arises from dynamic local anion rearrangements rather than static structural transformations. By correlating NPDF-resolved short-range structural evolution with ionic conductivity and activation energy, the work demonstrates that lithium-ion transport in halide electrolytes is controlled by collective anion motion at the local scale, providing a quantitative structure-transport relationship grounded in experimentally resolved short-range structure.

2.6 Challenges and Future Directions

Neutron-based characterization techniques, particularly ND and neutron total scattering combined with PDF analysis, have become increasingly important in the study of energy storage materials. Unlike X-rays and electrons, which interact primarily with electron density, neutrons scatter from atomic nuclei, endowing them with unique sensitivity to light elements such as Li, H, O, and F, as well as strong isotope and magnetic contrast. This fundamental difference makes neutron techniques a powerful and complementary tool for resolving structural features that are often inaccessible or ambiguous in X-ray and electron-based measurements.

ND has been extensively used to determine long-range crystal structures, phase evolution, and site occupancies in electrodes and solid electrolytes. In comparison with X-ray diffraction, neutrons can more reliably distinguish elements with similar atomic numbers and directly quantify lithium positions and occupancies, providing essential information on bulk structural evolution during electrochemical cycling. However, many key processes in energy storage materials—such as ion transport, oxygen redox, defect formation, and structural degradation—originate from local structural rearrangements that are averaged out in conventional diffraction. Here, neutron total scattering and PDF analysis play a critical complementary role. By capturing both Bragg and diffuse scattering, neutron PDF directly probes short- and intermediate-range atomic correlations, revealing local bonding environments, SRO, and lattice distortions that are invisible to average-structure probes.

Compared with electron microscopy, which offers atomic-resolution imaging but is inherently surface-sensitive and often limited by beam damage and small sampling volumes, neutron PDF provides statistically averaged bulk local structural information under realistic conditions. In disordered and complex systems—including Li-excess disordered rock salt cathodes, high-entropy materials, amorphous electrodes, and solid electrolytes—neutron PDF has proven essential for linking local structure to lithium-ion percolation, migration pathways, redox mechanisms, and mechanical stability. These insights complement real space imaging and reciprocal-space X-ray analysis, forming a more complete structural picture across length scales.

Looking forward, the impact of neutron-based techniques in energy storage research will increasingly depend on their integration with other characterization methods. Multimodal approaches combining ND and PDF with synchrotron X-ray scattering, electron microscopy, spectroscopy, and electrochemical analysis will enable a comprehensive understanding of structure–property relationships from the atomic to the device scale. Furthermore, the combination of neutron experiments with atomistic modeling, including density functional theory, AIMD, and machine-learning-assisted structural reconstruction, will be crucial for translating experimentally observed local structures into quantitative predictions of transport and stability.

Advances in operando and in-situ neutron techniques, enabled by high-flux spallation sources and improved instrumentation, are expected to further expand the capability of neutrons to probe energy materials under realistic working conditions. Together, ND and neutron total scattering/PDF will continue to serve as indispensable complements to X-ray and electron techniques, playing a central role in the rational design and optimization of next-generation energy storage materials.

References

1 J. Wei, H. Chen, Y. Chen, Y. Chen, Y. Chi, C. Deng, H. Dong, L. Dong, S. Fang, J. Feng, S. Fu, L. He, W. He, Y. Heng, K. Huang, X. Jia, W. Kang, X. Kong, J. Li, T. Liang, G. Lin, Z. Liu, H. Ouyang, Q. Qin, H. Qu, C. Shi, H. Sun, J. Tang, J. Tao,

C. Wang, F. Wang, D. Wang, Q. Wang, S. Wang, T. Wei, J. Xi, T. Xu, Z. Xu, W. Yin, X. Yin, J. Zhang, Z. Zhang, Z. Zhang, M. Zhou, T. Zhu, China spallation neutron source: Design, R&D, and outlook, Nuclear Instruments and Methods in Physics Research Section A: Accelerators, Spectrometers, Detectors and Associated Equipment 600(1) (2009) 10–13. https://doi.org/10.1016/j.nima.2008.11.017.

2 G.E. Pérez, J.M. Brittain, I. McClelland, S. Hull, M.O. Jones, H.Y. Playford, S.A. Cussen, P.J. Baker, E.M. Reynolds, Neutron and muon characterisation techniques for battery materials, Journal of Materials Chemistry A 11(20) (2023) 10493–10531. https://doi.org/10.1039/d2ta07235a.

3 Z.H. Lu, R.A. Donaberger, J.R. Dahn, Superlattice ordering of Mn, Ni, and Co in layered alkali transition metal oxides with P2, P3, and O3 structures, Chemistry of Materials 12(12) (2000) 3583–3590. https://doi.org/10.1021/cm000359m.

4 M. Bianchini, F. Fauth, E. Suard, J.-B. Leriche, C. Masquelier, L. Croguennec, Spinel materials for Li-ion batteries: New insights obtained by operando neutron and synchrotron X-ray diffraction, Acta Crystallographica Section B Structural Science, Crystal Engineering and Materials 71(6) (2015) 688–701. https://doi.org/10.1107/s2052520615017199.

5 M. Bianchini, E. Suard, L. Croguennec, C. Masquelier, Li-Rich Li1+xMn2–xO4 spinel electrode materials: An operando neutron diffraction study during Li+ extraction/insertion, The Journal of Physical Chemistry C 118(45) (2014) 25947–25955. https://doi.org/10.1021/jp509027g.

6 S. Liang, Y. Sun, Y.J. Cheng, S.V. Roth, P. Müller-Buschbaum, L. Cao, J. Peng, S. Wang, X. Sun, C. Wang, Probing all-solid-state batteries with real-time synchrotron and neutron techniques, Advanced Energy Materials (2025). https://doi.org/10.1002/aenm.202504045.

7 T.U.o.M.T. Heinz Maier-Leibnitz research neutron source (FRM II), Reactor, 2025. (Accessed 18 December 2025). https://www.frm2.tum.de/en/frm2/the-neutron-source/reactor/.

8 C. Ye, China Advanced Research Reactor (CARR): A new reactor to be built in China for neutron scattering studies, Physica B: Condensed Matter 241-243 (1997) 48–49. https://doi.org/10.1016/S0921-4526(97)00509-7.

9 H. Chen, Y. Chen, S. Fu, L. Ma, S. Wang, F. Chen, Y. Chen, H. Dong, L. Dong, G. Feng, J. Gu, L. He, K. He, W. He, C. Hu, J. Huang, Q. Ji, X. Jia, D. Jin, L. Kang, W. Kang, T. Liang, G. Lin, H. Liu, J. Li, H. Ouyang, F. Qi, X. Qi, H. Qu, H. Sun, Z. Sun, L. Shen, J. Tang, J. Tao, F. Wang, L. Wang, P. Wang, Q. Wang, Y. Wu, J. Xi, T. Xu, W. Yin, B. Zhang, J. Zhang, J. Zhang, S. Zhang, J. Zhao, L. Zhao, Y. Zhao, M. Zhou, T. Zhu, J. Zhuang, Design and construction of the China spallation neutron source, Nuclear Instruments and Methods in Physics Research Section A: Accelerators, Spectrometers, Detectors and Associated Equipment 1078 (2025) 170431. https://doi.org/10.1016/j.nima.2025.170431.

10 L. He, S. Deng, F. Shen, J. Chen, H. Lu, Z. Tan, H. Zheng, J. Hao, D. Zhao, Q. Ma, The performance of general purpose powder diffractometer at CSNS, Nuclear Instruments and Methods in Physics Research Section A: Accelerators, Spectrometers, Detectors and Associated Equipment 1054 (2023). https://doi.org/10.1016/j.nima.2023.168414.

11 B. Qiu, Y. Zhou, H. Liang, M. Zhang, K. Gu, T. Zeng, Z. Zhou, W. Wen, P. Miao, L. He, Y. Xiao, S. Burke, Z. Liu, Y.S. Meng, Negative thermal expansion and oxygen-redox electrochemistry, Nature 640(8060) (2025) 941–946. https://doi.org/10.1038/s41586-025-08765-x.

12 A. Zeng, N. Li, H. Zhang, S. Qiu, Z. Xu, Y. Zhang, W. Yin, E. Zhao, X. Xiao, Anchoring ligand electron enables robust metal-oxygen coordination toward 4.5 V O3-type sodium-ion battery cathodes, Angewandte Chemie International Edition 64 (2025) e202514220. https://doi.org/10.1002/anie.202514220.

13 B. Zhang, Y. Zhang, X. Wang, H. Liu, Y. Yan, S. Zhou, Y. Tang, G. Zeng, X. Wu, H.G. Liao, Y. Qiu, H. Huang, L. Zheng, J. Xu, W. Yin, Z. Huang, Y. Xiao, Q. Xie, D.L. Peng, C. Li, Y. Qiao, S.G. Sun, Role of substitution elements in enhancing the structural stability of Li-Rich Layered Cathodes, Journal of the American Chemical Society 145 (2023) 8700–8713. https://doi.org/10.1021/jacs.3c01999.

14 M.T. Dove, G. Li, Review: Pair distribution functions from neutron total scattering for the study of local structure in disordered materials, Nuclear Analysis 1(4) (2022) 100037. https://doi.org/10.1016/j.nucana.2022.100037.

15 Z. Song, T. Wang, H. Yang, W.H. Kan, Y. Chen, Q. Yu, L. Wang, Y. Zhang, Y. Dai, H. Chen, W. Yin, T. Honda, M. Avdeev, H. Xu, J. Ma, Y. Huang, W. Luo, Promoting high-voltage stability through local lattice distortion of halide solid electrolytes, Nature Communications 15(1) (2024) 1481. https://doi.org/10.1038/s41467-024-45864-1.

16 M. Yang, T. Yang, M. Dong, Z. Huang, Y. Pu, L. Jin, R. Wang, Y. Luo, T. Zeng, Y. Tang, J. Chen, R.E. Dunin-Borkowski, Z. Chen, Y. Qiao, Y. Xiao, Long-range cation disorder enhances comprehensive performance in Mn-rich layered sodium cathodes, Advanced Materials 37(30) (2025) e2500984. https://doi.org/10.1002/adma.202500984.

17 C.U.o.H.K. Centre for neutron scattering, Multi Physics Instrument (MPI), 2025. https://www.cityu.edu.hk/cns/crf/mpi.htm.

18 L. Yang, T. Kang, J. Xu, W. Yin. Unveiling local magnetic correlations: the development of magnetic pair distribution function at CSNS, Nuclear Instruments and Methods in Physics Research Section A: Accelerators, Spectrometers, Detectors and Associated Equipment 1069 (2024) 169967. https://doi.org/10.1016/j.nima.2024.169967.

19 P. Debye, Zerstreuung von Röntgenstrahlen, Annalen der Physik 351(6) (1915) 809–823. https://doi.org/10.1002/andp.19153510606.

20 S. Huang, K. Xiang, J. Mi, Recent advances in synchrotron X-ray studies of the atomic structures of metal alloys in liquid state, Journal of Materials Science & Technology 203 (2024) 180–200. https://doi.org/10.1016/j.jmst.2024.02.038.

21 Y. Ren, X. Zuo, Synchrotron X-ray and neutron diffraction, total scattering, and small-angle scattering techniques for rechargeable battery research, Small Methods 2(8) (2018). https://doi.org/10.1002/smtd.201800064.

22 K.W. Chapman, Emerging operando and X-ray pair distribution function methods for energy materials development, MRS Bulletin 41(3) (2016) 231–240. https://doi.org/10.1557/mrs.2016.26.

23 Y. Niu, Z. Zhuo, J. Hao, T. Yang, T. Cui, C. Gao, Y. Yu, J. Jiao, T. Luo, Y. Zhou, H. Ren, Y. Xiang, L. He, B. Li, Designing co-free medium-Ni layered oxide cathodes via additional li substitution, Advanced Materials 37 (2025) e07530. https://doi.org/10.1002/adma.202507530.

24 C. Li, W.H. Kan, H. Xie, Y. Jiang, Z. Zhao, C. Zhu, Y. Xia, J. Zhang, K. Xu, D. Mu, F. Wu, Inducing favorable cation antisite by doping halogen in Ni-rich layered cathode with ultrahigh stability, Advanced Science (Weinh) 6(4) (2019) 1801406. https://doi.org/10.1002/advs.201801406.

25 C. Li, W.H. Kan, H. Xie, Y. Jiang, Z. Zhao, C. Zhu, Y. Xia, J. Zhang, K. Xu, D. Mu, F. Wu, Lithium-ion batteries: Inducing favorable cation antisite by doping halogen in Ni-rich layered cathode with ultrahigh stability (Adv. Sci. 4/2019), Advanced Science 6(4) (2019) 1970021. https://doi.org/10.1002/advs.201970021.

26 S.-T. Myung, S. Komaba, K. Kurihara, K. Hosoya, N. Kumagai, Y.-K. Sun, I. Nakai, M. Yonemura, T. Kamiyama, Synthesis of Li[(Ni0.5Mn0.5)1-xLix]O2 by emulsion drying method and impact of excess li on structural and electrochemical properties, Chemistry of Materials 18(6) (2006) 1658–1666. https://doi.org/10.1021/cm052704j.

27 D. Mohanty, A. Huq, E.A. Payzant, A.S. Sefat, J. Li, D.P. Abraham, D.L. Wood, C. Daniel, Neutron diffraction and magnetic susceptibility studies on a high-voltage Li1.2Mn0.55Ni0.15Co0.10O2 lithium ion battery cathode: Insight into the crystal structure, Chemistry of Materials 25(20) (2013) 4064–4070. https://doi.org/10.1021/cm402278q.

28 P.E. Pearce, A.J. Perez, G. Rousse, M. Saubanere, D. Batuk, D. Foix, E. McCalla, A.M. Abakumov, G. Van Tendeloo, M.L. Doublet, J.M. Tarascon, Evidence for anionic redox activity in a tridimensional-ordered Li-rich positive electrode beta-Li2IrO3, Nature Materials 16(5) (2017) 580–586. https://doi.org/10.1038/nmat4864.

29 N. Li, E. Zhao, Z. Zhang, W. Yin, L. He, B. Wang, F. Wang, X. Xiao, J. Zhao, Gradient and De-clustered anionic redox enabled undetectable O(2) formation in 4.5 V sodium manganese oxide cathodes, Advanced Materials 36(48) (2024) e2408984. https://doi.org/10.1002/adma.202408984.

30 K. Wu, E. Zhao, P. Ran, W. Yin, Z. Zhang, B.T. Wang, K. Ikeda, T. Otomo, X. Xiao, F. Wang, J. Zhao, Localizing oxygen lattice evolutions eliminates oxygen release and voltage decay in all-Mn-based Li-rich cathodes, Small 19(18) (2023) 2300419. https://doi.org/10.1002/smll.202300419.

31 A. Urban, D.-H. Seo, G. Ceder, Computational understanding of Li-ion batteries, npj Computational Materials 2(1) (2016). https://doi.org/10.1038/npjcompumats.2016.2.

32 J. Lee, D.A. Kitchaev, D.H. Kwon, C.W. Lee, J.K. Papp, Y.S. Liu, Z. Lun, R.J. Clement, T. Shi, B.D. McCloskey, J. Guo, M. Balasubramanian, G. Ceder, Reversible Mn($^{2+}$)/Mn($^{4+}$) double redox in lithium-excess cathode materials, Nature 556(7700) (2018) 185–190. https://doi.org/10.1038/s41586-018-0015-4.

33 J. Liao, H. Chen, Y. Xie, Z. Li, S. Tan, S. Zhou, L. Jiang, X. Zhang, M. Liu, Y.B. He, F. Kang, Z. Lun, S. Zhao, T. Hou, Modeling short-range order in high-entropy cation-disordered rocksalt-type cathodes, Advanced Energy Materials 15(32) (2025). https://doi.org/10.1002/aenm.202501857.

34 C.-X. Zhao, Z. Wang, D. Jacobson, Y. Li, B. Khaykovich, S. Fayfar, L. Zheng, J. LaManna, X. Chen, D.S. Hussey, F. Chen, G.M. Veith, C. Wang, Electrolytes that reduce electro-osmotic drag improve fast charging of lithium-ion batteries, Science 390(6774) (2025) 745–750. https://doi.org/10.1126/science.adv1739.

35 Z. Liu, P.-H. Chien, S. Wang, S. Song, M. Lu, S. Chen, S. Xia, J. Liu, Y. Mo, H. Chen, Tuning collective anion motion enables superionic conductivity in solid-state halide electrolytes, Nature Chemistry 16(10) (2024) 1584–1591. https://doi.org/10.1038/s41557-024-01634-6.

3

X-ray Diffraction and Scattering Techniques: From Crystal Structure to Multiscale Structural Analysis

3.1 Fundamentals and Techniques of X-ray Diffraction

The discovery and development of X-ray diffraction (XRD) are deeply intertwined with the history of modern physics and crystallography. In 1895, Wilhelm Conrad Röntgen observed an unknown form of penetrating radiation while experimenting with cathode ray tubes. He named it X-rays for their mysterious nature, and their ability to pass through matter and produce photographic images quickly led to their application in medicine and materials science.

A breakthrough occurred in 1912 when Max von Laue and his colleagues demonstrated that crystals could diffract X-rays, providing the first direct evidence of their wave nature and revealing that crystal lattices act as three-dimensional diffraction gratings. This achievement opened the door to structural determination of solids at the atomic scale. In 1913, William Lawrence Bragg and William Henry Bragg formulated the relationship now known as Bragg's law, linking the angles of diffraction to the spacing between lattice planes. Their work laid the foundation for the quantitative analysis of crystal structures and earned them the 1915 Nobel Prize in Physics.

Shortly after, Henry Moseley established a systematic correlation between X-ray frequencies and atomic numbers, now called Moseley's Law. This provided a firm basis for the modern periodic table and allowed precise identification of elements.

Over the following decades, XRD techniques evolved from early photographic film methods to modern diffractometers equipped with precise goniometers, monochromators, and advanced detectors. Today, XRD is not only a standard tool for phase identification but also an indispensable technique for determining lattice parameters, quantifying phase fractions, studying microstructure, and probing materials in-situ under real operating conditions.

3.1.1 Fundamental Principles of XRD

XRD is governed by the constructive interference of X-ray waves scattered by the periodic arrangement of atoms in a crystal lattice. The condition for constructive interference is expressed by Bragg's law:

$$n\lambda = 2d_{hkl} \sin \theta \tag{3.1}$$

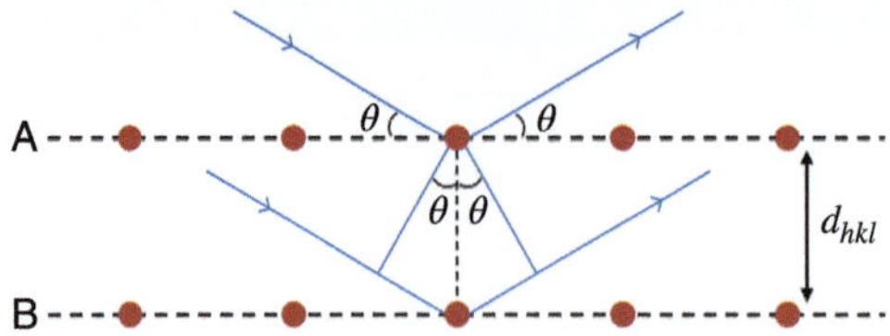

Figure 3.1 Bragg geometry and path difference construction.

where n is the diffraction order, λ is the incident X-ray wavelength, d_{hkl} is the spacing between crystallographic planes indexed by Miller indices, and θ is the Bragg angle. This relationship directly connects the positions of diffraction peaks with the crystal's geometric structure.

Figure 3.1 illustrates the real space geometry behind this relation for two parallel planes (A and B) separated by d_{hkl}. The incident and reflected angles are equal (θ), and the ray scattered from the lower plane travels an extra distance of $2d_{\text{hkl}} \sin\theta$. When this path difference equals an integer multiple of the wavelength ($n\lambda$), the reflected waves interfere constructively and a diffraction maximum is observed. In practice, we most often consider first-order diffraction ($n = 1$), but higher orders are permitted when the geometry and spectral bandwidth allow.

Bragg's law determines where diffraction peaks appear, but the structure factor determines their intensities. The structure factor for a reflection with indices is:

$$F_{\text{hkl}} = \sum_{j=1}^{N} f_j \exp\left[2\pi i\left(hx_j + ky_j + lz_j\right)\right] \tag{3.2}$$

Here, f_j is the atomic scattering factor of the j-th atom in the unit cell, and (x_j, y_j, z_j) are its fractional coordinates. The measured intensity of a diffraction peak is proportional to $|F_{hkl}|^2$, modified by experimental factors such as polarization, Lorentz corrections, and absorption.

Interference between waves scattered by different atoms can be constructive or destructive, sometimes resulting in systematic absences—peaks that are forbidden by the crystal's symmetry. These absences, together with intensity patterns, are critical for determining the space group of a crystal.

In powder diffraction, the sample contains an enormous number of randomly oriented crystallites. This randomness ensures that for any set of lattice planes, some fraction of the crystallites satisfy the Bragg condition, producing diffraction in all azimuthal directions. Each (hkl) reflection therefore forms a complete cone of scattering in reciprocal space, recorded as a ring in 2D images or as a sharp peak in 1D scans.

This powder averaging is a practical advantage—it removes the need to orient the sample precisely as in single-crystal diffraction. However, it also introduces the problem of peak overlap. Different reflections may have similar d_{hkl} spacings, especially in low-symmetry crystals, large unit cells, or when peak broadening occurs due to small crystallite size or microstrain. Overlap complicates manual peak identification and can obscure weaker reflections entirely.

To overcome these limitations, modern analysis relies on whole-pattern fitting methods, such as Rietveld refinement, which simultaneously model all peaks using a structural and instrumental profile function. Recognizing the inevitability of peak overlap helps beginners avoid overinterpretation of individual peaks and encourages adoption of statistically robust refinement approaches from the outset.

3.1.2 Diffraction Modes: Bragg, Laue, and Powder

With the interference principles now established, XRD can be realized experimentally in three canonical modes that differ by spectrum (monochromatic vs. polychromatic), sample state (single crystal vs. polycrystalline powder), and the degrees of freedom scanned during data collection. Each mode interrogates reciprocal space in a distinct way and therefore yields different types of information and data products; understanding these differences clarifies why later sections choose particular instruments and geometries for particular scientific questions.

Bragg (rotating-crystal/single-crystal) mode employs a monochromatic beam incident on a single crystal that is rotated with respect to the X-ray wavevector and the Ewald sphere. As the crystal is rocked (ω), spun (φ), and/or tilted (κ), successive reciprocal lattice points intersect the Ewald sphere, producing discrete reflections at precisely defined scattering vectors. Because each reflection can be measured in isolation with minimal peak overlap, single-crystal Bragg experiments provide exceptionally clean intensity sets for structural solution and refinement, precise unit cell determination, and accurate orientation matrices. Contemporary diffractometers use goniometers offering multiple rotational axes and area detectors to accelerate full reciprocal space mapping, while historical camera methods (e.g., Weissenberg, precession) achieved similar separation of reflections by recording curved sections of reciprocal space on film. The principal limitations are the need for a suitable single crystal (size, perfection, mosaicity) and the susceptibility of intensities to dynamical diffraction in very perfect crystals—factors that motivate powder methods when crystals are unavailable or small.

Laue mode uses a polychromatic ("white") spectrum on a stationary or minimally rocked single crystal. Because a continuum of wavelengths is present, many Bragg conditions are satisfied simultaneously, yielding a characteristic Laue spot pattern. The geometry is particularly powerful for (i) rapid crystal orientation and alignment (e.g., cutting wafers, mounting specimens for other probes), (ii) mapping lattice orientation and elastic strain fields when combined with micro-focused beams (Laue microdiffraction), and (iii) ultrafast studies that require sub-millisecond exposures, where time-resolved wavelength calibration is impractical. Indexing a Laue pattern typically requires an approximate unit cell (or an energy-dispersive detector to assign wavelengths ex post facto), and absolute structure factor extraction is less straightforward than in monochromatic single-crystal work; nonetheless, the mode excels wherever orientation/strain information and speed dominate over full-intensity accuracy.

Powder (Debye–Scherrer/Bragg–Brentano) mode targets polycrystalline or powdered specimens comprising myriad randomly oriented crystallites. Under these conditions, each set of lattice planes generates a diffraction cone; 2D detectors register concentric rings and 1D scans record sharp peaks at 2θ positions determined by d_{hkl}. Two laboratory implementations dominate: transmission through a capillary (Debye–Scherrer), which improves powder averaging and mitigates preferred orientation and absorption, and reflection from a flat specimen (Bragg–Brentano), which offers high throughput and straightforward sample handling. Powder diffraction is the workhorse for phase identification, quantitative phase analysis, lattice parameter metrology, and microstructural characterization (size/strain, residual stress, texture) via whole-pattern approaches. Its principal trade-off is peak overlap in complex/low-symmetry structures or when peaks are broadened by small coherent domain sizes and microstrain—precisely the scenario addressed later by total scattering/Pair Distribution Function methods when local order is of interest.

Taken together, the three modes form a practical decision tree. If a well-formed single crystal is available and the goal is full three-dimensional structural refinement with unique reflections and minimal overlap, the Bragg rotating-crystal mode is unmatched. If rapid orientation determination, strain mapping, or ultrafast snapshotting is required, Laue is the natural choice. If the specimen is polycrystalline, multiphase, or available only as a powder—or if statistical representativeness across grains is desired—powder diffraction provides the most robust and widely applicable pathway, with an analysis ecosystem that supports both qualitative and quantitative outcomes.

3.1.3 Powder XRD Instrumentation and Geometries

A modern powder diffractometer generates a quasi-monochromatic X-ray beam, delivers it to the specimen under controlled angular conditions, and records the diffracted intensity with high precision. Laboratory sources most commonly employ sealed tubes (e.g., Cu Kα, Co Kα, Mo Kα), while rotating-anode generators provide higher flux for demanding applications. Wavelength selection and beam conditioning are achieved with crystal monochromators (graphite, Ge(111)) or multilayer optics together with divergence/receiving slits and Soller collimators, which jointly define beam footprint, divergence, and the instrument resolution function. These optical choices directly trade intensity against peak width and background.

Two measurement geometries dominate powder work. In Bragg–Brentano reflection geometry, the incident and diffracted beams are symmetric with respect to the specimen surface, and the source–specimen–detector trio lie on the focusing circle. As illustrated in Figure 3.2, the Bragg–Brentano geometry is the most commonly employed configuration in laboratory X-ray diffractometers: the source and detector execute a coupled θ–2θ scan while the specimen is held at the goniometer axis on the focusing circle, ensuring that the Bragg condition is satisfied throughout the scan and enabling accurate intensity recording. This configuration is well-suited to flat, compacted powders and offers high throughput for routine phase analysis and refinement. By contrast, Debye–Scherrer transmission geometry mounts the

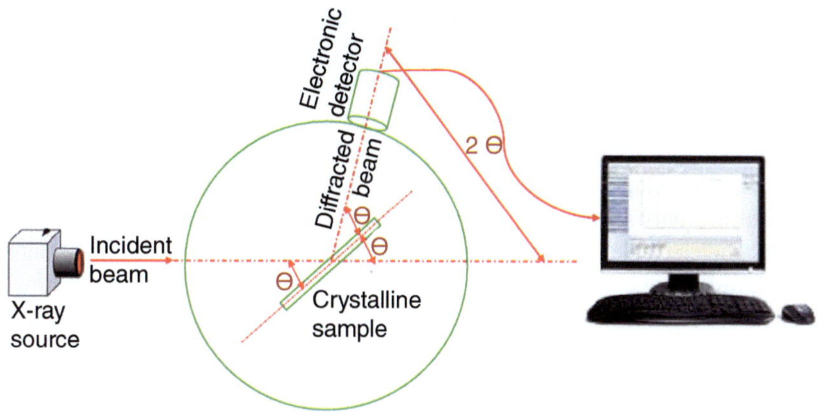

Figure 3.2 XRD instrument schematic [1] / with permission of John Wiley & Sons.

powder in a thin capillary and transmits the beam through the specimen; improved powder averaging mitigates preferred orientation and microabsorption, which is advantageous for quantitative analysis and low-symmetry phases, albeit at the cost of more involved sample preparation.

Detector technology continues to shape data quality and speed. Point scintillation counters remain the benchmark for resolution and instrument function metrology; linear position-sensitive detectors accelerate routine scans but require geometric-defocusing corrections at low angles; two-dimensional area detectors capture complete Debye rings in a single exposure and are therefore essential for time-resolved and in-situ measurements as well as azimuthal integrations for texture analysis. Choice of detector should be made in concert with optics and geometry so that the instrumental profile is appropriate for the intended analysis (e.g., high-resolution lattice parameter metrology vs. fast phase mapping).

In practical terms, the combination of source, optics, geometry, and detector defines the resolution–intensity–speed envelope of an instrument. Bragg–Brentano reflection geometry coupled with slit-defined optics and a linear detector is a robust "workhorse" configuration for polycrystalline specimens, whereas transmission geometry with capillaries and parallel-beam optics is often preferred when minimizing preferred orientation, absorption bias, or sample surface artifacts. With the instrumental grammar established, we turn next to specimen preparation and data collection strategies, which ultimately determine whether the achievable instrumental performance translates into traceable, quantitative diffraction data.

3.1.4 Sample Preparation and Data Collection

Accurate powder diffraction measurements require careful sample preparation to minimize artifacts such as preferred orientation, microabsorption, and surface roughness. Samples are generally ground to a fine, homogeneous powder with particle sizes below 10–20 μm, ensuring random orientation of crystallites.

Overgrinding, however, can induce strain or partial amorphization, affecting peak shapes and intensities.

In reflection geometry, the specimen surface should be flat and positioned precisely at the goniometer axis to avoid sample displacement errors. The thickness must be sufficient to achieve "infinite" absorption for the chosen X-ray wavelength, ensuring that all incident photons interact with the sample rather than penetrating to the substrate. In transmission geometry, uniform packing of the capillary and optimal diameter (typically 0.3–1 mm) help reduce microabsorption and ensure uniform illumination.

Data collection strategies involve selecting an appropriate angular range, step size, and counting time. The angular range should cover all reflections of interest, typically from 10 ° to 90 ° 2θ for Cu Kα radiation. Step sizes on the order of 0.01–0.02 ° 2θ balance resolution with total scan time, while counting time per step determines statistical precision. For quantitative or Rietveld analysis, higher counting statistics and stable environmental conditions are crucial.

Environmental controls, such as temperature stages or gas flow cells, allow non-ambient measurements, while maintaining mechanical and thermal stability is essential to avoid peak shifts or broadening during long scans. Calibration using a standard reference material (e.g., NIST SRM 640c silicon) is recommended to correct for zero shift, instrument broadening, and systematic errors in peak position.

3.1.5 Phase Identification

One of the primary applications of powder XRD is qualitative phase identification—determining which crystalline phases are present in a sample. This process relies on comparing the measured diffraction pattern to reference data from comprehensive databases such as the International Centre for Diffraction Data Powder Diffraction File (ICDD PDF).

Phase identification begins with extracting peak positions (2θ) and relative intensities from the experimental pattern. Modern software automates peak searching and matches the data against database entries using search–match algorithms. Candidate phases are selected based on the best statistical match and the chemical plausibility given the sample's composition and preparation history.

The matching process must account for possible peak overlaps, differences in relative intensities due to preferred orientation, and small shifts in peak positions caused by lattice strain or compositional variations. Where multiple phases are present, a composite match is required, ensuring that all observed peaks are explained without introducing redundant or unsubstantiated phases.

After initial identification, refinement of lattice parameters against the observed peak positions provides further confirmation and can reveal compositional trends (e.g., solid solution effects). In complex cases, advanced methods such as whole-pattern fitting or Rietveld refinement can combine phase identification with quantitative analysis, improving reliability.

Accurate phase identification depends critically on high-quality data: well-resolved peaks, accurate 2θ calibration, and minimal background interference. The

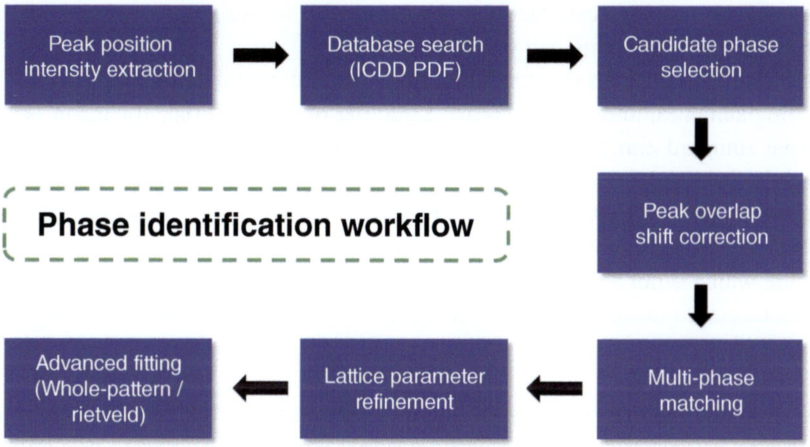

Figure 3.3 Illustration of the process of qualitative phase identification in powder XRD.

general workflow of qualitative phase identification in powder XRD is schematically summarized in Figure 3.3. The combination of robust experimental practices and comprehensive reference databases ensures that powder XRD remains the gold standard for crystalline phase analysis in materials science.

3.1.6 Quantitative Phase Analysis

Quantitative phase analysis (QPA) in powder XRD aims to determine the weight fractions of crystalline phases present in a multiphase material. The simplest approach is the Reference Intensity Ratio (RIR) method, where the integrated intensity of a chosen reflection is compared to that of a known standard, typically corundum (α-Al_2O_3). The RIR constant for each phase is tabulated in databases such as the ICDD PDF. The phase weight fraction W_i can be calculated from:

$$W_i = \frac{\frac{I_i}{RIR_i}}{\sum_j \frac{I_j}{RIR_j}} \tag{3.3}$$

where I_i is the integrated peak intensity and R_i is the RIR for phase i.

While RIR is straightforward, it is limited by peak overlap, preferred orientation, and microabsorption effects. A more robust approach is **whole-pattern fitting**, particularly **Rietveld refinement**, which models the entire diffraction profile using structural parameters, instrumental functions, and background models. In Rietveld QPA, the scale factor for each phase, obtained through least-squares refinement, is directly related to its weight fraction:

$$W_i = \frac{S_i(Z_i M_i V_i)}{\sum_j S_j(Z_j M_j V_j)} \tag{3.4}$$

where S is the refined scale factor, Z the number of formula units per cell, M the molar mass, and V the unit cell volume.

Accurate QPA requires careful calibration of instrumental parameters, adequate counting statistics, and representative sampling of the material. Inclusion of an internal standard can help correct for sample absorption and improve accuracy, especially for highly absorbing phases or poorly crystalline materials. In complex systems, combining QPA with complementary methods such as scanning electron microscopy (SEM) or thermal analysis can validate results and reveal amorphous content, which is not directly detectable by conventional XRD.

3.1.7 Microstructure, Defects, and Stress/Texture

XRD provides valuable information about microstructural features such as crystallite size, microstrain, lattice defects, residual stress, and preferred orientation (texture). These factors influence peak positions, widths, and shapes, and can be extracted from careful line profile analysis. The main XRD-based analysis routes for extracting grain size, microstrain, residual stress, and texture information are schematically illustrated in Figure 3.4.

Crystallite size refers to the size of coherently diffracting domains, which may be smaller than the physical particle size. Peak broadening due to finite crystallite size can be estimated using the Scherrer equation:

$$D = \frac{K\lambda}{\beta \cos \theta} \tag{3.5}$$

where D is the domain size, K a shape factor ($\approx$0.9), λ the wavelength, β the peak broadening (FWHM) corrected for instrumental effects, and θ the Bragg angle. However, size-induced broadening is often accompanied by microstrain, which also broadens peaks. The Williamson–Hall method plots $\beta \cos \theta$ versus $\sin \theta$ to separate

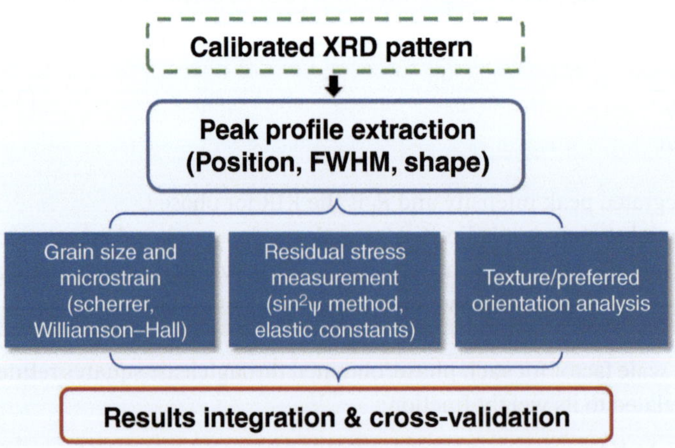

Figure 3.4 Schematic diagram of XRD analysis routes for grain size and microstrain, residual stress, and texture determination.

these contributions, though more sophisticated approaches using whole-pattern refinement provide greater reliability.

Residual stress in polycrystalline materials can be measured by the $\sin^2\psi$ method, which tracks the shift in lattice spacing with specimen tilt. This requires knowledge of the material's elastic constants and careful control of experimental geometry to avoid introducing artifacts.

Texture—the preferred orientation of crystallites—alters relative peak intensities and can bias phase quantification. It is commonly assessed via pole figure measurements, which map the distribution of a specific crystallographic direction in sample coordinates. In Rietveld analysis, texture can be incorporated through models such as the March-Dollase function or spherical harmonics.

Microstructural analysis via XRD is powerful when coupled with complementary characterization methods such as transmission electron microscopy (TEM) or electron backscatter diffraction, which can validate and extend the interpretation derived from diffraction profiles.

3.1.8 Total Scattering and PDF Analysis

While conventional powder diffraction techniques are highly effective in quantifying crystalline phases and probing microstructural features such as grain size, strain, and texture, they are fundamentally limited to information derived from long-range periodicity. Many technologically important materials—nanocrystals, disordered solids, amorphous phases, and heterogeneous composites—exhibit significant structural motifs that do not manifest as sharp Bragg reflections. In such cases, a comprehensive understanding of material behavior requires going beyond the Bragg paradigm to include diffuse scattering arising from short-range order. This broader approach, known as total scattering, captures both the long-range and local structural correlations within a single dataset, enabling analysis in real space through the pair distribution function (PDF).

Total scattering collects both the sharp Bragg reflections from long-range order and the broad diffuse component from local or short-range order. Data are measured over a wide momentum transfer range $Q = 4\pi \sin\theta/\lambda$ to capture as much structural information as possible. After background subtraction and corrections for absorption, polarization, fluorescence, Compton scattering and multiple scattering, the intensity is normalized by scattering factors to yield the total structure function $S(Q)$. From $S(Q)$, one defines $F(Q) = Q\left[S(Q) - 1\right]$, whose sine Fourier transform gives the real-space pair distribution function $G(r)$:

$$G(r) = \frac{2}{\pi}\int_0^{Q_{max}} F(Q)\sin Qr dQ \tag{3.6}$$

Here, peaks in $G(r)$ correspond to interatomic distances, their areas to coordination numbers, and widths to static or thermal disorder.

High real-space resolution requires large Q_{max} (often 20–25 Å^{-1} or more), which is best achieved with high-energy synchrotron X-rays (60–100 keV) or, for neutron PDF, time-of-flight sources. Laboratory XRD can perform PDF with lower Q_{max} using Mo or Ag anodes and photon-counting detectors, trading resolution for accessibility. Corrections and normalization are critical: sample container scattering, instrument resolution, and diffuse background can be of similar magnitude to the diffuse signal of interest, especially in disordered or nanocrystalline systems.

PDF analysis can be carried out using "small-box" models—refining crystallographic or nanoparticle structural models against $G(r)$—or "large-box" approaches such as Reverse Monte Carlo or molecular dynamics to capture complex disorder and medium-range order. X-ray PDFs weight elements by atomic form factors, making them more sensitive to heavier atoms, while neutron PDFs weight by scattering lengths, giving better contrast for light elements like Li, O, or H. Combining both provides complementary sensitivity.

Applications include:

1. Nanocrystals: determining domain size distributions, particle shape, and surface relaxation via Debye scattering equation modeling.
2. Disordered and mixed-order systems: identifying local motifs in glasses, fast-ion conductors, and amorphous-crystalline composites.
3. Molecular and hybrid materials: constraining both intramolecular geometry and intermolecular packing in MOFs, polymers, and molecular solids.
4. Operando studies: tracking short-range structural changes during electrochemical cycling or under thermal/pressure stimuli, often revealing distortions or amorphization preceding macroscopic phase transitions.

Best practice involves maximizing Q_{max} and counting statistics, measuring blanks under identical conditions, verifying corrections for Compton and fluorescence, and reporting all processing parameters to ensure reproducibility. Properly executed, total scattering/PDF extends the reach of diffraction beyond average long-range order, providing a bridge between conventional powder methods and the in-situ/operando techniques that follow.

3.1.9 In-situ and Operando XRD

In-situ XRD is a powerful extension of conventional XRD that enables the direct observation of structural evolution in materials under operating or processing conditions. Unlike ex-situ measurements, which require the sample to be removed from its native environment—often interrupting or altering the physical or chemical state—in-situ XRD captures diffraction patterns in real time while the sample is subjected to controlled external stimuli. This capability is crucial for understanding dynamic processes such as phase transitions, crystallization, lattice expansion/contraction, stress development, and reaction intermediates, which cannot always be inferred from static end-point measurements. The motivation for developing in-situ XRD stems from the need to link structure and function more directly, reduce ambiguities caused by sample handling, and reveal transient states that may govern performance or failure in practical applications.

The fundamental principle of in-situ XRD remains the same as in conventional diffraction: incident monochromatic X-rays interact with the periodic atomic planes of a crystalline material, producing constructive interference according to Bragg's law. However, in-situ configurations integrate additional instrumentation to apply stimuli and maintain a measurement-compatible environment. These setups can include temperature-controlled stages for heating or cooling, mechanical rigs for applying stress or strain, electrochemical cells for battery studies, environmental chambers for controlling gas composition and humidity, and flow reactors for monitoring catalysts under reaction conditions. The challenge lies in designing a measurement geometry that accommodates the in-situ apparatus without compromising diffraction resolution, angular range, or signal-to-noise ratio.

In-situ XRD systems can be broadly categorized according to the nature of the stimulus or environment. Typical in-situ XRD cell configurations and the corresponding diffraction signatures of different reaction mechanisms are illustrated in Figure 3.5:

1. Thermal in-situ XRD tracks structural changes during heating or cooling, enabling the study of thermal expansion, phase transformations, and reaction kinetics in ceramics, metals, semiconductors, and polymers.
2. Mechanical in-situ XRD monitors strain-induced lattice distortions and defect evolution under tensile, compressive, or cyclic loading.
3. Electrochemical in-situ XRD—one of the most developed subfields—follows structural changes in battery electrodes during charge/discharge, fuel cell catalysts under operation, and electrochromic materials during switching.
4. Environmental in-situ XRD examines reactions under specific gas atmospheres or pressures, such as oxidation/reduction, catalytic activation, or corrosion processes.

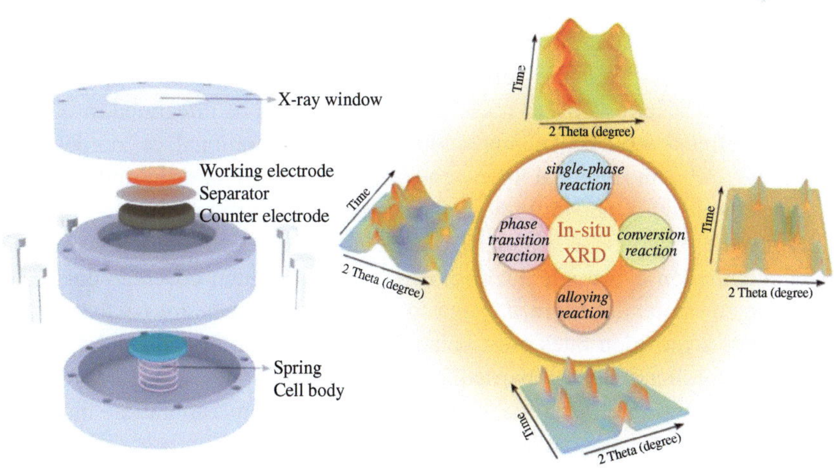

Figure 3.5 Schematic diagram of in-situ cell and schematic diagram of in-situ XRD patterns about four typical reaction mechanisms: single-phase reaction, phase transition reaction, alloying reaction, and conversion reaction [2] / with permission of John Wiley & Sons.

From an instrumentation perspective, in-situ XRD can be implemented on laboratory diffractometers or at synchrotron radiation (SR) facilities. Laboratory systems are more accessible and suitable for slower processes, but synchrotron sources, with their high brilliance and tunable wavelengths, provide the temporal resolution and penetration power needed for rapid or bulk-sensitive measurements. High-energy synchrotron X-rays can probe deep into dense materials or through complex sample environments, making them ideal for industrially relevant conditions.

The design of in-situ measurement platforms is closely linked to the application field. For instance, an electrochemical in-situ cell for lithium-ion batteries must be X-ray transparent while maintaining proper sealing, electrical contacts, and electrolyte containment. Kapton windows, Be domes, or thin glass capillaries are often used to allow X-ray transmission. Similarly, high-temperature stages employ materials such as Pt or sapphire to withstand extreme conditions without interfering with the diffraction signal. In gas reaction studies, reaction cells incorporate gas flow control, heating elements, and safety interlocks, while preserving the alignment and stability of the sample.

The analytical output of in-situ XRD experiments extends beyond simple phase identification. By performing Rietveld refinement or Le Bail fitting on sequential diffraction patterns, one can quantify phase fractions, lattice parameters, crystallite sizes, and microstrain as functions of time or stimulus. This enables the extraction of kinetic parameters for phase transformations, identification of metastable intermediates, and mapping of structural pathways that dictate macroscopic performance. For example, in the context of shape memory alloys, in-situ XRD can capture the martensitic transformation process during thermal cycling, revealing transformation temperatures and hysteresis behavior. In heterogeneous catalysis, the technique can track the formation and consumption of active phases under reaction feeds, providing insight into catalyst stability and regeneration strategies.

An important advantage of in-situ XRD is its compatibility with complementary techniques. Simultaneous measurements—such as X-ray absorption spectroscopy (XAS), Raman spectroscopy, or mass spectrometry—can be integrated into in-situ setups to correlate structural changes with electronic, vibrational, or compositional information. This multimodal approach enhances the mechanistic understanding of complex systems where multiple processes occur concurrently.

Despite the progress in in-situ XRD, several technical considerations remain. Signal attenuation by sample environments, limited angular coverage due to cell geometry, and potential beam damage to sensitive materials are persistent challenges. Data acquisition rates must balance the need for temporal resolution with sufficient counting statistics, particularly for transient states with low scattering intensity. Furthermore, accurate temperature or pressure calibration inside the measurement cell is essential for quantitative analysis.

In-situ XRD has found widespread applications across diverse fields. In metallurgy, it is used to study recrystallization, precipitation, and phase stability under service-like conditions. In ceramics, it helps optimize sintering protocols by tracking phase evolution during heating. In semiconductors, it monitors epitaxial growth, strain relaxation, and phase segregation during thin film deposition. In pharmaceuticals, it assists in understanding polymorphic transformations during formulation or storage. In geology and planetary science, it supports high-pressure/high-temperature experiments to replicate deep Earth or extraterrestrial environments.

The continued development of in-situ XRD will benefit from advances in X-ray source brilliance, detector speed, and data analysis algorithms. Real time Rietveld refinement, machine learning-based pattern recognition, and automated cell control are making it possible to conduct more complex experiments with higher throughput and interpret data more effectively. As these improvements converge, in-situ XRD is evolving from a specialized research tool into a more routine technique, capable of providing time-resolved structural information critical for material innovation and process optimization.

3.2 Synchrotron Radiation XRD

SR-XRD is one of the most important tools in materials science for obtaining crystallographic information. Its working principle is identical to that of conventional X-ray powder diffraction, both relying on the Bragg diffraction condition to determine unit cell parameters, crystal structure, and microstructural characteristics. Compared with laboratory X-ray sources, SR offers extremely high brilliance (up to 10^8–10^{10} times higher), excellent collimation, narrow energy bandwidth, and tunable wavelength. These properties lead to significant advantages in diffraction signal intensity, angular resolution, and temporal resolution, enabling SR-XRD to excel in applications involving small sample quantities, complex multiphase systems, and in-situ/real time measurements.

In powder diffraction, peak broadening originates from both instrumental resolution and sample-related effects. Laboratory diffractometers, limited by relatively large X-ray source sizes and non-negligible beam divergence, typically achieve 2θ resolutions of only 0.05 °–0.2 °, leading to broader peaks and poor separation of closely spaced reflections, which constrains Rietveld refinement. In contrast, synchrotron facilities, equipped with high-precision collimation systems and multiple optical elements, can reduce the instrumental contribution to peak widths down to 0.002 °–0.05 °, dramatically improving the ability to resolve closely spaced peaks.

3.2.1 SR-XRD Instrumentation

These performance gains arise from several key design features: (1) Source and beam control: High-brilliance X-rays are generated in an electron storage ring using undulators or wigglers, with upstream collimating optics to control beam shape and divergence. (2) Monochromatization and focusing: A double-crystal monochromator precisely selects the wavelength, while multilayer mirrors or bent crystal mirrors provide horizontal and vertical focusing. (3) Collimation system: Cox-type arrangements combine vertical and horizontal Soller slits to suppress divergence while maintaining flux; Parrish-type systems place high-precision crystal analyzers

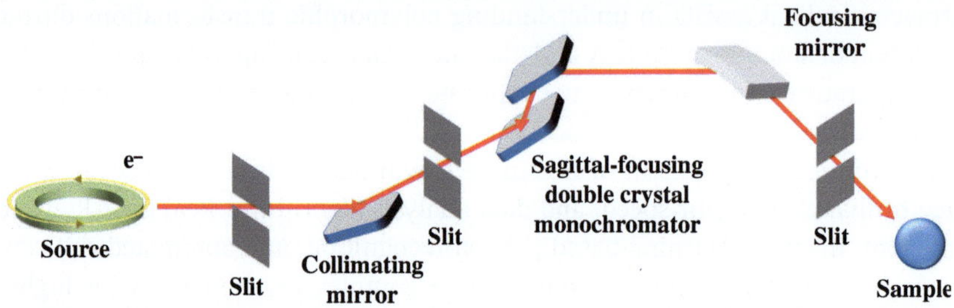

Figure 3.6 Schematic diagram of SR-XRD powder diffraction experiment.

downstream of slits, achieving peak resolutions of 0.01 °–0.05 °, suitable for Rietveld refinement and microstructural analysis. (4) Detection modes: High-resolution measurements often employ multianalyzer crystal arrays with point detectors for strict energy discrimination, whereas rapid measurements use two-dimensional pixel detectors for millisecond-level acquisition.

The typical SR-XRD optical path is: source → collimating mirror → monochromator → focusing mirror → sample position → analyzer crystal/detector, with optional heat filters (e.g., Be windows or Kapton foils) to suppress higher-order harmonics and thermal effects. Both reflection and transmission geometries can be used by simply rotating the sample without rearranging the instrument, and coupled scans are supported (a schematic layout of a typical SR powder XRD experiment is shown in Figure 3.6). Diffractometers designed for synchrotron powder diffraction and four-circle single-crystal diffraction can share goniometers. Depending on the experimental objective, SR-XRD enables in-situ battery cycling tests, high-pressure/variable-temperature structure analysis, quantitative multiphase analysis, and nanostructure studies.

The development of SR-XRD dates back to the early 1970s with first-generation synchrotron sources, when its potential for diffraction experiments was initially explored. In the 1980s, second-generation facilities delivered substantial improvements in stability and beam brilliance, leading to the establishment of dedicated powder diffraction beamlines and the introduction of Soller slits, Cox-type optics, and Parrish-type analyzers, greatly enhancing angular resolution. Since the 1990s, third-generation synchrotrons have provided ultra-high brilliance and low beam divergence, making multianalyzer arrays, precision goniometers, and in-situ environments standard. This era also saw the emergence of microbeam, rapid detection, and full-pattern recording capabilities. In recent years, advances in detector technology and optical design have driven SR-XRD toward the combined goals of high angular and temporal resolution, multimode operation, and multidetector integration, offering unprecedented capabilities for real time, in-situ, and multifield-coupled structural studies.

3.2.2 Example of a SR-XRD Beamline

As an example, the ID22 high-resolution powder diffraction beamline at the European SR Facility (ESRF) is shown schematically in Figure 3.7. High-brilliance

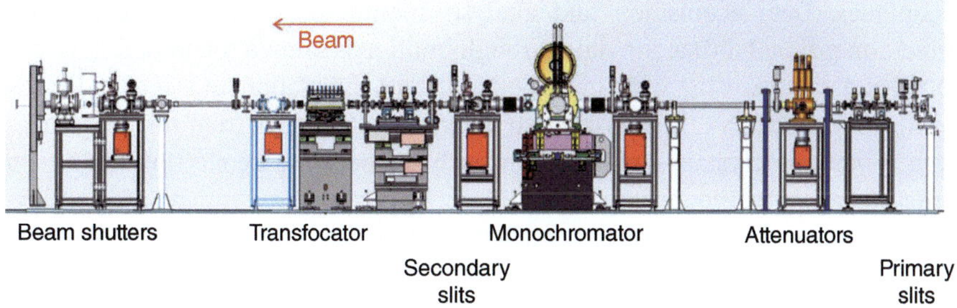

Figure 3.7 Arrangement of the ESRF ID22 beamline optics Adapted from [3].

X-rays generated by the storage ring first pass through slit S1, then enter a liquid-nitrogen-cooled channel-cut Si(111) monochromator (4 mm channel width) for wavelength selection. The beam may then be directed through a focusing system (185 parabolic compound refractive lenses, radius of curvature 200 μm) to adjust the beam size, or left unfocused for optimal angular resolution. In high-resolution mode, the beam is transmitted to the experimental hutch to directly illuminate the sample mounted on a precision goniometer, with diffraction signals collected by a multianalyzer crystal array and pixel detector. Depending on experimental needs, the beam can remain unfocused for maximum resolution or be focused to tens of micrometers for micro-area measurements.

The beamline operates in the energy range of 6–75 keV ($\lambda \simeq 2.1$–0.165 Å) and is equipped with high-precision coaxial horizontal rotation axes (2θ circle accuracy 5×10^{-6o}, ω axis accuracy 10^{-5o}). The analyzer system consists of 13 Si(111) channels, with a scanning range from $-52°$ to $152°$. In unfocused mode, at 35 keV and 200 mA, the beam intensity is approximately 1.1×10^{13} photons mm^{-2} s$_{-1}$; in focused mode, the minimum spot size can reach 50 μm × 50 μm. Equipped with large-area imaging detectors, the beamline can also perform high-energy pair distribution function (PDF) and texture analysis. It supports a broad range of experiments from Rietveld refinement to high-energy PDF and texture studies, with particular suitability for in-situ tracking of phase transitions and microstrain evolution during battery cycling.

With its high brilliance, excellent collimation, and flexible optical design, SR-XRD delivers a substantial enhancement in angular resolution and detection sensitivity over laboratory powder diffraction. In battery research, this enables real time, precise characterization of phase transitions, microstrain accumulation, and interface structure evolution during charge–discharge cycles—capabilities far beyond the reach of conventional X-ray sources.

3.2.3 SR-XRD: Single-crystal Diffraction Methods

In single-crystal diffraction (SCD) studies, SR-XRD (SR SCD) enables the precise determination of atomic positions, thermal vibration parameters, and absolute configurations, providing richer structural information than many other diffraction techniques. Data acquisition and analysis require the collection of complete three-dimensional diffraction data through multi-axis sample rotation, followed by complex data processing steps including indexing, integration, phasing, and model refinement. Under SR, it is possible to determine structures from microcrystals—even submicron crystals—while achieving high-speed data collection and in-situ structural analysis under extreme conditions.

Owing to its high brilliance, excellent collimation, and tunable wavelength, SR SCD offers distinct advantages over laboratory X-ray sources. High brilliance ensures sufficient signal-to-noise ratio even for extremely small sample volumes or very weak diffraction signals, allowing high-precision data collection. Superior collimation and low beam divergence not only improve angular resolution but also reduce peak shape distortions, which is beneficial for refining complex structures. Wavelength tunability offers flexibility in mitigating absorption effects, optimizing reflection conditions, or performing anomalous dispersion experiments. These capabilities make SR particularly powerful for the structural determination of microcrystals, biological macromolecular crystals, and in-situ measurements under extreme conditions, often yielding complete and high-quality structures in cases where laboratory instrumentation is inadequate.

The analytical workflow for SCD generally comprises several stages:

1. **Sample selection and data collection**: An appropriate crystal is selected, and experimental conditions such as morphology, temperature, and exchange medium are defined. Diffraction patterns are recorded using a four-circle diffractometer, Laue method, or other photographic techniques. Indexing of the diffraction data yields the unit cell parameters.
2. **Preliminary data processing**: The number of atoms or molecules in the unit cell is determined, diffraction intensities (I) are converted to structure factor amplitudes (|F|), and the space group is identified.
3. **Structure solution (phasing)**: Common methods include direct methods, multiple-solution methods, reflection-reconstruction methods, isomorphous replacement, heavy-atom methods, and isomorphous heavy-atom techniques. In special cases, the Patterson method may be used to locate atomic positions.
4. **Structure model building and validation**: Phases (α_hkl) are determined, and |F| values combined with phases are used to compute the electron density distribution ρ_0, from which an initial structural model is derived. Structure factors calculated from the model are compared with observed values, and discrepancies are used to refine the model.
5. **Structure refinement**: Least-squares refinement is employed to optimize model parameters until the difference function (difference electron density) converges, yielding precise structural parameters including atomic coordinates, bond lengths, and bond angles.
6. **Interpretation and application**: The relationship between structure and physical or chemical properties is analyzed, and the results are applied in theoretical modeling or practical applications.

3.2.4 SR-XRD: Polycrystalline Diffraction Methods

In polycrystalline diffraction, SR-XRD exploits the high brilliance, low divergence, and tunable wavelength of synchrotron sources to obtain high-quality diffraction data in extremely short acquisition times. Unlike SCD, polycrystalline diffraction imposes minimal requirements on sample morphology—powders or polycrystalline aggregates can be measured directly—making it widely applicable for phase identification, refinement of unit cell parameters, microstructural analysis (grain size, microstrain), and the study of phase transformation kinetics. SR-XRD can be combined with in-situ environments such as mechanical loading, heating/cooling, and electrochemical cycling, enabling real time tracking of structural evolution under external stimuli and providing a powerful tool for research in materials science, chemistry, and energy-related fields.

Compared with laboratory XRD, SR-XRD offers multiple advantages in polycrystalline diffraction analysis. High brilliance significantly shortens acquisition times while maintaining high signal-to-noise ratio and achieving high angular resolution, which is particularly important for in-situ experiments on dynamic processes. Wavelength tunability allows optimization of penetration depth, reduction of absorption effects, or enhancement of anomalous dispersion signals. Low divergence yields sharp, accurately positioned peaks, improving refinement precision and detection sensitivity. Furthermore, SR-XRD, when equipped with high-throughput two-dimensional detectors and fast readout systems, can achieve millisecond temporal resolution, enabling the study of rapid phase transitions, reaction kinetics, and transient intermediate phases—capabilities rarely attainable with most laboratory instruments.

The procedure for determining crystal structures from polycrystalline diffraction is similar to that for SCD and typically includes the following stages:

1. Peak indexing and unit cell determination: Using high-resolution synchrotron powder diffraction patterns, peaks are precisely indexed, and unit cell parameters and crystal system are determined, often by comparison with reference materials or known structures.
2. Data acquisition and intensity calculation: At the selected wavelength, high signal-to-noise diffraction data covering a wide angular range are collected, and peak intensities are converted into structure factor amplitudes $|F|$.
3. Peak deconvolution and initial structure solution: Overlapping peaks are separated to obtain as many independent structure amplitudes as possible; Patterson function analysis or other phasing methods may be used to propose atomic positions and infer possible space groups.
4. Initial structure model building: Based on deconvoluted peaks and phase solutions, an initial crystal structure model is constructed and refined for chemical and crystallographic plausibility.
5. Structure refinement and parameter optimization: Full-pattern Rietveld refinement is performed to optimize the model, yielding high-precision atomic coordinates, bond lengths, bond angles, thermal displacement parameters, and statistical indicators such as R-factors, thus establishing the final structure.

3.3 Small-angle/Wide-angle X-ray Scattering Techniques

Small-angle and wide-angle X-ray scattering (SAXS/WAXS) techniques play a critical and complementary role in elucidating the multiscale structural characteristics of functional materials. SAXS is particularly sensitive to electron density fluctuations over length scales ranging from approximately 1 to 100 nm, enabling quantitative analysis of nanoscale heterogeneities, including particle size distributions, pore structures, phase separation, and domain evolution during dynamic processes. In contrast, WAXS probes atomic-scale ordering and crystallographic information, providing direct insight into crystal structure, phase composition, lattice parameters, and long-range structural coherence. When combined, SAXS and WAXS offer a powerful, hierarchical view of material structure spanning from the nanoscale to the atomic scale, allowing correlations to be established between mesoscale morphology and long-range crystallographic order. As such, SAXS/WAXS techniques have become indispensable tools for establishing robust structure–property relationships in advanced materials research.

3.3.1 Introduction to SAXS/WAXS

SAXS refers to the detection of coherent scattering signals produced by electron density inhomogeneities in a sample when probed with incident X-rays, recorded in the small-angle range of approximately 2 °–5 °. SAXS primarily probes nanostructures in the 1–100 nm size range, corresponding to long-range variations in electron density within the material. It is highly sensitive to internal microstructural features such as morphology, size distribution, orientation, porosity, and electron density contrast, and is therefore widely applied in the structural analysis of polymers, colloids, nanoparticles, biomacromolecules, and porous media. SAXS data are characterized by strong scattering intensity in the low-q region, reflecting overall particle morphology and structural periodicity. The theoretical foundations of SAXS were established and developed by Guinier, Porod, Kratky, Hosemann, Debye, and others, who demonstrated that the scattering pattern and intensity distribution are independent of the atomic composition of the scatterers and of their crystallinity, but depend solely on the shape, size distribution, and surrounding medium's electron density. In essence, SAXS arises from electron density differences within the system.

WAXS is used to probe larger scattering angles (typically $2\theta > 5\,°–10\,°$, corresponding to $q \approx 1 - 40\ \text{nm}^{-1}$), providing information on interatomic or intermolecular spacings. For highly crystalline materials, WAXS data primarily display Bragg diffraction peaks, enabling precise lattice parameter refinement, phase composition analysis, strain, and texture characterization. For partially ordered or semicrystalline systems, WAXS contains both diffraction peaks and diffuse scattering, revealing medium- and short-range order. The high brilliance, collimation, and energy tunability of synchrotron sources enable WAXS to achieve higher q-resolution, shorter

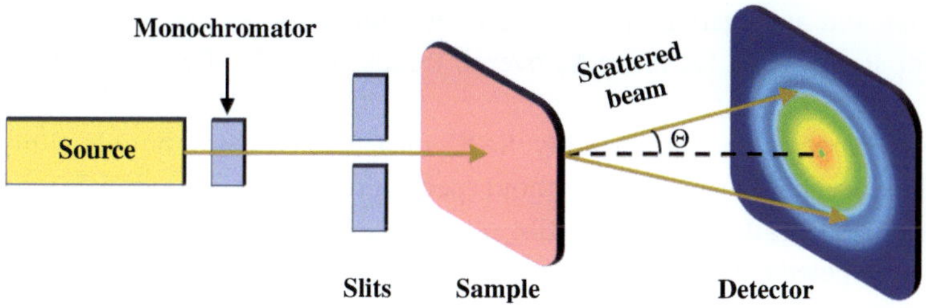

Figure 3.8 Schematic diagram of synchrotron-based SAXS/WAXS experiment.

acquisition times, and greater experimental flexibility, along with enhanced penetration for measurements on thick or encapsulated samples. A schematic configuration of a synchrotron-based SAXS/WAXS experiment is shown in Figure 3.8.

It is important to note that both SAXS and WAXS fundamentally depend on high-quality X-ray sources and precise measurement systems, with data quality and experimental capability directly tied to source performance. On laboratory-based X-ray systems, limitations in brilliance, collimation, and energy tunability restrict measurement sensitivity, q-space coverage, and time resolution. For SAXS, an optimal beamline requires a focused, monochromatic hard X-ray beam, produced by monochromatizing the synchrotron beam with a crystal monochromator and focusing it with mirrors. The beam must be focused (both horizontally and vertically) at the detector to meet the resolution and sensitivity requirements.

3.3.2 Complementarity of SAXS/WAXS with Other Characterization Techniques

SAXS and WAXS are frequently used together, with SAXS providing nanometer-scale, long-range structural information and WAXS probing atomic- or molecular-scale order. Together, they allow multiscale structural characterization from the nanometer to the atomic level on a single experimental platform. For example, in polymer research, SAXS can reveal lamellar structures between crystallites, while WAXS analyzes molecular packing within crystallites, producing a complete structural picture.

SAXS, TEM, and SEM are all essential tools for investigating structures from the nanometer to sub-micrometer scale. They differ in the type of information provided, spatial and temporal resolution, and sample requirements, and are thus complementary in materials microstructure research.

1. Information complementarity: SAXS yields statistical parameters—particle size distribution, morphology, surface area, porosity—over the 1–100 nm range from bulk scattering, while TEM/SEM provide direct images of local morphology, interfaces, and defects. Combined, they enable global-to-local, multiscale analysis.
2. Spatial and temporal resolution complementarity: Synchrotron SAXS achieves millisecond–microsecond time resolution for dynamic processes such as phase

transitions and aggregation. TEM/SEM offer nanometer–atomic spatial resolution for static detail. SAXS tracks overall trends; TEM/SEM validate critical states.

3. Sample preparation and applicability: SAXS measures powders, liquids, films, bulk solids, and porous materials without special preparation, including in-situ. TEM requires ultrathin sections; SEM often needs conductive coating. SAXS can guide targeted electron microscopy preparation.
4. Data interpretation and validation: SAXS relies on model-based fitting with possible non-uniqueness; TEM/SEM directly test model plausibility, improving reliability and accuracy.

Thus, SAXS and TEM/SEM are complementary: global statistical analysis paired with local high-resolution imaging. On synchrotron platforms, SAXS offers high brilliance, high flux, wide q-range coverage, and nondestructive measurements, enabling rapid, statistically representative nanostructural characterization beyond the reach of TEM or SEM.

On synchrotron platforms, WAXS and XRD (SXRD) share overlapping functionalities. SXRD is a core method for determining crystal structures by measuring and analyzing elastic scattering signals over a large-angle range. This "wide-angle scattering" includes not only traditional diffraction peaks but also diffuse scattering from partially disordered systems. In this range, WAXS experiments are closely related to SXRD in principle and data type, but have broader applicability, extending from crystallographic refinement to medium- and short-range order analysis in semicrystalline and amorphous materials. Synchrotron WAXS, like XRD, is based on Bragg's law, but benefits from the high intensity, superior collimation, and energy tunability of synchrotron beams, enabling higher q-resolution, shorter data acquisition times, and more flexible geometries. These capabilities allow WAXS to detect weak diffraction, probe heterogeneous systems, and perform in-situ and time-resolved experiments. High-energy synchrotron WAXS can simultaneously provide wide q-range coverage and high penetration for measurements on thick or encapsulated samples.

In practice, synchrotron XRD can be regarded as a specific form of WAXS optimized for highly crystalline systems, serving tasks such as lattice parameter refinement, phase identification, strain, and texture analysis. WAXS has broader applicability, fulfilling these roles and combining with SAXS for multiscale structural characterization from the nanometer to atomic level. For example, in polymers and composites, SAXS captures long-range morphology while WAXS provides crystallite size and packing information, forming a complete structural description. Thus, in synchrotron-based materials research, WAXS is not merely an extension of XRD but a bridge between crystallography and amorphous structure analysis.

3.3.3 SAXS/WAXS Instrumentation Overview

WAXS/SAXS and synchrotron XRD share a high degree of similarity in their source and optical systems, but differ markedly in beamline layout, detector placement, and experimental environment. The main similarities and differences are as follows:

1. Source and monochromatization: All three rely on high-brilliance synchrotron sources, with double-crystal monochromators (e.g., Si(111), Si(220)) for precise energy selection; the energy range and monochromatization principles are essentially identical.
2. Focusing and collimation: All employ bent cylindrical mirrors, Kirkpatrick–Baez (KB) mirrors, or multilayer optics for beam focusing and collimation, with comparable slit and aperture configurations, although SAXS typically demands higher collimation precision in the low-angle region to suppress parasitic scattering.
3. Sample stage and environment: All use high-precision, multi-axis stages and can be equipped with in-situ devices for temperature, humidity, and mechanical loading control; however, SAXS/WAXS require a wider sample-to-detector distance range to cover different q-regions compared with XRD.
4. Detector placement: XRD detectors are typically fixed to the diffractometer arm to cover a wide 2θ range for complete diffraction profiles; WAXS detectors are positioned at short distances (tens of centimeters) to capture high-q scattering; SAXS detectors are placed at long distances (meters to over ten meters) to record low-q scattering, with some beamlines employing dual-detector setups for simultaneous acquisition.
5. Beam path and background control: XRD often operates in air, whereas SAXS requires long evacuated or helium-filled flight tubes to minimize air scattering; WAXS, with its shorter path, has less stringent requirements.
6. Data acquisition and analysis: All support automated control and online processing; however, SAXS/WAXS data are usually two-dimensional scattering images requiring azimuthal integration, normalization, and model-based fitting to extract structural parameters, while XRD data are typically one-dimensional diffraction profiles

3.3.4 Grazing-incidence Geometry and Surface-sensitive Scattering

Grazing incidence refers to a measurement geometry in which X-rays illuminate the sample surface at a very small incident angle, typically less than 1 °, close to or slightly above the material's critical angle. Under these conditions, the X-ray penetration depth into the sample is only on the order of tens to hundreds of nanometers, meaning that the detected signal originates predominantly from the thin film or near-surface region, while contributions from the bulk substrate are strongly suppressed. Grazing-incidence conditions give rise to refraction and total-reflection-related phenomena, such as the Yoneda peak, which enhance surface sensitivity but require refraction corrections to the scattering vector q.

On synchrotron platforms, the fundamental distinction between grazing-incidence modes (GIXRD, GIWAXS, GISAXS) and conventional measurement modes (transmission or standard reflection geometries) lies in the control of the incident angle and the associated probing depth. (1) Incident angle control: Grazing-incidence modes set the incident angle α_i close to or slightly above the sample's critical angle (typically <1 °), whereas in conventional modes the incident angle is determined naturally by the diffraction condition (Bragg angle) or experimental geometry. (2) Probing depth and sensitivity: In grazing incidence, the X-ray penetration depth is restricted to tens or hundreds of nanometers, providing signals primarily from thin films or surface structures, while conventional modes penetrate from micrometers to millimeters, producing bulk-sensitive information. (3) Signal composition: Grazing incidence suppresses bulk substrate scattering, thereby enhancing surface, interface, and film diffraction or scattering signals; conventional modes generally contain combined contributions from both substrate and surface.

When the film is very thin and its lattice spacing large, small-angle incidence can excite diffraction or scattering from the film at extremely low 2θ values. Well-crystallized films produce sharp diffraction peaks, whereas disordered or semi-crystalline films exhibit a combination of diffuse scattering and Bragg reflections. As the incident angle increases, the X-ray path length within the film and the sampled volume both increase, leading to stronger surface signals; however, once the angle significantly exceeds the critical angle, substrate contributions rise, requiring a trade-off between signal-to-noise ratio and surface selectivity [4].

3.3.5 GIXRD, GIWAXS, and GISAXS on Synchrotron Platforms

1. GIXRD / GIWAXS (wide-angle regime)
   These techniques record diffraction signals from the surface or thin films over a large scattering-angle range, enabling phase identification, lattice constant refinement, crystallite size analysis, and texture/orientation studies. Two-dimensional detectors can simultaneously capture complete diffraction rings and orientation information. By adjusting the incident angle and photon energy, it is possible to "depth profile" different layers within the film. The high brilliance and collimation of SR allow bidirectional in-plane/out-of-plane measurements for ultrathin films (<10–50 nm), weakly diffracting samples, and in-situ processes such as annealing, tensile deformation, or vapor deposition growth.
2. GISAXS (small-angle regime)
   This method records low-q (10^{-3}–10^{-1} Å^{-1}) scattering from nanostructures and periodicities in surfaces or films, including quantum dot arrays, phase-separated morphologies, pores, and roughness. The grazing-incidence geometry decouples q_z from $q_{||}$, facilitating the distinction between in-plane order and out-of-plane layering or thickness fluctuations. By combining with analytical models, one can extract statistical quantities such as size distributions, correlation lengths, and specific surface areas. The combination of high flux and long sample-to-detector distances enables joint GISAXS/GIWAXS configurations to cover length scales from the nanometer to the atomic regime in a single setup.

3. **Advantages and implementation considerations**
   High brilliance and high flux provide strong surface signals and short exposure times, suitable for rapid mapping and time-resolved measurements. Energy tunability allows control of penetration depth, optimization of contrast, and even near-edge resonant scattering. Superior collimation and fine beam spots enable micro-area measurements and in-plane texture resolution in thin films and patterned devices. Multidetector layouts permit parallel acquisition of long-distance SAXS (GISAXS) and short-distance WAXS (GIWAXS) data. Data processing must account for refraction/reflection effects, Yoneda enhancement, footprint effects, and instrument function corrections.

The experimental geometry and scattering configuration of a typical synchrotron-based GIWAXS measurement on thin films are schematically illustrated in Figure 3.9.

3.4 Application Cases

XRD is extensively applied across a wide range of scientific and engineering fields. In materials science and solid-state chemistry, XRD is used to identify crystalline phases, determine crystal structures, and evaluate long-range order. In energy-related research, such as batteries and catalysts, XRD enables monitoring of phase evolution and structural stability during operation. Additionally, XRD plays an

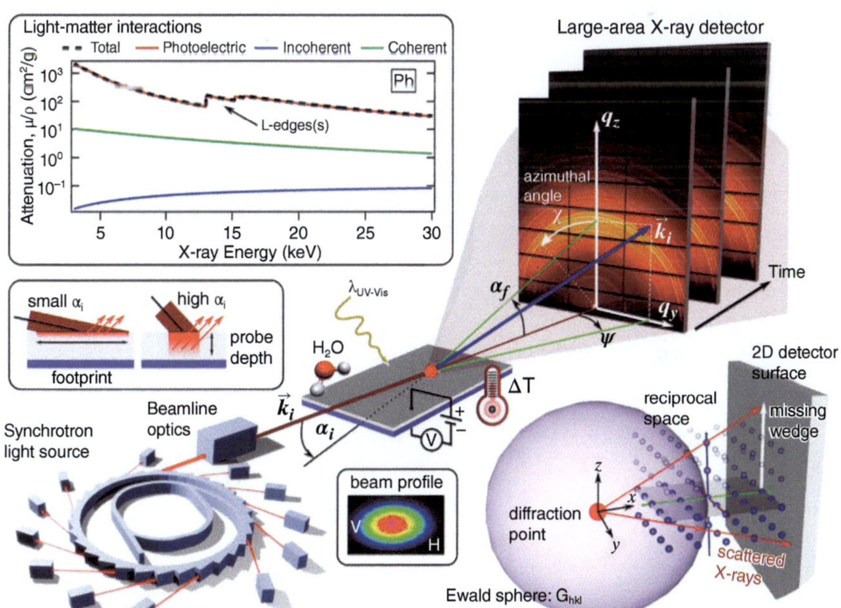

Figure 3.9 Schematic diagram of synchrotron-based GIWAXS experiment on thin film materials [4] / with permission of John Wiley & Sons.

important role in geology, metallurgy, pharmaceuticals, and semiconductor industries for quality control, stress analysis, and polymorph identification.

3.4.1 Rietveld + PDF Reveals Complex Phase Composition, Phase Ratios, and Short-range Order

SR-based pair distribution function (PDF) analysis is a powerful structural characterization technique that utilizes total scattering data to extract real space information, enabling detailed investigation of both short-range and long-range ordering in materials. This method is particularly effective in probing local structural features such as defects, distortions, and disorder, which are difficult to detect by conventional XRD. When combined with Rietveld refinement, which provides precise atomic structural parameters, the PDF method offers comprehensive insights into structural evolution during the operation of lithium-ion battery (LIB) materials.

In LIB research, PDF analysis has been extensively applied to study phase transitions and local structure changes during electrochemical cycling. For example, studies on layered O_2-$LiMn_{0.75}O_2$ cathodes have shown that PDF can reveal subtle local structural distortions and oxygen slab thickness changes in initial charge-discharge progress (Figure 3.10).

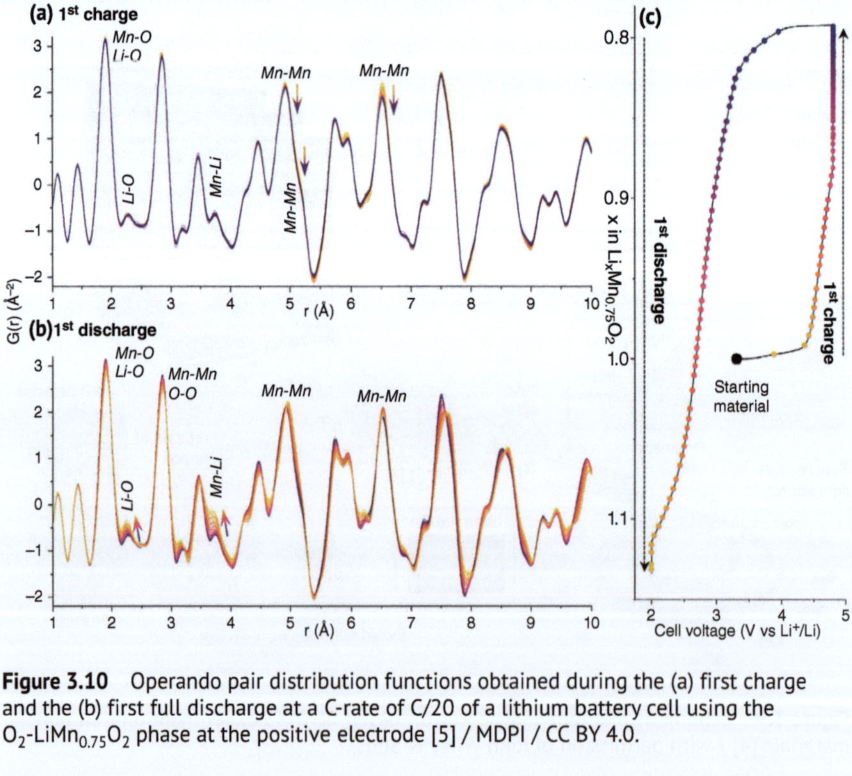

Figure 3.10 Operando pair distribution functions obtained during the (a) first charge and the (b) first full discharge at a C-rate of C/20 of a lithium battery cell using the O_2-$LiMn_{0.75}O_2$ phase at the positive electrode [5] / MDPI / CC BY 4.0.

Another representative application is in the study of Li-rich layered oxides. PDF analysis has demonstrated that, during the initial charge, local TM–O bond length changes and TM migration toward the lithium layer can be detected, providing direct evidence for the early-stage formation of spinel-like domains. This is essential for understanding voltage fade and structural degradation. In anode materials, synchrotron PDF has been used to track the lithiation/delithiation mechanisms in Si, Sn, and Ti-based anodes [6]. By monitoring local coordination changes, researchers have been able to determine whether structural evolution follows an amorphization-crystallization pathway or occurs through a solid solution reaction. This information is crucial for designing materials with high cycle stability. When PDF is combined with Rietveld refinement, researchers can simultaneously capture local distortions and average crystallographic parameters. For example [7], in $LiFePO_4$ and its doped derivatives, combined analysis has quantified the coexistence of ordered and disordered phases and revealed the influence of dopants on Li-ion migration channels. Similarly, in high-voltage spinel $LiNi_{0.5}Mn_{1.5}O_4$, this approach has provided precise information on Mn oxidation states, cation ordering, and oxygen vacancy formation (Figure 3.11).

Overall, synchrotron-based PDF analysis, especially when paired with Rietveld refinement, enables multiscale structural characterization, from atomic-level distortions to long-range ordering. This dual capability is essential for correlating structure–property relationships, diagnosing degradation mechanisms, and guiding the rational design of next-generation LIB materials.

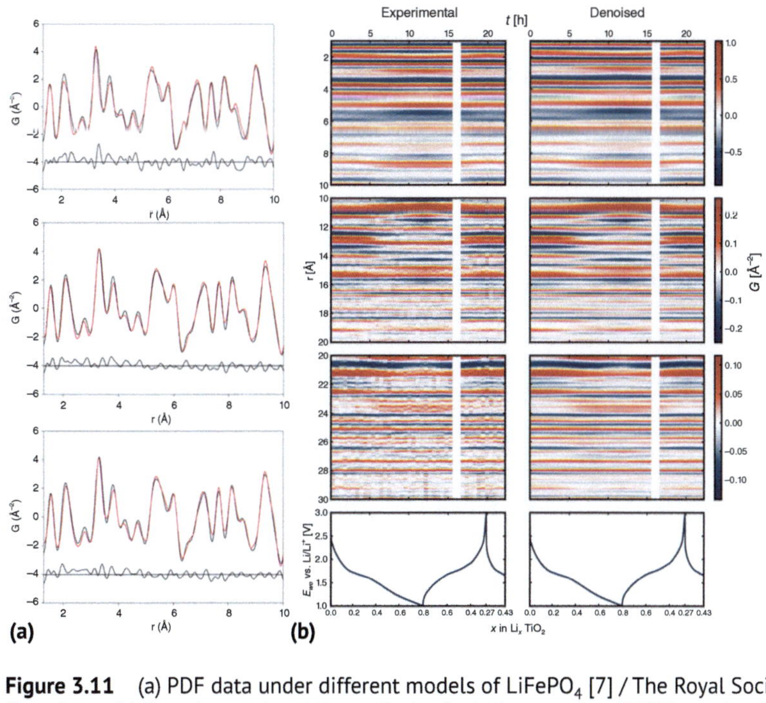

Figure 3.11 (a) PDF data under different models of $LiFePO_4$ [7] / The Royal Society of Chemistry; (b) Experimental and theoretical calculation results of $LiTiO_2$ under different charge and discharge states Adapted from [7].

3.4.2 Application of XRD in Phase Transition and Structural Evolution of LIB Materials During Charge and Discharge Processes

In LIB research, electrode materials undergo complex phase transitions and structural evolution during charge–discharge processes, which directly influence the specific capacity, cycle life, and rate performance of the battery. XRD, as a classical technique for analyzing crystal structures, provides critical information on phase composition, lattice parameter variations, and changes in crystallinity. However, conventional laboratory XRD has certain limitations in temporal resolution, spatial resolution, and signal-to-noise ratio, making it difficult to capture transient microstructural changes occurring during rapid electrochemical reactions. SR-XRD, with its advantages of high brightness, high coherence, and tunable wavelength, enables in-situ or quasi-in-situ structural analysis with high temporal and angular resolution. This capability allows researchers to reveal essential features such as phase boundary migration, phase coexistence regions, phase transition kinetics, and microstrain evolution during lithium insertion and extraction. SR-XRD has been widely applied to investigate phase transition mechanisms in layered oxides, spinels, and olivines, as well as to evaluate the structural stability of silicon, graphite, and metal oxide anodes. These insights provide a solid experimental foundation for designing high-performance electrode materials and optimizing electrochemical reaction pathways.

In-situ high-energy XRD was employed to resolve the electrode-specific contributions to overcharge-induced failure in LIBs [8]. By tracking the lattice dilation of the aluminum and copper current collectors on the cathode and anode sides, respectively, the local temperatures within commercial 18650 cells were quantitatively inferred during electrochemical abuse. This approach enables the thermal evolution of the two electrodes to be independently distinguished, providing direct insight into their asymmetric responses under overcharge conditions (Figure 3.12).

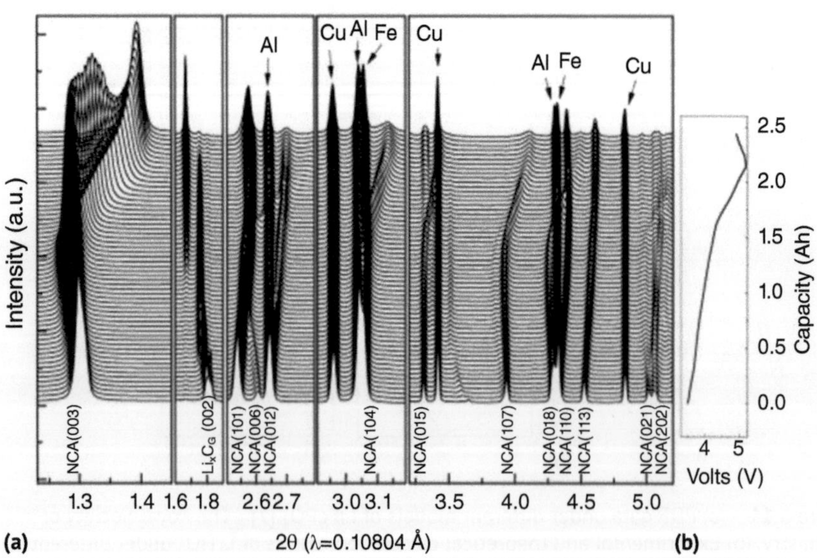

Figure 3.12 (a) In-situ XRD patterns and (b) the electrochemical curve of MCMB/NCA 18650 lithium-ion cell undergoing normal charge and overcharge abuse [8] / with permission of Elsevier.

The cathode temperature exhibits an early and continuous increase once the cell voltage reaches 4.16 V, indicative of progressive thermal activation upon deep delithiation. By contrast, the anode temperature decreases slightly during the initial charging stage and rises only after the voltage exceeds 4.58 V. This pronounced divergence in temperature evolution underscores the dominant role of cathode instability in governing overcharge degradation for the chemistry investigated. These observations suggest that improving the electrochemical robustness of delithiated cathode materials is critical for enhancing the intrinsic tolerance of LIBs to overcharge abuse.

Operando XRD studies by Jürgen Janek et al. [9] reveal that the structural degradation of $LiNiO_2$ is intrinsically linked to lithium content–dependent phase evolution. While moderate delithiation ($x(Li) \geq 0.26$) proceeds through symmetry lowering and Li-vacancy ordering with limited lattice perturbation, deep delithiation triggers a catastrophic H2–H3 transition marked by abrupt *c*-axis collapse and substantial volume contraction. Crucially, the most severe contraction of the Li interslab spacing occurs not during the phase transition itself, but upon further lithium extraction within the H3 phase, identifying deep delithiation as the primary driver of irreversible structural instability in high-Ni layered cathodes (Figure 3.13).

Yoon et al. [10] systematically investigated the thermal-structure evolution of charged $Li_{0.33}Ni_{1/3}Co_{1/3}Mn_{1/3}O_2$ cathodes and its correlation with thermal stability using time-resolved XRD (TR-XRD), as shown in Figure 3.14. Measurements were performed over a broad temperature range from 25 °C to 600 °C, both in the presence and absence of electrolyte, with $Li_{0.27}Ni_{0.8}Co_{0.15}Al_{0.05}O_2$ serving as a reference material. The $Li_{0.33}Ni_{1/3}Co_{1/3}Mn_{1/3}O_2$ cathode exhibits a distinct sequence of phase transitions upon heating. In the absence of electrolyte, the layered structure

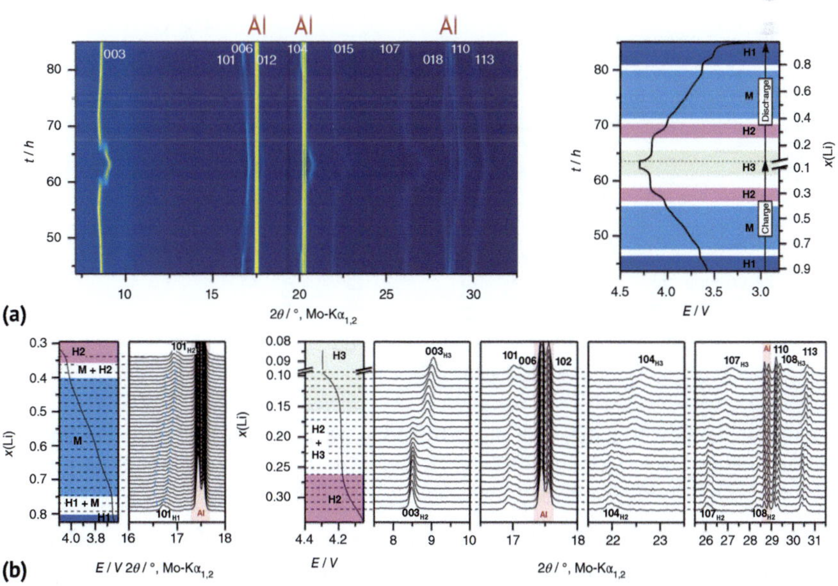

Figure 3.13 Operando XRD of an LNO/Li cell at C/20 in the second cycle. (a) Contour plot. (b) Selected diffraction patterns showing the splitting of 101 reflection in the monoclinic phase region and selected diffraction patterns of the H2–H3 transformation region [9] / with permission of John Wiley & Sons.

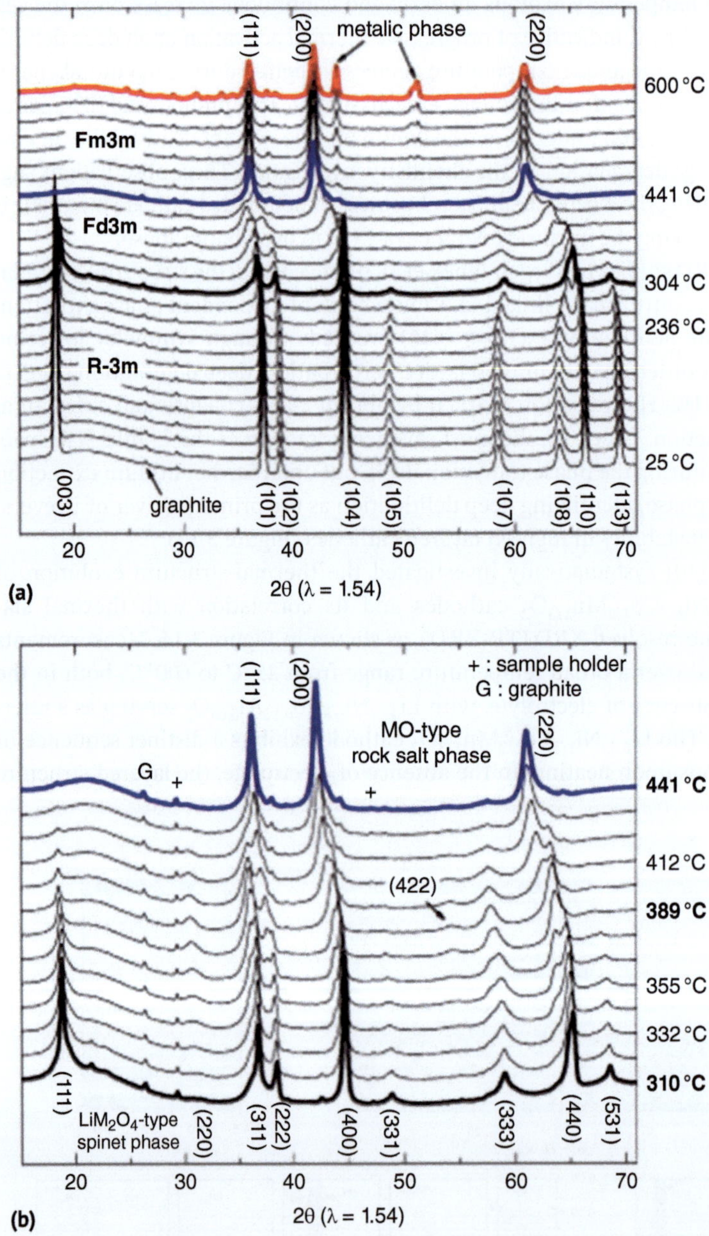

Figure 3.14 (a) TR-XRD patterns of the charged $Li_{0.33}Ni_{1/3}Co_{1/3}Mn_{1/3}O_2$ in the presence of electrolyte heated from 25 °C to 600 °C. (b) TR-XRD patterns as in temperatures between 310 and 441 °C with smaller temperature step [10] / with permission of Elsevier.

initially transforms into a LiM_2O_4-type spinel phase, followed by conversion to a M_3O_4-type spinel, which remains stable up to 600 °C.

When electrolyte is present, the thermal decomposition pathway is further modified. An additional transformation from the M_3O_4-type spinel to an MO-type rock-salt phase occurs in the temperature interval of approximately 400–441 °C,

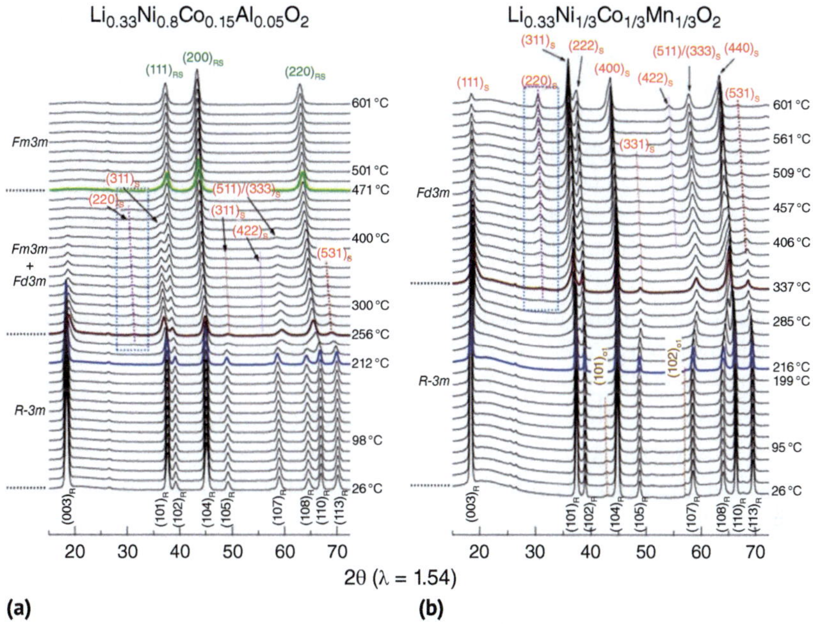

Figure 3.15 TR-XRD patterns of the overcharged (a) $Li_{0.33}Ni_{0.8}Co_{0.15}Al_{0.05}O_2$, and, (b) $Li_{0.33}Ni_{1/3}Co_{1/3}Mn_{1/3}O_2$ during heating up to 600 °C Adapted from [11].

accompanied by the emergence of metallic phases at around 460 °C. Compared with $Li_{0.27}Ni_{0.8}Co_{0.15}Al_{0.05}O_2$ under identical conditions, $Li_{0.33}Ni_{1/3}Co_{1/3}Mn_{1/3}O_2$ displays a markedly delayed spinel-to-rock-salt transition, effectively extending the thermal stability window of the spinel phases from ~20 °C to ~140 °C.

Yang et al. [11] employed time-resolved TR-XRD during controlled heating to systematically probe the structural evolution of overcharged $Li_xNi_{0.8}Co_{0.15}Al_{0.05}O_2$ and $Li_xNi_{1/3}Co_{1/3}Mn_{1/3}O_2$ cathode materials, as shown in Figure 3.15. This operando thermal analysis enabled a comprehensive comparison of the two systems in their overcharged states, encompassing differences in phase distribution after overcharge, as well as phase nucleation, growth, and propagation behavior upon heating.

In addition, TR-XRD provided insight into the preferred crystallographic sites and migration pathways of Ni, Co, and Mn cations, allowing their distinct roles in governing thermal stability to be disentangled. These structural transformations were further correlated with the onset and extent of oxygen release during heating, establishing a direct link between cation migration, phase evolution, and lattice oxygen instability. Collectively, these findings offer mechanistic guidance for the rational design of next-generation cathode materials with enhanced thermal robustness and improved safety under abusive conditions.

3.4.3 Application of GIXRD and GISAXS/GIWAXS to Inspect Thin Film Samples in LIB Systems

Synchrotron X-ray and neutron scientific facilities are the most advanced resources with a large variety of state-of-the-art instrumentations and powerful tools for material research. X-rays and neutrons interact with matter in different ways and offer complementary views of materials at various levels. This section summarizes the principles and applications of synchrotron X-ray- and neutron-based diffraction, total scattering, and small-angle scattering techniques in rechargeable battery research. All of these approaches fall within the framework of elastic scattering and enable structural characterization across multiple length scales. Conventional diffraction methods are primarily employed to identify crystalline phases and to resolve average atomic arrangements with high precision. In contrast, total scattering combined with pair distribution function (PDF) analysis provides access to local atomic correlations, irrespective of long-range crystallinity, and is therefore applicable to disordered, amorphous, or liquid-like systems.

Small-angle scattering techniques complement these methods by probing nanoscale features, yielding quantitative information on particle size, morphology, and hierarchical organization. In addition, the unique advantages of synchrotron-based high-energy X-rays are highlighted, including their deep penetration depth and extended momentum transfer range, which make them particularly well-suited for PDF measurements and operando investigations of battery materials.

Chen et al. [12] developed a novel lamellar interphase featuring 2D directional channels and confinement-regulated dielectric environments to modulate Li^+ desolvation and deposition. The directional orientation of copolymer P(VDF-TrFE-CTFE) noted as GPVTC was elucidated through grazing incidence WAXS (GIWAXS) patterns. As depicted in Figure 3.16a, b, measurements of GPVTC and G/PVTC were conducted in two X-ray beam directions (parallel and perpendicular) relative to the membrane plane, respectively. A highly oriented alignment was observed in GPVTC, where the PVTC is uniformly distributed in parallel within the GO-confined channel. Compared to the G/PVTC membrane, the GPVTC membrane exhibited a narrower and more robust (001) peak after the signal normalization of grazing incidence GIWAXS (Figure 3.16a, c). The azimuthal angle (φ) plot (Figure 3.16d) further illustrates that GPVTC displays a distinctive orientation, as calculated by the anisotropic scattering mode. This orientation was quantified through the Hermann orientation factor, yielding a value of 0.69. In contrast, the G/PVTC displayed a disordered distribution of GO nanosheets, with no observable aligned channels. These results effectively demonstrated that the confined GO channel guides the orientation of PVTC chains, forming highly directional confined channels. The SEM, XRD and solvent selection schematic diagrams further reveal the structural advantages of GPVTC (Figure 3.16e–g).

Vogt et al. [13] used operando grazing-incidence small-angle X-ray scattering/XRD (GISAXS/GIXD) technology to study the structural changes of a model ordered mesoporous nickel cobalt oxide ($NiCo_2O_4$) thin film anode during battery operation (Figure 3.17). The study found that the size of the mesopores has a significant effect on the stability of the ordered nanostructure during cycling. The anode with small mesopores (about 9 nm) undergoes structural collapse in the first two charge and discharge cycles, while the anode with large mesopores (17–28 nm) basically maintains its nanostructure in the first two cycles, thereby significantly improving the capacity retention rate. However, as more charge and discharge cycles continue, the

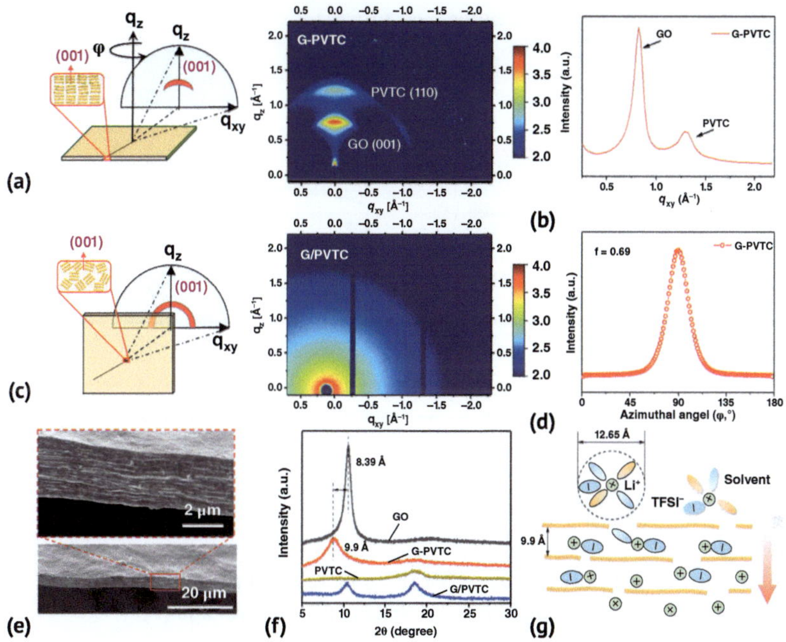

Figure 3.16 GIWAXS image of (a) GPVTC and (b) G/PVTC. The qxy axis corresponds to the direction parallel to the sample surface; the qz axis represents the scattering perpendicular to the surface. (c) Diffractograms and (d) azimuthal plot of the GPVTC obtained by integration of GIWAXS. (e) Cross-sectional SEM image of the GPVTC. (f) XRD patterns of GO, PVTC, G/PVTC, and GPVTC. (g) The schematic diagram illustrates the selective sieving of solvent molecules by the lamellar interphase [12] / with permission of the American Chemical Society.

ordered nanostructure is gradually lost, resulting in a decline in battery performance. These multiscale tests provide insights into how atomic-scale changes are converted into nanostructural changes during battery operation.

3.4.4 Dimensional, Strain, and Defect Analysis of LIB Materials Using XRD

LIBs have been widely employed in portable electronics, electric vehicles, and large-scale energy storage systems owing to their high-energy density, long cycle life, and excellent rate capability. The microstructural features of electrode materials—including grain size, strain state, and defect distribution—play a decisive role in determining their electrochemical performance and cycling stability. For instance, nano-sized grains can shorten lithium-ion diffusion paths and enhance rate capability, but excessively small grain sizes may lead to increased surface side reactions. Moreover, strain accumulation and defect formation during cycling can induce lattice distortion, trigger phase transitions, or even cause structural collapse, thereby accelerating capacity fading. Therefore, the development of precise structural characterization and analytical methods is of great significance for elucidating failure mechanisms, optimizing material design, and improving battery performance.

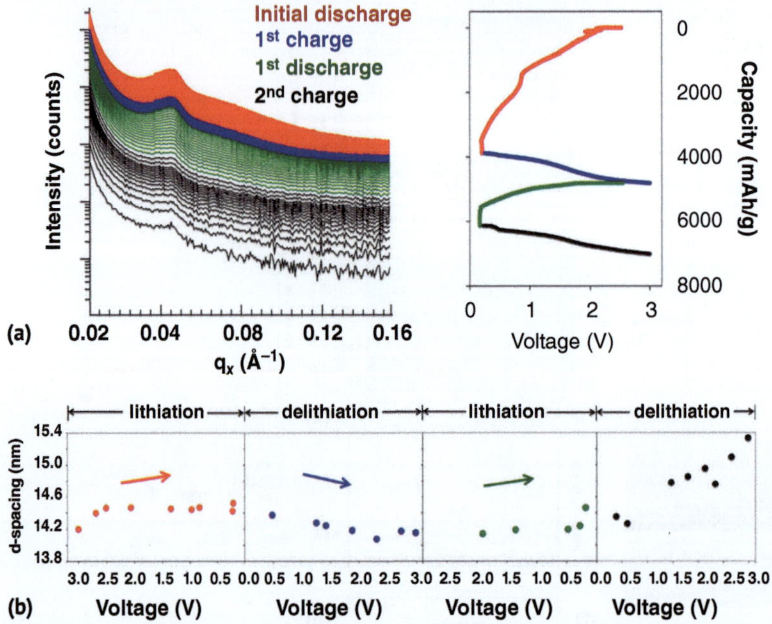

Figure 3.17 (a) 1D (in plane) operando GISAXS profiles from galvanostatic cycling of m-NC9 film anode. (b) Quantitative analysis of change in d-spacing of m-NC9 from GISAXS during successive lithiation and delithiation [13] / with permission of the American Chemical Society.

XRD, a nondestructive and multiscale structural characterization technique, provides essential information on crystal structure, phase composition, and preferred orientation. By analyzing diffraction peak positions, widths, and shapes, XRD enables the quantitative determination of grain size, strain, and defects. For example, the Scherrer equation is commonly used to estimate nanoscale grain sizes; the Williamson–Hall method separates the contributions of grain refinement and microstrain to peak broadening; and Rietveld refinement combined with peak shape analysis can yield dislocation density, stacking fault probability, and other defect-related parameters. Particularly, the advent of high-resolution synchrotron XRD has made it possible to track grain growth, strain evolution, and defect formation under in-situ or quasi in-situ electrochemical cycling conditions.

In recent years, XRD-based multiscale structural analysis strategies have evolved from static measurements to dynamic process tracking, which not only deepens the understanding of structure–performance relationships but also provides solid experimental and theoretical foundations for the rational design of high-performance LIB materials.

Guo et al. [14] employed in-situ XRD to systematically investigate the phase evolution of lithium-rich cathode materials under high average-voltage conditions, thereby elucidating the structural origin of the voltage increase. In-situ XRD patterns were collected during a reannealing process following partial chemical delithiation, allowing direct tracking of temperature-dependent structural changes. Figure 3.18 presents contour maps of both diffraction peak intensity and peak

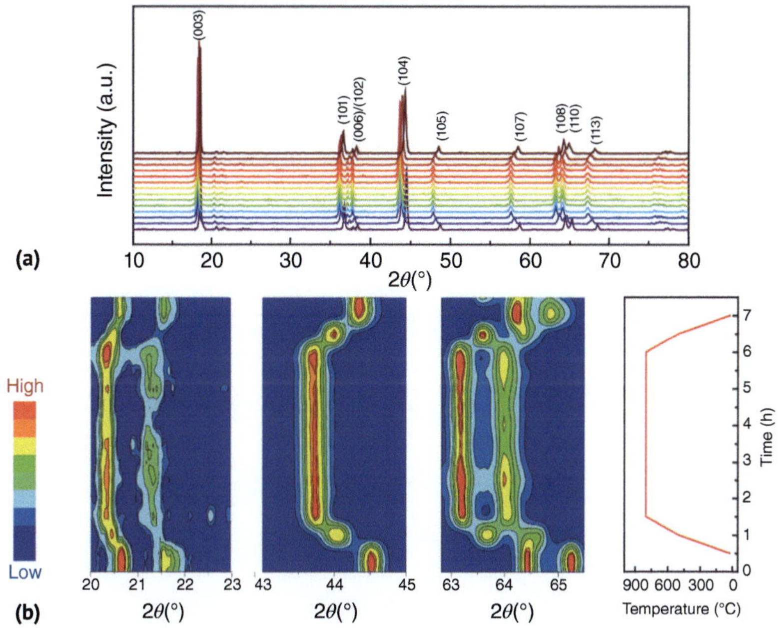

Figure 3.18 (a) In-situ XRD patterns of the LL-811 material during the first reannealing treatment process. (b) Contour plot of the evolution of the characteristic diffraction peaks during the first reannealing treatment process [14] / with permission of John Wiley & Sons.

position as a function of temperature, which clearly capture the continuous evolution of the crystal structure.

The pristine LL-811 material displays the characteristic R3m layered framework with a relatively low degree of cation disorder. After partial lithium extraction in solution followed by reannealing, the superlattice reflection associated with the Li_2MnO_3 component, located in the 20–23° range, shows a pronounced reduction in intensity. This attenuation indicates a diminished Li_2MnO_3 contribution, in agreement with the observed electrochemical behavior. In addition, the intensity of the (104) reflection near 43–45° decreases after treatment, leading to an increased $I_{(003)}/I_{(104)}$ ratio. This change signifies a suppression of cation mixing during reannealing and suggests the relocation of Ni^{2+} ions from the 3b octahedral sites in the lithium layer to the 3a octahedral sites within the transition metal slab.

During the reannealing process, a gradual enhancement of the (108) reflection is observed, accompanied by a concurrent attenuation of the (110) peak, further indicating a reduction in the Li_2MnO_3 component associated with the C2/m phase. As the temperature increases, all diffraction reflections shift systematically toward lower diffraction angles, reflecting thermal lattice expansion. Correspondingly, both the a and c lattice parameters increase, a change that is predominantly governed by the elevated temperature. Upon cooling to ambient conditions, the lattice constants of the reannealed sample (1TLL-811) remain slightly larger than those of the pristine LL-811 material. This irreversible lattice expansion is attributed to high-temperature-induced redistribution of transition metal cations, particularly the migration of Ni^{2+} ions during the calcination process, which alters the average crystallographic framework.

Wang et al. [15] employed in-situ XRD to identify intermediate phases and elucidate the dynamic structural evolution of ($0.5Li_2MnO_3{\cdot}0.5LiMn_{1/3}Ni_{1/3}Co_{1/3}O_2$) (denoted as LHMNCO TBA) during a complete heating–cooling cycle, as illustrated in Figure 3.19. The temperature-resolved diffraction patterns reveal that a predominantly spinel-type phase becomes stabilized at approximately 500 °C. The

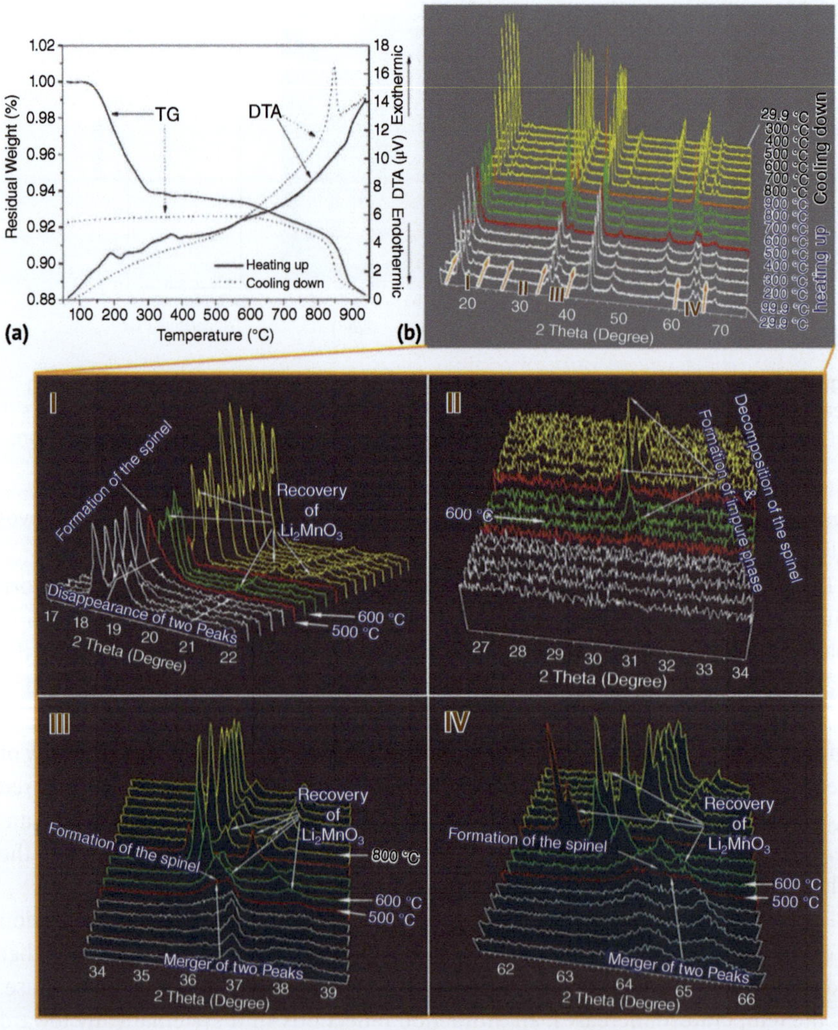

Figure 3.19 In-situ characterizations to identify phase transition of LHMNCO TBA during heat treatments. (a) TG/DTA curves when LHMNCO TBA is heated from room temperature to 950 °C (solid lines) and cooled back to ambient temperature (dashed lines) in air. (b) In-situ XRD patterns of LHMNCO TBA powders when heated to 900 °C and subsequently cooled down to room temperature, with selected portions shown from I to IV [15] / with permission of John Wiley & Sons.

continuous nature of the phase evolution is more clearly visualized in the magnified 2θ regions (I–IV).

In region I, the two reflections originally located at $2\theta \approx 18–20°$ progressively merge into a single diffraction peak at around 300 °C, corresponding to the (111)s plane of a cubic spinel structure. A slight shift of this peak toward lower diffraction angles with increasing temperature is attributed to thermal expansion of the lattice during in-situ heating. Concurrently, the gradual attenuation of the (020)M reflection at $2\theta \approx 20.8°$ up to 500 °C indicates the continuous consumption of the layered Li_2MnO_3 component.

The initially well-ordered layered $LiMn_{1/3}Ni_{1/3}Co_{1/3}O_2$ phase is evidenced by distinct peak splitting in the (006)R–(012)R and (108)R–(110)R doublets. Upon heating to 500 °C, these characteristic layered reflections collapse into broad (222)S and (440)S peaks, as shown in regions III and IV, signifying a complete transformation from the layered structure to the spinel phase.

Further increasing the annealing temperature beyond 600 °C leads to undesirable degradation of the newly formed spinel phase. This process is accompanied by the emergence of $MnCo_2O_4$ as an impurity phase and partial reappearance of layered Li_2MnO_3, consistent with the continuous mass loss observed in the thermogravimetric (TG) profiles during heating. The formation of $MnCo_2O_4$ is clearly identified at approximately 600 °C by reflections in the 2θ range of 30–31 ° (region II), while the recovery of Li_2MnO_3 is confirmed by the increased intensity of its characteristic peaks in regions I, III, and IV.

After cooling to room temperature, the final in-situ XRD pattern indicates coexistence of three phases, namely a spinel-structured compound, residual layered Li_2MnO_3, and $MnCo_2O_4$ impurity. The partial mass regain observed in the TG curve during the cooling stage is likely associated with reformation of the layered Li_2MnO_3 phase, supported by the presence of a pronounced exothermic peak in the differential thermal analysis (DTA) signal near 850 °C during cooling.

3.5 Conclusions and Perspectives

XRD together with WAXS has progressed well beyond its traditional role in phase identification, evolving into a powerful and flexible toolbox for interrogating structural, chemical, and microstructural evolution in functional materials and operating devices. In LIBs and other electrochemical energy storage systems, the combination of SR, high-energy diffraction, and advanced laboratory-based WAXS has enabled detailed tracking of crystalline phase transitions, lattice strain responses, cation redistribution, oxygen redox–induced structural changes, and their coupling to electrochemical performance. Owing to their nondestructive character and sensitivity across a broad range of length scales, XRD and WAXS continue to play indispensable roles not only in battery research but also in soft matter physics, catalysis, geoscience, and biological materials. Continuous advances in X-ray sources, detector speed, and computational analysis are rapidly transforming XRD from a predominantly ex-situ structural probe into a time-resolved, operando, and increasingly data-driven platform capable of examining materials under realistic working environments.

Future progress in XRD-based techniques will be strongly influenced by demands for enhanced temporal and spatial resolution, deeper integration with complementary characterization methods, and broader accessibility of operando capabilities. State-of-the-art synchrotron facilities, coupled with fast two-dimensional detectors, now routinely achieve millisecond- and even microsecond-scale time resolution, while sub-micrometer beam sizes enable spatially resolved mapping of structural gradients within individual particles or across full electrochemical cells. Such capabilities are essential for capturing short-lived intermediates, kinetic asymmetries, localized strain accumulation, and heterogeneous degradation processes. In parallel, multimodal experimental strategies—integrating XRD/WAXS with XAS, fluorescence mapping, tomography, and small-angle scattering—are becoming standard on modern beamlines. These approaches provide concurrent information on crystallography, oxidation states, elemental distributions, and hierarchical morphology, allowing quantitative insight into the coupled chemical–mechanical phenomena governing battery cycling, catalytic transformations, and polymer self-organization. At the same time, significant advances in laboratory diffractometers—featuring microfocus X-ray sources, hybrid photon-counting detectors, and tailored operando environments—are narrowing the gap between synchrotron and in-house measurements. This trend substantially expands access to quasi-operando studies and high-throughput screening without exclusive dependence on limited synchrotron beamtime.

The scope of XRD/WAXS applications is expected to expand further, with LIBs and emerging electrochemical technologies remaining major beneficiaries. For cathode materials, particularly structurally complex systems such as Li-rich layered oxides, synchrotron XRD continues to elucidate phase transition sequences, transition metal migration pathways, oxygen-redox-driven lattice distortions, and the origins of voltage fade. High-energy diffraction additionally enables penetration through realistic cell architectures, facilitating in-situ monitoring of structural evolution during prolonged cycling or aggressive fast-charging protocols. Complementary WAXS and SAXS measurements capture nanoscale morphological changes, including particle fracture, pore development, and interphase formation, which are central to understanding mechanical degradation. Similar methodological frameworks are rapidly extending to solid-state batteries, where grazing-incidence diffraction, depth-resolved scattering, and operando tomography enable direct observation of buried interfaces, interphase growth, and stress accumulation within solid electrolytes. Beyond electrochemical systems, WAXS and SAXS remain foundational in soft matter and biomolecular research, enabling time-resolved studies of polymer crystallization, protein conformational changes, and nanoparticle nucleation. In geoscience and cultural-heritage investigations, high-energy diffraction, microbeam scanning, and portable instruments support non-invasive analysis of minerals, pigments, and archaeological artifacts, offering quantitative structural information where destructive sampling is not feasible.

Despite their versatility, XRD and WAXS possess inherent limitations that must be carefully considered. Conventional powder diffraction yields volume-averaged structural information, potentially obscuring local heterogeneities, minority phases, or nanoscale defects unless supplemented by microbeam mapping, tomography, or total scattering approaches. Bragg diffraction alone is also relatively insensitive to amorphous or highly disordered components, necessitating pair distribution function (PDF) analysis or complementary spectroscopic techniques when investigating complex electrode composites or soft materials. Operando measurements introduce additional challenges, including background contributions from cell components, attenuation by electrolytes, and geometric constraints, all of which demand careful experimental design and data treatment. Furthermore, restricted access to synchrotron facilities and the massive data volumes generated by modern detectors underscore the need for robust, automated workflows for peak analysis, phase quantification, and model-based interpretation. These pressures are accelerating the adoption of machine learning and data-mining approaches for pattern recognition, automated refinement, and structure–property correlation.

Looking forward, the continued evolution of XRD/WAXS will be driven by the convergence of advanced instrumentation, operando methodologies, and data-centric analysis strategies. Establishing standardized operando cell designs, metadata protocols, and open-access databases will be crucial for improving reproducibility and enabling meaningful cross-platform comparisons. Deeper integration of diffraction with complementary synchrotron techniques will further illuminate the interplay between structure, chemistry, and mechanics that dictates performance in batteries, catalysts, and soft matter systems. Meanwhile, increasingly capable laboratory instruments will democratize access to advanced diffraction measurements, enabling rapid feedback during material synthesis, processing, and device optimization. Ultimately, combining XRD with PDF analysis, atomistic modeling, and machine learning frameworks will provide a unified view of long-range order, local disorder, and dynamic processes. In this way, XRD is poised to function not merely as a characterization method, but as a central mechanistic tool guiding the rational design and optimization of next-generation functional materials and electrochemical technologies.

References

1 H. Khan, A.S. Yerramilli, A. D'Oliveira, T.L. Alford, D.C. Boffito, G.S. Patience, Experimental methods in chemical engineering: X-ray diffraction spectroscopy—XRD, The Canadian Journal of Chemical Engineering 98(6) (2020) 1255–1266. https://doi.org/10.1002/cjce.23747.

2 M. Xia, T. Liu, N. Peng, R. Zheng, X. Cheng, H. Zhu, H. Yu, M. Shui, J. Shu, Lab-scale in situ X-ray diffraction technique for different battery systems: Designs, applications, and perspectives, Small Methods 3(7) (2019) 1900119. https://doi.org/10.1002/smtd.201900119.

3 A. Fitch, C. Dejoie, E. Covacci, G. Confalonieri, O. Grendal, L. Claustre, ID22– the high-resolution powder-diffraction beamline at ESRF, Journal of Synchrotron Radiation. 30(5) (2023) 1003–1012. https://doi.org/10.1107/S1600577523004915.

4 J.A. Steele, E. Solano, D. Hardy, D. Dayton, D. Ladd, K. White, P. Chen, J. Hou, H. Huang, R.A. Saha, L. Wang, F. Gao, J. Hofkens, M.B.J. Roeffaers, D. Chernyshov, M.F. Toney, How to GIWAXS: Grazing incidence wide angle X-ray scattering applied to metal halide perovskite thin films, Advanced Energy Materials 13(27) (2023) 2300760. https://doi.org/10.1002/aenm.202300760.

5 G. Zhao, V. Saïbi, L. Castro, A. Guéguen, S. Belin, M. Duttine, D. Carlier, P. Salles, C. Delmas, M. Guignard. A Complete Operando Structural and Redox Perspective on the O_2-Type Lithium-Rich Manganese Oxide $LiMn_{0.75}O_2$, Advanced Energy Materials, 16(9) (2026) e05540. https://doi.org/10.1002/aenm.202505540.

6 M.A. Karlsen, J. Billet, S. Tao, I. Van Driessche, S.J.L. Billinge, D.B. Ravnsbaek, Operando pair distribution function analysis of nanocrystalline functional materials: The case of TiO(2)-bronze nanocrystals in Li-ion battery electrodes, Journal of Applied Crystallography 57(Pt 4) (2024) 1171–1183. https://doi.org/10.1107/S1600576724005624.

7 M. Bini, S. Ferrari, D. Capsoni, P. Mustarelli, G. Spina, F.D. Giallo, M. Lantieri, C. Leonelli, A. Rizzuti, V. Massarotti, Pair distribution function analysis and Mössbauer study of defects in microwave-hydrothermal LiFePO4, RSC Advances 2(1) (2012) 250–258. https://doi.org/10.1039/c1ra00525a.

8 C.-K. Lin, Y. Ren, K. Amine, Y. Qin, Z. Chen, In situ high-energy X-ray diffraction to study overcharge abuse of 18650-size lithium-ion battery, Journal of Power Sources 230 (2013) 32–37. https://doi.org/10.1016/j.jpowsour.2012.12.032.

9 L. de Biasi, A. Schiele, M. Roca-Ayats, G. Garcia, T. Brezesinski, P. Hartmann, J. Janek, Phase transformation behavior and stability of LiNiO(2) cathode material for li-ion batteries obtained from in situ gas analysis and operando X-ray diffraction, ChemSusChem 12(10) (2019) 2240–2250. https://doi.org/10.1002/cssc.201900032.

10 K.-W. Nam, W.-S. Yoon, X.-Q. Yang, Structural changes and thermal stability of charged LiNi1/3Co1/3Mn1/3O2 cathode material for Li-ion batteries studied by time-resolved XRD, Journal of Power Sources 189(1) (2009) 515–518. https://doi.org/10.1016/j.jpowsour.2008.10.130.

11 K.W. Nam, S.M. Bak, E. Hu, X. Yu, Y. Zhou, X. Wang, L. Wu, Y. Zhu, K.Y. Chung, X.Q. Yang, combining in situ synchrotron X-ray diffraction and absorption techniques with transmission electron microscopy to study the origin of thermal instability in overcharged cathode materials for lithium-ion batteries, Advanced Functional Materials 23(8) (2012) 1047–1063. https://doi.org/10.1002/adfm.201200693.

12 R. Lv, C. Luo, K. Wang, W. Tang, B. Liu, Y. Liu, Y. Li, L. Li, F. Wu, R. Chen, Confinement-regulated lamellar interphase with enhanced dielectric environments for lithium metal batteries, ACS Nano 19(22) (2025) 21007–21019. https://doi.org/10.1021/acsnano.5c04491.

13 S.M. Bhaway, Z. Qiang, Y. Xia, X. Xia, B. Lee, K.G. Yager, L. Zhang, K. Kisslinger, Y.M. Chen, K. Liu, Y. Zhu, B.D. Vogt, operando grazing incidence small-angle

X-ray scattering/X-ray diffraction of model ordered mesoporous lithium-ion battery anodes, ACS Nano 11(2) (2017) 1443–1454. https://doi.org/10.1021/acsnano.6b06708.

14 J.L. Shi, D.D. Xiao, M. Ge, X. Yu, Y. Chu, X. Huang, X.D. Zhang, Y.X. Yin, X.Q. Yang, Y.G. Guo, L. Gu, L.J. Wan, High-capacity cathode material with high voltage for Li-ion batteries, Advanced Materials 30(9) (2018) 1705575. https://doi.org/10.1002/adma.201705575.

15 J. Zhao, R. Huang, W. Gao, J.M. Zuo, X.F. Zhang, S.T. Misture, Y. Chen, J.V. Lockard, B. Zhang, S. Guo, M.R. Khoshi, K. Dooley, H. He, Y. Wang, An ion-exchange promoted phase transition in a Li-excess layered cathode material for high-performance Lithium ion batteries, Advanced Energy Materials 5(9) (2015) 1401937. https://doi.org/10.1002/aenm.201401937.

4

Transmission Electron Microscopy for Atomic Scale and Dynamic Characterization of Battery Materials

4.1 Electron Beam Imaging and Basic Configuration of Transmission Electron Microscopy (TEM)

4.1.1 Electron Beam Imaging

The resolution of optical microscopy is limited by the wavelength of the illuminating light. Traditional visible light microscopes have a resolution limit of 200 nm, while near-field optical microscopes can achieve a resolution of 50 nm. In comparison, TEM uses a high-voltage accelerated electron beam as the illuminating light source. Benefiting from the ultra-short wavelength of the high-speed electron beam (the electron beam wavelength corresponding to a 100 kV accelerating voltage is approximately 0.0037 nm), it is theoretically capable of achieving extremely high resolution. However, the wavelength of the electron beam is not the primary factor affecting TEM resolution; instead, factors such as spherical aberration, chromatic aberration, and astigmatism of the electromagnetic lenses in TEM impose significant limitations. Currently, with the advancement of aberration-corrected TEM technology, the highest resolution of scanning TEM (STEM) can reach 0.06 nm.

In 1931, Ernst Ruska and his supervisor Max Knoll successfully fabricated an imaging device consisting of two electromagnetic lenses, achieving a magnification of $3.6 \times 4.8 = 17.3$. Although the magnification was relatively low, it marked the first imaging accomplished using an electron beam. The magnification of optical microscopy is restricted by the light wavelength, and they hoped that electron beams could overcome this resolution limitation. Despite Louis de Broglie's wave-particle duality theory having been proposed as early as 1924 and experimentally verified by the electron diffraction experiments conducted by Davisson and Thomson in 1927, Ruska, an engineer, only learned about it after completing electron beam imaging in 1931. Initially, they were quite disappointed because the resolution of electron beam imaging is also constrained by wavelength. However, after calculating the electron wavelength, they became excited—electron wavelengths are five orders of magnitude smaller than those of visible light. Clearly, electron beam imaging holds great promise, and this has been proven by subsequent developments. As an engineer, Ruska maintained a persistent pursuit of electron microscopy technology over the years, made significant contributions to the field of TEM, and thus shared the 1986 Nobel Prize in Physics.

The imaging principle of convex lenses has found widespread applications. For instance, old-fashioned film projectors use transmitted light and convex lenses to magnify slides into real images projected onto a screen. The imaging logic of TEM is similar to that of slide projectors. The only differences are that glass lenses are replaced by electromagnetic lenses (which perform functions analogous to glass lenses, such as converging parallel electron beams to the focal plane) and visible light waves are replaced by the de Broglie waves of accelerated electron beams. With such a short wavelength "ruler", imaging of micro-objects becomes feasible.

There is more than one imaging electromagnetic lens in a transmission electron microscope. The real image formed by the previous electromagnetic lens can be used as the object for the next lens, enabling repeated magnification. Consequently, samples at the nanometer or even atomic scale can be magnified into observable images on a screen. By exploiting the diffraction phenomenon of electron beams in crystals and manipulating the direct beam and diffracted beams, TEM offers various imaging modes, endowing it with profound theoretical foundations and rich imaging capabilities. Examples include diffraction using electron beams, imaging by manipulating direct and diffracted beams, or scanning transmission imaging, which will be elaborated on in this chapter.

4.1.2 Basic Configuration of Modern TEM

The structure of modern TEM is highly complex, but from a user's perspective, it is sufficient to understand its main working logic without delving into the specific hardware structure and technology. A transmission electron microscope consists of an illumination system, an imaging system, and an imaging recording system. Meanwhile, the entire process requires various operations on the electron beam, so the instrument system must operate under high-vacuum conditions.

4.1.2.1 Illumination System

It comprises an electron gun and condenser lenses. The condenser lens system is a set of lenses that can converge the accelerated electrons and irradiate them onto the sample. Depending on the operating mode, it can generate parallel electron beams and convergent electron beams. Parallel electron beams are mainly used for TEM imaging and selected area electron diffraction (SAED), while convergent electron beams are primarily employed in STEM, X-ray energy dispersive spectroscopy (EDS), electron energy loss spectroscopy (EELS), and convergent beam electron diffraction. Deflection coils can adjust the translation and incident angle of the electron beam for imaging in different modes.

4.1.2.2 Imaging System

It includes (from top to bottom) a sample holder (where the sample is placed), an objective lens (for imaging or diffraction), an intermediate lens (for forming a magnified image), and a projection lens (for further magnifying the image). High-magnification transmission electron microscopes may be equipped with multiple intermediate lenses or projection lenses. The imaging system also incorporates an objective aperture and a selected area aperture. The principle of the aperture is to block part of the electron beam and select another part, which is a selective operation on the transmitted electron beams (direct beam and diffracted beams) (specific uses are described in the corresponding sections). The objective aperture is located on the back focal plane of the objective lens, where parallel light beams converge. The objective aperture can capture different convergent electron beams for subsequent imaging. The selected area aperture is situated on the image plane of the objective lens and can be used to select a specific area of the image.

4.1.2.3 Imaging and Recording System

The carrier of imaging information in TEM is electrons, so the role of the observation system is to display and record the position and quantity (intensity) of the transmitted electron beam. The observation screen in TEM is coated with a material that can be excited by electrons to emit fluorescence, such as ZnS-doped material, which emits green light with a wavelength of 550 nm. Charge-coupled device (CCD) cameras are also commonly used in TEM for digital recording of electron information. CCD cameras convert incident electrons into light signals, which are then transformed into electrical signals and recorded as pixel brightness.

The various lenses in a TEM refer to electromagnetic lenses, which generate magnetic fields through current coils to control the trajectory of electron movement, and can also be understood as the "light path" for electron imaging. The effect of electromagnetic lenses on electrons is similar to that of convex lenses on light. Unlike optical lenses, electromagnetic lenses have adjustable focal lengths by varying the coil current.

Currently, electromagnetic lens technology is still not perfect and can produce spherical aberration and chromatic aberration (Figure 4.1). Spherical aberration is caused by the different convergence effects of the central paraxial region and the peripheral off-axis region of the electromagnetic lens on the electron beam. When the electron beam passes through the electromagnetic lens, the electron beam in the central axis region of the lens deflects at a relatively small angle, while the electron beam at the edge of the lens (i.e., the region far from the central axis of the lens) deflects at a larger angle under the action of the magnetic field. Ideally, both the electron beams in the central axis region and the edge region should converge at the focal point of the electromagnetic lens. However, due to the deflection angle error of the electron beam in different regions of the lens, the focusing position of the electron beam is dispersed, affecting the imaging resolution, which is called spherical aberration.

The deviation of the magnetic field axial symmetry of the electromagnetic lens leads to differences in the refraction ability of the electromagnetic lens for incident electron beams in different directions, which is called astigmatism. Imaging deviation caused by the

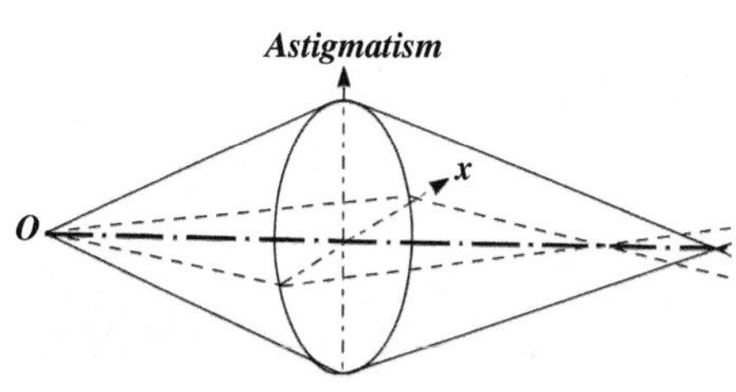

Figure 4.1 Diagram of chromatic aberration, spherical aberration, and astigmatism.

uneven kinetic energy of electrons in the light source electron beam (insufficiently monochromatic wavelength) is known as chromatic aberration. After the electron beam passes through the lens, wavelength dispersion results in the dispersion of the convergence point, which is essentially a problem of the illuminating electron beam. Chromatic aberration, spherical aberration, and astigmatism can all be eliminated and corrected through improvements in hardware design, including spherical aberration correctors, stigmators, and enhancements in the stability of the electron beam light source, thereby improving the resolution of the TEM. Nevertheless, even the best-performing electromagnetic lenses have an imaging quality equivalent to a simple optical lens made from the bottom of a cola bottle. In other words, when comparing the precision of electromagnetic lenses to human vision, their effect is theoretically equivalent to partial blindness. Therefore, the resolution of current transmission electron microscopes is still far from reaching the order of magnitude of the electron beam wavelength, or in other words, there is significant room for improvement in the resolution of transmission electron microscopes relative to the electron beam wavelength.

4.1.3 Operating Modes of TEM

The electron beam penetrates the sample and reaches the objective lens of the imaging system. Through the convergence of the objective lens, a diffraction pattern is formed on the focal plane of the objective lens (if the sample is a thin crystal), and an image is formed on the image plane of the objective lens (Figure 4.2). The operating modes of TEM include imaging mode and diffraction mode.

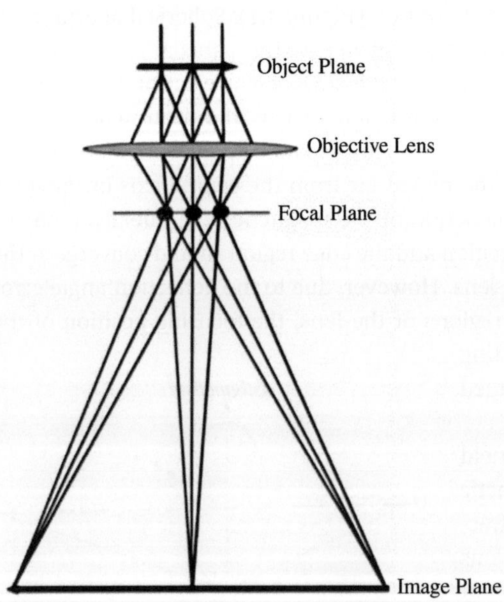

Figure 4.2 Schematic diagram of the positions of the object plane, objective lens, objective lens focal plane and image plane.

4.1.3.1 Imaging Mode

To observe an image, the image plane of the objective lens is used as the object plane of the intermediate lens, and the image can be magnified and projected onto a fluorescent screen or a CCD camera. TEM imaging utilizes parallel incident electron beams that pass through the sample and are scattered by it, forming an image on the image plane of the objective lens. In imaging mode, the object plane of the intermediate lens is the image plane of the objective lens, and then the intermediate lens magnifies it to obtain a microscopic image of the sample.

4.1.3.2 Diffraction Mode

When the electron beam passes through a thin crystal sample, diffraction can occur, resulting in a diffraction pattern on the focal plane. To observe the diffraction pattern, the focal plane of the objective lens is used as the object plane of the intermediate lens, that is, the diffraction pattern is magnified as an image and projected onto a fluorescent screen or a CCD camera. At this time, the diffraction pattern of the electron beam passing through the crystal sample is observed. Both diffraction mode and imaging mode are achieved by adjusting the parameters of the objective lens and the intermediate lens, and they have been set as two switchable basic modes on the TEM. TEM images are grayscale images, and their brightness contrast is the change in electron beam intensity collected by the fluorescent screen (or CCD camera). In the TEM imaging process, different contrasts can be selected for imaging, such as mass-thickness contrast, diffraction contrast, phase contrast, and Z contrast. We first discuss the most easily understandable mass-thickness contrast, and other contrast images will be covered later. Mass-thickness contrast arises from the differences in the thickness, density, or mass of the sample at different positions, leading to variations in electron scattering ability. Regions with higher density or greater thickness can scatter more electrons and form larger scattering angles. The objective aperture behind the objective lens can block some electrons with large scattering angles from participating in imaging, resulting in differences in the intensity of the transmitted electron beam and forming image contrast. Regions with high density, large thickness, and few transmitted electrons appear dark in the image, and vice versa. Mass-thickness contrast depends on the composition, density distribution, and thickness of the sample. It is the most typical imaging contrast in TEM, very similar to the imaging of film by a projector.

TEM analysis has high requirements for sample preparation. Since the detected objects are transmitted electrons and scattered electrons, the most important requirement for the sample is to be as thin as possible. Generally, the thickness should be less than 100 nm, and the thickness of samples for high-resolution TEM testing needs to be approximately 10 nm, also as thin as possible. For general powder samples, if the particles are at the micron scale, they need to be ground as much as possible to reduce the particle size. After ultrasonic dispersion in a solvent (such as anhydrous ethanol), the suspension is dropped onto a supporting grid and allowed to dry. Copper grids are commonly used, which are equipped with support films. Depending on the size and morphology of the observed sample, ordinary carbon films, microgrids, and ultrathin carbon films can be selected. If the particles are too large to be easily ground, the powder sample can be embedded in resin for curing and then ultrathin sectioned. Bulk samples need to be cut (for ceramic samples) or punched (for metal samples) into thin sheets with a diameter of 3 mm and subjected to ion beam thinning (similar to surface argon ion etching in X-ray photoelectron spectroscopy (XPS)). Sample preparation for TEM directly affects imaging and diffraction results. Therefore, for some special samples, it is recommended to consult TEM technicians before testing to ensure the test effect.

4.2 Electron Diffraction

Among the analytical methods discussed earlier, those using electron beams as probes typically leverage the ability of electron beams to be focused to the nanometer scale, enabling higher spatial resolution structural information compared to X-ray sources. Electron diffraction in TEM can be used to distinguish amorphous and crystalline structures at the micro-nano scale, analyze the structural characteristics of micro-nano scale crystals, and the orientation relationship between grains in polycrystalline aggregates. For single crystals with specific morphologies, the zone axis orientation and the exposed crystal plane indices can be analyzed through electron diffraction.

The physical basis of electron diffraction is the wave nature of electrons. Of course, there are significant differences between electron beams and X-rays. Firstly, the wavelength of electron beams is much shorter than that of X-rays, so the diffraction angle of electron beams is very small. Secondly, the Coulomb interaction between electron beams and atoms is stronger, and the scattering effect is much more intense than that of X-rays. Therefore, the electron diffraction signal is stronger than that of XRD. Parallel incident electron beams that are not scattered during their passage through the sample remain parallel and are called direct beams. After passing through the objective lens, they converge at the focal point, which is referred to as the direct spot. If the sample is a crystal, the incident electron beam undergoes diffraction at a certain crystal plane, forming a parallel beam in the diffraction direction, known as the diffracted beam. After passing through the objective lens, it converges at a point on the focal plane, called the diffraction spot. If the object plane of the intermediate lens is placed at the focal plane of the objective lens, the image formed on the fluorescent screen is the diffraction pattern of the focal plane, i.e., the diffraction mode of TEM imaging.

In the XRD section, the diffraction spectra of powder samples were mainly discussed, and the diffraction of different crystal planes in the crystal was analyzed based on Bragg's equation, so the content of reciprocal lattice was not involved. In the electron diffraction section, to better understand electron diffraction patterns, it is necessary to understand diffraction in reciprocal space. The key issue to understand is how a two-dimensional diffraction pattern corresponds to a three-dimensional crystal structure.

4.2.1 Ewald Sphere

After learning about reciprocal space and the Laue equation, electron diffraction can be understood in the reciprocal space. The experimental scenario is that an electron beam irradiates a crystal sample. Instead of focusing on the lattice point arrangement in the crystal, the reciprocal lattice corresponding to the crystal structure is constructed. As discussed earlier, only the direction of the wave vector can be marked in the real space, but its length cannot be drawn because the dimension of the wave vector is the reciprocal of length, which cannot be expressed in the real space. In the reciprocal space, the wave vector can be fully expressed with both direction and magnitude. As shown in Figure 4.3 select a lattice point in the reciprocal lattice as the origin O, and use the selected origin as the front end of the wave vector. Since it is in the reciprocal space, both the direction and magnitude of the wave vector can be drawn. Taking the end of the wave vector as the center (marked as center C), draw a sphere with a radius of $1/\lambda$ (the magnitude of the wave vector), which is called the Ewald sphere (or Ewald reflection sphere). The vector CO is the incident wave vector k0, and the vector from C to any point on the sphere can be the scattered wave vector k. When a certain reciprocal lattice point exactly falls on the Ewald sphere, k − k0 = g, satisfying the Laue equation. In other words, the reciprocal lattice points falling on the Ewald sphere correspond to crystal planes that satisfy the diffraction condition, and diffraction can occur. The diffraction direction is the direction of the scattered wave vector k.

4.2.2 Deviation Parameter

In the XRD section, to detect the diffraction signals of different crystal planes, it is necessary to continuously change the θ angle of the detector. In electron diffraction, however, the diffraction spots of several crystal planes can be observed simultaneously without adjusting the angle of the fluorescent screen or the detector. One

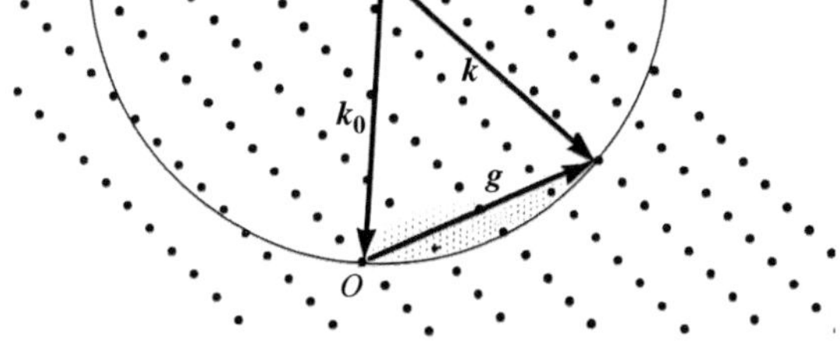

Figure 4.3 Reciprocal lattice and Ewald sphere.

of the reasons is that the wavelength of X-rays is much longer than that of electron beams. According to Bragg's equation, the diffraction angle of XRD is also much larger, while the diffraction angle of electron beams is very small, usually not exceeding 1°. If the incident electron beam is strictly parallel to a certain zone axis of the sample, that is, the direction of the electron beam is strictly parallel to the crystal plane, obviously, according to the Laue condition, theoretically, no diffraction will occur (this conclusion can also be obtained from Bragg's equation). However, clear diffraction patterns can be observed in actual experiments, which indicates that in the electron diffraction of transmission electron microscopes, a certain intensity of diffracted beams can still be observed even if the Bragg condition is not strictly satisfied. Due to the very short wavelength of electrons, the radius of the Ewald sphere is large. Compared with the crystal reciprocal lattice, the Ewald sphere surface is close to a plane. At the same time, since the transmitted sample is very thin, the corresponding reciprocal point is not an ideal point but presents an elongated rod shape (see below). In this way, the Ewald sphere surface can cut through many reciprocal rods, forming diffraction spots of different intensities.

If the reciprocal vector g does not intersect the Ewald sphere, a vector s is introduced, called the deviation vector. Since the electron diffraction Bragg angle is very small, the deviation vector s can be approximately regarded as along the direction of the incident wave vector, and understood as a scalar s, i.e., the deviation parameter.

This indicates that for a thin crystal with thickness T, the diffraction intensity is a function of the deviation parameter s. Figure 4.4 shows the functional relationship between diffraction intensity and deviation parameter. The strongest diffraction can be observed at s = 0, and a certain diffraction intensity can still be observed when s is not zero. The above results can be understood as that for thin samples, the lattice points in the reciprocal space are no longer pure points but elongated

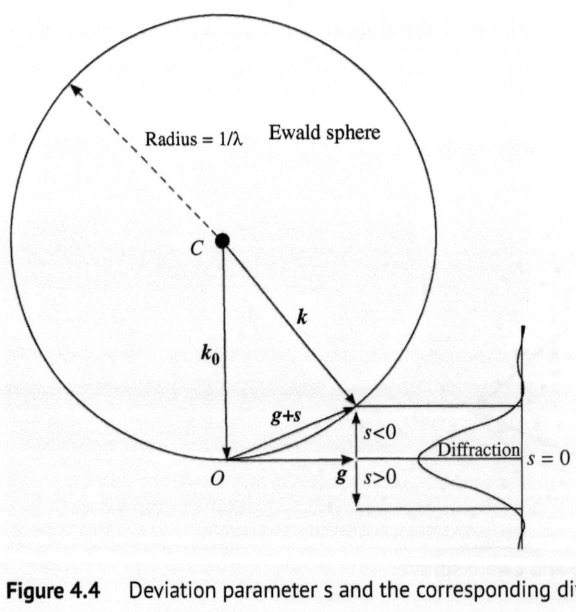

Figure 4.4 Deviation parameter s and the corresponding diffraction intensity distribution.

rods, called reciprocal rods. Obviously, even if the center of the reciprocal rod does not intersect the Ewald sphere, as long as it is within the deviation parameter range, the sphere can intersect with a part of the rod, or in other words, the Bragg condition is partially satisfied, so weak diffraction spots can be observed (this is also a phenomenon in experiments). There are two ways to change the deviation parameter s in experiments: one is to tilt the sample, which is equivalent to the Ewald sphere remaining stationary and the reciprocal lattice moving, and the deviation parameter is selected by observing the brightness of the diffraction spots; the other is to tilt the direction of the electron beam, which is equivalent to the reciprocal lattice remaining stationary and the deviation parameter is selected by moving the Ewald sphere.

4.2.3 SAED and Diffraction Pattern Calibration

Selecting an area of interest within the image range of a TEM (such as grains of different crystal phases) and then performing electron diffraction on the selected area is called SAED. The standard method is to use a selected area aperture. The selected area aperture is located on the image plane of the objective lens. An imaging area is constrained by the selected area aperture (at this time, the object plane of the intermediate lens is on the image plane of the objective lens, i.e., imaging mode). Then, the object plane of the intermediate lens is placed on the focal plane of the objective lens, and the electron microscope is switched to diffraction mode to observe the diffraction pattern on the focal plane of the objective lens. Due to the constraint of the selected area aperture, only the direct beam and diffracted beam participating in the imaging of the selected area can pass through the selected area aperture, thereby obtaining the diffraction pattern corresponding to the selected image area.

The selected area aperture is equivalent to adding a virtual aperture to the sample (the parallel electron beams outside the target area are not blocked and still hit the sample, but outside the selected area. After passing through the objective lens, this part of the electron beam is blocked by the selected area aperture and thus cannot participate in imaging). Note that regardless of whether the object plane of the intermediate lens is selected on the focal plane or the image plane of the objective lens, due to the blocking of the selected area aperture, only the electron beam that can pass through the selected area aperture can be diffracted or imaged. By selecting the image area on the image plane, that is, selecting the electron beam participating in imaging, the diffraction pattern of this part of the electron beam on the focal plane also corresponds to the selected image area. The selected area aperture is also called the intermediate lens aperture, and the selected area range on the sample is in the micron scale.

4.2.4 Single Crystal Electron Diffraction Pattern Calibration

When the electron beam is incident along the zone axis of a single crystal, the diffraction pattern is actually the lattice arrangement on the zero-order reciprocal plane of the reciprocal lattice. Assigning the diffraction spots on a diffraction pattern, or identifying which crystal plane's diffraction each diffraction spot comes from, is called diffraction pattern calibration. Even if the camera constant is not known, at least there is an inverse proportional relationship between R and d. Since the diffraction spots are lattice points on the reciprocal plane, the reciprocal vectors corresponding to the diffraction spots satisfy the vector addition relationship. For samples with known crystal structures, the zone axis can be determined through diffraction patterns, and the orientation of the crystal can be judged, such as determining the exposed crystal plane of the crystal nanosheet through diffraction patterns.

4.2.4.1 Polycrystalline Diffraction Patterns

If the sample for SAED is polycrystalline powder, it can be regarded as each randomly oriented single crystal particle undergoing diffraction. According to the electron diffraction formula, for each single crystal particle, if the (*hkl*) crystal plane can undergo diffraction, the distance between the corresponding diffraction spot and the direct spot is $R = K/d$. The connection direction between the direct spot and the diffraction spot (i.e., the *hkl* diffraction vector) is perpendicular to the (*hkl*) crystal plane. Since the orientation of the (*hkl*) crystal plane of the powder grains is random, as long as the (*hkl*) crystal plane undergoes diffraction, the diffraction spots will form a circular ring with a radius R from the direct spot. Due to the different distances of the diffraction spots, the diffraction patterns formed by all participating diffraction crystal planes are concentric rings around the direct spot, called polycrystalline diffraction rings. The ratio of the radii of the diffraction rings = the ratio of the reciprocals of the corresponding diffracted interplanar spacings.

Using samples with known structures, the camera constant of the transmission electron microscope can be determined. First, measure the radius of the diffraction ring of the polycrystalline sample to obtain the ratio of R^2. For the cubic crystal system structure, the crystal plane family index {*hkl*} corresponding to each diffraction ring can be determined through the increasing ratio law, and the corresponding interplanar spacing d can be found according to the PDF card of the sample. Then, the camera constant is calculated for each diffraction ring according to $K = Rd$, and the average value is taken as the camera constant. If the camera constant is known, for unknown structure polycrystalline powders, the ratio of R^2 can be calculated from the radius of the diffraction ring to infer the crystal lattice structure of the material. With the help of the camera constant, the interplanar spacing d corresponding to each diffraction ring is calculated, and the PDF card is searched using the d value corresponding to the diffraction ring with strong diffraction intensity to find the phase corresponding to the sample.

4.2.5 Secondary Diffraction

For very thin crystal samples, it is usually considered that the electron beam undergoes only one diffraction. Secondary diffraction refers to the diffracted beam undergoing a second diffraction in the sample before leaving the sample.

The probability of the second diffraction is approximately the square of the probability of the first diffraction. Therefore, if the sample is very thin, only the first diffraction needs to be considered. However, a very thin crystal sample is an idealized state. For relatively thick crystal samples, the electron beam can also undergo secondary diffraction or even multiple diffractions. If the crystal is thicker, the diffracted beam is stronger. Since the electron diffraction Bragg angle itself is very small, the direction of the diffracted beam is close to the direction of the incident beam. The secondary diffraction of the diffracted beam can produce a similar diffraction pattern. However, the diffraction spots of the first diffraction and the secondary diffraction are combined, and the diffraction spots that should have been extinct will appear. Secondary diffraction spots can also coincide with primary diffraction spots, resulting in abnormal brightness of certain diffraction spots.

4.3 Different Imaging Modes and Contrast Imaging

The electron beam in TEM is understood as an electron wave. The amplitude and phase of the electron wave scattered (diffracted) by the sample will change, resulting in image contrast. Earlier, we discussed the most easily understandable image contrast in TEM imaging—mass-thickness contrast, that is, the brightness contrast of the transmitted electron beam intensity caused by different thicknesses or compositions of the sample, which belongs to amplitude contrast. For crystal samples, when there is a strong diffraction effect, the contribution of mass-thickness contrast to image contrast can be ignored. After learning about electron diffraction, we can continue to discuss other modes of TEM imaging. After all, crystal materials are the most common samples in transmission electron microscope analysis, and it is only after understanding electron diffraction that it is easy to understand different imaging modes.

When the electron beam penetrates the crystal film, according to the angle between the incident direction of the electron beam and certain crystal planes in the film, if there are crystal planes that strictly satisfy or are very close to the Bragg diffraction condition, strong diffraction will be formed. If there is a certain deviation in the diffraction condition of the crystal plane (the deviation parameter s is not zero), diffractions of different intensities will be formed. The image contrast formed by using the direct beam for imaging or the diffracted beam for imaging and utilizing the different diffraction conditions at different positions of the sample is called diffraction contrast. Diffraction contrast is a direct reflection of the crystallographic characteristics of different parts of the sample. It is also a kind of amplitude contrast caused by the difference in the degree of satisfying the Bragg condition due to the different orientations or structures of the crystal everywhere.

The conditions for forming diffraction contrast include:

1. The difference in orientation between two grains causes them to deviate from the Bragg diffraction to different degrees, forming contrast.
2. The existence of defects or strain fields causes local distortion of the crystal, leading to changes in the Bragg condition and forming contrast.

3. The enrichment of micro-area elements or the existence of second-phase particles may cause changes in their interplanar spacing, leading to changes in the Bragg condition and forming contrast, including changes in the structure factor caused by different phase compositions, resulting in contrast.
4. The contrast displayed with the change of thickness in the perfect crystal forms thickness fringes.
5. The contrast caused by the different degrees of bending (deviation vector) in the perfect crystal forms equal-inclination fringes.

4.4 Diffraction Contrast Images

The essence of TEM image contrast is the change in the electron wave function after the interaction between the electron beam and the sample. If the exit wave function of the electron beam passing through the sample can be directly calculated, it will be very useful for understanding the TEM image. The formation of diffraction contrast is due to the diffraction of the electron beam in the crystal, so it is necessary to calculate the diffraction intensity. The kinematic theory of diffraction contrast is a simplified method for discussing diffraction intensity, which can basically explain diffraction contrast images. The kinematic theory of diffraction contrast requires the following approximate assumptions.

4.4.1 Two-beam Approximation

Diffraction contrast images are imaged with direct beams and diffracted beams respectively. By tilting the crystal and selecting the appropriate crystal plane direction, only one set of crystal planes (*hkl*) is very close to satisfying the Bragg diffraction condition, while other crystal planes are far from their respective diffraction positions. In this way, on the focal plane of the objective lens, in addition to the direct spot, there is only one strong diffraction spot, and the intensity of other diffraction spots is very weak, which can be ignored. From the perspective of wave vectors and reciprocal space, only one reciprocal point intersects the Ewald sphere, and other reciprocal points have a large deviation. For relatively thick crystals, the two-beam condition is easier to satisfy because the reciprocal rods of relatively thick crystals are shorter, and it is easier to satisfy the deviation condition (except for one reciprocal point that satisfies the diffraction condition). By tilting the sample, a specific (*hkl*) diffraction plane can be selected to form a strong diffracted beam (instead of forming a symmetric diffraction pattern by incident along the zone axis), and the other beam is a direct beam. Such a diffraction condition is called the two-beam condition. Imaging with the (*hkl*) diffracted beam as a dark-field image, only the position where the (*hkl*) crystal plane undergoes diffraction is bright. Bright-field images and dark-field images are two ways to achieve amplitude contrast images. When electrons are incident along a low-index crystal direction, the number of diffracted beams is large, the diffraction intensity is high, and the transmission intensity is weak, which becomes dark in the corresponding transmission bright-field image, and vice versa. Imaging under the two-beam condition, only one diffraction spot participates in imaging [corresponding to the (*hkl*) crystal plane of the crystal]. Therefore, whether it is bright-field imaging or dark-field imaging, the diffraction contrast on the image mainly comes from the hkl diffraction spot, and the reciprocal vector corresponding to this diffraction spot is usually marked as ghkl.

Under the ideal two-beam condition, the intensities in the bright- and dark-field images should be complementary. If the two-beam condition is not satisfied, that is, there are other diffracted beams at the same time, the contrast of the bright- and dark-field images cannot be completely complementary, and the diffraction contrast will also be reduced. This is the significance of the two-beam condition in diffraction contrast images.

4.4.2 Kinematic Approximation

The kinematic approximation assumes that the intensity of the diffracted beam is much lower than that of the incident beam, and the direct beam and the diffracted beam are independent of each other without interaction, and multiple reflections of the electron beam in the sample are ignored. For very thin crystals, the above assumptions are approximately valid under the condition of high-speed electron beams.

4.4.3 Column Approximation

It is assumed that the electron beam in the sample propagates in a small crystal column parallel to the electron beam, and the small crystal column can be regarded as a stack of n unit cells. Assuming that the propagation of the electron beam in the small crystal column is not affected by the adjacent small crystal columns, the calculation of the diffracted beam intensity does not consider transverse scattering, but only the intensity of the electron beam propagating out of the crystal along the small crystal column.

On the basis of the above assumptions, the intensity change of the diffracted beam after passing through the small crystal column can be simply calculated. Since the kinematic approximation ignores multiple reflections of the electron beam, the diffraction intensity of the electron beam by each unit cell in the small crystal column can be summed. In addition, according to the two-beam approximation, only the diffraction of a certain (*hkl*) crystal plane needs to be considered.

4.4.4 Crystal Defect Analysis

Structural defects in crystals cause lattice sites to deviate from strict lattice periodic positions, thereby changing the angle between specific crystal planes in the defect region and the electron beam. Compared with the diffraction conditions of perfect crystals, the Bragg conditions for diffraction in the defect region change, forming different diffraction contrasts. For defect analysis, the sample bending should be avoided as much as possible, and the analysis should be carried out in a relatively thin and uniform thickness region to avoid the interference of thickness fringes and bending effects. Diffraction contrast is mainly used for microstructure analysis (grain shape, grain size, grain boundaries, precipitates, etc.) and crystal defect analysis (dislocations, twins, stacking faults, etc.). Obviously, different crystal defects cause different distortions of the perfect crystal, thus different phase differences, and the resulting diffraction contrast images are also different. When the selected diffraction vector g is perpendicular to the Burgers vector ($g \cdot b = 0$), the defect cannot form contrast, and this relationship can be used to determine the Burgers vector b of the dislocation. The dislocation image is related to the sample tilting direction, because tilting the sample will change the Bragg condition for dislocation diffraction. Attention should be paid to this problem when observing the dislocation density in the sample.

4.5 High-resolution TEM (HRTEM) Images and Phase Contrast

HRTEM is slightly different from the previous TEM imaging modes discussed. Bright-field images can be formed with direct beams, and dark-field images can be formed with diffracted beams, while HRTEM uses both direct beams and diffracted beams for imaging. HRTEM has a higher magnification, which can easily observe the lattice period of crystals (usually on the order of several angstroms). To obtain HRTEM images, higher requirements are put forward for the hardware of the transmission electron microscope, and the quality of sample preparation is also the key to obtaining atomic-scale resolution images.

The direct beam and diffracted beam of electrons travel different paths in the crystal and undergo different scattering processes. Therefore, under the action of the crystal potential field, the exit electron wave functions of the electron beams in different directions and paths have different phases. HRTEM uses both direct beams and diffracted beams for imaging. During the imaging process, the phase difference of the objective lens and the over-focus or under-focus degree can also change the phase of the exit wave, leading to changes in imaging brightness. The direct beam and diffracted beam can interfere on the image plane. Due to their different phases, intensity brightness changes and contrasts are generated in the image. This contrast is caused by the phase change of the direct beam and diffracted beam under the action of different crystal potential fields in the sample, so the contrast of HRTEM images is called phase contrast.

HRTEM uses direct beams and more than one diffracted beam for imaging. Due to the different phases of the direct beam and diffracted beams, interference can be formed, and the image will show phase contrast. Assuming that imaging is performed with a direct beam and a diffracted beam, the direct spot and the diffraction spot on the focal plane can be regarded as two point light sources. For easy understanding, the direct spot and the diffraction spot light source can be compared with the Young's double-slit interference experiment. In the Young's double-slit experiment, light waves passing through two slits form bright and dark interference fringes on the screen. The spacing of bright (dark) fringes is $D\lambda/W$, where D is the distance between the slits and the screen, and W is the distance between the slits. According to the results of the Young's double-slit interference experiment, the spacing of the interference fringes of the direct beam and the diffracted beam is inversely proportional to the distance R between the diffraction spot and the direct spot. According to the electron diffraction formula, R is inversely proportional to the interplanar spacing d. Therefore, combined with the parameters of the transmission electron microscope (camera constant, objective focal length, etc.), the information of the interplanar spacing d can be obtained from the interference fringe spacing of the high-resolution image. The above discussion is only to facilitate the understanding of the formation of high-resolution images, but it is not strict. The images formed by phase contrast are difficult to explain intuitively because phase contrast is sensitive to many factors, including sample thickness, crystal orientation, and spherical aberration of the objective lens. In HRTEM imaging, selecting direct beams and multiple diffracted beams for imaging can form images containing lattice fringes in multiple directions. Note that although lattice fringe images can provide interplanar spacing information, they are not equivalent to direct imaging of atomic structures. The complete interpretation of the image, such as the possible atomic arrangement or structural defects in the image, needs to be analyzed in combination with computational simulation, and the atomic structure cannot be directly inferred from the image.

HRTEM of crystal samples generally requires the sample to be crystalline and stable, sufficiently thin, and the electron beam to be parallel to the zone axis. There are two interference (diffraction) processes in the HRTEM imaging process. The first is the diffraction of the electron beam by the lattice (resulting in phase change), and the second is the interference between the direct beam and the diffracted beam. The spacing of interference fringes is proportional to the interplanar spacing. The interplanar spacing can be obtained from the interference fringe spacing obtained from the image combined with the magnification of the electron microscope. Note that the bright or dark areas in HRTEM images cannot be directly identified as corresponding to atomic positions. Usually, high-resolution atomic images are not the positions of atoms, but interference phase contrast images formed by the exit electron wave of the sample after passing through various lenses such as the objective lens and intermediate lens. The electron wavelength λ (affected by the accelerating voltage), the defocus (over-focus or under-focus) amount Δf, and the spherical aberration coefficient Cs will all affect the interference conditions between the direct beam and the diffracted beam, thereby affecting the imaging quality of HRTEM. Only under very harsh conditions, that is, the crystal sample is very thin (2 nm or even thinner), the zone axis is strictly parallel to the incident electron beam, and at a specific defocus amount [called Scherzer defocus condition], the high-resolution image can correspond to the structure. At this time, the high-resolution image is called a structure image. In general, the above conditions cannot be satisfied, so the high-resolution image cannot be simply corresponding to the crystal structure (atomic arrangement). The bright and dark areas in the image (bright and dark areas of interference) can be converted into dark and bright areas with focusing and sample thickness. The correct interpretation of high-resolution images requires computational simulation.

4.6 STEM Images

As mentioned above, it is relatively difficult to obtain HRTEM structure images. HRTEM uses parallel electron beams to transmit samples. If electromagnetic lenses are used to converge the electron beam into a small spot and incident on the sample in scanning mode, it is called convergent beam mode or probe mode. The convergent electron beam in SEM scans the sample surface, but now the convergent electron beam has to pass through the sample for imaging. This mode is the STEM mode, and the image is a STEM image.

STEM uses a condenser lens to converge the electron beam into a probe and scan it on the sample. Imaging analysis is performed by collecting signals generated by the interaction between the electron probe and the sample. The imaging mode of STEM is essentially different from that of TEM. TEM uses parallel electron beams to transmit a sample area and selects direct beams or (and) diffracted beams to project onto a fluorescent screen for imaging (similar to optical imaging, it is a parallel imaging). STEM uses a convergent electron beam spot to scan point by point and line by line in a sample area, and records the transmission electron situation at each point. This is a serial imaging mode, so it is much slower than TEM imaging. TEM images are magnified by lenses, while STEM images do not require magnification by imaging lenses. In addition, it should be noted that the incident angle of the convergent beam electrons at each scanning point must be consistent to ensure the same scattering and diffraction directions. Eliminating the spherical aberration of the condenser lens to obtain a smaller electron beam spot is the key to improving image resolution.

STEM imaging includes Annular Bright-Field (ABF), Annular Dark-Field, and High Angle Annular Dark-Field (HAADF). Different annular detectors behind the sample are used for different imaging modes. Since the various imaging modes collect scattered electron signals at different acceptance angles, different images of the same position can be obtained at one time in the experiment, showing different material structure information. Light elements have weak scattering on electrons, while heavy elements have relatively strong scattering ability on electrons. Therefore, in the STEM image obtained with a HAADF detector (Figure 4.5), elements of different weights can form images of different brightness. The contrast is approximately proportional to the square (about 1.8 power) of the atomic number, so it is also called Z contrast image.

Both HRTEM and high-resolution STEM rely on spherical aberration correction technology. The main factor affecting HRTEM resolution is the spherical aberration of the objective lens, while the main factor affecting high-resolution STEM is the spherical aberration of the light source condenser lens. Therefore, spherical aberration electron microscopes include double spherical aberration (equipped with both objective spherical aberration corrector and condenser

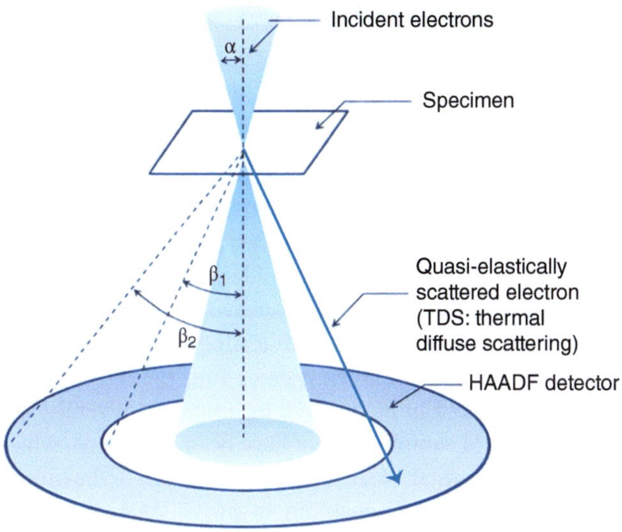

Figure 4.5 HAADF STEM detection.

spherical aberration corrector) and single spherical aberration (objective or condenser correction), which need to be selected according to experimental needs. At present, field emission STEM/TEM with aberration correctors can achieve sub-angstrom scale resolution. STEM imaging with condenser spherical aberration correction combined with spectroscopic analysis methods such as EDS and EELS (see Chapter 8) has become a characterization method with extremely high spatial resolution.

4.7 Introduction to New TEM Technologies

4.7.1 iDPC-STEM Imaging Technology

The HAADF probe of STEM can detect high scattering angle electron signals, which is beneficial for the observation of heavy elements. However, the signal intensity of low scattering angles for light elements is low, making it difficult to observe light elements such as carbon, nitrogen, and oxygen. The centroid of the convergent beam diffraction pattern shifts under the influence of the atomic potential field in the sample. Based on this principle, the integrated differential phase contrast (iDPC) technique utilizes a segmented detector for imaging and correlates the centroid of the diffraction pattern with atomic positions through computational simulation, thereby enhancing the ability to observe atomic position information. The iDPC-STEM technique can detect all scattered electrons within a specific region, enabling imaging even under ultra-low electron beam dose conditions. This makes it suitable for high-resolution imaging of materials susceptible to structural damage from electron beam irradiation, such as zeolite molecular sieves and metal-organic frameworks. Compared with HAADF STEM, iDPC-STEM exhibits higher sensitivity to light elements.

4.7.2 In-situ TEM

TEM offers extremely high spatial resolution for observing microstructures. When combined with in-situ conditions, it can provide critical information about the dynamic evolution of material microstructures. However, TEM testing imposes strict requirements on sample preparation: the sample holder has very limited operable space, and a high-vacuum environment is necessary to avoid scattering interference with the electron beam. These factors make in-situ testing highly challenging. One of the key technical difficulties is introducing gases into the high-vacuum system, a challenge similar to that faced in in-situ XPS technology. Currently, various types of commercial in-situ TEM sample holders have been developed, which can apply different external fields (thermal, electrical, magnetic, etc.) to the sample and introduce liquid or gaseous reactants, enabling in-situ TEM characterization of diverse material systems. One in-situ technical solution employs multiple differential pumps to maintain a moderate gas pressure (typically below 20 Torr) only in the region near the sample, while the rest of the sample chamber remains under high vacuum. Similar to the technical scheme of near-atmospheric pressure XPS, this setup is also known as environmental TEM (ETEM). In an ETEM system, the electron beam directly irradiates the sample; as the gas pressure increases, the imaging resolution decreases due to increased electron scattering by gas molecules. The technical development route is shown in Figure 4.6.

4.7.3 Cryogenic TEM (Cryo-TEM)

Compared with inorganic materials, biological materials (e.g., proteins, viruses) typically have high water content and require an aqueous environment to maintain their specific structures. However, TEM observations must be performed under high vacuum, and biological samples are far more sensitive to electron beams than inorganic materials—these factors make studying biological materials with electron microscopy particularly challenging. XRD can be used to study the structure of biological macromolecules, but it requires growing crystals of the macromolecules. Many proteins are difficult to crystallize, and the structure of crystallized samples may differ from their biologically active state. The method of studying the solution conformation of biological macromolecules using multidimensional nuclear magnetic resonance won the 2002 Nobel Prize in Chemistry, but this technique is only suitable for proteins with relatively small molecular weights. In 2017, the Nobel Prize in Chemistry was awarded to Jacques Dubochet, Joachim Frank, and Richard Henderson in recognition of their outstanding contributions to studying the structure of biological molecules using cryo-TEM. The key to cryo-TEM sample preparation is rapidly freezing aqueous biological samples to form an amorphous (glass-like) state of water, rather than ice crystals—ice crystals would destroy the structure of biological molecules.

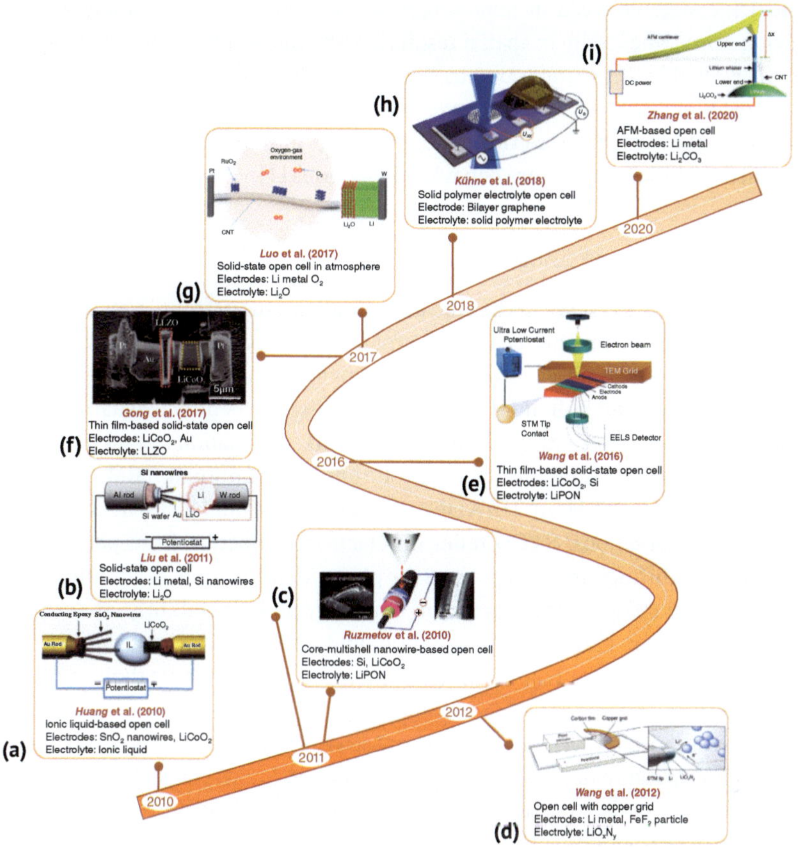

Figure 4.6 The development of open cells [1] / with permission of John Wiley & Sons.

Ice crystals also produce diffraction signals that interfere with the imaging of biological molecules. After cryogenic preparation, biological macromolecules embedded in glassy water are transferred to the TEM for observation under cryogenic conditions. To avoid electron beam-induced damage to biological samples, cryo-TEM typically uses low-dose electron beams for imaging. Despite this limitation, cryo-TEM has revolutionized structural biology by enabling high-resolution imaging of non-crystalline, biologically active macromolecules and complexes, such as membrane proteins and virus particles, which were previously inaccessible to traditional structural analysis techniques.

4.8 Application Cases

Transmission electron microscopy–based techniques represent one of the most central characterization platforms for probing microscopic mechanisms in Li- and Na-ion battery materials. This platform includes conventional TEM, SAED, aberration-corrected

TEM, EELS, cryogenic electron microscopy, and in-situ TEM. These methods provide strong complementarity in spatial resolution, chemical sensitivity, and temporal access. They enable structural, compositional, and dynamic interrogation of electrode and electrolyte materials. They reveal intrinsic features and interfacial reaction mechanisms. They establish a critical experimental foundation for understanding performance evolution and failure behavior in batteries [1].

4.8.1 TEM Imaging and SAED

TEM imaging and SAED serve as the core and complementary techniques for microstructural studies in electrochemical energy materials. They operate across cathodes, anodes, and solid or liquid electrolytes, and provide structural information from both surface regions and bulk domains. TEM imaging offers direct visualization of morphology, particle size, grain boundaries, defects, dislocations, stacking faults, surface reconstruction, and interfacial evolution. Conventional TEM, HRTEM, aberration-corrected STEM in HAADF or ABF modes, and dark-field imaging support these observations. SAED relies on electron diffraction from periodic lattices and records diffraction spots or rings for phase identification, orientation analysis, domain evaluation, phase-transition tracking, and crystallinity assessment. TEM imaging delivers real-space information, and SAED delivers reciprocal-space information. TEM focuses on structural features, and SAED reveals their crystallographic nature.

In layered Li-ion cathodes such as $LiCoO_2$, NCM, Ni-rich oxides, and Li-rich materials, TEM imaging detects dislocations, cracking, interfacial sliding, and local reconstruction driven by transition metal migration. SAED identifies phase transitions through spot splitting, spot weakening, spot disappearance, or the appearance of extra reflections. It distinguishes layered-to-rock salt or layered-to-spinel reconstructions and resolves H1–H2–H3 evolution. In Ni-rich cathodes, SAED confirms rock salt reconstruction or Li-deficient layered phases after cycling, and TEM imaging reveals their surface-to-bulk propagation. In Li-rich materials, TEM imaging captures surface reconstruction, defect accumulation, and lattice distortion triggered by oxygen redox, and SAED separates layered–monoclinic coexistence or layered–spinel duality and tracks symmetry changes.

In Na-ion layered cathodes, TEM imaging visualizes interlayer gliding, Na-layer rearrangement, and distortion from phase transitions. SAED determines P2–O2 or O3–P3 gliding pathways with high precision, and diffraction-spot shifts reveal the crystallographic trajectory of each transition. Frequent gliding in Na systems produces characteristic streaks and periodicity changes in SAED patterns, and these features reflect structural reversibility and cycling stability.

In anodes such as graphite, silicon, and lithium metal, TEM imaging highlights Solid Electrolyte Interphase (SEI) formation, interface evolution, cracking, and void formation. SAED presents diffuse rings for amorphous or short-range-ordered structures and confirms crystallization events such as $Li_{15}Si_4$ formation. It resolves crystallographic orientation and grain size in lithium dendrites and identifies Li–Si phases in lithiated silicon. In solid electrolytes including LLZO, LPS, and NASICON,

TEM imaging shows grain boundary phases, phase separation, pores, and cracks, and aberration-corrected STEM visualizes ion-transport channels. SAED identifies crystal structures in multiphase systems, evaluates phase purity, and detects cubic-to-tetragonal transitions. In liquid-electrolyte systems, TEM imaging focuses on SEI or CEI layering and organic–inorganic distribution. SAED confirms the crystallinity of inorganic SEI phases such as LiF, Li_2O, and Li_2CO_3.

TEM imaging and SAED form a complementary framework in all electrochemical systems. TEM provides direct structural morphology, defects, and interfaces. SAED determines phase transitions, lattice distortion, and orientation changes. The combination supports atomic-scale interpretation of structure–property relationships and guides the design of high-stability and high-energy-density electrodes and electrolytes.

TEM imaging and SAED act as central tools for structural analysis in Li-ion cathodes, including $LiCoO_2$, NCM, Ni-rich layered oxides, and Li-rich transition metal oxides. In $LiCoO_2$, HRTEM visualizes the layered framework, octahedral stacking, lattice periodicity, and distortion at high delithiation states. SAED distinguishes H1–H2–H3 evolution and identifies O1 stacking or O3 variants through spot splitting and symmetry changes. In NCM and related layered oxides, TEM imaging reveals local defects, cation mixing, lattice distortion, and surface reconstruction toward rock salt. SAED detects spot diffusion, extra reflections, and coexisting layered–rock salt signatures. In Ni-rich materials such as NCM811, TEM identifies surface residual phases, instability, and microcracks, and SAED separates rock salt surface layers from layered cores and tracks interface thickness and coherence. In Li-rich oxides, TEM imaging captures active-region evolution, transition metal migration, and activation behavior. SAED differentiates monoclinic and layered reflections, identifies transitions toward defect-rich layered or spinel/rock salt structures, and tracks structural order evolution during cycling. TEM supplies visual details of local structure and degradation, and SAED supplies crystallographic criteria. Their combination establishes a complete structure–degradation framework for layered oxides.

Researchers evaluated the structural stability of $Li(Ni_{0.95}Co_{0.04}Al_{0.01})O_2$ at deep-charge states near 4.6 V. Increasing delithiation produces surface Li depletion, structural reconstruction, and mechanical instability. TEM imaging shows Li-deficient surface regions and microcracks (Figure 4.7). Crack propagation begins at voltages above 4.3 V. SAED identifies structural transitions from the H3 layered phase to O1 stacking in the surface region. O1 stacking appears at ~4.4 V. Diffraction-spot broadening and streak formation confirm the layered-to-O1 reconstruction and indicate its irreversible nature. These observations reveal the origin of initial capacity loss. Surface Li depletion triggers H3 → H4 (O1) transition and initiates crack formation. Both effects weaken structural integrity and Li-ion transport. The results highlight the role of doping or Ni^{2+} migration control in suppressing H3 → H4 transition and mitigating crack formation. TEM imaging visualizes morphological degradation, and SAED identifies crystallographic pathways. The combination establishes a clear failure mechanism for NCA95 under deep-charge conditions [2].

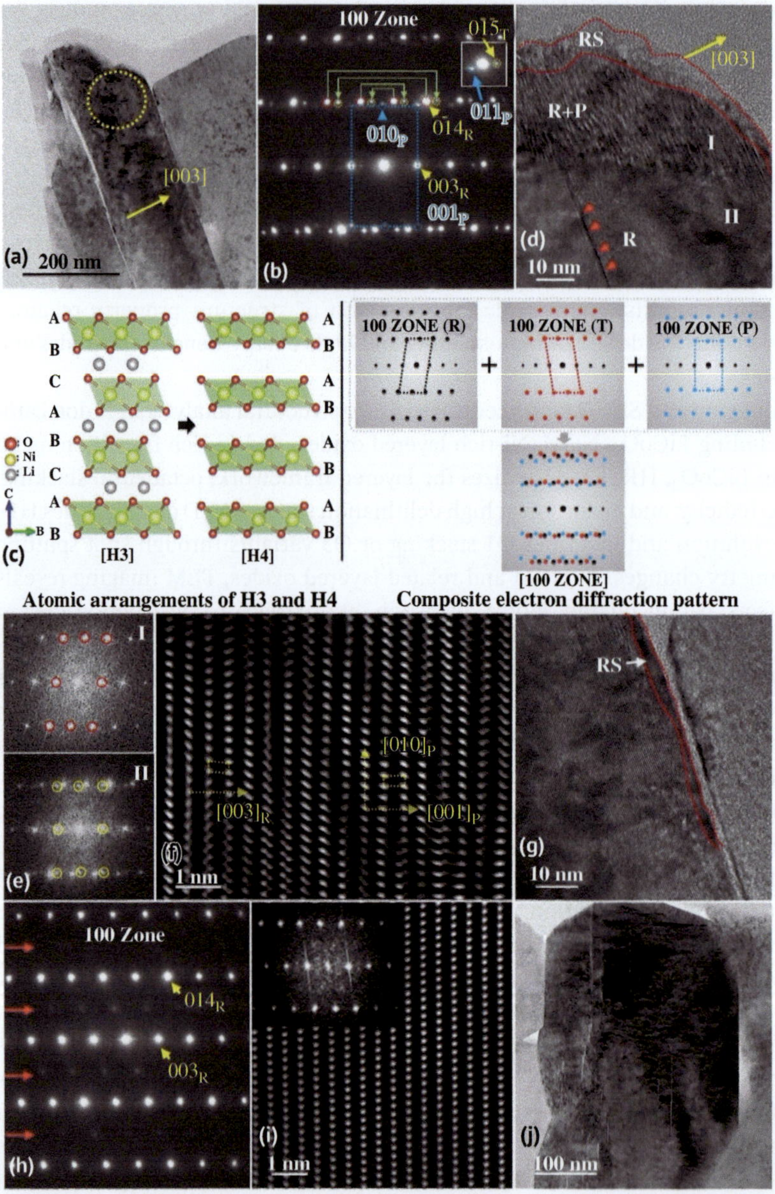

Figure 4.7 (a) TEM bright-field image from a primary particle of the NCA95 cathode charged to 4.6 V, (b) SAED pattern from the circled area in (a) with the inset showing the $(0\bar{1}4)$ spot, (c) atomic arrangements of H3 and H4 phases along the [100] direction illustrating the stacking sequence and illustration of composite electron diffraction pattern of the three diffraction structures, (d) HR-TEM image near the surface of the primary particle in (a), (e) Fourier transform of the regions I and II in (d), (f) HR-TEM image of the two-phase region showing the H3 and H4 phases, (g) HR-TEM image along the interparticle boundary in (a) showing a thin layer of rock salt phase along the boundary, (h) SAED pattern from the particle interior, (i) HR-TEM image of the particle interior with the inset showing the Fourier transform of the image, (j) TEM bright-field image from a primary particle with fine hairline intraparticle cracks. (R: H3, P: H4, T: twin, RS: rock salt) [2] / with permission of John Wiley & Sons.

TEM imaging and SAED play central roles in sodium-ion battery research. They support analysis of stacking sequences, phase-transition mechanisms, distortion driven by ion migration, and microstructural degradation during electrochemical cycling. They also form a core platform for constructing structural evolution maps in Na-ion systems. In layered Na cathodes such as P2-$Na_{0.67}MO_2$ (M = Ni, Mn, Co), TEM imaging provides direct visualization of P2 → O2 and P2 → OP4 transitions triggered by gliding after Na extraction. TEM also reveals interlayer misalignment, slab distortion, local collapse, and crack formation under high-rate or deep-charge conditions. SAED records shifts, splitting, or new periodic streaks in diffraction patterns and captures stacking changes from ABBA to ABCD or other sequences. These features mark the crystallographic reconstruction pathway during Na extraction.

In O3-type Na cathodes such as O3-$NaTMO_2$, TEM imaging detects slab bending, interlayer-spacing changes, and surface reconstruction toward rock salt-like phases. SAED distinguishes O3 → P3 or O3 → O1 transition sequences and identifies gliding directions through changes in streak periodicity. SAED also reveals whether the structural reconstruction remains reversible. In Prussian blue and its derivatives such as $Na_2Fe[Fe(CN)_6]$, TEM imaging highlights vacancy distribution, octahedral distortion, and water content effects on framework stability. SAED detects symmetry breaking during cubic-to-rhombohedral or cubic-to-monoclinic transitions and provides key evidence for water-dependent structural reversibility.

In hard-carbon anodes, TEM imaging shows micropores, graphitic nanodomains, or onion-like motifs. SAED produces diffuse rings, and intensity evolution reflects the formation of short-range order or Na clusters during insertion. In solid-state Na electrolytes such as NASICON or Na_3SbS_4, TEM imaging reveals grain boundary phases, nanopores, and interfacial contact quality. SAED confirms phase purity in multiphase mixtures and identifies local phase transitions or lattice distortion after cycling. Overall, TEM imaging provides morphology, defects, and interface evolution under practical operating states. SAED offers crystallographic criteria for reconstruction, phase-transition sequences, and gliding pathways. Their combination enables atomic-scale interpretation of structural evolution in Na-ion cathodes, anodes, and electrolytes and supports the advancement of Na-ion materials design and mechanism studies.

The authors developed a medium-entropy NASICON-type Na electrode material, NVAMP-0.3, through the introduction of multiple metal species including Al, Mg, Co, Ni, Cu, and Zn. This strategy increases configurational entropy and enhances voltage and capacity. Structural characterization involved HRTEM imaging, which revealed microstructural continuity, lattice integrity, and particle-level crystallinity. TEM-EDS elemental mapping confirmed homogeneous distribution of metal species such as V, Al, Mg, and Co at the nanoscale. For phase-transition analysis, the study relied on in-situ X-ray diffraction to track lattice-parameter evolution and to identify solid-solution behavior or phase-transition modes during Na insertion and extraction. The use of TEM (HRTEM and EDS mapping, Figure 4.8) aimed to validate nanoscale structural integrity and doping homogeneity in the medium-entropy NASICON framework. These features strongly influence Na-ion transport and electrochemical stability. HRTEM resolved lattice fringes and enabled assessment of defects such as

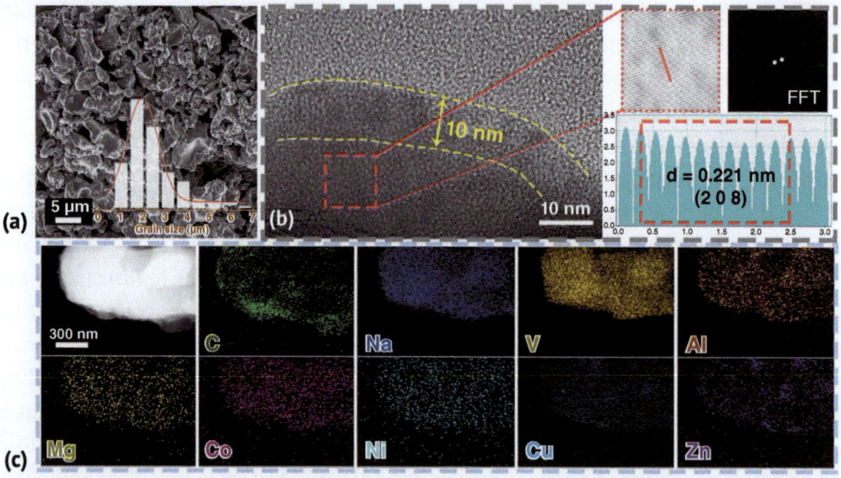

Figure 4.8 (a) SEM image of NVAMP-0.3 (the inset is the particle size distribution curve). (b) HRTEM image of NVAMP-0.3. (c) TEM-EDS elemental mappings of NVAMP-0.3 materials [3] / with permission of John Wiley & Sons.

dislocations or interfacial discontinuities. EDS mapping verified successful multielement incorporation through the entropy-engineering strategy. The combined TEM evidence and in-situ XRD results support the conclusion that NVAMP-0.3 exhibits a high average operating voltage near 3.33 V, a high capacity of 138.1 mAh g^{-1}, and strong low-temperature performance with significant retention at −40 °C. TEM confirms nanoscale structural stability, and homogeneous elemental distribution provides a microscopic basis for its excellent electrochemical behavior [3].

TEM imaging and SAED serve as essential microstructural tools for studies of Li-ion anode materials. They support analysis of graphite-based carbons, hard carbon, Si-based materials, metal oxides such as TiO_2 and Fe_3O_4, alloy-type anodes, and sulfide anodes. In graphite and hard-carbon anodes, TEM imaging provides direct access to layered stacking, interlayer spacing, defect distribution, and disordered domains. SAED presents characteristic hexagonal diffraction rings for graphite and enables evaluation of graphitization and structural order. In hard carbon, TEM identifies turbostratic layers, microporous regions, and graphitic nanodomains. SAED produces diffuse rings, and their intensity profiles distinguish multiple scales of disorder and local short-range order. These features form the basis for understanding Li- or Na-storage mechanisms.

In Si-based anodes such as nanosilicon or silicon oxide, TEM imaging captures volume expansion, crack formation, and interfacial evolution. TEM also visualizes pulverization of Si nanoparticles, SEI coverage, and particle aggregation during cycling. SAED detects structural evolution of Si, including amorphization and formation of Li–Si alloy phases such as $Li_{15}Si_4$, and provides key evidence for degradation pathways in Si anodes. In metal-oxide anodes such as TiO_2, Fe_3O_4, and Co_3O_4, TEM imaging resolves particle size, morphology, lattice distribution, and interface structure. HRTEM identifies lattice fringes and exposed facets. SAED presents well-defined diffraction spots, confirms phase purity and structural stability, and reveals possible phase reconstruction after cycling, such as lattice distortion in lithiated TiO_2 or conversion-driven phase evolution in Fe_3O_4.

In alloy-type anodes such as Sn, Sb, or Bi, TEM and SAED track phase evolution during alloying and dealloying. TEM captures size variation of alloy nanoparticles, amorphization events, and stress-release behavior. SAED records transient diffraction features of intermediate alloy phases and provides direct evidence for reaction pathways. In sulfide anodes such as MoS_2 or VS_4, TEM imaging analyzes interlayer spacing, exfoliation degree, defect distribution, and structural changes during intercalation. SAED monitors diffraction ring evolution and spot variation to identify phase transitions, structural rearrangement, or disordering during lithiation and delithiation. Overall, TEM imaging reports morphology, lattice features, and interface structure, while SAED supplies crystallographic signatures of phase evolution. Their combination offers indispensable microscopic insight for mechanistic studies, stability evaluation, and failure analysis of diverse anode systems.

In studies of Si-based anodes for all-solid-state batteries, TEM imaging reveals nanoscale structural features, including Si particles, metal-alloy or metal-framework interfaces, voids, and deformation modes during cycling. TEM confirms the integrity and continuity of the Si alloy or Si framework composite at the nanoscale (Figure 4.9). This evidence supports the conclusion that the hybrid conductive and mechanical framework strengthens anode stability. Overall, the reported TEM analysis demonstrates the advantages of the Si alloy composite design in morphology retention and interfacial robustness [4].

In both solid-state and liquid-electrolyte systems, TEM imaging and SAED serve as indispensable microstructural probes for elucidating the intrinsic structure of electrolytes, the evolution of interfacial chemistry, phase transformations, defect distributions, and reaction pathways at electrode–electrolyte interfaces. For liquid-electrolyte systems, where SEI/CEI layers typically possess thicknesses of only a few to several tens of nanometers and consist of mixed amorphous and nanocrystalline domains, TEM imaging enables direct visualization of their stratified morphology, multilayered architectures, spatial distributions of organic–inorganic components, and the formation of cracks or voids. High-resolution TEM further resolves lattice fringes of crystalline SEI species such as Li_2CO_3, LiF, and Li_2O, while SAED provides essential information for distinguishing amorphous from crystalline domains via diffuse rings or weak diffraction spots. These

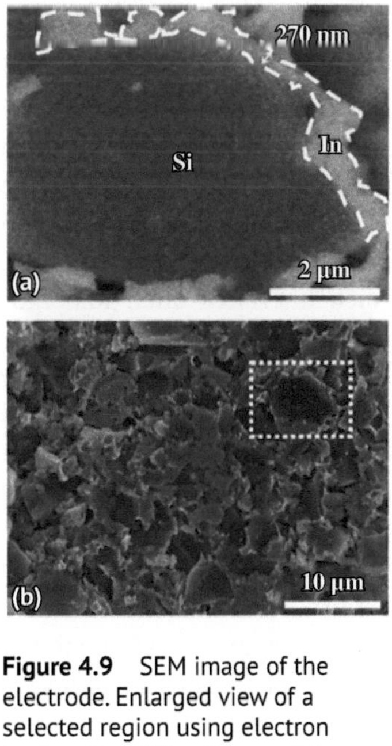

Figure 4.9 SEM image of the electrode. Enlarged view of a selected region using electron backscatter diffraction model.

signatures allow the crystallinity and structural heterogeneity of SEI to be correlated with electrolyte composition, additive chemistry, or applied electrochemical conditions. In fluorine-rich or B/P-containing electrolytes, TEM-EELS offers an additional capability to map the distribution of heteroatoms, complementing SAED-based identification of nanocrystalline phases. In solid-state electrolytes (SSEs), including oxides (e.g., LLZO), sulfides (e.g., LGPS), and polymer electrolytes, TEM imaging is widely employed to assess lattice integrity, microcracking, grain boundary features, defect clustering, and phase distribution. Particularly in oxide electrolytes, HRTEM provides atomically resolved views of dendrite penetration pathways, grain boundary chemistry, and stress-induced sub-grain distortions. SAED is crucial for rapid verification of crystal structures, assessment of phase purity, and identification of lattice distortions or impurity phases induced by dopants. In sulfide electrolytes, which are sensitive to air exposure and prone to partial amorphization, TEM reveals the presence of glassy and nanocrystalline domains, whereas the diffuse SAED rings help evaluate whether the material retains long-range order or undergoes disordering during cycling. At solid-solid interfaces, TEM represents one of the few techniques capable of directly revealing interfacial contact quality, void formation, chemical stratification, reaction-layer thickness, and crack propagation—parameters central to understanding interfacial resistance in solid-state batteries. SAED further identifies newly formed interphase compounds such as Li_2S, Li_3P, or Li-Nb-O species and evaluates whether interfacial reactions intensify or stabilize upon cycling. In Li metal/SSE interfaces, TEM directly captures dendrite nucleation sites, tip morphology, propagation trajectories, and localized failure arising from electric-field concentration, while SAED determines whether dendrites are crystalline Li or partially amorphous. Collectively, TEM provides spatial, morphological, and interfacial evolution information, whereas SAED supplies crystallographic phase identification, and their synergy enables a comprehensive understanding of electrolyte bulk structure and interface chemistry, thereby guiding the design of stable interfaces, controlled SEI/CEI compositions, improved interfacial contact, and enhanced electrolyte stability.

Focused on the thermal stability of the interface between the lithium-manganese spinel cathode ($LiMn_2O_4$, LMO) and the garnet-type solid electrolyte (LLZTO), such interfaces often undergo high-temperature processing during co-sintering or experience elevated temperatures during operation, and their chemical/structural instability can critically impact interfacial resistance and ionic transport. To investigate these processes, the study employs in-situ TEM imaging combined with high-resolution HRTEM to track the microstructural evolution of the interface from room temperature to 500 °C (Figure 4.10). Cross-sectional TEM–STEM specimens were analyzed at various temperatures using HAADF and HRTEM imaging, accompanied by EELS mapping to monitor the thermally driven migration of Li, Mn, La, and Ta. Parallel in-situ XRD and in-situ Raman spectroscopy further confirmed the formation and transformation of interfacial phases. The results show that around 300 °C, Mn begins to diffuse from LMO into LLZTO, forming an intermediate $LiMn_3O_4$ phase; upon heating to ~500 °C, this phase further transforms into MnO, accompanied by substantial reconstruction of the interfacial lattice. HRTEM images

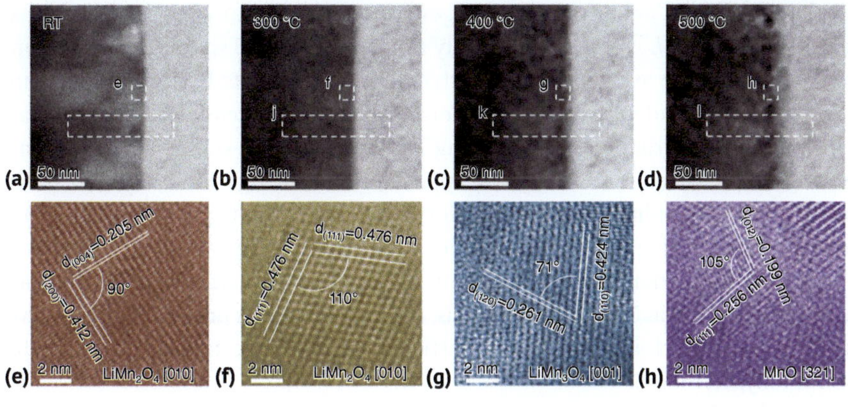

Figure 4.10 In-situ TEM investigation of LMO/LLZTO interface during heating. (a)–(d) HAADF images of LMO/LLZTO interface at RT, 300, 400, and 500 °C, respectively. (e)–(h) HRTEM images for LMO/LLZTO interface at marked regions in (a)–(d), respectively [5] / John Wiley & Sons / CC BY 4.0.

clearly capture the lattice fringes of newly formed phases and the associated interfacial distortions, while SAED signatures (extracted from HRTEM regions) unambiguously identify the crystallographic transitions from $LiMn_3O_4$ to MnO. These structural transformations significantly impede ionic transport, manifested as a sharp increase in interfacial resistance in EIS measurements. The authors conclude that although the bulk phases of LMO and LLZTO exhibit reasonable thermal stability, their solid-solid interface becomes chemically reactive and structurally unstable at relatively moderate temperatures, and this thermally induced degradation is the primary origin of elevated interfacial resistance. The work highlights the critical need for careful temperature control during the fabrication and thermal treatment of solid-state batteries and provides mechanistic guidance for designing protective coatings or buffer layers to stabilize cathode–electrolyte interfaces [5].

4.8.2 Electron Energy Loss Spectroscopy

EELS, an advanced characterization technique integrated within TEM, plays an irreplaceable role in lithium- and sodium-ion battery materials research due to its ultrahigh energy resolution (0.1 eV level) and superior spatial resolution (sub-angstrom level). Possessing significant advantages over traditional X-ray EDS, EELS excels particularly in detecting light elements (Li, B, C, N, O, F) and resolving electronic structures, including band structures, oxidation states, and chemical bonding environments. Across cathode, anode, and electrolyte characterization, EELS applications share commonalities while exhibiting distinct differences driven by specific electrochemical properties.

In lithium- and sodium-ion battery cathode research, EELS focuses primarily on probing transition metal (TM) valence variations and oxygen anionic redox activities. During charge/discharge processes involving layered oxides (e.g., NCM ternary materials), spinels, or polyanionic cathodes, TM ion oxidation/reduction occurs fundamentally. By analyzing TM *L*-edges (corresponding to 2p-to-3d transitions), specifically the L_3/L_2 white-line intensity ratio and chemical shifts, researchers precisely quantify the average valence distribution of elements like Ni, Co, Mn, and Fe in both bulk and surface regions. For instance, EELS frequently reveal surface "rock salt" reconstruction layers causing TM reduction and subsequent impedance growth. Lithium and sodium cathode applications share high similarity, with differences stemming mainly from the complex phase transitions (e.g., P2-O3) and severe lattice distortions in sodium-ion batteries caused by the larger Na-ion radius; here, EELS track local coordination environment changes. Furthermore, in high-voltage cathodes like Li-rich Mn-based materials, the oxygen *K*-edge pre-edge feature serves as direct evidence for lattice oxygen participation in charge compensation mechanisms (oxygen redox).

In lithium-ion battery cathode research, EELS find widespread use in $LiCoO_2$, NCM ($LiNi_xCo_yM_zO_2$), general layered oxides, Ni-rich materials (Ni >0.8), and Li-rich systems ($Li_{1+x}M_{1-x}O_2$) due to its high spatial resolution, elemental sensitivity, and quantitative analysis capabilities for TM valence and local coordination environments. For $LiCoO_2$, the Co $L_{2,3}$-edge sensitively reflects Co^{3+}/Co^{4+} ratio changes, making the Co L_3/L_2 intensity ratio a standard indicator for judging Co redox degrees and potential cation migration (e.g., Co migration to Li layers) during high-voltage formation or cycling; the O *K*-edge pre-peak determines TM–O covalency variations, revealing lattice oxygen reaction trends. Regarding NCM materials, energy loss features at transition metal *L*-edges (Ni $L_{2,3}$, Co $L_{2,3}$, Mn $L_{2,3}$) facilitate the resolution of oxidation state changes, specifically Ni^{4+} formation, Mn redox reversibility, and cycling-induced local reduction (increased Ni^{2+}) or oxygen defect generation. O *K*-edge analysis further characterizes oxygen local density of states changes, typically linking closely with structural degradation and rock salt phase formation in Ni-rich systems. In general layered cathodes, EELS analyze intraparticle composition gradients, surface reduction enrichment, and interfacial redox inhomogeneity, often employing STEM-EELS line-scans to observe oxidation state gradient evolution from surface to bulk. In Ni-rich materials (Ni > 0.8), prone to surface rock salt formation (NiO-like) and internal oxygen vacancy generation, EELS proves critical: Ni *L*-edge white-line intensity determines rock salt phase thickness and Ni^{2+} distribution, while O *K*-pre-peak attenuation indicates oxygen vacancy accumulation and TM–O bond weakening, identifying core sources of capacity decay and thermal instability.

Introducing EELS analysis to resolve a long-standing core controversy: the true participation of Co in the redox process within Li-rich layered oxides (LRLO) and the reversibility of Co valence changes during charge/discharge. Due to resolution limitations of traditional XAS or XPS in distinguishing subtle local coordination changes, combined with the sensitivity of EELS at the TM *L*-edge regarding white-line intensity and energy shifts for determining Co electronic structure and coordination environment, the authors employ STEM-EELS to conduct point-by-point analysis of local Co electronic states in both pristine and electrochemically cycled materials (Figure 4.11). During the actual analysis, they prioritize comparing Co L_3/L_2 white-line energy positions and intensity ratios to explore Co-O

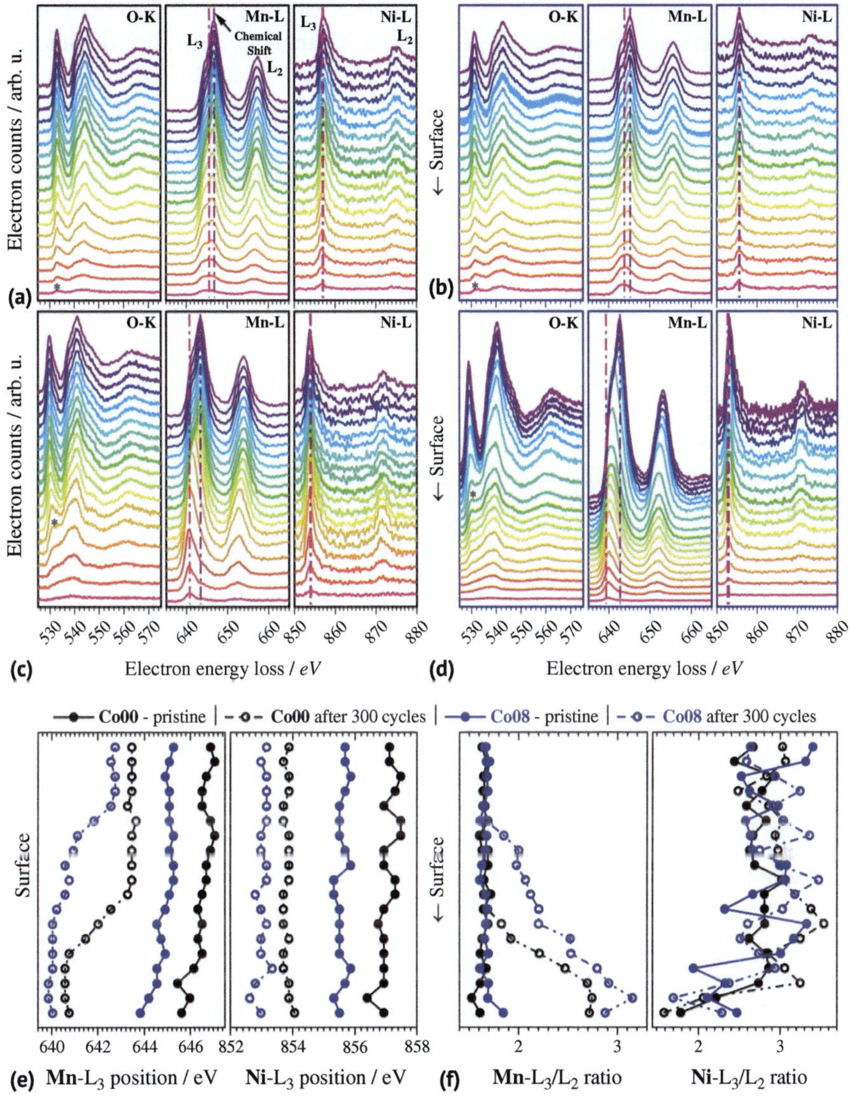

Figure 4.11 Series of EELS spectra of the O K-edge, Mn L-edge, and Ni L-edge of Co00 (a, c) and Co08 (b, d) powders as made (fresh) (a, b) and extracted from cycled electrodes (c, d). The electrodes' powders were collected after 300 cycles at 1 C (discharged state). Depth profile shifts of (e) the L_3 peak position and (f) the ratio of L_3/L_2 intensities at the Mn L-edges and Ni L-edges [6] / John Wiley & Sons / CC BY 4.0.

bond covalency, Co 3d electron occupancy, and local oxygen environment variations; simultaneously, they assess transition metal-oxygen electronic reconstruction and oxygen electrochemical participation via O K-edge pre-peak changes. Research results demonstrate the absence of significant changes in Co L-edge energy positions and white-line ratios before or after cycling, indicating the maintenance of a stable Co valence state without major participation in charge compensation processes; conversely, obvious changes in the O *K*-edge pre-peak further support an oxygen anionic redox-dominated mechanism. Ultimately, deep EELS characterization clearly confirms the non-central role of Co as a reversible redox center in this Li-rich material, providing direct in-situ evidence for clarifying the electronic chemical role of Co in LRLO [6].

EELS serves as the most electronically sensitive spectroscopic technique under TEM for lithium-ion battery anode materials. It acts as a pivotal tool for resolving valence evolution, bonding structure reconstruction, local chemical environment variations, and the spatial distribution of SEI chemical compositions across various anode systems. For graphite anodes, EELS primarily analyzes the π-σ peak intensity ratio and energy shifts of the Carbon *K*-edge. This distinction differentiates amorphous carbon, graphitic phases, carbon coating layers, and carbonaceous species within the SEI. It provides critical evidence for revealing local structural evolution during fast lithiation, surface film formation, and graphite exfoliation. Regarding battery anode materials, EELS characterization priorities shift from TM electronic states to carbon hybridization states and SEI fine chemical composition analysis. For graphite, hard-carbon, or silicon-carbon anodes, the Carbon *K*-edge Energy Loss Near-Edge Structure clearly distinguishes the ratio between sp2 hybridization (graphitic conductive networks) and sp3 hybridization (diamond-like structures or defects), thereby evaluating graphitization degrees and structural degradation after cycling. Crucially, EELS serves as a powerful tool for SEI component resolution. Given the light-element composition of SEI films (LiF, Li2CO3, alkyl lithium carbonates), EELS differentiates organic and inorganic layer distributions at the nanoscale via F *K*-edge, O *K*-edge, and Li *K*-edge fingerprint peaks. Specifically, the Li *K*-edge chemical shift distinguishes metallic lithium (Li0) from ionic lithium (Li^+), a capability vital for investigating dead lithium formation and dendrite growth mechanisms in lithium metal anodes. Regarding silicon-based anodes (Si, SiO_x, Si alloys), EELS utilizes the Si $L_{2,3}$-edge white-line structure to judge the local silicon coordination environment (ranging from tetrahedral Si^0 to partially oxidized Si^+/Si^{4+}). Variations in the O *K*-edge pre-peak quantitatively describe SiO_x oxidation degrees and silicate SEI components formed during cycling. Furthermore, EELS tracks the covalent-to-metallic bond transition induced by Si–Li alloying at the nanoscale. For alloy-type anodes (tin, germanium, phosphorus-based), EELS tracks valence changes during lithiation via energy shifts in Sn *M*, Ge *L*, and P *L*-edges. This reveals metal–lithium alloy formation mechanisms and their reversibility. Specifically, in tin-based systems, EELS distinguishes the interfacial distribution of Sn^0, Sn^{2+}, and Sn^{4+}, thereby evaluating SEI electronic insulation and structural stability. In metal-oxide anodes (e.g., TiO_2, Fe_3O_4, CoO), spectral changes in TM *L* and O *K*-edges finely resolve reduction–oxidation processes during electrochemical reactions. Examples include the reversible Ti^{4+}\to$Ti^{3+}$ transition and oxygen vacancy generation contributing to lithium diffusion. Crucially, combining EELS with STEM imaging achieves atomic-scale interfacial chemical analysis. This directly visualizes anode-SEI spatial heterogeneity, including SEI thickness gradients, inorganic/organic layer distribution, and F-containing species (like LiF) identification.

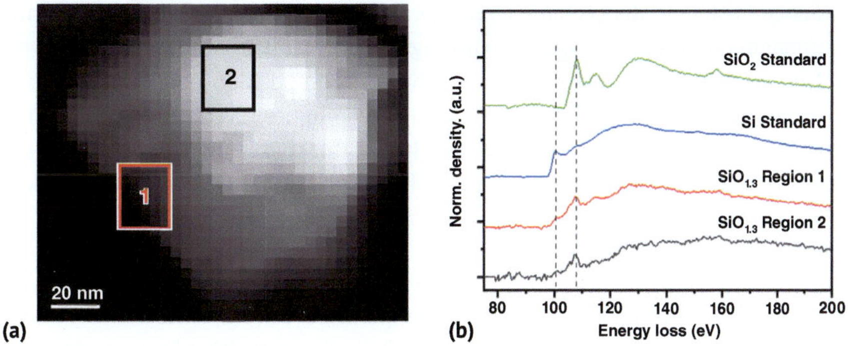

Figure 4.12 Microstructure information of pristine $SiO_{1.3}$ powder using STEM-EELS. (a) HAADF image of pristine $SiO_{1.3}$ powder and (b) corresponding EELS spectra of surface and bulk region together with the Si *L*-edge from the standard powder samples Si and SiO_2 [7] / with permission of John Wiley & Sons.

Si-based anodes, including SiO_x, form interfacial phases and Li–Si or Li-Si-O compounds during prelithiation or the first cycle. These phases differ from products formed at ambient conditions. They determine the initial Coulombic efficiency and influence long-term cycling stability. Conventional ex situ characterization often alters the chemical states and microstructures of these sensitive products during sample preparation or air exposure. This limitation motivated the use of low-temperature and low-damage analytical methods for accurate interfacial identification. Cryogenic and low-dose TEM imaging was conducted and combined with localized spectroscopy, including STEM-EELS mapping (Figure 4.12). This approach distinguishes lithium silicates, lithium silicides, and conventional SEI products at the nanoscale. Interfacial spectral differences between non-prelithiated and prelithiated samples were compared to evaluate the chemical impact of prelithiation. Local chemical state and elemental mapping based on EELS confirms the formation of a mosaic-like interfacial phase after prelithiation. This phase contains lithium silicate or silicate species with favorable electrical conductivity and mechanical robustness. These components exhibit distinct O *K*- and Si *L*-edge features under cryo-EELS, in contrast to carbonate- or organic-rich SEI formed without prelithiation. The study demonstrates that prelithiation compensates Li inventory and suppresses irreversible side reactions through interfacial-phase regulation. These effects improve initial efficiency and cycling retention [7].

In solid and liquid-electrolyte systems, EELS has become a core technique for resolving bulk electrolyte structures, interface reaction product compositions, ion conduction network configurations, and solid-solid interface chemical evolution due to its ultra-high energy resolution and chemical state sensitivity toward light elements (especially O, N, F, S, P). In liquid-electrolyte systems, EELS primarily identifies differences between light-element inorganic phases (e.g., LiF, Li_2CO_3, Li_2O, Li_3N, $Li_xPF_yO_z$) and organic phases ($ROCO_2Li$, PEO-like polycarbonates) within the SEI/CEI. Through pre-peak changes, white-line intensity ratios, and energy shifts of F *K*, O *K*, and C *K*-edges, it quantitatively judges the composition and spatial distribution of interface decomposition products. For instance, the LiF

F *K*-edge exhibits a sharp characteristic peak, whereas organic fluorine-containing components show broader spectra with slight energy shifts. The presence of π(C═O) and σ(C—O) in the C *K*-edge reflects solvent decomposition depth. Additionally, utilizing STEM-EELS line-scan and mapping techniques reveals SEI layered structures (inorganic inner layer/organic outer layer), thickness inhomogeneity, and surface enrichment regions. This quantitatively portrays interfacial electronic blocking performance and stability sources.

Interfacial products at Li metal-electrolyte interfaces, including SEI and reaction layers, contain light elements such as Li, O, C, F, S, and P. Many components readily reconstruct or decompose under room temperature TEM or vacuum conditions. Reliable identification of Li_2CO_3, LiF, or Li_2O distributions therefore remains challenging. The combination of cryo-TEM and EELS avoids these artifacts and enables direct access to interfacial chemistry (Figure 4.13). Cryo-TEM imaging of cycled

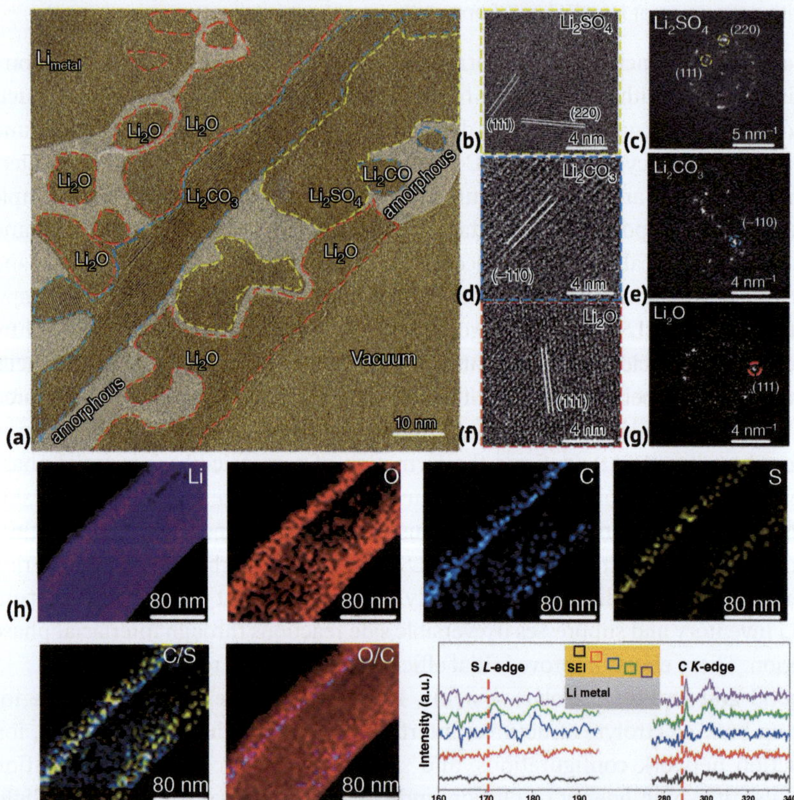

Figure 4.13 (a) Cryo-HRTEM images of SEI in (EC/DEC+ 2%DTD) electrolyte. (b), (c) HRTEM and FFT of Li_2SO_4 crystal. (d), (e) Li_2CO_3 crystal. (f), (g) Li_2O crystal. (h) EELS elemental mapping of lithium (purple) in the lithium dendrites. Oxygen (red), carbon (blue), and sulfur (yellow) in the lithium dendrites. (i) Composite map of carbon and sulfur; oxygen and carbon. (j) The EELS of S and C *K*-edge scanning from Li Metal to SEI surface. Please note that the HRTEM images are acquired using the same electron dosage of $\approx$100 e Å^{-2} [8] / with permission of John Wiley & Sons.

or deposited Li was performed under ultra-low dose and cryogenic conditions and combined with EELS mapping and edge analysis. This strategy locates Li_2CO_3, LiF, and Li_2O within the SEI and resolves their chemical states. EELS signals correlate with cryogenic imaging and diffraction results. This correlation enables interpretation of interfacial chemistry without thermal or beam-induced decomposition. Cryo-TEM and EELS together reveal that Li_2CO_3 does not always act as a stable inner-layer component. It often resides in the outer reactive region of the SEI and continues to react with electrolyte to release gas, leading to SEI porosity. In contrast, LiF- and Li_2O-rich nanocrystalline or amorphous mixed phases form a denser inner layer. This layer suppresses further parasitic reactions. These spectroscopic and imaging results indicate that S- or F-containing additives promote a favorable inorganic inner SEI. This strategy significantly enhances Li metal cycling efficiency [8].

In SSEs (sulfides, oxides, halides), EELS reveals local coordination structures, defect concentrations, anion displacements, and bonding characteristics via spectral changes in S $L_{2,3}$, O K, Cl K, and P L-edges. For example, in sulfide electrolytes (Li_6PS_5Cl, $Li_{10}GeP_2S_{12}$), the S L-edge pre-peak reflects P—S and Ge—S bond covalency. The Cl K-edge energy shift tracks anionic site structural changes caused by doping regulation. The O K-edge pre-peak identifies vacancies and polyhedral distortions in oxide electrolytes (LLZO, LATP). At solid-solid interfaces (e.g., solid electrolyte / cathode, solid electrolyte / anode contact interfaces), EELS provides in-situ revelations of reaction-layer spatial thickness, gradient composition, and electronic structure changes. For example, at the sulfide electrolyte and Ni-rich cathode contact interface, EELS identifies Ni^{2+}/Ni^{3+} valence changes and tracks S–O and P–O bond formation, thereby revealing interfacial oxidation reactions. At the lithium metal / solid electrolyte interface, EELS determines the formation and distribution of reduction products like Li_2S, Li_3P, and LiCl, aiding in the analysis of interfacial resistance growth roots during cycling.

In summary, the physical essence of EELS in battery component characterization involves inferring atomic electronic structures by measuring energy loss from incident electrons exciting inner-shell electrons. All applications aim to establish structure-performance relationships (e.g., capacity fading, voltage hysteresis) at the microscopic/atomic scale. EELS targets distinct energy ranges and elements for different battery parts: cathodes require focusing on TM L-edges and oxygen K -edges in the 400–1000 eV range, emphasizing multivalence analysis; conversely, anodes and electrolyte interfaces demand attention to light elements like Li, C, and F (K-edges) in the 50–300 eV range, emphasizing chemical bond type differentiation (e.g., C=O vs. C—H). Furthermore, the physical perspective differs: cathodes prioritize electronic density of state changes caused by charge transfer and crystal field splitting, whereas anodes and interfaces prioritize phase composition and amorphous/crystalline structural identification. Thus, EELS constitutes a critical bridge connecting macroscopic electrochemical performance with atomic-level electronic behavior in modern battery material analysis, offering profound insights into interface reaction mechanisms and light-element evolution unmatched by other spectroscopic means. The core advantage of EELS in electrolyte and interface research lies in its ability to simultaneously provide elemental distribution, coordination structure, and valence information with sub-nanometer resolution while accurately distinguishing different chemical environments of light elements. It serves as a key characterization means for understanding electrolyte stability, interface engineering strategies, and solid-state battery failure mechanisms.

4.8.3 Cryogenic Electron Microscopy

In lithium-ion and sodium-ion battery research, cryogenic electron microscopy (cryo-EM) has emerged as a core technique for revealing metastable intermediate phases in energy materials. This capability stems from its ability to rapidly freeze electrodes and electrolytes in their working state at low temperatures, thereby maximally avoiding structural damage commonly induced by electron beam irradiation, vacuum desolvation, and component migration in traditional TEM/SAED characterization.

In lithium-ion battery cathode systems, many high-energy-density materials (e.g., Ni-rich NCM, LRLO, high-voltage $LiCoO_2$) form oxygen-rich isolated bodies, undergo in-situ TM–O reconstruction, or generate oxygen-containing intermediate phases during charge and discharge. These intermediate phases typically exhibit extremely high electron beam sensitivity. Conventional TEM imaging often exacerbates oxygen defects, causes lattice collapse, or directly induces irreversible phase transitions. Conversely, cryo-EM, by rapidly freezing samples to liquid nitrogen or liquid helium temperatures, allows for the observation of these intermediate phases in a near-intrinsic state. This has enabled the first direct observation of the "rock salt reconstruction layer" on Ni-rich cathode surfaces and peroxide/superoxide short-range ordered structures (e.g., O–O dimers) during the charging of Li-rich materials.

In sodium-ion battery cathodes, such as P2- and O3-type layered oxides, Na-vacancy-ordered phases, Na-striped phases, or short-range distorted interlayer sliding structures are easily generated during sodium deintercalation. These structures are highly unstable and similarly sensitive to electron beams. Cryo-EM preserves the authentic distorted morphology of their layered frameworks, enabling researchers to track the initiation sites of interlayer sliding and local TM–O coordination changes caused by sodium deintercalation, thus more accurately constructing sodium cathode phase-transition pathways. Cryo-EM was employed to capture interfacial penetration of Na or Li and CEI morphology at high potentials (Figure 4.14). Conventional TEM fails to preserve native CEI and light-element-enriched phases formed at electrolyte–electrode contacts. The experimental workflow includes rapid freezing under inert atmosphere using liquid nitrogen or helium stages, cryo-FIB preparation or cold transfer into TEM, and low-dose imaging with spectroscopic analysis. This protocol reveals the true spatial distribution of layered, particulate, F-rich, or O-rich CEI components. Results show that cryo-TEM directly identifies CEI densification or solvent-residue differences induced by specific electrolytes or additives. These microscopic features correlate with cathode phase transitions, oxygen release, and capacity fading at high voltage. The findings highlight the decisive evidential role of cryo-TEM in elucidating cathode surface degradation mechanisms [9].

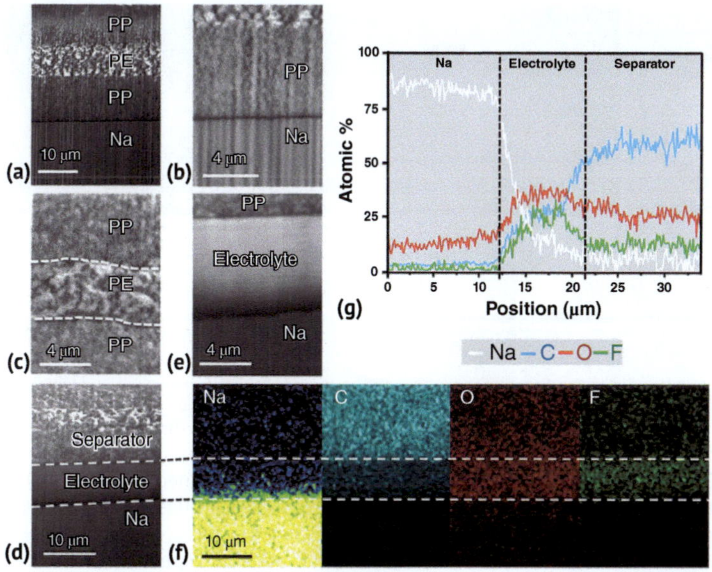

Figure 4.14 (a) Cross-sectional SEM image of an uncycled symmetric Na||Na cell with solid interfacial contact between the separator and electrode. (b) A higher magnification SEM image of the separator-electrode interface seen in (a). (c) A high-magnification cross-sectional SEM image of the tri-layer polymer separator. (d) Cross-sectional SEM image of an uncycled symmetric Na||Na cell with a layer of electrolyte between the separator and electrode. (e) A higher magnification cross-sectional SEM image of the electrolyte layer between the separator and electrode seen in (d). (f) EDS maps of the Na, C, O, and F signal from the same region shown in (d). (g) Line plot of the normalized intensity of the atomic % of Na, C, O, and F EDS signal as a function of vertical position along image d. All milling, imaging, and EDS were done at cryo-temperatures [9] / with permission of John Wiley & Sons.

In anode material systems, particularly lithium metal, sodium metal, silicon-based anodes, and tin-based anodes, extremely sensitive intermediate phases arise during lithiation/delithiation. These include Li_2O/LiOH nanophases, amorphous Li_xSi alloy phases, short-range ordered structures during the initial stages of metal–lithium alloying, and inorganic-rich nanodomains within the SEI/AEI (anion-derived interphase). Conventional TEM often leads to the evaporation of SEI organic components, decomposition of inorganic phases, or thermally induced crystallization of Li_xSi. In contrast, cryo-EM allows for the observation of the native multilayer structure of the SEI, organic/inorganic component distribution, LiF-rich/Li_2CO_3-rich spatial gradient layers, and initial nanovoid and grain boundary configurations of lithium metal dendrites under frozen conditions.

At liquid-electrolyte system interfaces (SEI/CEI), cryo-EM stands as the sole technique capable of authentically capturing initial intermediate products of solvent decomposition. Examples include soft phases rich in $ROCO_2Li$ and polycarbonate segments formed in EC/DEC electrolyte systems, as well as multilayer structures like $RO–PO_2F_2$ and LiF_xO_y on high-voltage cathode surfaces. These components are typically completely destroyed and rendered unidentifiable by the electron beam in conventional TEM.

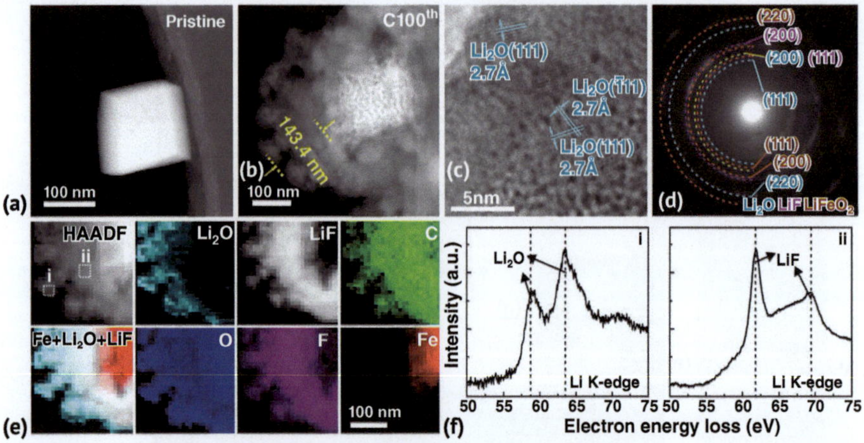

Figure 4.15 Identical-location cryo-S/TEM characterization of SEI formed on bare Fe_2O_3 quasi-cube after 100 cycles. (a)–(f) HAADF-STEM images on an Fe_2O_3 quasi-cube before (a) and after 100 times full charge (C100th) (b), and the corresponding HRTEM image (c), SAED pattern (d), HAADF-STEM image and EELS maps (e), and Li *K*-edge EEL spectra (from the two selected rectangular regions in [e]) after 100 times full charge (C100th) (f) [10] / John Wiley & Sons / CC BY 4.0.

Guo and co-workers applied an identical-location strategy combined with cryo-TEM and low-temperature spectroscopy to compare reversible changes in SEI thickness and composition at the same electrode region under different states of charge (Figure 4.15). Cryo-TEM was selected due to severe solvent evaporation, organic damage, and inorganic reorganization under room temperature TEM conditions. These artifacts obscure true SEI breathing behavior during cycling. Samples were rapidly frozen under inert atmosphere and transferred under cryogenic conditions. Low-dose imaging and cryo-EELS enabled quantitative comparison of spatial redistribution of organic and inorganic components such as Li_2O, LiF, and carbonates. The results demonstrate SEI breathing during lithiation and delithiation. SEI expands during lithiation with increased incorporation of insertion products and contracts locally during delithiation. Improper mixing of SEI components, including strong intermixing of Li_2O with organic species, causes irreversible thickening and increased interfacial resistance. These observations establish interfacial engineering strategies, including carbon coatings or interfacial isolation layers, as effective routes to extend cycling life [10].

In solid-state electrolyte systems—whether sulfides, oxides, or halides—contact interfaces with cathodes or anodes form extremely thin but highly active solid-solid intermediate layers (e.g., Li_2S/Li_3PS_4/LiCl mixed layers or oxide–metal reduction layers at LLZO/Ni-rich cathode interfaces). These intermediate phases usually decompose rapidly under electron beams, whereas cryo-EM preserves their authentic chemical composition, nanostructural hierarchy, and defect distribution, providing critical evidence for understanding solid-state battery interfacial impedance growth mechanisms.

Overall, the common advantage of cryo-EM across different systems lies in preserving authentic intermediate phases, avoiding beam damage, resolving light-element compositions, and providing in-situ phase-transition pathways. The differences arise from the distinct structural stability, chemical sensitivity, and interfacial complexity of various material systems, directing cryo-EM more towards capturing oxygen redox intermediate phases in cathodes, revealing SEI/AEI structures in anodes, and resolving the chemical evolution of solid-solid interface reaction layers in SSEs. Cryogenic electron microscopy (cryo-TEM / cryo-STEM) provides authentic structural views of critical interfaces in lithium-ion batteries (e.g., lithium metal SEI, solid-state electrolyte interfaces) that are difficult to achieve with traditional electron microscopy. The analytical rationale is primarily that these interfacial layers are extremely sensitive (prone to electron beam damage, reconstruction under vacuum or high temperature) and difficult to preserve intact via room temperature TEM. The analytical process typically involves rapid freezing (vitrification), cryo-stage sample preparation (e.g., cryo-FIB), low-dose imaging, and spectroscopic (EELS or EFTEM) analysis. Analytical conclusions reveal the authentic SEI microstructure (composition, 3D network, or particles), reaction-layer thickness, interfacial connectivity, and the spatial distribution of interfacial chemical states (e.g., redox products, coordination structures). These provide a molecular/atomic-scale mechanistic basis for interface engineering (e.g., SEI optimization, interface passivation). With the maturation of cryo-EM sample preparation and imaging technologies (e.g., cryo-FIB, low-dose probes, cryo-transfer), its importance in battery interface research will further increase, promising to provide direct structural guidance for the design of high-performance, high-safety next-generation lithium batteries (e.g., lithium metal, solid-state batteries).

4.8.4 In-situ TEM

In-situ TEM (in-situ TEM) serves as a core tool in battery material science for directly observing electrochemical processes, phase transformation kinetics, and interfacial evolution. Its applications in surface and bulk phase studies cover lithium-ion battery cathodes (layered oxides, NCM, Ni-rich, Li-rich), sodium-ion battery cathodes, various anodes (silicon, graphite, alloy/conversion types), and solid/liquid electrolytes with distinct yet complementary focuses. Regarding cathode materials, in-situ TEM enables direct nanoscale visualization of interlayer gliding, phase-boundary migration, surface reconstruction, and crack propagation triggered by lithium (or sodium) deintercalation. Phenomena such as layered-to-spinel/rock salt reconstruction during deep deintercalation, microcrack initiation from lattice strain concentration, and interface migration in multiphase coexistence regions allow "real-time" capture within the TEM via electrochemical in-situ microbatteries or MEMS-based nanodevices. In comparison, sodium-ion cathodes exhibit complex short-range distortions arising from interlayer gliding (P2-O2), vacancy serialization, and Na-vacancy ordering. In-situ TEM reveals the germination of these glides on single particles and their propagation along grain boundaries, directly associating macroscopic voltage plateaus with local structural reconstruction. For anodes, in-situ TEM holds particular importance for alloy anodes like silicon. Researchers observe crystalline-to-amorphous transitions, two-phase-front propagation rates, surface crack formation/propagation, and intraparticle strain relaxation or pulverization at the single-particle level. Early in-situ TEM work directly visualized the two-phase lithiation interface advancement mechanism in silicon nanospheres, altering the understanding of "uniform corrosion or brittle fracture" mechanisms. Regarding carbon-based anodes like graphite and hard carbon, in-situ TEM monitors the reversibility of graphite interlayer spacing upon lithiation/delithiation, exfoliation caused by local expansion, and nucleation of SEI-like products in liquid microcells despite the milder nature of most processes.

Regarding electrolytes and interfaces, in-situ TEM operates through two complementary approaches. First, liquid or quasi-liquid microcells facilitate direct observation of early-stage electrolyte/electrode interface reactions and CEI/SEI layer formation, capturing solvent decomposition product locations and nucleation kinetics. Second, MEMS-based solid batteries or cryo-FIB prepared cross-sectional samples in solid-state battery models allow direct visualization of solid-solid interface intermediate phases (e.g., Li_2S, Li_3P, oxide reduction products) growing during cycling or heating and affecting ion conduction channels. Despite the unprecedented spatiotemporal resolution of in-situ TEM, experimental designs, observation focal points, and interpretability differ significantly across systems. Cathodes and most inorganic solids (e.g., layered oxides, NASICON-type) generally exhibit higher electron beam tolerance, permitting high-resolution imaging and atomic-scale spectroscopy (EELS/EDS) for quantitative analysis of precipitated TM valence states and oxygen vacancy generation. Conversely, sulfur-containing or organic-rich systems (e.g., liquid SEI or organic polymer electrolytes) suffer easily from beam damage, necessitating low-dose strategies, high-frame-rate cameras, or cryo-TEM integration to maintain the pristine state. Liquid in-situ cells display authentic solvent decomposition pathways but suffer from spatial resolution and spectroscopic sensitivity limits due to liquid window membranes, thickness, and beam scattering [11]. Solid-state microbatteries and MEMS chips offer higher resolution under quasi-in-situ electrochemical conditions but potentially introduce boundary conditions (e.g., non-ideal electrode contact, local geometric effects) affecting conclusion universality.

Advanced in-situ TEM was employed to directly capture real-time Li-ion intercalation and deintercalation dynamics in graphite and graphene anodes during electrochemical cycling. Conventional static characterization fails to resolve microscopic structural changes during lithiation, including interlayer-spacing variation and local stacking rearrangement. In-situ TEM provides direct nanoscale visualization of these processes (Figure 4.16). Controlled potentials were applied on an in-situ electrochemical nanobattery platform. Li ions gradually entered graphite or graphene interlayers. Time-resolved bright-field and dark-field TEM images were recorded. Electron diffraction and spectroscopic analysis confirmed structural evolution mechanisms. Results show progressive interlayer expansion, local dislocation formation, and layer gliding during lithiation. These processes correlate strongly with nonuniform Li insertion. This behavior directly influences reversibility and cycling stability. In-situ TEM provides quantitative and visual evidence of nanoscale

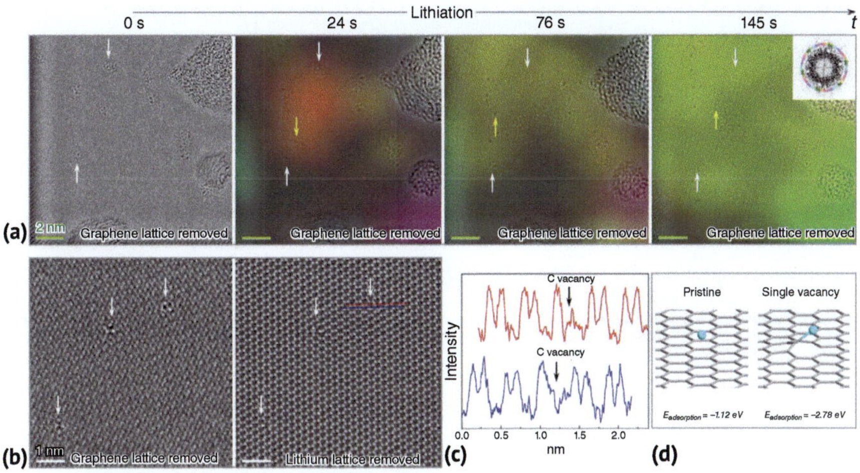

Figure 4.16 Observing lithiation at atomic scale. (a) A series of CC/CS-corrected HRTEM images acquired during a lithiation experiment, with bilayer graphene lattice removed by Fourier filter. The white arrows mark the C vacancy defects in graphene. The yellow arrows mark some foreign O atoms inside the Li crystal. Colored orientation maps of Li crystals are obtained and overlaid on the images by applying a colored bandpass mask to the second order reflections of Li in the Fourier transforms (see the inset Fourier transform). (b-left) Another graphene removed HRTEM image containing the contrast of the C vacancy defects marked by the arrows. (b-right) The same image as the left but with Li lattice removed, leaving the graphene lattice. (c) Extracted line profiles from (b). (d) Calculated configurations of interaction between graphene and an isolated Li atom (left: Li adsorption on pristine graphene, right: Li adsorption on single vacancy) [12] / with permission of John Wiley & Sons.

structural evolution. These insights guide the design of high-rate and long-life carbon-based anodes [12].

Lu et al. discuss the roles and challenges of in-situ TEM techniques in investigating lithiation/delithiation (or sodiation) electrode materials from a cathode mechanism perspective (Figure 4.17). Analytically, the development of high-energy-density batteries (e.g., layered oxides, phosphates, spinels, and other complex cathodes) makes understanding phase-transition mechanisms, structural reconstruction, phase-boundary migration, and defect formation in working states critical for optimizing cycle performance and life. Traditional ex situ (post-cycling sampling) methods provide structural snapshots but fail to reveal transient in-situ phase structural evolution, internal particle interface motion, and grain boundary behavior during reactions [13]. In-situ TEM offers a unique perspective combining high spatial (nanometer or atomic level) and temporal resolution (real-time observation of multiple cycle stages), enabling direct recording of structural kinetics in cathode materials within "working battery (or near-real electrode) models."

Overall, in-situ TEM stands as a fundamental technology for understanding secondary rechargeable cathode mechanisms (especially high specific energy density systems), exerting profound influence on the development of next-generation battery cathode materials.

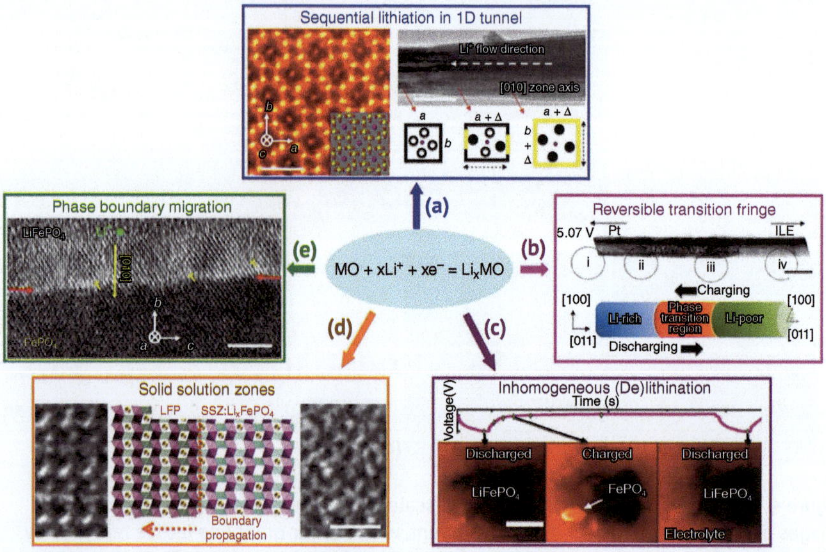

Figure 4.17 (a) Sequential Li+ filling inside α-MnO2 tunnels leads to crystal expansion asymmetry. The left scale bar, 1 nm, and the right bar, 50 nm. (b) Reversible movement of a phase-transition region (red) between Li-rich and Li-poor phases during cycling of a single LiMn2O4 nanowire-based battery was captured. Scale bar, 200 nm. (c) Inhomogeneous (de)lithiation among LFP particles revealed by time-sequential energy-filtered TEM in an electrochemistry liquid cell. Scale bar, 200 nm. (d) The movement of a solid-solution zone was captured dynamically during delithiation of a LFP crystal. Scale bar, 1 nm. (e) The migration of a LFP/FP phase boundary was captured during lithiation with its migration direction identified as well. Scale bar, 5 nm [13] / Springer Nature / CC BY 4.0.

4.8.5 Aberration-corrected TEM

In the field of Li- and Na-ion battery research, aberration-corrected TEM (AC-TEM) has emerged as a key tool for resolving intrinsic microstructures in complex electrode and electrolyte systems. This capability arises from sub-angstrom spatial resolution and high sensitivity to light elements and local lattice distortions. AC-TEM shows complementary yet distinct roles in studies of cathodes, anodes, and solid or liquid-electrolyte surfaces and bulk phases.

In Li- and Na-ion battery cathodes, the primary value of AC-TEM lies in direct visualization of atomic column arrangements, local dislocations, antiphase boundaries, and surface reconstruction layers in layered oxides, spinels, and multiphase structures. In $LiCoO_2$, NCM, and Ni-rich layered cathodes, AC-TEM resolves atomic occupancies in TM layers and Li layers. The method identifies cation mixing, stacking faults, and cycling-induced nucleation of rock salt or spinel phases. These observations provide decisive insight into structural degradation and capacity fading at high voltage. In Li-rich Mn-based cathodes, atomic-scale imaging of oxygen-related defects, short-range ordering, and surface disorder delivers key structural evidence for anionic redox activity and voltage decay.

High-voltage oxides such as $LiNi_{0.8}Co_{0.15}Al_{0.05}O_2$ exhibit phase transitions, local defects, cation disorder, and stacking faults during deep delithiation and high-voltage cycling. These features drive capacity loss and resistance growth. Atomic-scale lattice and interfacial information becomes essential. Conventional non-corrected TEM fails to provide clear contrast due to aberration and image delocalization. Wan and co-workers applied aberration-corrected high-resolution TEM combined with HAADF imaging and diffraction modes to pristine and cycled cathodes [14]. AC-TEM resolved TM and Li layer occupancies at sub-angstrom resolution. The method visualized ultrathin surface reconstruction layers with rock salt or spinel character and atomic configurations at grain boundaries and mismatch interfaces (Figure 4.18). Aberration correction reduced image distortion and improved Z contrast reliability. HAADF STEM images distinguished Co, Ni, and Mn disorder and Li-site defects. AC-TEM evidence revealed pronounced layered-to-rock salt reconstruction at particle surfaces, increased cation mixing, and atomic-scale mismatch between structural slabs after high-voltage cycling. These defects represent the fundamental origin of cycling instability. Control of surface and grain boundary stability emerges as a key design principle for high-voltage cathodes.

In Na-ion battery cathodes such as layered $NaMO_2$ and polyanionic compounds, AC-TEM primarily reveals stronger interlayer gliding, phase-boundary formation,

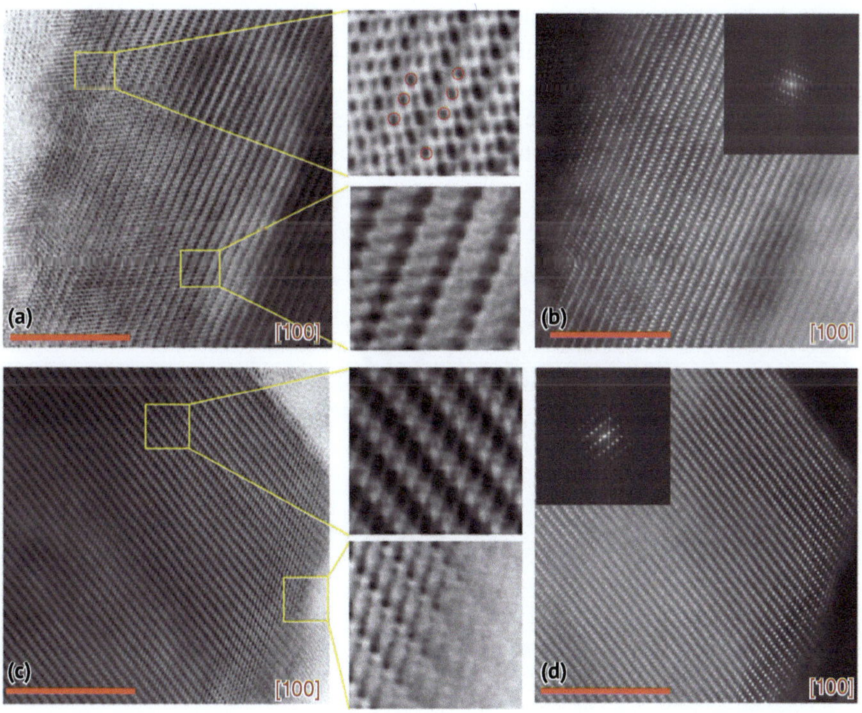

Figure 4.18 (a) ABF image of pristine Li-excess cathode material LL-811, the Ni2+ ions occupied at the Li site have been highlighted with a red circle. (b) HAADF image of pristine Li-excess cathode material LL-811 and its fast Fourier transformation image. (c) ABF image of three times treated Li-excess cathode material 3TLL-811. (d) HAADF image of three times treated Li-excess cathode material 3TLL-811 and its fast Fourier transformation image. All scale bars are 5 nm [14] / with permission of John Wiley & Sons.

and large-scale strain accumulation during Na^+ extraction. The conceptual framework resembles Li systems. The larger Na^+ radius and higher structural flexibility place greater emphasis on spatial correlations among phase interfaces and defect networks. Layered Na-ion cathodes such as $NaNi_xCo_yMn_{1-x-y}O_2$ undergo pronounced layered-to-rock salt transitions, Na-layer vacancy formation, and extended dislocation networks during desodiation. These changes strongly affect Na^+ diffusion pathways and cycling performance. Direct observation of light-element positions and phase-boundary formation becomes essential. Conventional TEM lacks sufficient contrast and resolution for these features. Kerstin Volz and co-workers employed aberration-corrected STEM combined with 4D STEM and STEM-EELS for high structural and chemical resolution [15]. Atomic-scale AC-STEM images revealed alternating Na and TM layers (Figure 4.19). 4D STEM reconstructed light-element positions and identified layered-to-rock salt transition regions. Aberration correction enhanced point resolution and contrast. Local Na-layer defects, stacking faults, and lattice distortions became quantifiable. Results demonstrated that Na and O rearrangement, Na-vacancy enrichment, and mismatch networks collectively determine the phase-transition pathway. AC-TEM and 4D STEM provided direct atomic-level evidence for degradation and cycling behavior in Na cathodes. These results highlight the irreplaceable role of AC-TEM in Na-based systems.

In battery anodes, AC-TEM shifts emphasis from ordered lattice analysis to capture of structural heterogeneity and interfacial details. In graphite anodes, aberration-corrected imaging resolves subtle interlayer-spacing variations, edge dislocations, and local disorder. These observations support atomic-scale interpretation of lithiation

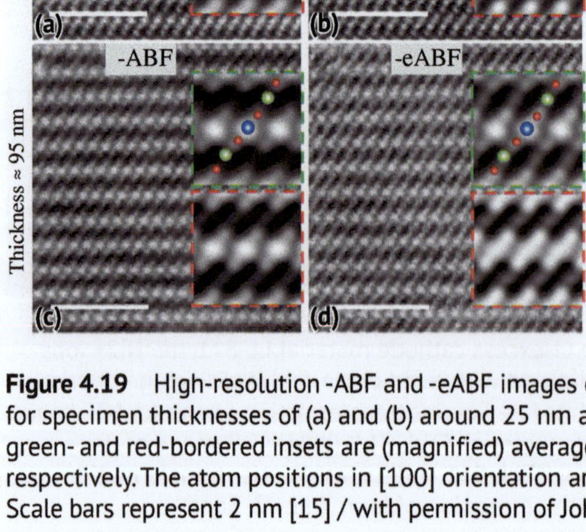

Figure 4.19 High-resolution -ABF and -eABF images of LNO viewed in the [100] orientation for specimen thicknesses of (a) and (b) around 25 nm and (c) and (d) around 95 nm. The green- and red-bordered insets are (magnified) averaged experimental and simulated images, respectively. The atom positions in [100] orientation are shown in the green-bordered insets. Scale bars represent 2 nm [15] / with permission of John Wiley & Sons.

kinetics and rate capability. In Si- and Sn-based alloy anodes, AC-TEM distinguishes residual short-range order during crystalline-to-amorphous transitions, nanocrystal embedding in amorphous matrices, and stress localization sites. These features reveal intrinsic origins of volume expansion and pulverization failure. In Li metal anodes, strong beam sensitivity persists. AC-TEM under low-dose or cryogenic conditions resolves Li deposition morphology, crystal orientation, and geometric relations with interfacial layers. The analysis links atomic-scale morphology to macroscopic dendrite behavior. Si-based anodes attract attention due to high theoretical capacity. Lithiation induces severe volume expansion and complex amorphous-crystalline coexistence. Conventional TEM fails to distinguish amorphous Si, nanocrystalline domains, and SEI sublayers. These features govern cycling degradation. Xue and co-workers applied aberration-corrected high-resolution TEM under low dose to track nanoscale evolution at different cycling states [16]. AC-TEM distinguished crystalline Si, amorphous Si, and atomic arrangements at grain boundaries and nested phases (Figure 4.20). STEM-EDS and STEM-EELS mapping quantified spatial distributions of Li-rich alloy phases and SEI components. AC-TEM images showed nonuniform crystalline-to-amorphous transformation and formation of complex phase-boundary networks. Local structural variations correlated with stress concentration and crack initiation. Atomic-scale SEI imaging resolved an inorganic-rich inner layer and an organic-rich outer layer. This layered structure strongly affected interfacial stability and cycle life. The study underscored the central role of AC-TEM in elucidating Si anode microstructure and failure mechanisms.

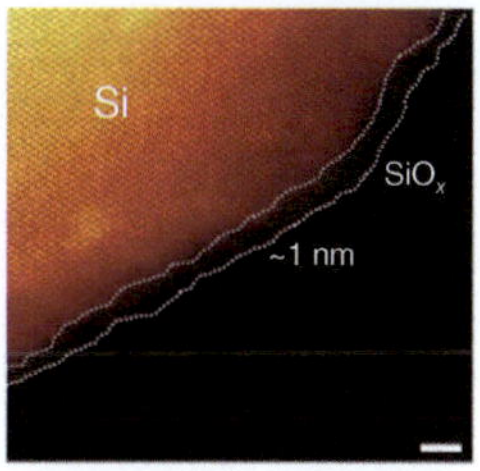

Figure 4.20 Interfacial charge landscape of Si (semiconductor) versus graphite (conductor) and electrolyte-guided formation of a self-limiting interphase. AC-HRTEM image of a Si particle. Scale bar, 2 nm [16] / with permission of John Wiley & Sons.

In solid and liquid-electrolyte systems, especially electrode–electrolyte interfaces such as SEI, CEI, and solid-solid contacts, AC-TEM excels in precise determination of atomic continuity, reaction-layer thickness, and local crystallization. In liquid electrolytes, AC-TEM combined with cryogenic or quasi-in-situ preparation identifies nanocrystalline inorganic phases such as LiF and Li_2O and their spatial relation to amorphous regions. In solid electrolytes based on sulfides, oxides, or polymers, AC-TEM reveals atomic matching at interfaces, crystallographic features of reaction products, and stress-induced defects. These studies emphasize interfacial continuity and localized chemistry effects on ion transport and interfacial resistance. Interfacial products in solid electrolytes or CEI layers are ultrathin, locally disordered, and rich in light elements. These characteristics impose stringent demands on structural and chemical contrast. AC-TEM improves atomic-scale resolution and contrast and enables direct imaging of interfacial atomic arrangements. Julia Maibach and co-workers combined AC-STEM with HAADF and ABF imaging to analyze surface-layer evolution in LNMO cathodes during cycling [17]. The study identified a transition from ordered layered structures to disordered or reconstructed phases and resolved defects and phase boundaries within boundary layers

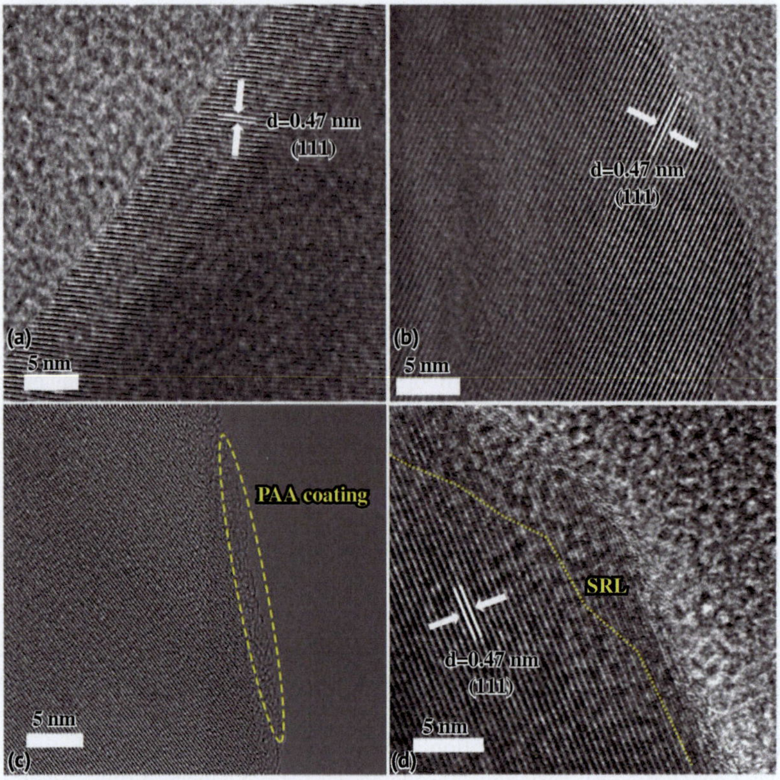

Figure 4.21 HRTEM images of (a) P-LNMO, (b) W-LNMO, (c) A-LNMO, and (d) H-LNMO. SRL: surface reconstruction layer. Pristine LNMO was treated with water, 3% poly (acrylic acid) (PAA), and 5% H_3PO_4, which were labeled with the prefixes P- for pristine, W- for water, A- for PAA, and H- for H_3PO_4 [17] / with permission of John Wiley & Sons.

(Figure 4.21). Low-aberration conditions clarified atomic columns and associated dislocations and stacking faults. AC-TEM images demonstrated progressive surface evolution from layered order to defect accumulation and reconstructed phases. These changes localized at surfaces and near interfaces correlated strongly with electrochemical degradation. Interfacial continuity and defect density emerged as key factors governing ion transport and cycling durability. The work highlighted the importance of AC-TEM for quantitative interface analysis.

Overall, AC-TEM applications across battery systems share common goals and show clear distinctions. The common objective centers on atomic-scale structural resolution for intrinsic understanding of phase transitions, defects, and interfaces. Differences arise from structural order, beam sensitivity, and research focus. Cathode studies prioritize lattice and phase transitions. Anode studies emphasize heterogeneous evolution and mechanical effects. Electrolyte and interface studies focus on reaction layers and atomic contact. Continued integration of low-dose imaging, cryogenic conditions, and multimodal spectroscopy is transforming AC-TEM from a high-resolution structural tool into a comprehensive platform for interfacial and reaction-mechanism analysis. AC-TEM will remain indispensable for next-generation batteries with high-energy density and enhanced safety.

4.9 Conclusion and Prospective

The above discussion demonstrates that a multidimensional characterization framework centered on transmission electron microscopy, including conventional TEM, SAED, aberration-corrected TEM and STEM (AC-TEM/AC-STEM), EELS, cryogenic electron microscopy (cryo-EM), and in-situ TEM, has formed a systematic, hierarchical, and highly complementary platform for Li- and Na-ion battery research. This framework has become a core scientific approach for revealing intrinsic structures, interfacial reaction mechanisms, and failure evolution in electrode and electrolyte materials.

From a structural perspective, TEM imaging and SAED constitute the foundational platform for battery materials characterization. These techniques identify crystal structures, phase compositions, and orientation relationships at the nanoscale. They reveal coexistence and transformation among layered, spinel, and rock salt phases in cathodes. They capture crystallization-to-amorphization evolution during lithiation or alloying in anodes. Such structural information provides essential prerequisites for understanding capacity fading, polarization growth, and phase instability. Conventional TEM and SAED show inherent limitations in resolution and contrast for complex defects, atomic-scale disorder, and light-element-dominated interfacial systems.

The introduction of aberration-corrected TEM overcomes these limitations. Sub-angstrom resolution enables direct imaging of atomic occupancies in TM and Li or Na layers. The method resolves cation mixing, stacking faults, antiphase boundaries, and surface reconstruction layers. Structural understanding advances from average structures to local atomic configurations. In Li- and Na-ion cathodes, AC-TEM reveals atomic-scale origins of layered-to-rock salt or spinel transitions under high voltage or deep deintercalation. In anodes, detailed analysis of amorphous-nanocrystalline coexistence, stress-induced defects, and interfacial morphology provides direct evidence for failure mechanisms in Si-based, carbon-based, and Li metal anodes. In solid and liquid-electrolyte interfaces, AC-TEM shows unique capability for resolving ultrathin reaction layers and interfacial atomic continuity. Complementarily, EELS represents the chemical dimension of electron microscopy. High sensitivity to light elements and valence states enables reliable nanoscale determination of composition and spatial distribution in SEI and CEI layers, lithium silicates, and LiF-rich interphases. This capability establishes a direct link between microscopic chemical structure and macroscopic electrochemical behavior.

For highly beam-sensitive intermediates and interfaces, cryogenic electron microscopy further expands the applicability of electron microscopy. Rapid freezing and low-dose imaging preserve native structures and chemical states at electrode–electrolyte interfaces. Cryo-EM enables faithful capture of SEI and CEI layering, solvent decomposition pathways, and interfacial reaction intermediates. These features remain unstable under conventional ex situ conditions. This capability proves critical for Li metal anodes, solid-state electrolyte interfaces, and systems containing organic–inorganic composites. Cryo-EM provides compelling experimental support for interface engineering and electrolyte design. In contrast, in-situ TEM adds a temporal dimension to electron microscopy. Simplified electrochemical or chemical environments inside the microscope enable real-time tracking of phase transitions, morphological evolution, and interfacial reaction kinetics. In-situ TEM provides dynamic validation for mechanisms previously inferred post mortem. Examples include phase-front propagation in cathode particles and nucleation and growth of Li dendrites in anodes. Interpretation requires caution due to model-system constraints.

Future development of electron microscopy for batteries will follow three major trends. The first trend involves deep integration of multiple techniques. Structural identification by TEM and SAED, atomic-scale analysis by AC-TEM, chemical state probing by EELS, intrinsic structure preservation by cryo-EM, and dynamic observation by in-situ TEM will form a unified structure-chemistry-kinetics characterization paradigm. The second trend involves continued progress toward low-dose, low-damage conditions and operation closer to realistic battery environments. This direction minimizes artifacts from beam effects and environmental simplification. The third trend involves increased reliance on quantitative analysis and multiscale theoretical modeling. This approach improves predictability in structure-performance correlations. With continued technical and methodological advances, electron microscopy will play an increasingly central and irreplaceable role in elucidating intrinsic structures, interfacial reaction mechanisms, and failure origins in Li- and Na-ion battery materials. This progress will provide a robust microscopic foundation for rational design of next-generation batteries with high-energy density, high safety, and long cycle life.

References

1 Z. Sun, J. Pan, W. Chen, H. Chen, S. Zhou, X. Wu, Y. Wang, K. Kim, J. Li, H. Liu, Y. Yuan, J. Wang, D. Su, D.-L. Peng, Q. Zhang, Electrochemical processes and reactions in rechargeable battery materials revealed via in situ transmission electron microscopy, Advanced Energy Materials 14(2) (2024) 2303165. https://doi.org/10.1002/aenm.202303165.

2 B. Cheol Lee, J. Kim, H.-S. Kim, G.-T. Park, Y.-K. Sun, C. Seung Yoon, Structural analysis of deeply charged Li(Ni0.95Co0.04Al0.01)O2 cathode for li-ion battery, ChemSusChem 18(8) (2025) e202401856. https://doi.org/10.1002/cssc.202401856.

3 C. Shi, D. Muhtar, X. Lu, F. Liu, X. Lu, Z. Sun, Z. Guo, A medium-entropy NASICON cathode for sodium-ion batteries achieving high energy density through dual enhancement of voltage and capacity, Advanced Energy Materials 15(26) (2025) 2500448. https://doi.org/10.1002/aenm.202500448.

4 R. Li, J. Zeng, P. Wang, T. He, L. Rao, L. Zhou, M. Hou, X. Zhang, T. Lin, Y. Zhang, N. Zhang, A high-performance silicon-based anode enabled by hybrid pathways for all-solid-state batteries, Advanced Energy Materials 15(37) (2025) e02913. https://doi.org/10.1002/aenm.202502913.

5 X. Lei, J. Wang, Y. Su, S. Guo, G. Li, P. Ji, X. Wang, L. Gu, C. Ge, Y. Zhang, D. Su, Thermal-induced structure evolution at the interface between cathode and solid-state electrolyte, Small Structures 5(3) (2024) 2300342. https://doi.org/10.1002/sstr.202300342.

6 H. Choi, A.R. Schuer, H. Moon, G. Melinte, G.-T. Kim, J. Asenbauer, A. Kazzazi, M. Kuenzel, S. Passerini, Is cobalt in li-rich layered oxides for li-ion batteries necessary? ChemElectroChem 11(17) (2024) e202400391. https://doi.org/10.1002/celc.202400391.

7 S. Bai, W. Bao, K. Qian, B. Han, W. Li, B. Sayahpour, B. Sreenarayanan, D.H.S. Tan, S.-y. Ham, Y.S. Meng, Elucidating the role of prelithiation in Si-based anodes for interface stabilization, Advanced Energy Materials 13(28) (2023) 2301041. https://doi.org/10.1002/aenm.202301041.

8 B. Han, Z. Zhang, Y. Zou, K. Xu, G. Xu, H. Wang, H. Meng, Y. Deng, J. Li, M. Gu, Poor stability of Li2CO3 in the solid electrolyte interphase of a lithium-metal anode revealed by cryo-electron microscopy, Advanced Materials 33(22) (2021) 2100404. https://doi.org/10.1002/adma.202100404.

9 K.C. Matthews, B. Rush, R. Gearba, X. Guo, G. Yu, J.H. Warner, Cryo-electron microscopy reveals Na infiltration into separator pore free-volume as a degradation mechanism in Na anode: Liquid electrolyte electrochemical cells, Advanced Materials 36(23) (2024) 2308711. https://doi.org/10.1002/adma.202308711.

10 X. Guo, X. Du, V. Nicolosi, B. Zhang, Y. Zhu, Tailoring breathing behavior of solid electrolyte interphases unraveled by cryogenic transmission electron microscopy, Advanced Energy Materials 13(21) (2023) 2300240. https://doi.org/10.1002/aenm.202300240.

11 Y. Li, T. Wu, Y. Yuan, J. Lu, In Situ TEM studies on electrochemical mechanisms of rechargeable ion battery cathodes, Small Structures 4(8) (2023) 2300001. https://doi.org/10.1002/sstr.202300001.

12 Y. Li, F. Börrnert, M. Ghorbani-Asl, J. Biskupek, X. Zhang, Y. Zhang, D. Bresser, A.V. Krasheninnikov, U. Kaiser, In situ TEM investigation of the lithiation and delithiation process between graphene sheets in the presence of atomic defects, Advanced Functional Materials 34(41) (2024) 2406034. https://doi.org/10.1002/adfm.202406034.

13 Y. Yuan, K. Amine, J. Lu, R. Shahbazian-Yassar, Understanding materials challenges for rechargeable ion batteries with in situ transmission electron microscopy, Nature Communications 8(1) (2017) 15806. https://doi.org/10.1038/ncomms15806.

14 J.-L. Shi, D.-D. Xiao, M. Ge, X. Yu, Y. Chu, X. Huang, X.-D. Zhang, Y.-X. Yin, X.-Q. Yang, Y.-G. Guo, L. Gu, L.-J. Wan, High-capacity cathode material with high voltage for li-ion batteries, Advanced Materials 30(9) (2018) 1705575. https://doi.org/10.1002/adma.201705575.

15 S. Ahmed, M. Bianchini, A. Pokle, M.S. Munde, P. Hartmann, T. Brezesinski, A. Beyer, J. Janek, K. Volz, Visualization of light elements using 4D STEM: The layered-to-rock salt phase transition in LiNiO2 cathode material, Advanced Energy Materials 10(25) (2020) 2001026. https://doi.org/10.1002/aenm.202001026.

16 L. Xu, X. Han, J. Sung, Y. Hu, J. Wang, Q. Han, R. Gao, Y. Li, W. Xue, 4.5-V-Class safe lithium-ion batteries with silicon-majority-graphite anodes enabled by self-limiting interphase, Advanced Materials n/a(n/a) (2025) e15562. https://doi.org/10.1002/adma.202515562.

17 J. He, G. Melinte, M.S.D. Darma, W. Hua, C. Das, A. Schökel, M. Etter, A.-L. Hansen, L. Mereacre, U. Geckle, T. Bergfeldt, Z. Sun, M. Knapp, H. Ehrenberg, J. Maibach, Surface structure evolution and its impact on the electrochemical performances of aqueous-processed high-voltage spinel LiNi0.5Mn1.5O4 cathodes in lithium-ion batteries, Advanced Functional Materials 32(46) (2022) 2207937. https://doi.org/10.1002/adfm.202207937.

5

Transmission X-ray Microscopy for Multiphysics Imaging of Batteries

5.1 Fundamental Principles and Theories

Transmission X-ray Microscopy (TXM) combines X-ray transmission imaging with high-resolution microscopy. It leverages the differential X-ray absorption of sample components: when a collimated X-ray beam penetrates a thin specimen, regions with higher density or atomic number absorb more X-rays. The transmitted photons are detected and converted into a magnified image, enabling submicron visualization of internal structures without destructive sample preparation. Its core advantage lies in characterizing both morphology and elemental composition of materials like battery electrodes in their near-native states.

5.1.1 Theories and Progress

The advent and development of the TXM are rooted in the understanding of the nature of X-rays and their interactions with matter. Its theoretical foundation was established within classical physics and evolved alongside the field. This journey began with Röntgen's discovery in 1895, when a new form of radiation with remarkable penetrating power—X-rays—was observed.

It was not until 1912 that Laue's X-ray crystal diffraction experiment demonstrated that X-rays are electromagnetic waves with extremely short wavelengths of approximately 0.01–10 nm. This scale, comparable to atomic spacings, gives X-rays a natural advantage in probing internal microstructures. According to classical electromagnetic theory, X-rays exhibit wave phenomena such as refraction, interference, and diffraction. When X-rays traverse a sample, spatially varying electron density induces phase retardation of the wavefront, forming the physical basis of phase-contrast imaging.

However, wave-based descriptions alone are insufficient. Einstein's explanation of the photoelectric effect and the Compton scattering experiment revealed the particle nature of X-rays, which can be regarded as photons with energy $E = h\nu$. From this perspective, X-ray–matter interactions mainly include the photoelectric effect, Compton scattering, and coherent scattering.

The photoelectric effect involves complete photon absorption and underpins absorption contrast imaging, following the Lambert–Beer law. Compton scattering,

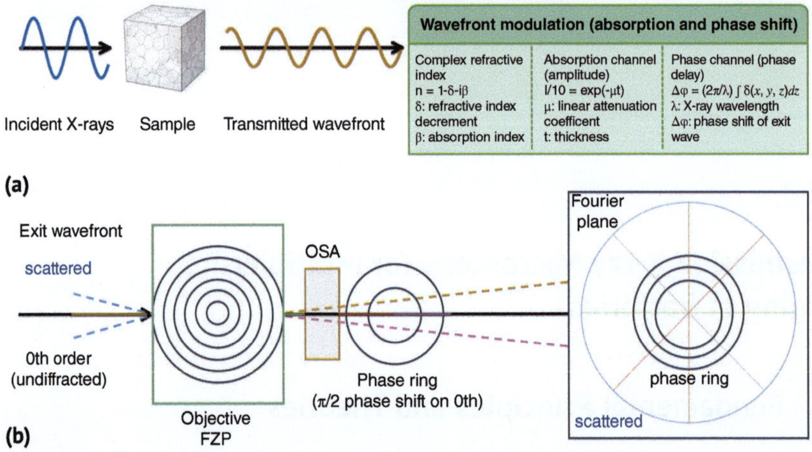

Figure 5.1 The working mechanism diagram of the TXM test.

an inelastic process, contributes mainly to imaging background. Coherent scattering is elastic and preserves photon energy, forming the basis of X-ray diffraction and wavefront coherence in phase-contrast imaging.

Therefore, contrast mechanisms in TXM directly reflect the wave–particle duality of X-rays: absorption contrast arises from particle-like interactions, while phase contrast originates from wave-like interference and diffraction. Overall, TXM imaging theory has evolved from absorption-based projection imaging toward high-resolution, high-contrast imaging that fully exploits X-ray wave–particle duality (Figure 5.1).

5.1.2 Phase Contrast and Zernike Phase Contrast Principle

Initial X-ray imaging relied on simple projection imaging based entirely on absorption contrast. While intuitive, this approach performs poorly for weakly absorbing specimens composed of light elements or with uniform thickness, where minimal absorption differences lead to extremely low image contrast. Phase contrast in TXM was developed to overcome this limitation. Its core principle is that, even when absorption is negligible, spatial variations in electron density cause differences in the real part of the refractive index, resulting in phase delays in the transmitted wavefront.

The Zernike phase contrast method introduces an annular phase-shifting plate at the back focal plane of the objective. This plate imparts a $\pi/2$ phase shift to the unscattered (zero-order) wave, altering its interference with scattered waves from cosine to sine dependence. For small phase variations, the image intensity becomes approximately proportional to the sample's phase projection, enabling linear conversion of phase information into intensity contrast.

In the TXM optical path, an annular illumination is formed by the condenser. After transmission through the sample, the objective lens generates a diffraction pattern at its back focal plane, where unscattered light concentrates at the annular region and scattered light distributes at higher spatial frequencies. The Zernike phase ring, precisely positioned at this location, selectively phase-shifts the unscattered wave. Interference between the phase-shifted unscattered wave and scattered waves at the image plane then produces a high-contrast phase image directly detectable by a camera.

Zernike phase contrast TXM derives its value from wavefront manipulation through interference and has become particularly important for imaging weakly absorbing materials. Its primary advantage is high-contrast imaging of light-element battery materials whose structures are nearly invisible in absorption contrast. By converting phase shifts into intensity variations, it reveals internal microstructures with rich detail, making it highly effective for studies of battery degradation mechanisms and interface evolution.

A second advantage is its ability to provide real time, direct imaging. Unlike phase-retrieval techniques requiring extensive postprocessing, Zernike phase contrast produces high-contrast images immediately at the detector, enabling in-situ observation of dynamic processes. Third, its spatial resolution is determined mainly by the Fresnel zone plate rather than the contrast mechanism itself, allowing high contrast to be combined with nanometer-scale resolution.

However, its interference-based principle also introduces inherent limitations. The most prominent are the "halo" and "shading" artifacts. Halo artifacts appear near sharp phase boundaries due to incomplete collection of high-frequency information, complicating edge interpretation. Shading artifacts arise from fabrication imperfections or misalignment of the phase ring, introducing nonuniform background contrast and reducing quantitative reliability.

Another limitation is the extremely stringent requirement on fabrication and alignment. The phase ring must coincide precisely with the annular illumination image, and its thickness must be controlled to generate an accurate $\pi/2$ phase shift. These demands greatly increase experimental complexity. In addition, although image intensity is approximately proportional to phase projection, accurate quantitative retrieval of the refractive index decrement δ remains difficult due to the influence of multiple coupled factors, often requiring additional scanning or phase-retrieval procedures. Finally, the method is best suited for relatively thin samples, as multiple scattering in thick specimens degrades image quality and invalidates the linear contrast assumption.

Despite these limitations, Zernike phase contrast remains the most widely used and successful phase-contrast method in TXM, enabling real time, high-contrast, and high-resolution nanoscale imaging.

5.1.3 X-ray Detection and Signal Acquisition System

The signal detection and collection system in TXM forms the critical bridge between X-ray–matter interactions and digital images, directly determining the system's ultimate performance.

1. Core functions and operational workflow
   The primary task of the TXM detection system is to convert invisible X-ray photon signals into digital electrical signals suitable for storage, processing, and

display. Incident X-ray photons first deposit energy in the detector's conversion layer, generating primary signals through interactions such as the photoelectric effect. These signals take the form of electron–hole pairs in direct detectors or visible/UV photons in indirect detectors.

The generated signals are then collected and amplified. In direct detection, charges are collected within pixel units by an electric field. In indirect detection, visible photons are coupled to photosensitive sensors via optical systems and converted into electronic charges. These charges are transformed into voltage signals, digitized by an analog-to-digital converter, and arranged into a two-dimensional image matrix according to pixel position. As each stage may introduce noise or distortion, the detection system is designed to maximize signal fidelity so that the final image accurately reflects the sample's intrinsic properties.

2. Principles of indirect and direct detection
TXM detectors are mainly divided into indirect and direct detection schemes, which differ fundamentally in working principle and performance.

Indirect detection employs a two-step conversion process. X-ray photons are first absorbed by a scintillator, producing visible photons, which are then transferred through an optical relay to a visible light sensor for signal readout. This approach benefits from mature visible light camera technology and enables a large field of view at relatively low cost. However, the spatial spreading of visible light within the scintillator causes image blur and limits resolution. Reducing scintillator thickness can improve resolution but at the cost of reduced X-ray absorption efficiency.

In direct detection, X-ray photons are absorbed directly in a pixelated semiconductor detector, generating electron–hole pairs that are collected by an internal electric field. This avoids scintillator-induced point spread, confining each photon event to a single pixel and enabling higher spatial resolution. Direct detectors also typically offer higher detective quantum efficiency and faster readout, making them suitable for high-speed imaging, though they are more expensive and usually provide a smaller field of view. The detector choice therefore involves trade-offs among resolution, speed, field of view, and cost.

3. Key performance parameters
Detector performance is defined by several interrelated parameters that collectively determine image quality.

Spatial resolution depends on the effective pixel size and the system point spread function. To satisfy Nyquist sampling, the detector pixel size must be at most half of the system's theoretical resolution.

Dynamic range describes the ratio between the maximum and minimum detectable signals in a single exposure and determines whether features in both thick and thin sample regions can be recorded without saturation or loss in noise.

Detective quantum efficiency (DQE) measures how efficiently a detector transfers signal-to-noise ratio from incident X-rays to the final image. A high DQE enables higher image quality at a given dose or, alternatively, lower radiation dose for the same image quality, which is essential for radiation-sensitive samples.

Frame rate determines acquisition speed and is critical for in-situ dynamic studies and fast tomography. Closely related is readout noise, an additive electronic noise that sets the lower detection limit for weak signals and often dominates image noise at low signal levels. Together, these parameters define a complex optimization space for detector selection.

4. Decisive impact of the detection system
The detection system actively defines the performance boundaries and application scope of TXM.

First, detector resolution and DQE set the effective information limits of the system, determining whether the theoretical performance of the X-ray source and objective can be practically realized.

Second, frame rate and dynamic range govern the feasibility and efficiency of experiments, particularly for capturing fast dynamic processes and resolving multiscale structures in a single exposure.

Third, detector DQE directly influences radiation dose requirements. High-DQE detectors enable high-resolution or long-term imaging of radiation-sensitive samples, expanding TXM applications in life and soft-matter sciences.

Finally, detector stability and radiation resistance are essential for long-term data reliability. Sustained exposure to intense synchrotron radiation can degrade detector performance, whereas robust detector design ensures stability, accuracy, and reproducibility.

In summary, the detection system is the cornerstone of TXM, jointly defining image quality in terms of resolution, signal-to-noise ratio, speed, and radiation dose.

5.2 Equipment and Analysis

A typical synchrotron radiation TXM beamline system primarily consists of an X-ray source, beamline optics, the main microscope assembly, detection and data acquisition systems, as well as beamline control systems and operational software (Figure 5.2).

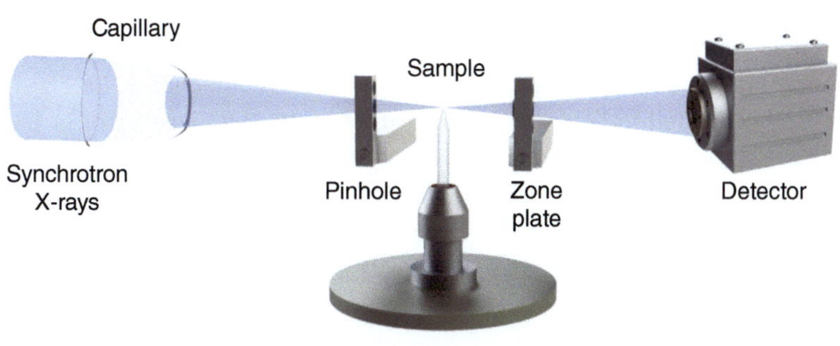

Figure 5.2 Schematic diagram of the synchrotron radiation TXM test device.

5.2.1 Typical Equipment

The source of a synchrotron radiation TXM beamline is an electron storage ring-a circular vacuum facility with a circumference of several hundred meters, where electrons are accelerated to nearly the speed of light (typically with energies of 3–5 GeV) and circulate continuously. When these high-energy electrons are deflected by bending magnets along the ring, they emit extremely bright, highly collimated, broadband electromagnetic radiation tangentially, in accordance with electrodynamic principles-this is synchrotron radiation. However, the core X-ray source for modern high-performance TXM beamlines is generally not the basic bending magnet, but rather more advanced periodic magnetic structures known as insertion devices, precisely installed in the straight sections of the storage ring. These primarily include undulators and wigglers. Taking the most commonly used undulator as an example, it consists of multiple pairs of periodically alternating north-south magnetic poles that force the electron beam to undergo numerous small-scale periodic transverse oscillations. Each oscillation emits X-rays, and due to the extremely high velocity of the electrons, the radiation wavefronts from different oscillation points coherently superimpose in the forward direction. This produces quasi-monochromatic light that is several orders of magnitude brighter than that from bending magnets. Key technical parameters include: the magnetic period length (which determines the characteristic energy), the number of periods (affecting spectral brightness and monochromaticity), and the K-value (a dimensionless parameter governing harmonic energy and bandwidth). It is the exceptional brightness, superior collimation, and high transverse coherence provided by these insertion devices that establish the indispensable physical foundation for TXM to achieve high-throughput, high-resolution, and high-contrast imaging.

Beamline serves as a precise bridge connecting the light source to the microscope main body. Its task is to "process" the raw, divergent, and broad-spectrum synchrotron radiation into a monochromatic, focused, and clean X-ray beam that meets the imaging requirements of TXM. It begins at the Front-End area, a critical region responsible for safely isolating the ultrahigh vacuum of the storage ring from the subsequent beamlines, and which contains beam defining apertures and safety interlocks. Afterwards, the beam enters the main optical components section. First is the Prefocusing Mirror(s), typically one or more curved reflective mirrors (such as ellipsoidal or cylindrical mirrors). It collects the divergent beam from a large solid angle via the principle of Total Reflection and performs initial focusing to increase the Photon Flux Density at subsequent optical components. Its Figure Error is typically required to be at the submicron level, with a Slope Error of less than 0.5 μrad. Next is the core monochromatizing component-the Double-Crystal Monochromator. It consists of two parallel crystals (commonly Silicon Si (111)). By precisely adjusting the relative angle (θ) between the two crystals and utilizing Bragg's Diffraction Law ($n\lambda = 2\ \mathrm{D}\ \sin\theta$), it selects a specific wavelength (energy) of X-rays with a narrow bandwidth from the continuous "white light." The typical Energy Resolution ($\Delta E/E$) for a Si (111) monochromator is in the order of 10^{-4}, meaning for 10 keV X-rays, the bandwidth is approximately 1 eV. Finally, there is the

Main Focusing Mirror System, comprised of a pair of vertically arranged elliptical reflecting mirrors (one for horizontal focusing, the other for vertical focusing). Using nanoprecision Piezoelectric Motors to control their orientation and bending radius, it ultimately focuses the monochromatized X-ray beam onto the sample. The Beam Spot Size can be reduced to below 1 μm, thereby injecting extremely high photon flux into the microscale sample area. This is key to achieving high signal-to-noise ratio and high spatial resolution imaging. All high-precision optical components operate within an ultrahigh vacuum or controlled atmosphere environment and are accompanied by a series of Slits, Beam Position Monitors, and Fluorescent Screens for beam positioning and diagnostics.

The microscope body is the component that performs the core imaging function, integrating precision mechanics, X-ray optics, and sample environment. The heart of its illumination system is the condenser, which functions to re-image the light source (or the image from the primary focusing mirror) onto the back focal plane of the objective lens, forming the (typically annular) illumination field required for TXM. The sample stage is the platform that supports and manipulates the specimen. It must be high-precision and multidegree-of-freedom, providing at least translations along the X, Y, and Z axes (with travel ranges up to several tens of millimeters and resolution better than 50 nm) and 360° continuous rotation (for tomography, with eccentricity better than 100 nm). It often incorporates a nanoprecision piezoelectric scanning stage for fine positioning. The core imaging element is the objective lens, which determines the theoretical resolution of the system. It is a circular diffraction lens comprised of concentric metal rings (typically gold) that alternate between transmitting (or phase-shifting) and opaque. Its outermost zone width, Δr, directly determines the diffraction-limited resolution (approximately equal to $1.22\Delta r$). Its focal length is related to the X-ray energy and the zone plate structure, exhibiting strong chromatic dispersion. This, in turn, demands that the illumination beam possesses a high degree of monochromaticity. To achieve Zernike phase contrast, an annular phase plate is precisely positioned at the back focal plane (Fourier plane) of the objective lens. Its ring is matched to the illumination aperture, and through specific material (e.g., nickel, gold) and thickness design, it introduces a precise $\pi/2$ phase shift and moderate attenuation to the zero-order diffracted light. The entire optical path (from the source to the sample and then to the detector) is typically maintained in a vacuum or helium-purged environment to minimize air absorption and scattering of soft X-rays.

The detection system is responsible for converting invisible X-ray images into quantifiable digital electrical signals. In zone-plate-based TXM, the detector is positioned at the image plane of the objective lens. As mentioned earlier, mainstream technologies are divided into indirect detection and direct detection. Indirect detection employs a scintillator to convert X-rays into visible light, which is then coupled to a sensor via relay optics. Its advantage is a large field of view, but its resolution is limited by the scintillator's point spread function. Direct detection, on the other hand, uses a pixel detector where X-ray photons directly generate charge within semiconductor pixels, avoiding resolution loss. This method offers high speed but has a relatively small field of view and is more expensive. The data acquisition

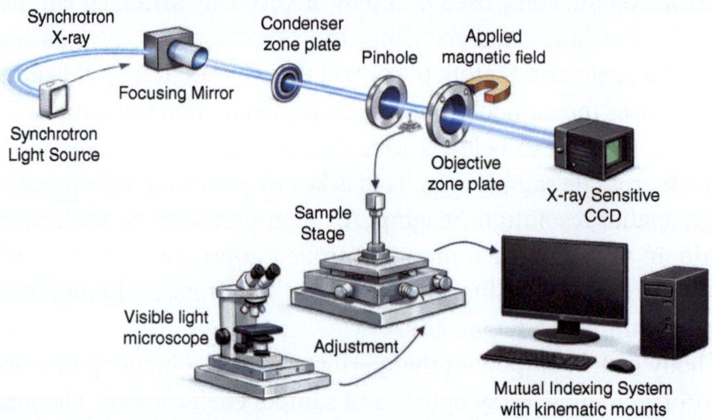

Figure 5.3 Beamline layout diagram of the SSRF TXM line-station.

system synchronously controls the rotation of the sample stage, the opening and closing of the shutter, and the exposure of the detector. It automatically collects thousands of projection images and transmits massive datasets (a single CT scan can reach several terabytes) via high-speed networks to computing clusters for storage and reconstruction.

The operation of the entire TXM beamline relies on a highly integrated and automated control system and software. Built on a distributed architecture, it utilizes input-output controllers and the Experimental Physics and Industrial Control System to achieve precise synchronization and real time monitoring of hundreds of hardware devices (such as motors, detectors, vacuum gauges, and temperature controllers). Users interact through a high-level graphical user interface, which encapsulates complex hardware commands into intuitive macro instructions. After data acquisition, specialized scientific software (e.g., Tomopy, TXM Wizard) performs pipeline processing, including projection alignment, phase retrieval, and 3D tomographic reconstruction (commonly using filtered back-projection or iterative algorithms). This ultimately generates analyzable 3D volumetric data, thereby completing the full cycle from physical measurement to scientific insight (Figure 5.3).

5.2.2 Acquisition Mode

TXM, as a powerful nanoscale characterization tool, achieves its functional diversity through several fundamental operational modes.

First is the most fundamental two-dimensional projection imaging mode. In this mode, the sample remains stationary, and the detector directly collects the two-dimensional transmission image of X-rays passing through the sample. This mode relies directly on the contrast generated by absorption and phase effects as X-rays traverse the sample. It offers fast imaging speed and can be used for rapid assessment of the sample's macroscopic morphology, uniformity of particle distribution, and precise localization of regions of interest for subsequent detailed measurements. It serves as the starting point for all TXM experiments and is the primary method for recording two-dimensional evolution processes during in-situ dynamic studies.

In lithium-ion battery research, conventional X-ray absorption contrast struggles to effectively resolve microstructures composed of low-atomic-number components such as carbon-based materials, polymer separators, and solid-state electrolytes. To address this limitation, Zernike phase-contrast imaging introduces an annular phase plate in the back focal plane of the objective lens, applying a $\pi/2$ phase shift to the direct (unscattered) transmitted wave. This enables interference with scattered waves modulated by the sample's refractive index gradient, thereby linearly converting phase retardation information into intensity contrast. This technique achieves breakthroughs in imaging critical features in battery materials, including interfacial phases, pore structures, and lithium deposition morphologies. It clearly visualizes the three-dimensional fiber network distribution in separators, grain-boundary structures in solid-state electrolytes, crack propagation in electrode materials, and dendritic lithium growth behavior at electrode-electrolyte interfaces. As such, it has become a core characterization method for analyzing failure mechanisms in battery materials and dynamically tracking interfacial evolution.

To obtain complete three-dimensional spatial information of electrode materials, the 3D nano-computed tomography mode must be employed. In this mode, the electrode sample is continuously rotated 180° around a vertical axis, while two-dimensional projection images are collected at hundreds to thousands of equally spaced angles. These projections are subsequently reconstructed into 3D volumetric data of components such as electrode particles, conductive agents, binders, and pores using computed tomography reconstruction algorithms (e.g., filtered back-projection or iterative reconstruction). This nondestructive "3D perspective" capability makes it an ultimate tool for investigating critical issues including pore connectivity (affecting electrolyte wetting and lithium-ion transport), 3D morphology of electrode-electrolyte interfaces, uniformity of active material distribution, crack formation and propagation in electrode particles during cycling, and lithium dendrite growth pathways. It enables precise quantification of key 3D parameters determining electrochemical performance-such as electrode porosity, specific surface area, tortuosity, and active material volume fraction-thereby providing direct evidence for optimizing electrode microstructure design and revealing battery failure mechanisms.

Furthermore, leveraging the unique advantage of tunable energy from synchrotron radiation sources, the X-ray spectroscopic imaging mode (particularly XANES-CT) has been realized (Figure 5.4). This mode involves scanning the X-ray energy across the absorption edge of specific elements in the sample and acquiring two-dimensional projections or completing full three-dimensional CT scans at each energy point. Due to the characteristic differences in X-ray absorption near-edge structure spectra of elements under different chemical states, by analyzing the absorption spectra of each voxel (three-dimensional pixel) in the image sequence, it is possible to reconstruct two-dimensional or even three-dimensional spatial distribution maps of the element in different valence states or coordination environments. This mode integrates morphological analysis with chemical-state identification,

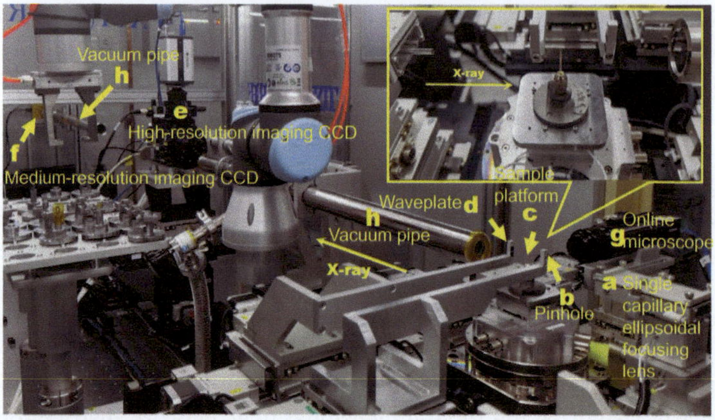

Figure 5.4 Beamline layout diagram of the SSRF TXM nano-CT line-station. Zhang et al. 2023/Springer Nature/CC BY 4.0

providing unparalleled microscopic insights into complex scientific issues involving changes in elemental chemical states, such as the valence distribution of transition metal ions in battery electrodes, the chemical states of active components in catalysts, and the speciation and migration of pollutants in environmental samples.

5.2.3 Qualitative Analysis

In the TXM characterization analysis, qualitative information regarding structure and properties is obtained through direct observation and interpretation of the acquired images, forming the basis for subsequent quantitative analysis. This process requires researchers to establish a profound correlation between two-dimensional grayscale images and the electrochemical behavior of electrode materials. When analyzing TXM images of electrode materials, the first step is morphological analysis of the active material particles: contrast variations in the images can directly assess the uniformity of secondary particle size distribution, the crystalline morphology of primary particles and the mixing homogeneity of conductive agents with the active material. For the porous structure of the electrode, the three-dimensional network formed by the binder PVDF and pores can be clearly distinguished, enabling the evaluation of ion transport channel connectivity (open or closed pores) and pore wall roughness. These structural characteristics directly influence electrolyte wettability and lithium-ion migration rates. In composite electrode observations, the distribution states of the three phases-active material, conductive agent, and binder-can be identified to assess whether there are "dead zones" caused by uneven conductive agent distribution or increased interfacial impedance due to binder aggregation. This qualitative analysis of microstructure provides direct visual evidence for understanding the degradation mechanisms of electrochemical performance, such as capacity fade, rate capability decline, and interfacial impedance growth, serving as a critical basis for optimizing material design and improving electrode processes.

Based on the difference in energy, it can be divided into transmission hard X-ray Microscopy (>5 keV) and transmission soft X-ray Microscopy (<2 keV). Through chemical contrast imaging with TXM, researchers can precisely resolve the spatial configuration relationships of key components in electrode materials. For the cathode materials, a hard X-ray imaging mode is usually employed for analysis. It enables direct visualization of whether the valence state distribution of active metal elements (e.g., nickel, manganese, cobalt) in cathode materials is homogeneous or exhibits local elemental segregation. It allows researchers to directly "see" the migration pathways of dissolved transition metals in the electrolyte and their deposition sites on the anode surface. For the anode materials and solid electrolyte interphase (SEI), a soft X-ray imaging mode is usually employed for analysis. It clearly distinguishes the distribution of lithium-silicon phase regions formed in silicon-based anodes during charge-discharge cycles, as well as the contact conditions between carbon conductive agents and active particles. By analyzing post-cycled electrodes, it becomes possible to trace whether crack propagation paths extend along grain boundaries or penetrate through the active particles themselves, and to determine the nucleation sites and growth morphology of lithium dendrites. This spatial correlation analysis can also be applied to study the distribution characteristics of electrolyte additives on electrode surfaces, observe the composition and thick distribution of the SEI, and analyze spatial variations in lithium-ion concentration gradients under different states of charge. These detailed spatial distribution insights into key components provide direct microscopic evidence for understanding battery capacity degradation mechanisms, interfacial side reaction pathways, and lithium-ion transport kinetics. Such information holds significant guiding value for optimizing electrode structure design and electrolyte formulations.

For TXM image sequences acquired during electrochemical in-situ or time-lapse experiments, the core of qualitative analysis lies in the accurate tracking of dynamic processes. This involves comparing image evolution of the same region under different cycle numbers, states of charge, or rate conditions to qualitatively resolve the microstructural evolution of electrode materials and interfaces. In lithium-ion battery research, this technique can directly capture dynamic processes of key degradation mechanisms: it enables visual observation of the evolution of the SEI on anode surfaces, transitioning from uniform deposition to localized thickening, cracking, and subsequent repair during cycling; it also reveals the entire process of lithium dendrite formation under low coulombic efficiency conditions-from nucleation and whisker-like growth to the eventual development of moss-like or dendritic deposits-while clearly identifying preferred nucleation sites (e.g., at defects on copper current collectors or pore edges of separators) and morphological evolution characteristics. For high-nickel cathode materials, it allows real time monitoring of mechanical failure processes in secondary particles under deep delithiation states, including microcrack initiation at grain boundaries, propagation along primary-particle interfaces, and eventual particle pulverization. In silicon-based anode studies, it tracks anisotropic volume expansion (exceeding 300%) and contraction behaviors of silicon particles during alloying/dealloying processes, as well as resulting particle fracture and conductive network disruption. In all-solid-state battery systems, it enables observation of pore formation, growth, and interfacial contact loss at the lithium metal/solid electrolyte interface, along with dendrite penetration pathways along electrolyte grain boundaries. This "visualization" capability of structural evolution in key battery components (cathode, anode, electrolyte interfaces) provides direct evidence for understanding performance degradation mechanisms such as capacity fade and internal resistance growth, offering irreplaceable guidance for optimizing electrode structure design, interface regulation strategies, and electrolyte formulations.

Furthermore, the identification of defects and abnormal structures is also a critical component of qualitative analysis. High-resolution TXM images can clearly reveal various inherent defects in electrode materials generated during preparation or cycling: within the material, microscopic pores and cracks caused by inadequate sintering or stress from lithium-ion intercalation/deintercalation, inclusions of inactive impurities, and local strain field contrast induced by lattice distortion can be observed; on the material surface, scratches from the electrode coating process, coating inhomogeneity, and foreign contaminant particles can be identified. Additionally, in high-throughput characterization, care must be taken to distinguish artificial artifacts such as bubbles formed by strong X-ray irradiation-induced electrolyte decomposition or material damage, as well as electrode surface blistering. Accurate identification of these microstructural features is essential for assessing the structural integrity of active material particles, evaluating the stability of conductive networks and interfaces, and is directly linked to battery capacity degradation, internal resistance increase, and safety performance. This analysis provides direct visual feedback and improvement guidance for optimizing material synthesis routes (e.g., solid-state sintering, coprecipitation methods) and electrode manufacturing processes (e.g., slurry dispersion, calendaring density control).

It must be emphasized that all such qualitative interpretations should be grounded in a thorough understanding of the imaging conditions (whether absorption contrast or phase contrast) and potential imaging artifacts (such as halo effects in phase contrast, dead pixels in the detector, projection overlap, etc.), to avoid misinterpreting artifacts as authentic structures. In summary, through keen observation and professional interpretation of TXM images, researchers can swiftly obtain rich qualitative insights into the microstructure of samples-including morphology, distribution, evolution, and defects-without the need for complex mathematical processing. This enables the formation of preliminary yet crucial scientific understanding regarding material properties, performance, and the underlying physicochemical mechanisms, thereby guiding subsequent precise quantification, model construction, and in-depth mechanistic investigations.

5.2.4 Quantitative Analysis

From the 3D structural data of lithium-ion battery electrodes obtained through TXM, extracting precise quantitative parameters to achieve the transition from qualitative observation to quantitative analysis constitutes a systematic data processing and information mining workflow. The core of this process lies in converting the 3D image matrix containing grayscale information of different components into physical quantities that characterize key electrode properties. This procedure begins with critical data preprocessing: first, through image segmentation techniques such as global/local thresholding, region growing, or advanced machine learning algorithms, different components within the electrode's 3D structure (including active material particles, conductive additive networks, binder phases, pore structures, and microcracks) are accurately separated and assigned distinct labels. The accuracy of this step directly determines the reliability of subsequent quantitative analysis. This is followed by data optimization processing, where advanced algorithms like anisotropic diffusion filters are applied to suppress noise while preserving critical interfacial structural integrity. Additionally, morphological repair is performed on anomalous regions caused by segmentation errors or image defects, ultimately yielding binarized or labeled 3D reconstruction models that genuinely reflect the electrode's microstructure. This establishes a solid foundation for the precise quantification of key parameters such as the electrode's pore network, active material distribution, and interfacial contact.

In the quantitative analysis of lithium-ion battery electrode materials, key geometric parameters can be extracted from three-dimensional volumetric data: firstly, the volume fraction of each phase is calculated based on image segmentation, enabling precise quantification of electrode porosity, active material mass fraction, and conductive additive distribution, which is crucial for evaluating electrode compaction density and theoretical capacity. Secondly, by approximating individual active particles or pores as spheres of equivalent volume to compute their equivalent diameter (D50), and statistically analyzing the particle/pore-size distribution across the entire electrode region, the uniformity of active material particles, the fragmentation degree of secondary particles, and the connectivity of the pore network can be effectively assessed. These parameters directly correlate with the lithium-ion transport efficiency and rate performance of the electrode. Furthermore, the specific surface area, obtained by calculating the ratio of the total solid-liquid interfacial area to the sample volume, significantly influences the electrode reaction interface size, the kinetics of SEI formation, and high-rate charge/discharge performance. Additionally, analyzing the aspect ratio of active particles can reveal their degree of anisotropy, while statistically evaluating the orientation distribution of graphite anode layers or pores within solid-state electrolytes allows for quantification of lithium-ion transport path tortuosity, providing critical data support for optimizing electrode microstructure.

At a more profound level of topological parameters, calculating the Euler characteristic enables precise quantification of the connectivity within the electrode's internal pore network. This determines whether lithium-ion transport channels form isolated closed pores or an efficiently interconnected three-dimensional continuous network, which is crucial for predicting electrolyte wetting efficiency and lithium-ion migration capability. By employing shortest-path algorithms or field simulations, the electrode's tortuosity can be computed. This parameter is defined as the ratio of the actual average path length traversed by lithium ions moving from one side of the electrode to the other, relative to the electrode's macroscopic thickness. It directly determines the battery's rate performance-higher tortuosity correlates with greater ionic transport resistance, more severe polarization during high-rate charging/discharging, and lower capacity retention. Simultaneously, it profoundly influences the utilization efficiency of active materials and the concentration gradient distribution during electrochemical cycling.

For dynamic process studies (e.g., battery cycling, material creep), by comparing binarized 3D data from corresponding regions under different states (before/after cycling, before/after loading), it is possible to precisely quantify the growth rate of total crack volume, the mass loss of active materials, the evolution curve of porosity, and changes in interfacial area. This enables direct correlation of microstructural evolution with macroscopic performance degradation. Ultimately, by establishing structure–property relationships between these accurately extracted microstructural parameters (such as tortuosity, specific surface area, crack volume) and the material's macroscopic properties (e.g., ionic conductivity, catalytic activity, mechanical strength, capacity retention), key structural factors affecting performance can be revealed. This provides precise, quantitative guidance and theoretical foundations for targeted material design and optimization, thereby completing the scientific closed loop from observation to measurement, and from cognition to prediction.

5.2.5 Beamline Introduction

Taking the National Synchrotron Radiation Laboratory BL07W soft X-ray micro-imaging beamline as a prime example, this beamline serves as a critical platform for in-situ comprehensive characterization at the nanoscale in fields such as materials science, earth science, environmental science, and life sciences in China (Figure 5.5). The BL07W at NSRL is devoted to cryo nano-tomography in the energy range of 284–530 eV. An ellipsoidal capillary used as condense to focus monochromatic light onto the sample. Two Ni zone plate (ZP) lenses made by Zeiss

Figure 5.5 Schematic diagram of NSRL BL07W line-station. X -ray Imaging Laboratory/ https://mems.ustc.edu.cn/About-44.html/last accessed on Feb 16, 2026

with 40 nm and 25 nm outer most zone widths, respectively, are available, giving spatial resolution in 2D of down to 40 nm and 30 nm, respectively. Unlike the soft X-ray microimaging of BL07W, the TXM nano-CT line-station of SSRF is based on hard X-ray microimaging. The picture of the line station is shown in Figure 5.3 and Figure 5.4. The beamline employs an undulator as the insertion device light source, delivering high-flux and highly collimated X-ray beams. Through a precision optical system-including a high-precision double-crystal monochromator, a Kirkpatrick-Baez mirror focusing system, and high-numerical-aperture Fresnel zone plate objectives-along with advanced Zernike phase contrast techniques, it ensures outstanding imaging performance within the X-ray energy range of 5–14 keV. The two-dimensional imaging spatial resolution is better than 30 nm, and the three-dimensional tomography resolution exceeds 50 nm. Moreover, it achieves ultrahigh data acquisition efficiency: capturing a single two-dimensional projection image takes only 0.1–10 seconds, while completing a full nano-computed tomography dataset (typically comprising over 1,000 projections) requires merely 10 minutes to several hours.

The distinguishing features and advantages of SSRF TXM beamline lie in its highly integrated multimodal imaging capabilities. Users can seamlessly switch between various imaging modes on the same experimental platform, including absorption contrast imaging, Zernike phase contrast imaging, nano-CT, and X-ray absorption spectroscopy imaging (XANES-CT). This allows for the acquisition of high-resolution three-dimensional morphological information while simultaneously resolving the chemical valence states and spatial distribution of specific elements. Particularly notable is its powerful in-situ characterization capability. The beamline is specifically designed with modularized in-situ sample environment integration interfaces, compatible with various in-situ sample chambers such as electrochemical cells, high-temperature heating stages, mechanical tensile/compression devices, and gas/liquid environmental cells. This enables researchers to observe dynamic phenomena in real time, including the initiation and evolution of cracks in electrode particles during the charge-discharge processes of lithium-ion batteries, structural reconstruction of catalytic materials under reactive conditions, phase transitions of alloy materials at high temperatures, and micro-crack propagation in geological materials under stress.

In the field of energy materials science, the NSRL BL07W beamline has achieved groundbreaking research outcomes addressing key scientific challenges in lithium-ion batteries, leveraging its high-resolution phase-contrast CT and soft X-ray microimaging (Figure 5.6). For the SSRF TXM beamline, it has accomplished the first four-dimensional spatiotemporal visualization of microcrack initiation and propagation pathways along primary-particle boundaries in high-nickel cathode materials, revealing the synergistic degradation mechanism between intergranular cracks and interfacial side reactions during cycling. Through nanoscale elemental valence state mapping, it precisely tracked the spatiotemporal evolution of lattice oxygen escape and transition metal migration in lithium-rich manganese-based cathode materials during the first charging cycle. In silicon-based anode studies, it successfully quantified the volume expansion coefficients and stress distribution in amorphous-crystalline transition regions during alloying/dealloying processes, providing direct

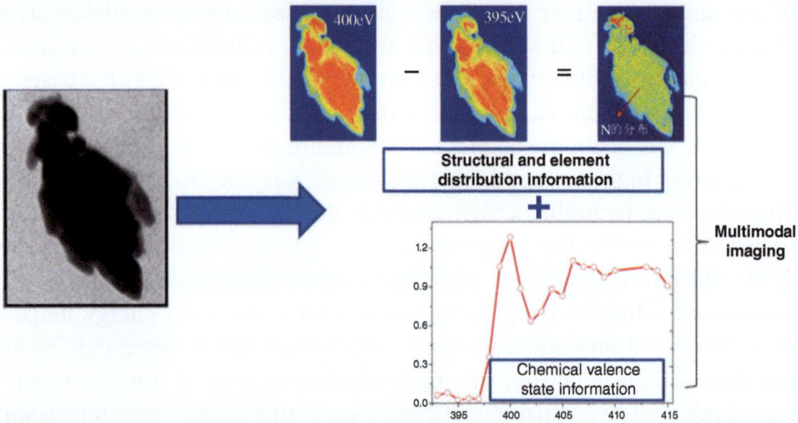

Figure 5.6 Schematic diagram of soft X-ray spectroscopy microscopy imaging technology of NSRL BL07W. X -ray Imaging Laboratory/https://mems.ustc.edu.cn/ProDetail-95.html/last accessed on Feb 16, 2026

data support for buffer layer design. For all-solid-state battery systems, it initially observed lithium dendrite penetration pathways along grain boundaries of solid electrolytes and interfacial ion transport bottlenecks, clarifying the regulatory role of interface modification on lithium metal deposition behavior. By integrating electrochemical workstations with specialized battery cells, this platform enables in-situ observation of critical dynamic processes-such as the evolution of cathode/SEI (CEI/SEI) and morphological transitions of lithium deposition-under realistic charge/discharge conditions, offering indispensable nanoscale insights into resolving the core conflict between energy density and cycle life in lithium-ion batteries.

In summary, the NSRL BL07W and SSRF BL18B and SSRF TXM beamlines have become indispensable national-level platform, supporting cutting-edge fundamental scientific research and addressing key technological challenges, thanks to its high-resolution, high-throughput capability, multimode integration, and powerful in-situ dynamic imaging performance.

5.3 Introduction to Emerging TXM Technology

The integration of TXM with electrochemical testing to achieve real time, in-situ observation of nano structural evolution in energy devices such as batteries during operation is a systematic engineering endeavor (Figure 5.7). This is accomplished through the careful design of specialized in-situ electrochemical cells, optimization of synchrotron radiation imaging parameters, and establishment of rigorous data acquisition and analysis protocols. The core of this approach lies in constructing a miniaturized reaction environment that enables precise control of electrochemical states while meeting the requirements for high-resolution X-ray imaging. The dedicated in-situ electrochemical cell must ensure good electrochemical performance while satisfying the geometric constraints for X-ray transmission.

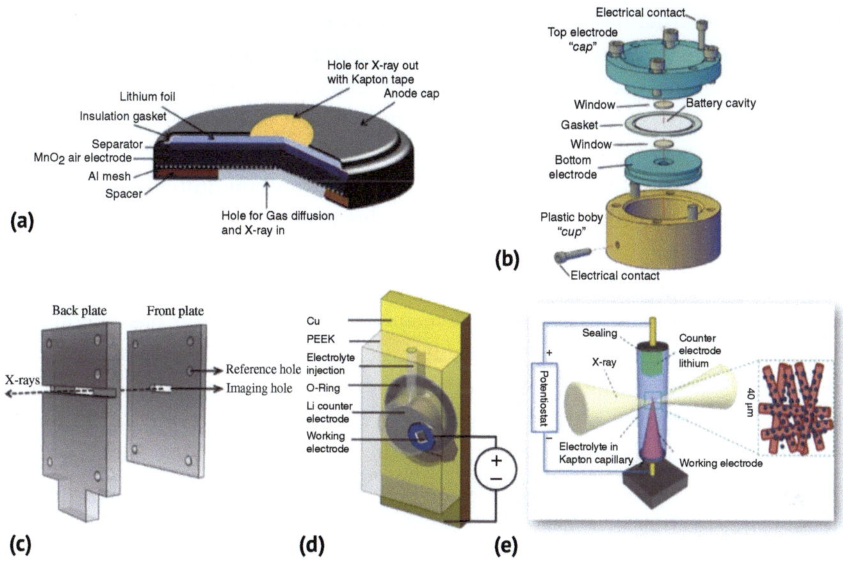

Figure 5.7 Schematic diagram of in-situ synchrotron radiation TXM [1] / with permission of John Wiley & Sons.

5.3.1 In-situ Electrochemical TXM

Typically, a three-electrode system (working electrode, counter electrode, and reference electrode) is employed to guarantee the accuracy of electrochemical measurements. The cell is sealed with X-ray-transparent window materials (such as high-purity beryllium sheets, ultrathin Kapton films, or silicon nitride membranes) to encapsulate the electrolyte and electrodes. A balance must be struck between window thickness and mechanical strength to optimize X-ray transmission and cell pressure tolerance. Electrode preparation also requires special consideration: the material under study (e.g., cathode or anode active materials for lithium-ion batteries) is typically coated onto conductive yet X-ray-transparent substrates such as carbon fibers or metal micro-grids to serve as the working electrode. This, along with the counter electrode (e.g., lithium metal foil or corresponding materials) and reference electrode, is precisely assembled into a cavity with dimensions on the millimeter or even micrometer scale. The interior of the cell must ensure sufficient electrolyte wetting and unobstructed ionic conduction pathways, while the entire assembly must be compatible with the geometric constraints and high-precision rotation (for tomography) of the TXM sample stage.

During the experiment, real time synchronized data acquisition is key to achieving dynamic observation. The TXM system (typically employing a high-flux and rapid synchrotron radiation source) continuously collects two-dimensional projections or fast tomography data of the sample at a preset temporal resolution (ranging from several seconds per frame to multiple frames per second). Simultaneously, the electrochemical workstation is strictly synchronized with the TXM control system, recording the applied voltage/current and the measured current/voltage response in real time, and precisely aligning timestamps with each X-ray image. This ensures that each microstructural image can be directly linked to a specific electrochemical state (such as a particular state of charge, charge/discharge cycle number, or overpotential). Such spatiotemporal correlation serves as the cornerstone for elucidating structure–property relationships. Through this coupled technique, researchers can directly observe numerous critical dynamic failure mechanisms. For instance, in lithium-ion battery studies, one can observe in real time the initiation and propagation of cracks in cathode material particles due to lattice volume changes during repeated lithium-ion intercalation and deintercalation, the loss of contact between active particles and conductive agents, as well as thickness fluctuations at the electrode level. In more hazardous lithium metal battery studies, it is possible to visually track the uneven nucleation of lithium dendrites on the anode surface, their subsequent growth into mossy or dendritic structures, and even the entire process of piercing the separator leading to short circuits, along with observing the formation and accumulation of "dead lithium." Additionally, this approach enables monitoring of unstable reactions at the electrode-electrolyte interface in solid-state batteries, the generation of interfacial pores, and the penetration behavior of lithium metal within the solid electrolyte.

To address the inherent challenges of this technique-such as radiolysis of organic electrolytes by X-rays, trade-offs between temporal and spatial resolution, and fluctuations in image contrast under complex operating conditions-it is generally necessary to adopt low-dose imaging techniques, optimize electrolyte composition, develop faster detectors, and apply advanced image processing algorithms. Ultimately, by analyzing these in-situ acquired image sequences, researchers can not only qualitatively describe the evolution pathways of microstructures, but also quantitatively extract parameters such as crack volume growth rate, fractal dimension of dendrite morphology, and interfacial contact loss area. This enables the establishment of quantitative structure–property relationships between electrode material structural evolution and electrochemical performance-such as capacity decay, increased internal resistance, and end-of-life behavior-from the nanoscale to the microscale. These insights provide unprecedented visual understanding and scientific basis for elucidating failure mechanisms in energy storage devices and guiding the design of next-generation high-performance batteries.

5.3.2 In-situ TXM with Coupled Thermal and Mechanical Fields

1. In-situ thermal TXM

   Thermal in-situ TXM enables researchers to observe the dynamic evolution of material microstructures from room temperature to over 1000 °C in real time under programmed temperature control by integrating micro-heating stages or laser heating devices. The core of this technology lies in constructing sample chambers using high-temperature-resistant materials (such as quartz capillaries, ceramics, or special alloys), and achieving precise temperature control with fast response through resistive or inductive heating, while ensuring that thermal expansion does not compromise the positioning accuracy of the sample relative to the X-ray beam. This technique is widely applied across multiple fields of materials science: in alloy studies, it directly visualizes dendritic growth during solidification, nucleation and growth of new phases during solid-state phase transformations, and grain-boundary migration at high temperatures; in ceramics and sintering research, it quantitatively tracks the formation and growth of sintering necks and the kinetics of pore shrinkage and coarsening, providing direct evidence for optimizing sintering processes; in catalysis studies, it observes nanoparticle migration and agglomeration under reactive atmospheres as well as the structural stability of support materials, thereby linking these to activity degradation mechanisms; in lithium-ion batteries, it monitors the interfacial separation between electrode materials and electrolytes during thermal cycling and the degradation of electrode microstructures. Thermal in-situ TXM offers unique dynamic visual evidence for understanding phase diagrams, reaction kinetics, and failure mechanisms of materials in high-temperature environments.

2. In-situ mechanical TXM
   In-situ mechanical TXM integrates miniaturized mechanical testing devices (such as tensile stages, compression stages, nano indenters, or microcantilevers) to perform high-resolution three-dimensional imaging of a material's internal structure while applying controlled mechanical loads. Its primary application value lies in establishing quantitative relationships between the microstructural evolution of electrode materials and their electrochemical performance. It enables direct observation of crack initiation (at grain boundaries/defects), propagation pathways, and particle pulverization in materials like silicon anodes and high-nickel cathodes during lithium-ion insertion/extraction processes. Furthermore, it reveals interfacial contact failure between active materials and conductive agents/binders, as well as the delamination process between electrode coatings and current collectors. The technique can simulate pore compression in separators and dendrite penetration behavior under stack pressure, and characterize the contact evolution and crack propagation at lithium metal/electrolyte interfaces in solid-state batteries. In-situ mechanical TXM provides a dynamic observational basis for deciphering battery mechanical failure mechanisms, optimizing electrode structural design, and enhancing safety.
3. Multiphysics-coupled In-Situ TXM
   The frontier of current in-situ TXM technology is advancing toward multi-physical field coupling. By simultaneously applying multiple external fields in a single experiment, it accurately simulates the multistress coupling environment faced by lithium-ion batteries under real service conditions. Typical coupling modes include electro-thermal coupling, mechanical-electrochemical coupling, and thermal-mechanical-electrochemical triple-field coupling. Electro-thermal coupling: performing charge-discharge cycles under precise temperature control to investigate the lattice oxygen release and phase-transition behavior of high-nickel cathode materials at elevated temperatures, the particle cracking and SEI film evolution dynamics of silicon-based anodes under thermal stress, and the decomposition process and thermal

runaway propagation paths of electrolytes under overheating conditions. Mechanical-electrochemical coupling: applying controlled static pressure or dynamic stress to electrodes during charge-discharge cycling to study the suppression mechanism of lithium dendrite growth in solid-state batteries under stack pressure, the strain distribution and plastic deformation behavior of graphite anodes during lithiation, and the electrode crack formation and conductive network degradation in flexible energy storage devices under repeated bending. Thermal-mechanical-electrochemical triple-field coupling: conducting electrochemical testing while controlling temperature and mechanical constraints to simulate the synergistic failure mechanisms of battery modules under real operating conditions, such as investigating the interfacial delamination of electrodes under combined high temperature and pressure, or the nucleation behavior of lithium plating under coupled temperature and stress fields. These multi-physical-field in-situ characterization techniques provide unprecedented insights into the failure mechanisms of lithium-ion batteries under coupled multi-factor interactions, significantly enhancing the optimization accuracy of battery safety, cycle life, and fast-charging performance.

These Multiphysics in-situ experiments have significantly enhanced the simulation and predictive capability of TXM technology. By replicating complex real-world operating conditions in a single experiment, they can reveal synergistic degradation mechanisms that cannot be induced by a single external field, thereby providing the most realistic testing and screening platform for designing and validating new materials for service under extreme conditions. In summary, through coupling thermal, mechanical, and other physical fields, in-situ TXM has evolved from a powerful static observation tool into a comprehensive research platform capable of dynamically reproducing and quantifying the structural responses and degradation behaviors of materials under real service environments. This has greatly expanded the frontiers of knowledge in materials science, chemistry, electrochemistry, and related fields.

5.3.3 Coupled with X-ray Fluorescence (XRF)

The integration of TXM and XRF essentially represents an in-depth combination of morphological imaging and elemental analysis techniques, achieving a leap in information dimensions. The core principle of TXM lies in utilizing transmitted X-rays that pass through the sample to generate high-resolution images reflecting the internal three-dimensional nanostructures (such as pores, particles, interfaces, cracks) of the sample via absorption contrast or phase contrast mechanisms. Its strength lies in providing precise geometric morphology, size distribution, and spatial arrangement information. However, it has limited capability in identifying elements, particularly being insensitive to trace or low-concentration elements. In contrast, XRF is a technique for analyzing elemental composition. Its principle involves high-energy X-rays exciting the sample, causing the ejection of inner-shell electrons from atoms, with outer-shell electrons transitioning to fill the vacancies and releasing characteristic-energy XRF. By detecting these characteristic X-rays, qualitative and quantitative analysis of elements in the sample can be performed. It exhibits extremely high sensitivity to trace elements (down to the ppm level), but its traditional spatial resolution is relatively low. Integrating the two, simultaneously collecting XRF signals while performing high-resolution structural imaging with TXM enables "seeing the structure clearly while knowing what elements it consists of." This approach is particularly valuable for addressing key scientific questions such as "where specific trace elements are located within a particular structural feature."

The TXM-XRF combined technique demonstrates irreplaceable value in mechanistic research and failure analysis of lithium-ion batteries. In cathode material development, it can not only clearly reveal the morphology, size distribution of primary particles within secondary spheres, and the three-dimensional pore network to ensure lithium-ion transport, but also precisely visualize the distribution uniformity of trace doping elements (such as key elements like cobalt, nickel, and manganese) inside and on the surface of particles. This helps determine whether elemental segregation occurs or if enrichment exists at grain boundaries, which directly relates to the structural stability and rate performance of the material, providing direct evidence for optimizing synthesis processes. In battery impurity analysis and source tracing, for contaminants on separators or electrode sheets, the combined technique can locate and qualitatively analyze trace metal impurities (such as iron, copper, and zinc) while revealing their microscopic morphology. It accurately determines whether these impurities introduced during production exist as isolated particles or have diffused to the surface of active materials. This is crucial for tracing contamination sources and evaluating their impact on battery self-discharge and cycling life. In advanced material design, it can reveal the true occurrence state of trace doping elements (such as aluminum and magnesium used for structural stabilization) within the cathode material lattice-whether they are successfully incorporated into the lattice or precipitate as inactive nano-oxides on the particle surface. This provides critical guidance for accurately understanding doping modification mechanisms. In interface and side reaction studies, this nondestructive technique can investigate the composition and distribution of the SEI film on cycled electrode surfaces. For example, it can observe whether dissolved transition metals (such as manganese) deposit on the anode surface and their spatial distribution or track the enrichment behavior of electrolyte additives (such as phosphorus or fluorine containing compounds) at specific interfaces. Simultaneously, it can monitor whether current collectors (such as copper) dissolve and migrate to the counter electrode under abuse conditions like overcharge or overdischarge, thereby revealing complex cross-contamination mechanisms. In battery failure analysis, it can precisely correlate microscopic defects (such as cracks in electrode particles or lithium dendrites on separator surfaces) with the co-localization of specific elements (such as fluorine and oxygen from electrolyte decomposition, or copper from corrosion). This enables accurate identification of the root causes of increased internal resistance, short circuits, or capacity degradation.

The core advantage of TXM-XRF integration lies in its ability to enable label-free, in-situ correlation analysis of structure and composition at the nanoscale.

It perfectly combines the "high-resolution structural imaging" of TXM with the "high-sensitivity elemental detection" of XRF, overcoming the limitations of individual techniques. However, this approach also faces several challenges. For instance, the spatial resolution of XRF is limited by the size of the X-ray probe beam spot, typically falling short of the highest resolution achievable with TXM. When performing three-dimensional XRF-CT, data acquisition and reconstruction become more complex and time-consuming. Additionally, quantitative analysis of XRF signals must account for factors such as matrix absorption effects and fluorescence yields. Nevertheless, with ongoing advancements in synchrotron radiation source brightness, improvements in detection efficiency, and progress in reconstruction algorithms, TXM-XRF integrated technology is emerging as an indispensable, cutting-edge characterization platform for comprehensively interpreting the relationship between material structure and composition at the nanoscale.

5.3.4 Coupled with X-ray Diffraction (XRD)

The combined technique of TXM and XRD is a powerful platform that correlates high-resolution morphological imaging in real space with crystallographic information in reciprocal space, enabling a leap from "seeing the structure" to "understanding the crystallographic essence." Its fundamental principle lies in the simultaneous acquisition of two complementary signals from the sample: the TXM channel records real space projections or 3D tomograms formed by the absorption and phase modulation of X-rays passing through the sample, with contrast primarily reflecting the density, composition, and morphological distribution of the sample. In contrast, the XRD channel collects X-rays diffracted by lattice planes within the sample using a 2D area detector (such as a flat-panel detector or pixel detector) placed behind the sample, generating Debye rings or diffraction spots. These diffraction patterns carry rich crystallographic information, including phase composition (identified by the positions of diffraction rings), grain orientation (determined by the azimuth of diffraction spots), lattice strain (precisely calculated from radial shifts of diffraction rings), and grain size (evaluated based on the broadening of diffraction rings). A typical implementation of this combined technique is scanning 3D XRD (S3XRD) or differential X-ray microscopy, which involves scanning the sample point-by-point or line-by-line using a focused X-ray nanoprobe. At each scanning point, a transmission image (for TXM reconstruction) and a corresponding 2D diffraction pattern are simultaneously recorded. By analyzing the diffraction pattern at each location (voxel), the crystal orientation matrix and full strain tensor at every point within the sample's 3D space can be reconstructed, thereby generating 3D orientation maps (i.e., 3D EBSD) and 3D strain field distributions.

This precisely correlated multimodal analysis capability at the nanoscale provides unprecedented insights into deciphering the failure mechanisms of key materials in lithium-ion batteries. In failure analysis of electrode materials, researchers can not only directly observe the microcrack networks formed in NMC ternary cathode secondary particles after cycling through TXM 3D imaging, but also immediately determine whether these cracks propagate along the grain boundaries of primary crystallites or penetrate directly through the grain interiors via correlated XRD data. Simultaneously, quantitative measurements of the local lattice strain distribution at crack tip regions offer critical experimental evidence for understanding how repeated lithium insertion/extraction-induced lattice volume changes and stress accumulation lead to mechanical failure. In studies of composite electrode systems, this technique can simultaneously visualize the three-dimensional distribution of conductive agents (such as Super P, carbon nanotubes-CNTs) and binders among active material particles, while XRD analysis reveals lattice distortions and residual stresses in active material particles resulting from interfacial interactions around carbon coating layers or dopant elements. This is crucial for optimizing electrode microstructure and improving interfacial ion/electron transport. In investigating structural evolution of electrode materials, it enables real time tracking of grain slippage, orientation rearrangement, and stress redistribution among polycrystalline particles caused by anisotropic volume changes during lithiation/delithiation of graphite anodes, intuitively revealing the heterogeneity of lithium intercalation reactions and its correlation with capacity fading. In phase transition mechanism studies, it allows observation of the moving two-phase interface in lithium iron phosphate (LiFePO4) while precisely determining the crystallographic orientation relationship between the olivine phase and iron phosphate phase, and quantitatively measuring the overall volume strain induced by anisotropic lattice parameter changes during phase transition, providing key data support for understanding the phase transition kinetics and volume stability of electrode materials.

In summary, the combination of TXM and XRD perfectly unifies the microscopic morphology, three-dimensional crystal structure, and local stress/strain fields at the nanoscale. This integration provides an unprecedented, comprehensive, and quantitative microcrystallographic perspective for understanding the processing-structure–property relationships of materials, thereby significantly advancing cutting-edge research in fields such as materials science, mechanical engineering, and condensed matter physics.

5.4 Application Cases

5.4.1 Two-dimensional TXM (2D TXM) Application Cases in Batteries

Two-dimensional TXM (2D TXM) occupies a distinct niche in battery characterization because it bridges the gap between bulk-averaged probes (diffraction/spectroscopy) and local, particle-resolved diagnostics while retaining statistically meaningful fields of view. In practice, 2D TXM in battery studies is rarely used as "simple imaging." Its most consequential role is as a quantitative mapper of reaction heterogeneity: spatially resolving where (and in what sequence) electrochemical redox, phase evolution, and damage initiation occur across particle ensembles and within individual particles. When operated in an energy-resolved mode (TXM-XANES), 2D TXM further converts transmitted intensity contrast into local chemical-state descriptors (e.g., transition metal valence proxies), allowing one to translate morphological or grayscale inhomogeneity into mechanistically interpretable redox maps. The studies summarized here collectively demonstrate that 2D TXM enables a rigorous treatment of "uniform reaction" assumptions—either validating uniformity under certain conditions or, more importantly, identifying the spatial fingerprints of nonuniformity that govern rate capability, degradation pathways, and cell-level lifetime.

A representative starting point is the explicit framing of heterogeneity as a design lever rather than a nuisance parameter. In bimodal cathodes designed for SiO/graphite || NCM full cells, 2D TXM-based mapping is used to track how different particle populations participate in redox at different depths of discharge, establishing a direct visualization of reaction partitioning across a heterogeneous electrode architecture. The central methodological point is not merely that a redox gradient exists, but that the gradient evolves with state-of-charge and can be rationally manipulated by engineering the cathode particle-size distribution and microstructure. Mechanistically, such spatially resolved evidence supports the idea that controlled heterogeneity can mitigate locally extreme operating conditions (e.g., overdriven subdomains), thereby extending longevity at the full-cell level. This "heterogeneity-aware" narrative is critical for 2D TXM: the technique's value is maximized when local maps are explicitly tied to performance-limiting phenomena rather than reported as descriptive images [2].

From this foundation, several works use 2D TXM to dissect how reaction heterogeneity emerges from intrinsic kinetics and transport constraints, particularly in materials where ion pathways and redox processes are structurally complex. A clear example is the disordered rock salt cathode study that interrogates particle-size-dependent redox kinetics and charge distribution. Here, 2D TXM is leveraged to visualize spatial variations in transition metal redox (as a proxy for local state-of-charge), enabling a direct comparison of "charged" regions across particles of different size classes. The key mechanistic insight is that particle size and internal pathway connectivity can produce markedly different charge distributions—often consistent with a reaction-front or core–shell-like progression in larger particles—highlighting that nominally identical electrochemical conditions may correspond to very different local redox states. Importantly, 2D TXM serves as the fast, high-throughput diagnostic that identifies whether heterogeneity is mild (supporting lumped-parameter descriptions) or severe enough to require explicit heterogeneity-aware interpretations and, potentially, 3D follow-up [3].

In high-nickel layered cathodes, 2D TXM becomes particularly powerful because microstructural and compositional engineering strategies are frequently deployed to stabilize high-voltage operation, yet their success depends on how effectively they redistribute local redox and suppress damaging gradients. A high-nickel heterostructured cathode with local stoichiometry control is a prototypical case in which 2D TXM-XANES is used (including operando implementations) to follow Ni redox evolution across multiple particles and, crucially, to resolve differential participation between surface-proximal regions and particle interiors. The mechanistic value of this approach is twofold. First, it provides direct evidence of whether surface regions “overreact” early during charge (a common precursor to surface reconstruction and impedance rise), and second, it quantifies how the core’s redox contribution evolves at higher voltage where transport limitations typically intensify. In this way, 2D TXM does not simply confirm that heterostructuring alters average behavior; it shows how the redox burden is redistributed spatially, which is the causal bridge between architectural design and improved stability [4].

A closely related, but conceptually distinct, use of 2D TXM is to establish when a system is sufficiently uniform that observed electrochemical improvements can be credibly attributed to lattice/site effects rather than to accidental compositional patchiness. In a P2-layered sodium cathode tuned by regulating sodium occupancy sites, TXM-XANES mapping is used to verify the spatial distribution of transition metal states across particles and to interpret which redox couples dominate in different voltage windows. Such evidence is not trivial: bulk XAS may indicate reversible average oxidation states, yet substantial inter/intraparticle heterogeneity can persist and dominate kinetics. By contrast, 2D TXM-XANES directly probes whether oxidized/reduced regions are uniformly distributed or whether subdomains lag behind, thereby strengthening mechanistic claims about how occupancy tuning modifies electrochemical behavior. More broadly, this class of studies illustrates an underappreciated role of 2D TXM: it can function as an assumption-checking tool that either validates the use of bulk-averaged descriptors or signals the need for spatially resolved mechanistic frameworks [5].

2D TXM is also well-suited to interrogating materials where the redox mechanism is not purely cationic and may involve coupled structural evolution. In hyperlithiated sacrificial cathodes based on Li_6CoO_4, Fe substitution is used to modulate anion redox reactions and structural evolution, and spatially resolved X-ray microscopy provides a critical avenue to test whether these complex processes occur uniformly or through localized initiation and propagation. While bulk signatures can suggest the presence or suppression of specific anion-redox-related pathways, 2D TXM-based mapping offers a means to identify whether reaction complexity manifests as spatially heterogeneous regions that could seed irreversible changes. In a book-chapter context, this example is valuable because it shows that 2D TXM is not limited to “mapping SOC gradients” in conventional intercalation cathodes; it can also provide mechanistic leverage in systems where oxygen activity and structural rearrangements are intertwined, and where spatial nonuniformity is often the root cause of degradation [6].

A major strength of 2D TXM is that it can connect reaction heterogeneity to chemo-mechanical failure, converting qualitative narratives about cracking into spatially anchored causal chains. In single-crystal $LiNi_{0.5}Mn1_{.5}O_4$, diffusion-induced intragranular cracking is mitigated by extending solid-solution behavior, and TXM-based chemical-state mapping is integrated with mechanics analysis to locate stress-concentration “hotspots” and correlate them with regions prone to cracking. This is a particularly instructive 2D TXM application case: rather than treating cracking as a postmortem artifact, the study positions local redox (and thus local concentration gradients) as a driver of internal stress fields, enabling a mechanism-led rationale for why extending the solid-solution regime reduces crack initiation. In effect, 2D TXM provides the spatial "state variable" field that makes transport–mechanics coupling testable, thereby elevating crack-mitigation strategies from empirical to mechanistic [7].

A similar transport–damage linkage emerges in polycrystalline sodium-ion cathodes designed for high-voltage fast charging via a grain-welding strategy. Here, 2D TXM-XANES is used after extended cycling to compare chemical-state distributions and infer how microstructural stabilization strategies influence the reversibility and uniformity of redox across the cathode (as illustrated in Figure 5.8). The mechanistic logic is consistent with the LNMO case but applied to a more complex polycrystalline architecture: if grain-boundary cohesion and intergranular connectivity are improved, reaction localization and associated stress concentrations should be reduced, leading to more spatially uniform redox and suppressed chemo-mechanical failure. The key contribution of 2D TXM in such studies is that it does not merely document that "the welded sample cracks less"; it provides a spatial chemical-state

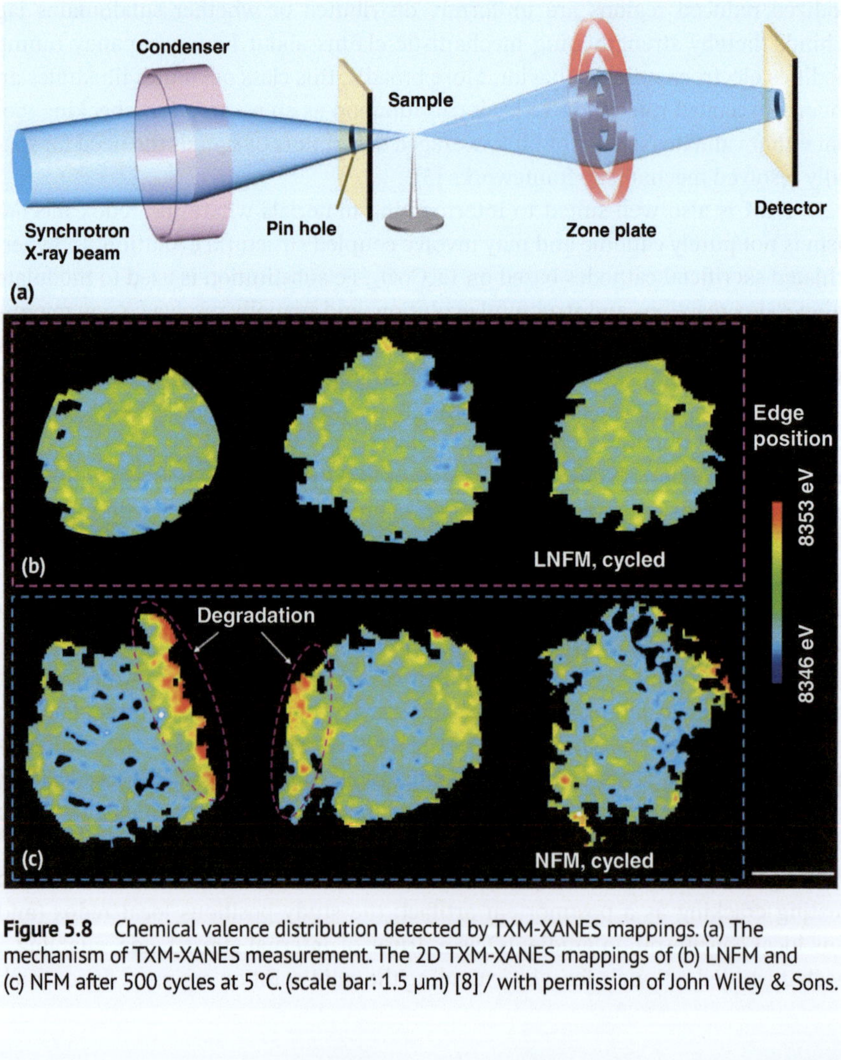

Figure 5.8 Chemical valence distribution detected by TXM-XANES mappings. (a) The mechanism of TXM-XANES measurement. The 2D TXM-XANES mappings of (b) LNFM and (c) NFM after 500 cycles at 5 °C. (scale bar: 1.5 μm) [8] / with permission of John Wiley & Sons.

perspective that is necessary to argue that damage suppression is achieved by mitigating heterogeneous reaction and its stress consequences [8].

While many canonical examples focus on positive electrodes, 2D TXM is equally important in solid-state architectures where interfacial and mesoscale transport bottlenecks dominate. In sulfide-based all-solid-state lithium metal batteries enabled by a stable catholyte interface, TXM-based mapping contributes to the mechanistic argument that interface design can stabilize practical operation by maintaining more uniform utilization and suppressing localized degradation in the composite cathode region. The essential message for 2D TXM is that solid-state systems amplify heterogeneity: contact loss, interphase formation, and tortuous ionic/electronic percolation can create strongly localized "dead zones" and overdriven regions. 2D TXM provides a direct route to visualizing such nonuniformity at relevant length scales (beyond what typical surface probes capture), making it an enabling tool for translating interface chemistry concepts into spatially resolved utilization and degradation evidence [9].

Another emerging theme is the use of 2D TXM to quantify how manufacturing and processing routes reshape the spatial distribution of degradation, which in turn informs what mitigation levers are most effective. In rapid Joule-heating densification of high-nickel cathodes, the microstructural evolution is studied in a time- and process-aware framework, and TXM-XANES mapping is used to compare spatial oxidation state distributions after cycling between conventionally processed and rapidly densified materials. This type of 2D TXM evidence is particularly impactful because it converts an intuitive processing hypothesis ("densification improves stability") into a spatially explicit mechanism: whether degradation is bulk-penetrating versus surface-confined, whether local redox inhomogeneity persists, and how pore evolution and microcracking correlate with chemical-state gradients. Such insights are precisely what is needed to prioritize subsequent engineering steps (e.g., whether coatings should target surface-dominant degradation or whether bulk transport limitations must be addressed) [10].

Finally, 2D TXM can play a decisive role in recycling and upcycling contexts where the central question is not only "what phases exist on average," but "how are they distributed spatially within regenerated particles?" In solid-state upcycling of NMC cathodes monitored by in-situ diffraction and ex-situ TXM, microscopy provides spatial evidence for phase/domain coexistence and mixing states that may not be fully resolved by bulk diffraction alone. In the logic of a 2D TXM subsection, this case is important because it demonstrates how 2D microscopy can serve as a spatial quality-control tool: identifying residual heterogeneity, incomplete transformation, or domain-level phase separation that will ultimately dictate transport paths, stress development, and long-term cycling stability in "recovered" cathode materials. It also creates a natural conceptual bridge to the subsequent 3D TXM subsection: once domain percolation, connectivity, and volumetric distributions become the performance-determining factors, tomography becomes the next methodological step [11].

Taken together, these application cases establish a coherent set of "2D TXM deliverables" for battery science. First, 2D TXM provides reaction heterogeneity maps that expose inter and intraparticle utilization differences under realistic cycling conditions (from engineered bimodal electrodes to DRX size effects and high-nickel heterostructures). Second, it functions as a mechanism validator: confirming when reactions are sufficiently uniform to attribute performance changes to intrinsic crystal-chemical design (e.g., sodium occupancy tuning) and, conversely, diagnosing when spatial nonuniformity dominates behavior. Third, it extends into chemically complex regimes (anion redox coupled to structural evolution), where spatial nonuniformity is often the hidden variable controlling irreversibility. Fourth, it provides the spatial field variables needed to connect transport gradients to stress localization and cracking, enabling more rigorous chemo-mechanical interpretation and design rules. Finally, it enables process- and sustainability-oriented frameworks—diagnosing how manufacturing routes and upcycling transformations rewrite the spatial distribution of redox and degradation, thereby guiding practical materials engineering, and it remains highly relevant to solid-state systems where mesoscale heterogeneity is often the primary limiter. In summary, 2D TXM's unique contribution is not just imaging capability; it is the ability to make "where the reaction occurs" and "where degradation initiates" experimentally measurable variables that can be engineered, modeled, and ultimately controlled.

5.4.2 3D TXM (Full-field Nanotomography) for Battery Materials and Cells

Three-dimensional TXM (3D TXM), most commonly implemented as full-field X-ray nanotomography, addresses a fundamental limitation of 2D projection imaging: in battery electrodes and interfaces, the performance-relevant objects—percolating pore networks, tortuous ion transport pathways, crack networks, and metal deposits—are intrinsically three-dimensional. As a consequence, the defining contribution of 3D TXM is not only higher information content but topological fidelity. Volumetric reconstruction makes it possible to quantify connectivity, interfacial area, phase contiguity, crack/void morphology, and thickness-direction gradients in a way that is not confounded by overlap in projection. These metrics are the natural inputs for multiscale transport, kinetics, and chemo-mechanical models, and they provide the "geometry truth" needed to connect microstructure to electrochemical observables with fewer hidden assumptions.

A first class of 3D TXM application cases emphasizes electrode architecture and manufacturability, where the central question is how processing determines mesoscale percolation of solids and pores. In scalable dry-printing manufacturing for lithium-ion batteries, 3D X-ray imaging is used to evaluate how the fabrication route organizes the active material/binder/carbon framework and porosity throughout the electrode volume. The value of 3D TXM-type reconstruction in this context is that it enables architecture to be expressed as quantitative descriptors rather than qualitative micrographs: pore-size distributions, spatial gradients across thickness, and the presence (or suppression) of isolated pore clusters that can trap electrolyte or impede ionic access. Such volumetric evidence supports a mechanistic manufacturing narrative—processing conditions reshape the 3D percolation landscape, which in turn governs local current distribution and utilization. Importantly, this is the regime in which 3D datasets become directly actionable: they can be converted into effective transport parameters (e.g., tortuosity factors) and used to rationalize why certain manufacturing choices improve life and energy density, rather than relying on empirical iteration [12].

A second and highly influential class of application cases focuses on cathode particle/secondary particle degradation, where crack and void networks evolve under coupled electrochemical and mechanical driving forces. In Ni-rich layered oxide cathodes, it is well established that microcracking can accelerate impedance growth and capacity fade by disrupting electronic/ionic pathways and exposing fresh reactive surfaces. The dual-modification study of Ni-rich layered oxides provides a representative demonstration of how 3D TXM complements chemistry-sensitive mapping: nanotomography captures the spatial distribution and topology of microcracks/voids, and when combined with spectroscopic contrast it becomes possible to interpret cracking not merely as "present/absent" but as a structural consequence of localized reaction and stress accumulation. The key advantage of 3D in such cases is that crack networks are rarely planar; what appears as a short crack in 2D may actually be a connected fracture surface that spans a significant fraction of a secondary particle in 3D. By resolving these networks volumetrically, 3D TXM provides a physically grounded basis for linking cathode modifications to suppressed crack connectivity and improved mechanical integrity [13].

A closely related, but mechanistically distinct, 3D TXM use case is oxygen-loss-driven damage and internal surface formation in layered oxides. In the $LiNiO_2$ study that constructs a thin disordered self-protective layer on primary particles, the central failure mode is framed around oxygen release and the associated structural reconstruction, which can generate internal voids and porous architectures. Here, the 3D aspect is decisive: oxygen-loss-induced porosity is not simply a surface phenomenon. Once pores nucleate and coalesce internally, they create new internal surfaces that can become electrochemically active (or parasitic), altering local chemistry and accelerating degradation. 3D spectrotomography and nanotomography provide a direct pathway to quantify these internal pore populations, their spatial distribution, and their evolution—information that is essential for evaluating whether a protective layer truly confines damage near the exterior or suppresses the generation of internal reactive surfaces throughout the particle volume. In other words, 3D TXM converts oxygen release from a bulk-averaged concept into a volumetric microstructural pathway with measurable signatures [14].

Whereas the preceding examples focus on cycling-induced evolution, 3D TXM is equally valuable upstream—during materials synthesis and processing—because many performance-limiting heterogeneities are "baked in" during calcination and phase formation. In probing dopant redistribution and phase propagation during layered oxide cathode synthesis, 3D TXM is used to visualize how oxidation state and chemical features propagate through particles as a function of processing stage (as shown in Figure 5.9). This class of evidence is difficult to obtain from bulk diffraction alone because averages can obscure persistent gradients; 3D mapping highlights that transformation fronts and oxidation can proceed nonuniformly,

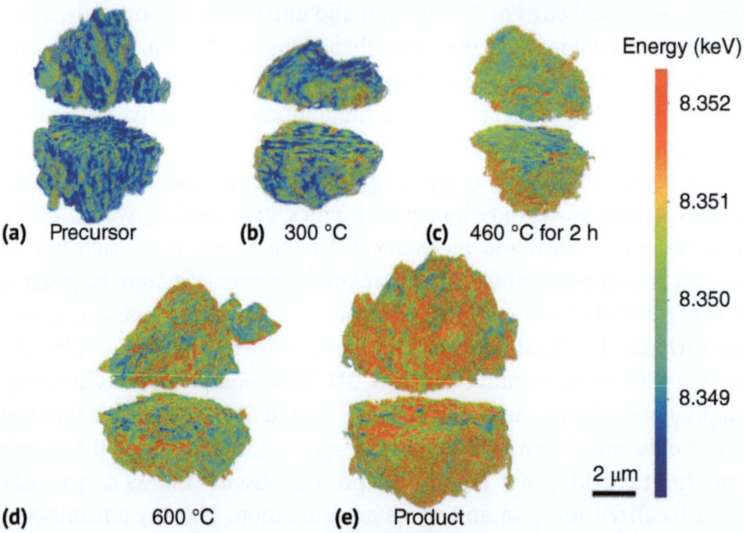

Figure 5.9 Ni *K*-edge 3D TXM results for the samples (a) at the pristine state; (b) calcined at 300 °C; (c) calcined at 460 °C for 2 h; (d) calcined at 600 °C; (e) complete calcination [15] / with permission of John Wiley & Sons.

producing surface-to-core differences that later manifest as spatially heterogeneous electrochemical activity. For a 3D TXM subsection, the broader point is that volumetric imaging enables a synthesis mechanism narrative: diffusion and reaction fronts in the solid state can produce enduring 3D chemical landscapes, and optimizing synthesis therefore requires controlling not only final phase identity but also the spatial uniformity of the transformed volume [15].

The importance of volumetric heterogeneity becomes even more apparent at the cell level, where operational protocols can drive coupled degradation across electrodes. In the study of model-based protective fast charging in commercial smartphone cells, TXM is used to diagnose cathode degradation that is interpreted in connection with anode lithium deposition phenomena. The mechanistic relevance of 3D-resolved cathode information in such work is that fast-charging protocols can induce nonuniform lithiation/delithiation histories, leading to local overoxidation, underutilization, and microstructural stress that may not be apparent from voltage curves alone. By providing spatially resolved evidence of cathode nonuniformity, TXM helps close the causal loop between protocol design, electrode-level heterogeneity, and full-cell degradation pathways (including crosstalk signatures associated with lithium deposition). This is an important theme to emphasize in a book chapter: 3D TXM can serve as the structural "ground truth" that validates or refines model-based interpretations of protocol-induced degradation [16].

Finally, 3D TXM is uniquely well matched to metal anodes and anode-free systems, where morphology and interphase are inherently three-dimensional and where stability is governed by coupled transport and interface fields. In the sodium metal anode study on support wettability and electrodeposition rate, 3D nanotomography is used to characterize deposit morphology and SEI microstructure under different interfacial conditions. The mechanistic role of 3D imaging here is to distinguish dense versus porous growth modes and to relate them to wettability-controlled nucleation and rate-dependent transport through the interphase. Such datasets support design rules that are difficult to justify with 2D snapshots alone: the stability of deposition is controlled by how 3D pathways for ion transport and local current focusing develop within (and beneath) the SEI [17]. In anode-free lithium metal batteries, where the negative electrode is formed in-situ, the coupling between host architecture and electrolyte formulation is similarly morphology-sensitive. The work achieving very high volumetric energy density highlights that advanced X-ray tomography can directly observe deposition/stripping behavior and associated structural evolution, enabling a mechanism-based assessment of how host/electrolyte synergy regulates deposit morphology and failure propensity [18]. Collectively, these studies underscore a general conclusion for 3D TXM: when the governing physics depends on percolation, tortuosity, and internal surface formation—as it does for metal growth and interphase evolution—volumetric reconstruction is not an incremental improvement; it is a prerequisite for credible mechanistic inference.

Across these application cases, the unifying logic is that 3D TXM transforms microstructure from a descriptive background into a quantitative, model-ready variable. It enables topology-aware metrics for manufactured electrodes, resolves the true connectivity and extent of crack/void networks in degraded cathodes, captures synthesis-stage phase propagation and chemical gradients that predetermine later heterogeneity, and provides volumetric evidence linking cell protocols to spatially nonuniform degradation. In metal and anode-free systems, 3D TXM is indispensable for interpreting deposition and interphase stability as coupled, three-dimensional transport problems.

5.4.3 Operando Electrochemical Testing Enabled by TXM

Operando TXM has matured into a uniquely incisive methodology for batteries because it transforms electrochemical cycling from a spatially averaged measurement into a resolved field problem: where reaction, transport, and degradation can be quantified as functions of position, time, and (when combined with spectroscopy) chemical state. In contrast to postmortem imaging, operando TXM can separate "what formed" from "what persisted," and can distinguish mechanisms that generate similar electrochemical signatures but arise from different spatial reaction modes. The core enabling step is that TXM provides not only morphological contrast but also energy-dependent contrast; thus, when the incident energy is scanned across an absorption edge, the evolving transmitted intensity can be converted into maps of oxidation state, phase fraction, and reaction pathways within real electrode architectures. This is why operando TXM has become most impactful when it is executed as TXM-XANES (or related spectromicroscopy), rather than as purely structural imaging.

At the platform level, the practical difficulty is that batteries are thick, multiphase, and dynamic—conditions that challenge the depth of focus, dose budget, alignment stability, and rotation geometry required for high-resolution computed tomography.

A representative solution is the "local in-situ nano-XCT" concept, which explicitly acknowledges that realistic cycling conditions and high-resolution tomography cannot always be achieved simultaneously across the full electrode area. Instead, a carefully prepared region-of-interest is isolated (for example, by laser shaping) so that the system can be rotated with minimized missing angles, while in-line phase contrast is exploited to access low-Z features that would otherwise be invisible in absorption contrast. The significance of this workflow is not merely instrumental: it establishes the operational regime in which operando TXM data are trustworthy enough to be used quantitatively—i.e., as time-resolved 3D volumes from which one can extract growth/decay rates, porosity evolution, tortuosity changes, and spatial correlation with electrochemical history. Such considerations also frame why many mechanistic advances have first been demonstrated in simplified or miniaturized in-situ cells before being translated toward more realistic formats [19].

Within intercalation cathodes, operando TXM's defining value is its ability to visualize reaction heterogeneity and nonequilibrium phase coexistence at the particle scale, thereby revealing kinetic pathways that are fundamentally obscured by ensemble-averaged probes. In single crystals of technologically relevant spinel chemistry, operando full-field TXM coupled with spectroscopy has been used to follow, in real time, the onset and crystallographic directionality of complex phase transitions and to connect these pathways to fracture events (as shown in Figure 5.10). The key mechanistic implication is that the phase transformation does not necessarily proceed along a single equilibrium trajectory; rather, multiple phases associated with substantially different electrochemical potentials can coexist at short length scales under dynamic conditions. This directly challenges interpretations that infer reaction mechanisms solely from voltage profiles or from equilibrium phase diagrams,

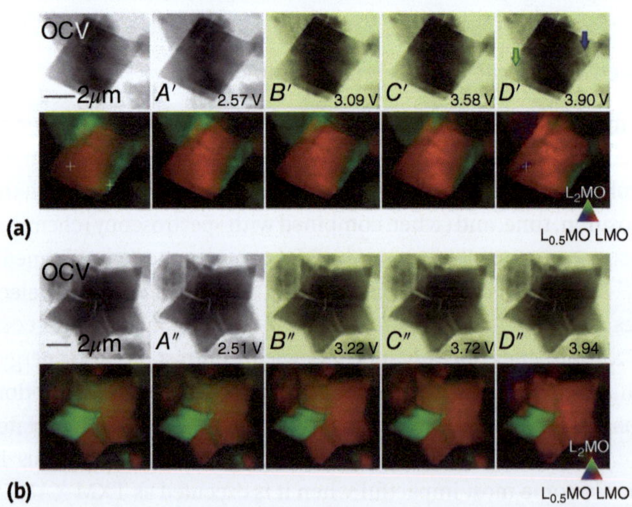

Figure 5.10 Chemical and morphological maps of FOVs of (a) II and (b) III during the tetragonal-to-cubic transition. Inverted optical density images collected at 6700 eV (upper row) and chemical phase maps (lower row) [20] / with permission of John Wiley & Sons.

and it clarifies how mechanical degradation can be triggered by spatially localized transformation sequences rather than by uniform lattice strain [20].

A complementary illustration arises in layered Na ion cathodes, where metastable structural evolution is tightly entangled with reaction inhomogeneity. By combining in operando diffraction with TXM-based chemical mapping, a non-equilibrium solid-solution regime (with coexisting $Na1 - \delta$ compositions) can be identified and tracked during cycling. Crucially, TXM-XANES imaging across element-specific absorption edges enables phase distributions to be mapped in multiparticle ensembles, demonstrating that chemical evolution can be spatially non-uniform even when electrochemical conditions are nominally homogeneous. The advantage of this approach is conceptual as much as technical: it creates a direct bridge between crystallographic descriptors (from diffraction) and spatial reaction fields (from TXM-XANES), allowing "metastability" to be interpreted not only as a bulk phase phenomenon but also as a spatially heterogeneous reaction mode [21].

Recent work on high-energy layered oxide cathodes further emphasizes that operando TXM-relevant measurements are essential when kinetics are nonlinearly coupled to state-of-charge. At the level of secondary particle agglomerates, delithiation has been observed to proceed through an initial latent stage followed by rapid, stochastic, and largely uniform delithiation of individual secondary particles, behavior that is strikingly inconsistent with the gradual increase in cell potential during charge. The mechanistic consequence is that electrode-level polarization cannot be straightforwardly mapped onto particle-level reaction rate. Modeling constrained by the operando observations requires an exchange current density that depends strongly on Li-site fraction and increases by orders of magnitude over a narrow composition window. For operando TXM-centered thinking, this is a pivotal message: spatially resolved operando probes are not only "diagnostics," but also boundary conditions for physics-based models of kinetics, because they reveal when the effective reaction law changes with composition in a way that bulk electrochemistry alone cannot reliably determine [22].

In conversion and alloying electrodes, operando TXM becomes even more mechanistically decisive because electrochemical performance is dominated by coupled phase conversion, interfacial transport, and chemo-mechanical reorganization. For alloying tin anodes, in-situ 3D synchrotron X-ray nanotomography has shown that severe morphological changes are concentrated in the earliest electrochemical events—specifically, the first delithiation and the subsequent second lithiation—after which the structure approaches a quasi-equilibrium with no further dramatic morphological evolution. This finding reframes "cycling stability" as an outcome that can be seeded (or lost) in the earliest cycles, implying that formation protocols and initial microstructural conditioning can dominate long-term mechanical integrity [23].

For Na-metal sulfide conversion chemistry, operando full-field TXM combined with XANES has enabled direct mapping of 2D chemical phase evolution and composition during repeated charge/discharge, while 3D nanotomography provides quantitative microstructural evolution and electrochemical intermittency measurements constrain transport. Mechanistically, first-cycle irreversibility can be attributed to partial Na trapping and large interfacial ion mobility resistance, whereas subsequent cycling exhibits improved reversibility after microstructural reorganization such as fracture and particle size reduction. Importantly, the same operando platform rationalizes large voltage hysteresis as a multifactor outcome—reaction overpotential and ohmic drop, amplified by the large Na ion radius and by diffusion resistance in the two-phase regime. The broader implication is that "conversion hysteresis" should not be treated as a purely thermodynamic penalty: operando TXM shows that it is strongly shaped by spatially evolving interfaces and their transport impedance, which can in principle be engineered [24].

Operando TXM can also arbitrate claims about structure-driven mitigation strategies, such as amorphization to suppress volume expansion. In amorphous $MoSnSe_{1.5}S_{1.5}$ electrodes, in-situ/operando measurements explicitly include TXM as a tool to interrogate real time phenomena and establish reaction mechanisms under both Li and Na storage. In that system, operando TXM is used to visualize volume expansion behavior (particularly contrasted against crystalline analogs), supporting the interpretation that amorphous structure can reduce mechanically deleterious expansion during cycling and that stability can differ markedly between Li-ion and Na ion operation under otherwise analogous conditions. This is a representative example of how operando TXM contributes not by supplying another "image," but by providing time-resolved evidence that links a design principle (amorphization) to its mechanically relevant consequence (expansion dynamics) under electrochemical control [25].

The interface challenge is most acute in solid-state batteries, where the electrochemical response is often governed by evolving solid–solid contact and localized mechanical failure. In-situ 3D XANES (spectroscopy tomography) has been used to track phase evolution inside active particles embedded in sulfide solid-state environments, revealing that conversion can initiate and propagate heterogeneously, accompanied by volume expansion and cracking, and may then slow dramatically or even terminate when interfacial changes induce contact loss. A further mechanistic insight is that dense metallic surface phases and strain-driven interfacial evolution can deteriorate reaction continuity, motivating interfacial tailoring strategies such as engineered electrolyte growth on particle surfaces to preserve contact. Here the "operando" aspect is essential: without tracking phase evolution as it occurs, one could easily misattribute capacity fade to intrinsic chemical irreversibility rather than to mechanically induced isolation at the interface [26].

Finally, operando TXM is increasingly relevant for metal anodes where direct optical observation can be obstructed by interfacial layers or scaffolds. In Zn metal systems employing a polyimide interface, the electrode surface can be visually hidden; in-situ TXM overcomes this by exploiting the negligible X-ray signal from the polymer relative to Zn, enabling real time observation of plating morphology during deposition within a purpose-built cell configuration. The mechanistic value is straightforward but powerful: morphological evolution—nucleation of spiky deposits, subsequent growth modes, and spatial localization—becomes observable under realistic electrochemical operation even when conventional imaging cannot access the interface. This reinforces a broader theme across chemistries: operando

TXM is often most valuable precisely where other operando diagnostics are blind, either because the interface is buried or because the relevant contrast is chemical rather than purely structural [27].

Collectively, these case studies show that operando TXM in batteries is best understood as an approach that quantifies spatial reaction fields—phase, composition, morphology, and damage—under electrochemical control. Its mechanistic leverage arises when it is used to connect (i) heterogeneity and nonequilibrium pathways, (ii) interfacial transport constraints, and (iii) chemo-mechanical evolution into a single causal narrative supported by time-resolved evidence, rather than inferred from averaged observables alone.

5.4.4 TXM with In-situ Heating

Thermal stimuli are intrinsic to essentially every stage of a battery material's lifecycle: precursor conversion during calcination, densification and annealing during electrode manufacturing, heat generation and runaway risk under abuse, and high-temperature relithiation/repair during direct recycling. What makes TXM with heating distinctive is not merely imaging "before/after" morphology, but its ability to transform thermal treatment into a spatially resolved reaction coordinate—mapping where oxidation/reduction, phase conversion, or interdiffusion proceeds first, how sharply reaction fronts develop, and which microstructural features act as kinetic bottlenecks. This spatial selectivity is particularly valuable because thermal processes in particles are often controlled by diffusion and interfacial accessibility, producing core–shell or patchy domains that ensemble probes compress into single averaged values.

A representative processing example is the use of TXM chemical imaging to map solid-state reaction heterogeneity during calcination of Ni-rich layered oxide cathodes. In the calcination pathway, multiple mass-transport channels compete: ingress of ambient oxygen, incorporation of lithium sources derived from Li2CO3 decomposition, and dehydration/oxidation of hydroxide precursors. TXM, especially in a spectroscopic mode (TXM-XANES), can spatially resolve transition metal oxidation states across partially reacted secondary particles, revealing reduced cores and oxidized shells at intermediate temperatures. A key mechanistic insight from such mapping is the identification of a slow oxygen-diffusion wavefront: near-surface domains oxidize first while inner regions remain reduced when oxygen diffusivity is insufficient at lower temperatures, yielding sharp redox boundaries that directly fingerprint transport limitation rather than intrinsic reaction thermodynamics [28].

This perspective elevates calcination from a nominal "time/temperature recipe" to a spatially heterogeneous process whose rate is dictated by where oxidant and lithium sources can reach. At higher temperatures, oxidation state maps can converge toward homogeneity, consistent with oxygen diffusion becoming fast enough to erase internal redox gradients, yet still requiring additional ambient oxygen to complete oxidation when internal oxygen is insufficient. In practical terms, TXM-derived reaction maps inform how particle size, porosity evolution, and gas accessibility can be tuned to suppress under-reacted cores and to minimize the coexistence of metastable intermediate phases—conditions often linked to performance scatter and safety variability in the final cathode.

The value of TXM-heating is even more pronounced when addressing safety-relevant thermal degradation of charged cathodes, where the onset of decomposition and the magnitude of heat release need not be governed by the same species or the same local environment. A clear example is the element-distinguished thermal behavior of lithium-rich Mn-rich layered oxides (LRMO), interrogated by combining diffraction with 2D TXM-XANES after controlled heat treatments (as illustrated in Figure 5.11). Because direct in-situ heating can be complicated by mechanical drift (e.g., softening/displacement of polymeric components), the study leveraged samples preheated to selected temperatures and then mapped Mn, Co, and Ni *K*-edge energies across individual particles, using edge energy shifts as pixel-level valence proxies [29].

Two mechanistic conclusions are particularly instructive for the TXM-heating toolbox. First, despite being a minority component, Ni reduction was observed to dominate the early stage of thermal decomposition (below ≈ 180 °C), implicating Ni as a primary controller of decomposition onset temperature. Second, Mn—often considered structurally stabilizing in conventional layered oxides—can, in LRMO, contribute to more intense heat release because oxygen-redox-derived lithium/oxygen vacancies promote MnO6 destabilization and Mn migration, altering the decomposition pathway.

Beyond element specificity, TXM also localized environment-dependent pathways: distinct thermal reduction behaviors emerged between regions covered by a solid polymer electrolyte (SPE) and exposed regions, effectively partitioning the particle surface into microenvironments with different dominant reduction chemistry

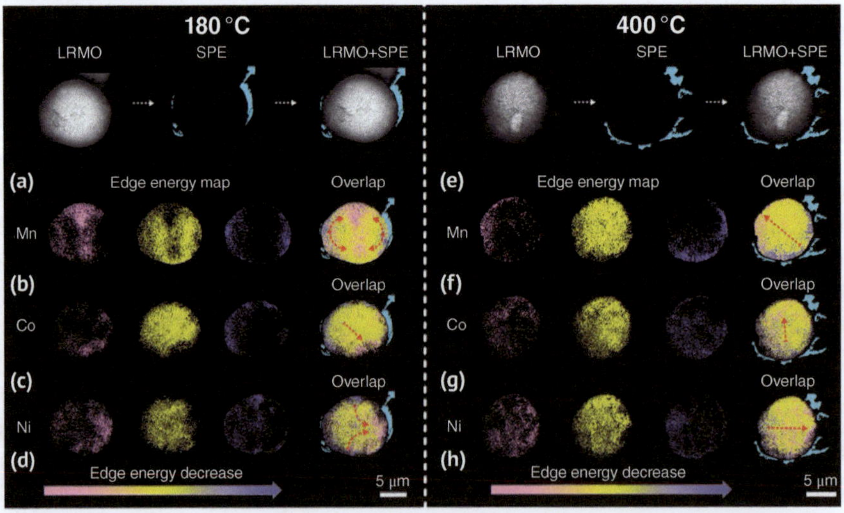

Figure 5.11 TXM projection images of 4.8 V LRMO@SPE after heat treatment at (a) 180 and (e) 400 °C, respectively. The (b, f) Mn, (c, g) Co, and (d, h) Ni *K*-edge energy maps segmented to relatively high, medium, and low edge energy sections are displayed to depict the valence distribution in spatial [29] / with permission of John Wiley & Sons.

(e.g., Mn reduction associated with covered regions, Ni reduction dominating exposed regions). This is a critical methodological message: TXM with thermal protocols can couple "where reactions occur" with "what reactions occur," enabling safety assessments that account for incomplete interfacial coverage and localized reactivity—features that bulk DSC/XRD cannot spatially discriminate.

Finally, thermal TXM is increasingly relevant to recycling and upcycling, where high-temperature relithiation and cation interdiffusion aim to restore or even upgrade legacy cathode compositions. In a precipitation-based upcycling approach that converts NMC111 toward higher-Ni NMC622, the essential question is whether diffusion truly homogenizes at all relevant length scales (secondary particle, primary particle, grain boundary). Tomographic TXM with dual-energy absorption contrast was used to quantify 3D elemental mixing—collecting tomograms above and below Ni and Co *K*-edges and using intensity differences to compute elemental ratios—providing nondestructive access to internal gradients that cross-sectional SEM-EDS can miss [30].

The TXM results showed that above ≈850 °C additional Ni can fully diffuse into polycrystalline secondary particles, yet diffusion limitations persist at the primary-particle scale, with Ni enrichment near grain boundaries even at 900 °C. This multiscale conclusion is precisely the kind of mechanistic clarification that thermal processing needs: "homogeneous by bulk composition" is not equivalent to "homogeneous where it matters," and TXM provides the quantitative bridge between these statements.

5.5 Conclusion and Prospective

TXM has evolved from a high-resolution imaging technique into a mechanism-oriented framework for studying batteries as spatially heterogeneous, multiscale, and strongly nonequilibrium systems. Two-dimensional TXM—particularly in its energy-resolved form (TXM-XANES)—enables rapid mapping of reaction heterogeneity and local redox states, allowing long-assumed "uniform utilization" behavior to be tested directly rather than inferred from bulk averages. Three-dimensional TXM (full-field nanotomography) removes the geometric ambiguity of projections and provides topology-faithful descriptors—connectivity, percolation pathways, crack and void networks, and internal surface generation—that can be quantified and linked to transport and chemo-mechanical responses. When these capabilities are combined with electrochemical control (operando TXM), structural and chemical evolution can be synchronized with current/voltage history to establish causality instead of postmortem correlation. Heating-enabled TXM extends the same philosophy to thermal coordinates, localizing diffusion-limited calcination fronts, environment-dependent decomposition pathways, and the length-scale limits of homogenization during upcycling. Across these modes, TXM's most distinctive contribution is to turn "where the reaction occurs" and "where degradation initiates" into measurable state variables that can be engineered, modeled, and validated.

This shift is important because many key performance-limiting and failure-driving phenomena—rate limitation, accelerated aging, cracking, contact loss, and thermal instability—are emergent outcomes of spatial nonuniformity coupled to evolving microstructure. Treating electrodes as uniform media often leaves mechanisms underdetermined: multiple hypotheses can reproduce the same voltage curve or average diffraction trend. TXM reduces this nonuniqueness by directly constraining the spatial organization of reaction and damage. The practical implication is that design principles can move from qualitative narratives (e.g., "stabilized interface" or "improved mechanical robustness") toward spatially anchored mechanisms supported by quantitative maps and statistically meaningful morphology metrics.

A central near-term opportunity is to elevate TXM from compelling visualization to reproducible, quantitative metrology. Achieving this will require standardizing calibration and analysis workflows so that derived heterogeneity metrics are comparable across facilities. For spectroscopic TXM, the priorities include consistent edge energy alignment, reference selection, fitting strategies, and uncertainty propagation from spectral noise to oxidation state maps. For tomography, transparent reporting is needed for acquisition geometry, dose, reconstruction parameters, segmentation thresholds, and sensitivity analyses that demonstrate how porosity, tortuosity, crack connectivity, and contact area vary with algorithmic choices. The outcome should be a community practice in which mechanistic claims are traceable to validated data quality and uncertainty bounds rather than to implicit processing assumptions.

At the same time, the greatest scientific leverage will come from increasing the information rate of operando measurements. Realistic electrochemical configurations impose severe constraints: thick multiphase electrodes, mechanically compliant stacks, and evolving interfaces challenge alignment stability and dose budgets. Progress will likely be driven by a combined strategy of instrumentation and computation. On the instrumentation side, higher coherent flux, faster detectors, and mechanically stable stages can expand the time–resolution window. On the computation side, sparse-angle and limited-dose reconstructions, drift- and deformation-aware algorithms, and phase-contrast methods that enhance sensitivity to low-Z components can reduce dose while preserving interpretability. The objective is not simply higher resolution, but higher mechanistic bandwidth: extracting reliable trajectories of phase fraction, reaction-front position, pore growth, and crack propagation under genuine cycling conditions.

Operando realism will increasingly depend on cell and fixture engineering as much as on optics. The next generation of in-situ platforms should be designed around the mechanistic observable of interest. If contact loss and interphase evolution dominate (as in solid-state architectures), pressure and compliance must be controlled and quantified while maintaining representative interfaces. If transport limitation in thick electrodes is central, geometries must preserve realistic thickness and electrolyte distribution rather than relying on overly thinned specimens that alter the physics. If thermal runaway precursors are the target, controlled heating with minimal drift and synchronized thermal readouts becomes essential. In this sense, "better TXM experiments" will be increasingly defined by faithful boundary conditions and well-characterized mechanical/thermal states.

As datasets grow in scale and complexity—multiscale, multimodal, and time-resolved—analytics will become the main differentiator. Manual interpretation will neither scale nor provide the uncertainty quantification required for rigorous mechanism identification. The most valuable role for machine learning is therefore physics-informed acceleration rather than black-box classification: automated segmentation of particles/pores/cracks, tracking of reaction fronts and damage evolution, co-registration across time/energy/modality, and data fusion with confidence estimates. Equally important is ensuring that analysis pipelines produce model-ready outputs, such as spatial reaction-rate fields, evolving boundary conditions for chemo-mechanical simulations, and topology-derived transport descriptors computed directly from 3D reconstructions.

The convergence of TXM with multiscale modeling offers a particularly promising route because it enables bidirectional closure. TXM can provide geometry truth and reaction-field constraints that reduce model nonuniqueness, distinguishing whether performance fade is driven primarily by localized underutilization, transport-induced reaction fronts, crack-mediated connectivity loss, or mechanically induced isolation. Conversely, models can guide experiments by identifying which observables and time windows are most discriminating among competing mechanisms, thereby improving experimental efficiency and maximizing the mechanistic content of beamtime-intensive studies. This tight coupling becomes especially important when operando observations reveal counterintuitive kinetics—such as abrupt transitions between latent and rapid delithiation behavior at the particle level—signaling that conventional reaction laws may require revision and that spatially resolved data should serve as decisive constraints.

Looking further ahead, TXM is positioned to expand from diagnosing electrochemical degradation into engineering the full battery lifecycle, including synthesis, manufacturing, safety, and recycling. Heating-enabled TXM already illustrates how thermal treatment can be reframed as a spatially heterogeneous reaction coordinate, localizing diffusion wavefronts, microenvironment effects, and incomplete homogenization that are invisible to bulk averages. This aligns directly with industrially relevant questions: how calcination protocols imprint persistent gradients; how densification modifies pore topology and internal surfaces; how element-specific decomposition initiates under thermal abuse; and how upcycling and relithiation progress across length scales rather than only in average composition. TXM will be most impactful where it yields actionable process rules—linking controllable parameters (temperature profiles, gas access, particle architecture, interfacial coverage, pressure history) to measurable spatial outcomes that predict performance scatter and safety margins.

In summary, TXM is becoming a comprehensive, mechanism-centric characterization framework that maps not only structure and chemistry, but the spatial organization of reaction and degradation under electrochemical and thermal driving forces. The next advances will be defined by quantitative rigor, operando realism, scalable analytics, and tight integration with predictive modeling so that TXM-derived spatial fields can serve as standardized inputs for design, optimization, and validation across the battery lifecycle.

References

1 L. Wang, J. Wang, P. Zuo, Probing battery electrochemistry with in operando synchrotron X-ray imaging techniques, Small Methods 2 (2018) 1700293.

2 H. Kim, C.M. Kim, S. Jo, S. Lee, S.J. Choi, H.J. Park, H. Yu, D. Kim, K. Kim, T.J. Shin, S.-M. Lee, Leveraging reaction heterogeneity in bimodal cathodes to enhance longevity of SiO/graphite | NCM full cells, Advanced Science 13 (2025) e18317.

3 Y. Zhang, A. Hu, J. Liu, Z. Xu, L. Mu, S. Sainio, D. Nordlund, L. Li, C.-J. Sun, X. Xiao, Y. Liu, F. Lin, Investigating particle size-dependent redox kinetics and charge distribution in disordered rocksalt cathodes, Advanced Functional Materials 32 (2022) 2110502.

4 C. Yin, M. Ge, Y. Chung, J. Bai, W.-K. Lee, K. Kisslinger, S.N. Ehrlich, X. Xiao, B. Qiu, S.E. Trask, A.R. Dunlop, A.N. Jansen, Z. Liu, Y. Shin, F. Wang, High-nickel heterostructured cathodes with local stoichiometry control for high-voltage operation, Small Structures 4 (2023) 2300236.

5 Q.-C. Wang, Z. Shadike, X.-L. Li, J. Bao, Q.-Q. Qiu, E. Hu, S.-M. Bak, X. Xiao, L. Ma, X.-J. Wu, X.-Q. Yang, Y.-N. Zhou, Tuning sodium occupancy sites in P2-layered cathode material for enhancing electrochemical performance, Advanced Energy Materials 11 (2021) 2003455.

6 W. Lee, H. Lee, Y. Byeon, J.H. Kim, W. Choi, M. Choi, M.-S. Park, W.-S. Yoon, Modulating anion redox reactions and structural evolution through Fe-substitution in Li_6CoO_4 hyper-lithiated sacrificial cathodes, Advanced Energy Materials 13 (2023) 2302316.

7 H. Shin, A. Ndukwe, T. Kim, J.H. Lee, G. Li, H.J. Lee, Mitigating diffusion-induced intragranular cracking in single-crystal $LiNi_{0.5}Mn_{1.5}O_4$ via extended solid-solution behavior, Angewandte Chemie International Edition 64 (2025) e202422726.

8 B. Peng, X. Huang, C. Xu, Z. Huang, R. Zhu, C. Jin, L. Wang, H. Zhou, Grain welding strategy suppressing chemomechanical failure in polycrystalline cathode for high-voltage fast-charging sodium-ion batteries, Advanced Materials 38(7) (2025) e14943.

9 H. Lee, S. Ko, S. Park, S.-M. Lee, Stable catholyte interface enables practical operation of sulfide-based all-solid-state Li metal batteries, Advanced Energy Materials 16(7) (2025) e03019.

10 M.-H. Kim, J. Seo, J. Shin, E. Park, A. Choi, M. Choi, E. Kim, S.Y. Lee, W. Jin, S.-C. Jeon, Y. Li, C. Song, R.S. Ruoff, W.K. Seong, S. Jin, H.-W. Lee, Microstructural evolution dynamics in rapid Joule heating densification of high-nickel cathodes, Advanced Materials 37 (2025) e08602.

11 M.L. Nisbet, D. Luong, E. Allen, S. Park, T.L. Kinnibrugh, J.E. Stubbs, P.J. Eng, J.T. Vaughey, T.T. Fister, In situ diffraction and ex situ transmission X-ray microscopy studies of solid-state upcycling for NMC cathodes, Advanced Energy Materials 15 (2025) 2500698.

12 J. Liu, B. Ludwig, Y. Liu, Z. Zheng, F. Wang, M. Tang, J. Wang, J. Wang, H. Pan, Y. Wang, Scalable dry printing manufacturing to enable long-life and high energy lithium-ion batteries, Advanced Materials Technologies 2 (2017) 1700106.

13 H. Yang, H.-H. Wu, M. Ge, L. Li, Y. Yuan, Q. Yao, J. Chen, L. Xia, J. Zheng, Z. Chen, J. Duan, K. Kisslinger, X.C. Zeng, W.-K. Lee, Q. Zhang, J. Lu, Simultaneously dual modification of Ni-rich layered oxide cathode for high-energy lithium-ion batteries, Advanced Functional Materials 29 (2019) 1808825.

14 J. Chen, Y. Yang, Y. Tang, Y. Wang, H. Li, X. Xiao, S. Wang, M.S.D. Darma, M. Etter, A. Missyul, A. Tayal, M. Knapp, H. Ehrenberg, S. Indris, W. Hua, Constructing a thin disordered self-protective layer on the $LiNiO_2$ primary particles against oxygen release, Advanced Functional Materials 33 (2023) 2211515.

15 Z. Yang, L. Mu, D. Hou, M.M. Rahman, Z. Xu, J. Liu, D. Nordlund, C.-J. Sun, X. Xiao, F. Lin, Probing dopant redistribution, phase propagation, and local chemical changes in the synthesis of layered oxide battery cathodes, Advanced Energy Materials 11 (2021) 2002719.

16 M.M. Besli, A. Subbaraman, F.R. Pour Safaei, C. Johnston, G. Schneider, N. Ravi, J. Christensen, Y. Liu, M.M. Doeff, M. Metzger, S. Kuppan, A study of model-based protective fast-charging and associated degradation in commercial smartphone cells: insights on cathode degradation as a result of lithium depositions on the anode, Advanced Energy Materials 11 (2021) 2003019.

17 C.-A. Lo, Y. Wang, V.R. Kankanallu, A. Singla, D. Yen, X. Zheng, K.G. Naik, B.S. Vishnugopi, C. Campbell, V. Raj, C. Zhao, L. Ma, J. Bai, F. Yang, R. Li, M. Ge, J. Watt, P.P. Mukherjee, D. Mitlin, Y. Karen Chen-Wiegart, Interdependence of support wettability-electrodeposition rate-sodium metal anode and SEI microstructure, Angewandte Chemie International Edition 64 (2025) e202412550.

18 D.-Y. Han, S. Kim, J. Son, S. Lee, S. Cho, D. Lee, J. Lee, T.K. Lee, N.-S. Choi, S. Park, Synergistic coupling of host and electrolyte achieving 1270 Wh L−1 in anode-free lithium metal batteries, Advanced Materials n/a (2025) e15906.

19 Z. Su, T.-T. Nguyen, C. Le Bourlot, F. Cadiou, A. Jamali, V. De Andrade, A.A. Franco, A. Demortière, Towards a local in situ x-ray nano computed tomography under realistic cycling conditions for battery research, Chemistry–Methods 2 (2022) e202100051.

20 Y.-S. Yu, C. Kim, Y. Liu, A. van der Ven, Y.S. Meng, R. Kostecki, J. Cabana, Nonequilibrium pathways during electrochemical phase transformations in single crystals revealed by dynamic chemical imaging at nanoscale resolution, Advanced Energy Materials 5 (2015) 1402040.

21 Y. Xie, H. Wang, G. Xu, J. Wang, H. Sheng, Z. Chen, Y. Ren, C.-J. Sun, J. Wen, J. Wang, D.J. Miller, J. Lu, K. Amine, Z.-F. Ma, In operando XRD and TXM study on the metastable structure change of $NaNi_{1/3}Fe_{1/3}Mn_{1/3}O_2$ under electrochemical sodium-ion intercalation, Advanced Energy Materials 6 (2016) 1601306.

22 M. Wolfman, B.M. May, V. Goel, S. Du, Y.-S. Yu, N.V. Faenza, N. Pereira, A. Grenier, K.M. Wiaderek, R. Xu, J. Wang, K.W. Chapman, G.G. Amatucci, K. Thornton, J. Cabana, Origin of rapid delithiation in secondary particles of LiNi0. 8Co0. 15Al0. 05O2 and $LiNi_yMn_zCo_{1-y-z}O_2$ cathodes, Advanced Energy Materials 13 (2023) 2300895.

23 J. Wang, Y.K. Chen-Wiegart, J. Wang, In situ three-dimensional synchrotron X-ray nanotomography of the (de) lithiation processes in tin anodes, Angewandte Chemie International Edition 53 (2014) 4460.

24 J. Wang, L. Wang, C. Eng, J. Wang, Elucidating the irreversible mechanism and voltage hysteresis in conversion reaction for high-energy sodium–metal sulfide batteries, Advanced Energy Materials 7 (2017) 1602706.
25 Y.-M. Chang, Y.-C. Wen, T.-Y. Chen, C.-C. Lin, S.-C. Huang, C.-S. Ni, A.-Y. Hou, C.-W. Hu, Y.-F. Liao, C.-H. Kuo, S.-F. Liu, W.-W. Wu, L.-J. Li, H.-Y. Chen, Understanding charge storage mechanisms for amorphous $MoSnSe_{1.5}S_{1.5}$ nanoflowers in alkali-ion batteries, Advanced Energy Materials 13 (2023) 2301125.
26 N. Sun, Q. Liu, Y. Cao, S. Lou, M. Ge, X. Xiao, W.-K. Lee, Y. Gao, G. Yin, J. Wang, X. Sun, Anisotropically electrochemical–mechanical evolution in solid-state batteries and interfacial tailored strategy, Angewandte Chemie International Edition 58 (2019) 18647.
27 C.-Y. Lai, Y.-S. Liao, H.-Y. Ku, W.-Y. Jao, S. Gull, H.-Y. Chen, J.-P. Chou, C.-C. Hu, Enhancing zinc electrode stability through pre-desolvation and accelerated charge transfer via a polyimide interface for zinc-ion batteries, Small 20 (2024) 2401713.
28 S. Jo, J. Han, S. Seo, O.-S. Kwon, S. Choi, J. Zhang, H. Hyun, J. Oh, J. Kim, J. Chung, H. Kim, J. Wang, J. Bae, J. Moon, Y.-C. Park, M.-H. Hong, M. Kim, Y. Liu, I. Sohn, K. Jung, J. Lim, Solid-state reaction heterogeneity during calcination of lithium-ion battery cathode, Advanced Materials 35 (2023) 2207076.
29 H. Pan, S. Jiao, Z. Xue, J. Zhang, X. Xu, L. Gan, Q. Li, Y. Liu, X. Yu, H. Li, L. Chen, X. Huang, The roles of Ni and Mn in the thermal stability of lithium-rich manganese-rich oxide cathode, Advanced Energy Materials 13 (2023) 2203989.
30 E. Allen, J. Macholz, M. Nisbet, N. Chowdhury, F. Wang, M. Porwal, D.T. Keane, W. Guise Jr., Z. Dundek, M. Caple, V. Nikitin, J. Cabana, Q. Dai, A. Lipson, Cathode upcycling for direct recycling of lithium-ion batteries using a precipitation approach, Advanced Energy Materials 15 (2025) 2500699.

6

Synchrotron-based Soft X-ray Absorption Spectroscopy for Electronic and Interfacial Characterization of Battery Materials

6.1 Fundamentals and Theory

When an assembly of atoms is irradiated with X-rays, part of the incident photons is absorbed by the material. At a specific photon energy, the absorption coefficient shows a sudden increase, known as the absorption edge. The position of this edge corresponds to the binding energy of a particular core level. At the absorption threshold, the photon energy matches the core-electron binding energy; in other words, the edge represents transitions from the ground state to the lowest unoccupied electronic states [1].

Kossel [2] first gave a classical explanation of X-ray absorption, suggesting that absorption takes place only when the photon energy is high enough to excite an electron into a free trajectory [2]. Petersen, Smoluchowski, and Kurylenko extended this understanding to molecular and crystalline systems, forming the foundation of modern X-ray absorption spectroscopy (XAS) theory [3, 4]. Since the advent of synchrotron radiation facilities and the progress of electronic structure theory in the 1960s, XAS has developed into a highly accurate and versatile analytical method.

The fundamental principle of soft XAS (sXAS) involves the excitation of core-level electrons into unoccupied states near the Fermi level using tunable synchrotron X-rays, enabling direct investigation of the local electronic and chemical environment. When the photon energy equals the binding energy of a core electron, a transition occurs, creating an excited state characterized by the presence of a core hole [5]. The detected absorption intensity is proportional to the transition probability, meaning that the resulting spectral line shape reflects the distribution of unoccupied electronic states and serves as a fingerprint of the valence and conduction band structures [6].

6.1.1 Fundamental Principles of X-ray Absorption

The distinction between "soft" and "hard" X-rays originates from their penetration depth and photon energy. The Hard X-rays, with photon energies ranging from a few keV to several tens of keV, are capable of penetrating thick samples, whereas soft X-rays, typically spanning several tens of electron volts up to about 2 keV, have

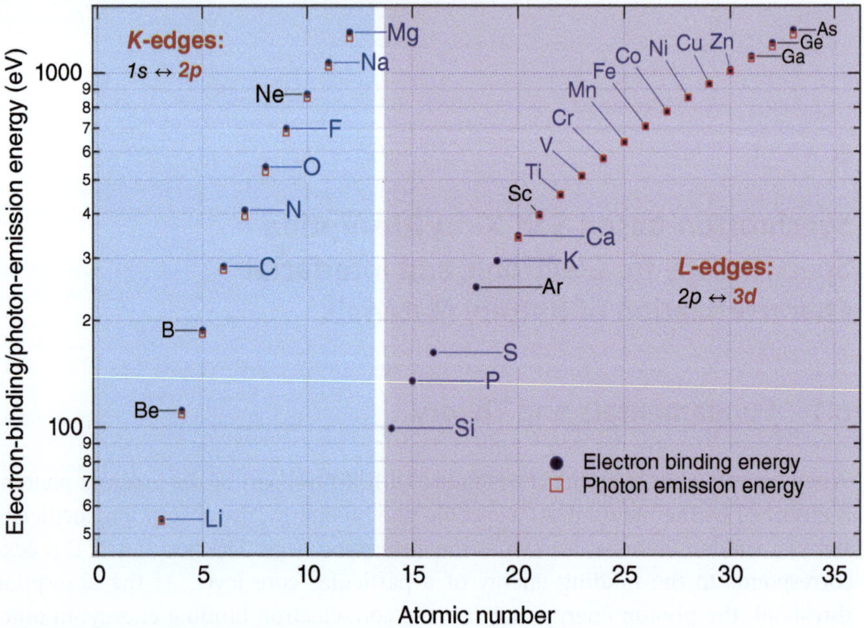

Figure 6.1 In the soft X-ray region (50 eV to 2 keV), spectroscopy provides access to core-level binding energies and corresponding primary X-ray emission energies associated with radiative decay processes [7] / with permission of IOP Publishing Ltd.

shallower penetration depths, which about tens to hundreds of nanometers [6]. This energy range conveniently encompasses the *K*-edges of light elements (such as O, N and F) as well as the *L*-edges of transition metals, as shown in Figure 6.1. Furthermore, the relatively long wavelength of soft X-rays also enables sensitive probing of long-range order and nanoscale morphology in materials.

In photon-in—photon-out (PIPO) soft X-ray methods, the probing depth depends on the incident photon energy, typically ranging from tens to several hundred nanometers. In core-level spectroscopy, when X-ray photons are absorbed and the excited core electrons remain within the solid (material) (instead of being emitted as photoelectrons), the excitation probability *W* is described by the Fermi Golden Rule:

$$W_{fi} = (2\pi/\hbar)\left|< \Phi_f|T|\Phi_i >\right|^2\delta\left(E_f - E_i - \hbar\Omega\right) \tag{6.1}$$

where Φ_i and final state Φ_f denote the initial and final states, the transition operator *T* is constrained by the electric dipole selection rule ($\Delta l = \pm 1$), which determines the allowed orbital transitions responsible for the observed absorption edges [1]. Specifically, transitions from s to p orbitals correspond to the *K*-edge (i.e., 1s → 2p excitations), while transitions from p to d orbitals correspond to the *L*-edge (i.e., 2p → 3d excitations). Quadrupole transitions correspond to final states whose angular momentum quantum number differs by two or remains the same as that of the initial state ($\Delta l = \pm$ 2 or 0, e.g., s → d, p → f or s → s, p → p, etc.). They are some hundred times weaker than the dipole transitions and can be neglected in most cases. These transitions define the characteristic edges that are experimentally detected in sXAS. For transition metal (TM) oxides, the soft X-ray *L*-edge originates from 2p to 3d electronic transitions, offering direct insight into the occupancy and hybridization of the 3d orbitals. In contrast, the oxygen *K*-edge corresponds to 1s → 2p excitations, which reveal the degree of hybridization between O 2p and TM 3d states. The intensity and shape of these absorption features therefore provide a direct indicator of the local electronic configuration and chemical bonding in the system. In lithium-ion battery cathode materials, for example, the spectral line shape of the TM L_3-edge varies systematically with lithiation level, directly reflecting the reconfiguration of 3d electronic states [8]. By analyzing the relative peak positions and intensities, one can quantify the redox states of Fe, Co, Ni, and other TMs.

Thus, sXAS not only differentiates chemical states and coordination environments but also elucidates orbital hybridization mechanisms and charge-transfer pathways. Supported by the high brilliance and tunability of synchrotron light sources, sXAS offers both high energy resolution and potential time-resolved capabilities. Consequently, it has become an indispensable technique for investigating the electronic structures of transition metal oxides and elucidating the key electronic mechanisms governing energy storage materials.

6.1.2 Self-absorption Effect and Multiplet Effect

In sXAS, the measured spectra are not only determined by the intrinsic electronic transitions of the absorbing atoms but also by several physical effects that influence the observed line shape and intensity. Among these, the self-absorption effect and the multiplet effect are particularly significant and must be considered for accurate data interpretation.

1. **Self-absorption Effect**
   The self-absorption effect occurs mainly in total-fluorescence-yield (TFY) detection when the fluorescence photons emitted after X-ray excitation are partially reabsorbed by the sample before escaping its surface [7]. This reabsorption process distorts the proportional relationship between the detected fluorescence signal and the actual absorption coefficient of the material. As a consequence, the recorded spectra exhibit a nonlinear response, characterized by an artificial suppression of the peak intensity and an alteration of the relative strength between the pre-edge and main-edge features. The magnitude of self-absorption depends strongly on the sample thickness, density, and photon energy. It becomes more pronounced in concentrated or thick specimens and at low photon energies where the attenuation length of soft X-rays is short. To minimize this effect, experimental geometries with shallow incident angles or the use of thin film samples are typically adopted. Properly correcting for self-absorption is essential for maintaining quantitative accuracy and preserving the true spectral features of the absorbing species.

2. Multiplet Effect

The multiplet effect is one of the most fundamental physical phenomena that influence the spectral characteristics of sXAS, especially at the TM *L*-edge. It arises from strong intra-atomic Coulomb and exchange interactions between the core hole created during the excitation process and the valence-shell electrons, leading to a manifold of final state configurations with distinct energies and symmetries. These interactions couple the spin and orbital angular momenta of the 2p core hole with those of the 3d electrons, producing a complex splitting pattern in the absorption spectrum that cannot be accounted for within a single-electron framework.

In TM oxides and coordination compounds, excitation of a 2p electron to the unoccupied 3d states generates a $2p^5 3d^{n+1}$ final configuration. Because the radial wave functions of 2p and 3d orbitals overlap strongly, their Coulomb and exchange interactions (quantified by the Slater-Condon parameters F and G) produce multiple coupled states with different total angular momenta and spin multiplicities. The magnitude of these splitting can reach several electronvolts, often exceeding the core hole lifetime broadening, thus becoming clearly visible as multiple peaks or shoulders in experimental L_3-and L_2-edge spectra [1]. The intensity and energy separation of these multiplet components depend sensitively on the oxidation state, crystal-field strength, spin configuration, and local symmetry of the absorbing atom.

The atomic multiplet model describes these features by considering the electron-electron repulsion, spin-orbit coupling, and exchange terms within the atom. However, to account for the influence of the surrounding ligand environment, the crystal-field multiplet model introduces an electrostatic potential that splits the 3d orbitals into t_{2g} and e_g subsets [1, 6]. The resulting spectral profile reflects both the intra-atomic multiplet coupling and the ligand-field symmetry. When charge-transfer interactions between metal and ligand orbitals are significant, the charge-transfer multiplet model extends this framework by mixing configurations such as 3dn and $3d^{n+1}\underline{L}$ (where $\underline{L}$ denotes a ligand hole) [1]. This inclusion allows the simulation of covalency effects and reproduces the experimental spectra of TM oxides and battery electrode materials with high fidelity.

The presence of multiplet effects in sXAS manifests most prominently in the TM L_3-edge, where the 2p → 3d transition gives rise to complex fine structures. In systems such as NiO or Fe-based cathode materials, the observed multiple peaks correspond to transitions into different spin and orbital configurations of the 3d shell [5]. These spectral features encode valuable information on the local spin state, oxidation number, and degree of metal-ligand hybridization.

Overall, the multiplet effect determines the fine structure and physical interpretability of sXAS spectra. While the self-absorption effect alters experimental intensities through photon reabsorption, the multiplet effect fundamentally reflects the intrinsic many-body nature of the electronic transitions. Understanding and properly modeling this effect through atomic, crystal-field, and charge-transfer multiplet theories is essential for the reliable elucidation of the local electronic structures of TM oxides, particularly those used as electrodes in electrochemical energy storage systems.

6.1.3 Signal Collection

The intrinsic absorption coefficient represents the extent to which the incoming X-rays are attenuated. However, because soft X-rays have a very limited penetration depth, direct attenuation measurements can only be performed on ultrathin specimens. In battery research, apart from scanning transmission X-ray microscopy, sXAS is generally measured by recording the intensity of decay products generated during core hole relaxation. As shown in Figure 6.2, the commonly employed detection channels include Auger electron yield (AEY), total electron yield (TEY) (Figure 6.2b), partial fluorescence yield (PFY), and TFY (Figure 6.2c). Because electrons have an extremely limited escape depth, AEY and TEY are strongly surface-sensitive methods, probing only a few nanometers below the surface. Conversely, PFY and TFY offer enhanced bulk sensitivity, with probing depths reaching hundreds of nanometers depending on the photon energy. Modern sXAS instruments are generally capable of simultaneously collecting TEY and TFY signals, which allows direct comparison between the electronic states of the surface and the bulk in battery materials [5]. Nevertheless, quantitative interpretation of TFY data requires caution, as significant line-shape distortions may arise from self-absorption effects. Such distortions arise from the abrupt variation in X-ray penetration depth as the photon energy is swept through an absorption edge. Under such circumstances, the inverse PFY (iPFY) method should be applied, as discussed in detail later.

X-ray transmission is the most direct approach for acquiring sXAS spectra at the *K*-edges of low-Z elements, such as the oxygen *K*-edge, allowing quantitative evaluation of the absorption cross section. However, for light elements like oxygen, the attenuation length of 530 eV X-rays at the oxygen *K*-edge in layered oxides is roughly 100 nm, requiring the preparation of ultrathin samples [6]. Transmission measurements are quantitatively reliable only when the specimen has uniform thickness, any variations in thickness will distort the spectral line shape, typically causing the intense peaks to appear artificially flattened. This phenomenon, commonly referred to as the pinhole effect, also affects measurements of inhomogeneous samples, such

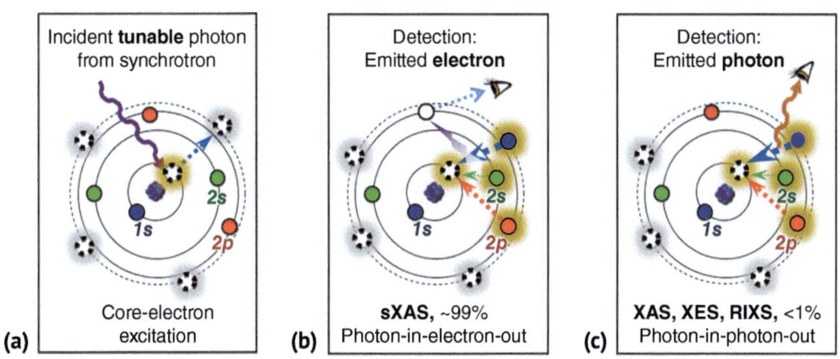

Figure 6.2 An atomic illustration of core-level soft X-ray absorption and emission spectroscopy, including (a) core-electron excitation, (b) photon-in-electron-out techniques such as sXAS, and (c) photon-in–photon-out methods like XAS, XES, and RIXS Adapted from [9].

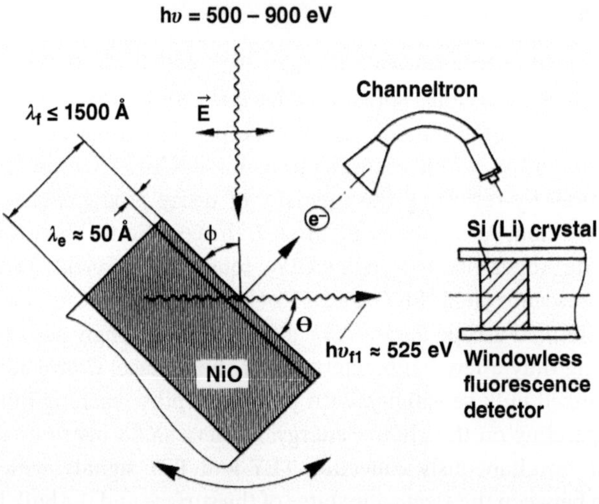

Figure 6.3 The schematic illustrates the experimental configuration for FY and TEY detection. In the fluorescence detection geometry, the angles φ and θ define the directions of the incident beam and the emitted fluorescence photons relative to the sample surface, respectively [10] / with permission of the American Physical Society.

as those containing nanoparticles or other nanostructured features detected with an area-averaged X-ray probe.

Alternative detection approaches are based on observing the decay pathways that occur after a core hole has been excited. As shown in Figure 6.3, the fluorescence yield (FY) method records the number of X-ray photons emitted during the relaxation process, while the electron yield (EY) method detects the emitted electrons. In the Auger process, the energy emitted as an electron fills the core hole is transferred to a different electron, which is subsequently ejected [10]. Additional relaxation steps can result in further electron emission, and EY quantifies the total number of electrons emitted through these processes.

1. **EY Detection**
   TEY detection, electrons escaping from the sample surface are recorded using electron analyzers. The detected signal may come from integrated counts in energy-dispersive detectors or, for conductive materials, from measuring the current that flows to the sample. When electrons of specific kinetic energies are selected, the technique is referred to as PEY; if a defined decay channel is monitored, it is known as AEY.

A distinct characteristic of EY detection is its very shallow probing depth: the escape length of electrons is typically only a few nanometers, which is considerably shorter than the penetration depth of the incident X-rays. As a result, the TEY signal generally avoids saturation. As a result, EY measurements make sXAS extremely sensitive to the surface, allowing signals from the surface and subsurface regions to be preferentially amplified. At the same time, this strong surface sensitivity implies that the TEY spectrum may not accurately reflect the bulk properties. In cases of surface oxidation or chemical modification, the obtained TEY spectrum differs from that of the bulk material. Even for atomically clean single-crystal surfaces, discrepancies arise because surface atoms have different coordination environments and thus distinct electronic structures. Given a probing depth of roughly 4 nm, the topmost atomic layers can contribute as much as 10% to the overall spectral intensity. Furthermore, many crystalline materials exhibit surface reconstructions that extend several layers deep, so TEY data often contain some degree of surface-induced influence.

When TEY measurements are performed in a gaseous environment, conversion EY (CEY) becomes an alternative detection option [11]. In such measurements, electrons emitted from the sample ionize surrounding gas molecules, and the generated ions are counted to reconstruct the sXAS spectral profile. Typically, two detectors are employed—one monitoring the gas phase alone, and another recording both the sample surface and gas phase contributions—allowing the separation of the two signals. The ionization efficiency is governed by the kinetic energy of the emitted electrons and the type of gas molecules involved, which makes careful calibration crucial for obtaining reliable CEY measurements.

A similar CEY approach can be used for liquid environments. In this configuration, sXAS spectra are obtained for surfaces in contact with a liquid by detecting ions generated within the liquid through electron-induced ionization, a method often termed ion current detection [11]. Regardless of the medium (gas or liquid), CEY techniques must be carefully calibrated and tested to ensure their linear response relative to the true sXAS absorption cross section.

2. **FY Detection**
   Once a core-level vacancy forms, an electron from an outer, higher energy shell transitions to occupy it, and the surplus energy is emitted as an X-ray photon. Detecting these fluorescent X-rays after radiative relaxation provides a signal that reflects the X-ray absorption spectral profile. In TFY detection, one detects all emitted X-ray photons, from all elements present in the sample. Assuming that nonresonant photoionization contributes only a constant background, the TFY signal can reproduce the sXAS spectral shape.

However, spectral distortions may arise from saturation and self-absorption effects. This occurs because off-resonant X-ray emission typically penetrates more deeply into a material than incident X-rays tuned near an absorption edge. As the photon energy is swept across the edge, variations in the absorption coefficient lead to corresponding changes in the X-ray penetration depth. If the background absorption is negligible, this variation can cause severe distortion of the observed spectral shape. In contrast, when the background absorption dominates, the overall attenuation length remains constant across the edge, and saturation does not occur. Consequently, TFY is applicable primarily to materials in which the absorbing element is present in low concentration. For highly dilute systems, the signal from other elements may dominate, leading to a weak signal-to-noise ratio for the element of interest. For dense materials such as bulk oxides, TFY spectra typically appear heavily saturated.

Beyond saturation, self-absorption (the reabsorption of emitted fluorescence photons within the sample) can further modify the spectral line shape. This effect becomes significant only in concentrated samples. Both saturation and self-absorption depend on the measurement geometry, making it possible to assess and correct their effects experimentally. For the oxygen *K*-edge, no spectral distortions arise from state-dependent decay processes. An alternative approach is PFY detection, in which only a specific portion of the emitted X-ray energy spectrum is measured. PFY can be advantageous when the emission from another element dominates the overall fluorescence, and it allows for the detection of XAS spectra even from extremely dilute elements, with concentrations as low as 100 ppm [6]. A particular variation of this technique is the iPFY method, where the fluorescence from a different element than the one under study is monitored.

Extended X-ray absorption fine structure (EXAFS) spectroscopy is now widely used as a powerful method to probe the local atomic surroundings of a particular element within a material. In most cases, the X-ray absorption of bulk materials is measured using the transmission method. However, this approach is generally impractical in the soft X-ray region, as it would require the preparation of extremely thin and homogeneous films, which is impractical for most studies [12]. In the case of concentrated (and thick) samples, the fluorescence EXAFS signal exhibits a systematic, energy-dependent decrease in amplitude known as self-absorption (or reabsorption). This attenuation arises from the absorption of both the incoming X-rays and the outgoing fluorescence photons as they traverse the sample. Under such conditions, applying standard EXAFS analysis without correction can introduce errors exceeding 50% in parameters derived from the EXAFS amplitude. This reabsorption process distorts the proportional relationship between the detected fluorescence signal and the actual absorption coefficient of the material. As a consequence, the recorded spectra exhibit a nonlinear response, characterized by an artificial suppression of the peak intensity and an alteration of the relative strength between the pre-edge and main-edge features. The magnitude of self-absorption depends strongly on the sample thickness, density, and photon energy. It becomes more pronounced in concentrated or thick specimens and at low photon energies where the attenuation length of soft X-rays is short. To minimize this effect, experimental geometries with shallow incident angles or the use of thin film samples are adopted. However, to broaden the range of materials and experimental conditions for which sXAS can be effectively applied, more advanced detection approaches have been developed [6]. One such method is the iPFY technique, which enables the measurement of the absorption coefficient with bulk sensitivity while avoiding both saturation and self-absorption effects. This approach uses an energy-resolving detector to measure the PFY. In contrast to conventional PFY, which tracks resonant emission from the element being excited, iPFY records the normal (nonresonant) X-ray emission spectrum (NXES) originating from an element other than the one whose absorption edge is being studied. The inverse of this NXES-based PFY signal-referred to as the iPFY has been demonstrated, both experimentally (for example, at the Cu *L*- and Nd *M*-edges of $La_{1.475}Nd_{0.4}Sr_{0.125}CuO_4$) and theoretically, to be directly proportional to the intrinsic XAS cross section and unaffected by self-absorption or saturation effects [12]. The fundamental principle of iPFY differs from that of traditional FY or EY detection. Whereas FY and EY measure the number of core hole excitations through an absorption edge, iPFY effectively probes the X-ray attenuation length, closely resembling transmission-based measurements. Given these advantages, the iPFY technique have broad applicability in sXAS across a wide variety of materials and experimental environments. In conclusion, properly correcting for self-absorption is essential for maintaining quantitative accuracy and preserving the true spectral features of the absorbing species.

6.2 Equipment and Analysis

As a core branch of advanced synchrotron radiation characterization techniques, sXAS serves as a powerful tool for probing the fundamental microscopic nature of matter. It leverages photon properties in the 200–2000 eV energy range and relies on highly integrated, precise experimental systems. The technical foundation of the method resides in the resonant absorption interaction between soft X-rays and the core-level electrons of atoms. When this intrinsic physical process is coupled with the high brightness and superior energy resolution characteristic of modern synchrotron radiation sources, it facilitates the establishment of a comprehensive characterization framework. This framework effectively bridges the gap from the macroscopic equipment chain to the extraction of detailed microscopic information.

The technique demonstrates particularly high sensitivity for light elements, such as carbon, nitrogen, and oxygen, as well as for the *L*-edges of TMs. Furthermore, its inherently shallow penetration depth, typically ranging from a few nanometers to several tens of nanometers, along with its inherent capacity for multimodal detection, secures its unique and indispensable position in contemporary research. This applies especially to studies focusing on surfaces and interfaces, low-dimensional structures, and complex functional materials. From a fundamental scientific perspective, sXAS provides comprehensive analytical coverage across several core dimensions. These include the identification of elemental composition, the quantitative analysis of chemical states, the investigation of electron orbital hybridization and spin configurations, and the determination of local atomic parameters such as bond lengths and coordination numbers.

Regarding its applied scientific value, the technique delivers precise and systematically obtained data that serves as critical scientific support for several key technological domains. These domains encompass the exploration of catalytic reaction mechanisms, the performance optimization of next-generation energy materials, the deliberate control and tuning of electronic structures in magnetic materials, and the rational design of interfaces in advanced nanodevices. A subsequent detailed discussion encompassing the composition of the experimental apparatus, the methodologies for multidimensional data analysis, and the specific configuration of dedicated beamlines will serve to fully elucidate the operational principles, inherent advantages, and the broad spectrum of application potential inherent to sXAS technology.

6.2.1 Typical sXAS Equipment

The complete experimental setup for synchrotron radiation-based sXAS is, in essence, a highly integrated and precision-optimized system combining photon science and vacuum technology, designed specifically for probing the electronic structure of matter. The entire process originates at the synchrotron radiation source itself. This source is not merely a luminous body; it generates an intense, continuous spectrum of electromagnetic waves emitted tangentially by relativistic electrons as they are deflected by strong magnetic fields within a storage ring. Here, an undulator utilizes its periodic magnetic field to induce small-amplitude oscillations in the electron beam. The radiation from successive oscillations interferes coherently, producing an exceptionally brilliant, well-collimated, and partially coherent photon beam, which forms the foundation for achieving high signal-to-noise ratios and rapid data acquisition. Meanwhile, a bending magnet provides broader-spectrum, highly stable radiation suitable for more general-purpose measurements.

Figure 6.4 illustrates a representative schematic layout of a sXAS beamline at a synchrotron radiation facility. The raw radiation enters a sophisticated beamline system, which functions as an integrated and precision-engineered assembly of optical components rather than a mere conduit. Its primary role is to isolate monochromatic light of a specific energy from the broadband synchrotron spectrum using a monochromator. Within the soft X-ray regime, this is predominantly accomplished with a grating monochromator, for instance of the variable-inclusion-angle type. In such devices, the desired wavelength is selected through precise rotation of the grating, thereby altering the diffraction angle. A defining performance parameter is the energy resolution, which can exceed 0.01%, enabling the detection of electronic-state shifts as subtle as 0.1 eV near absorption edges. Following this spectral purification, the beam is typically focused by a mirror system based on the principle of total external reflection, such as a Kirkpatrick–Baez mirror pair. This focusing step compresses the beam spot to micrometer or even nanometer dimensions, permitting spatially resolved spectromicroscopy. Concurrently, dedicated harmonic-rejection mirrors within the beamline filter out higher-order harmonics. This ensures that the photons ultimately incident on the sample are spectrally pure and free from stray radiation.

The processed beam finally arrives at the multifunctional experimental endstation, where it interacts with matter and its effects are decoded. The core interaction region must be maintained under an ultrahigh vacuum environment, typically better

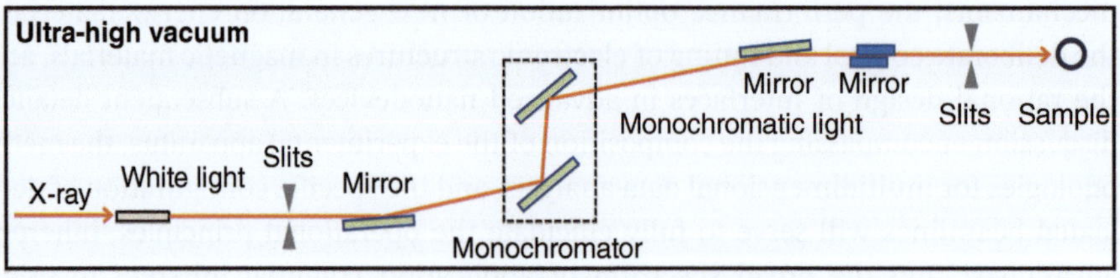

Figure 6.4 Schematic layout of a sXAS beamline at a synchrotron radiation facility.

than 10^{-8} Pascal. This is an absolute prerequisite both to prevent severe absorption of soft X-rays by air and to preserve clean sample surfaces for authentic surface science studies. The sample is mounted on a precision-manipulable stage and can be integrated with various sample environments tailored to research needs. These may include cryogenic temperatures down to liquid helium (approx. 4.2 K), high temperatures up to several hundred Kelvin, magnetic fields of several Tesla, or in-situ gas/liquid reaction cells simulating real chemical reaction conditions. The absorption event is captured through complementary detection modes, including the surface-sensitive TEY, the bulk-sensitive FY, and the chemically specific AEY. Furthermore, by raster-scanning the sample under the focused beam, two-dimensional spatial mapping of chemical composition and valence states can be achieved.

Therefore, an advanced sXAS setup represents a complete scientific continuum that extends from the intrinsic properties of the synchrotron source, through the spectral shaping and spatial conditioning of the beamline, to the final transformation of photon-matter interactions into quantitative electronic structure information at the endstation via multimode detection within sophisticated sample environments. Every component within this continuum is engineered with meticulous intent, driven by the pursuit of ultimate performance in sensitivity, spatial or temporal resolution, and environmental realism required to address frontier scientific questions across disciplines including catalysis, energy materials, and magnetism.

6.2.2 Analysis of Elemental Composition and Valence State

Elemental identification is achieved based on the characteristic energy positions of absorption edges in an X-ray absorption spectrum, enabling the precise qualitative determination of the elemental species present within a sample [13]. Each element possesses a distinct “fingerprint” in terms of its absorption edge energy—for instance, the carbon (C) *K*-edge is located near 284 eV, the iron (Fe) L_3-edge around 711 eV, and the oxygen (O) *K*-edge at approximately 532 eV. By comparing the edge energies observed in the experimental spectrum against a standardized database of elemental energies, one can definitively identify the major constituents, trace dopants, and impurity elements. This foundational step is crucial for subsequent, more detailed analysis of valence states and local structure.

The oxidation state of an element exhibits a direct and quantifiable correlation with several characteristic spectral features observable in its X-ray absorption spectrum. These diagnostic features encompass the energy position and shift of the absorption edge, the characteristic asymmetry of the edge line shape, the spectral intensity and full-width at half-maximum of pre-edge peaks, which often originate from dipole-forbidden transitions such as those from a 1s to a 3d orbital, as well as the peak intensity of the white line. The white line is defined as the main absorption peak located directly at the edge onset. The fundamental physical principle governing these correlations follows a well-established trend. As the oxidation state of an element increases, the electron density surrounding its atomic nucleus correspondingly decreases. This reduction in shielding leads to a stronger effective nuclear charge experienced by the core electrons. The increased nuclear attraction, in turn, raises the binding energy required to eject a photoelectron from the core level. Consequently, the absorption edge undergoes a systematic shift towards higher energy, a phenomenon widely recognized as a chemical shift. Additionally, the intensity of pre-edge features frequently shows a marked enhancement with rising oxidation state, a behavior that is especially prominent and diagnostically valuable for TM elements. By performing a quantitative analysis of these spectral parameters using established analytical methods, including linear combination fitting and spectral deconvolution, the specific valence state of an element can be determined with a high degree of precision. This capability allows for the accurate distinction between species such as Fe^{2+} and Fe^{3+}. This detailed spectral analysis furnishes essential insights into critical material properties, serving to elucidate complex redox reaction mechanisms and to track the dynamic evolution of the electronic state in active components under operational conditions.

6.2.3 Analysis of Electronic Structure Information

X-ray Absorption Near-Edge Structure (XANES) serves as a sensitive probe for analyzing orbital hybridization and the symmetry of the local atomic environment, with spectral features directly revealing the fundamental nature of chemical bonding. For example, carbon atoms with sp^3 hybridization, such as those in diamond, and those with sp^2 hybridization, as found in graphite, exhibit marked differences in their *K*-edge XANES spectra, particularly in pre-edge peak intensity and the splitting pattern of the main absorption feature. The symmetry of the coordination geometry, including tetrahedral, octahedral, or square planar arrangements, leads to distinct splitting patterns at the absorption edge. TM ions in octahedral coordination, such as Mn^{3+}, typically display a well-defined double-peak structure in their *L*-edge spectra, whereas this splitting is less pronounced under tetrahedral coordination. By systematically interpreting spectral parameters such as peak energy shifts, the number of split features, and relative intensity ratios, it is possible to infer key bonding characteristics. These include the relative contributions of σ and π bonding, the degree of orbital hybridization, and the symmetry of the coordination environment. This analysis, in turn, helps clarify the relationships between a material's electronic delocalization, its charge-transfer capability, and its resulting electrochemical activity.

For 3d transition metal elements, including Fe, Co, and Ni, sXAS, particularly in the *L*-edge region, is an effective technique for distinguishing their electron spin states, such as high-spin, low-spin, or intermediate-spin configurations. Differences in spin state induce systematic variations in the spectral features, specifically in the intensity ratio and energy separation of the spin-orbit split peaks. For instance, high-spin Fe^{3+}, as present in Fe_2O_3, shows a higher main peak intensity at the L_3-edge compared to low-spin Fe^{3+} found in compounds like $Fe(CN)_6^{3-}$. Through the quantitative analysis of these spin-sensitive spectral parameters, combined with supporting calculations based on crystal-field theory, the specific electronic spin configuration of an element can be determined conclusively. This provides critical support for understanding the mechanisms behind magnetic interactions, electron transport kinetics, and the intrinsic nature of catalytically active sites in various functional materials.

6.2.4 Analysis of Local Structural Information

The oscillatory features observed in the EXAFS region originate from the interference of photoelectrons scattered by neighboring atoms surrounding the absorbing species. The frequency and amplitude of these oscillations are directly correlated with key local structural parameters, which include the identity of the coordinating atoms, their bond distances, and the coordination number [14]. To decode this information, the EXAFS spectrum undergoes a series of analytical steps, including Fourier transformation, phase correction, and fitting with theoretical scattering models. This rigorous quantitative analysis allows for the precise extraction of structural parameters, such as the bond length between the central atom and its neighbors, with a typical accuracy near 0.01 Å, the coordination number, generally determined within an uncertainty of approximately 0.5 atoms, and the elemental type of the coordinating ligands. Utilizing these refined parameters, a detailed model of the short-range atomic arrangement around the absorbing atom can be constructed, thereby clarifying its local order. This methodology proves particularly valuable for monitoring dynamic changes in the coordination environment of functional elements within materials. A pertinent example is tracking the contraction or expansion of metal-oxygen bonds, such as Ni-O, during processes like lithium intercalation in battery electrodes, providing essential mechanistic insights into ion diffusion pathways and structural stability.

Furthermore, EXAFS analysis quantitatively characterizes the degree of structural disorder within the local environment. This encompasses both static disorder, which refers to inherent variations in atomic positions, and dynamic disorder arising from thermal vibrations. The magnitude of disorder is represented by the Debye-Waller factor obtained from the fitting procedure. Additionally, when multiple distinct coordination environments coexist for the same element within a material, the EXAFS spectrum reflects a superposition of contributions from each environment. Advanced fitting methodologies allow for the deconvolution of these contributions, determining the proportional distribution, specific bond lengths, and coordination numbers for each distinct site. This analytical power is particularly valuable for studying complex systems such as defective structures, amorphous solids, and interfacial regions, offering profound understanding of how local structural heterogeneity governs material properties including ionic conductivity, mechanical integrity, and long-term electrochemical performance.

6.2.5 Analysis of Surface and Interface Information

The analysis of surface chemical states is a key strength of sXAS. Due to the shallow penetration depth of soft X-rays, typically ranging from a few to several tens of nanometers, the technique is inherently surface-sensitive. It selectively probes the chemical information from the topmost layers of a sample, effectively avoiding interference from the bulk material. Through sXAS, direct information can be obtained regarding the valence state distribution of surface atoms, the evolution of their coordination environment, and the identity and bonding configuration of adsorbed species such as hydroxyl groups, water molecules, or organic functional groups. This capability is critically important in fields like catalysis and energy materials. For example, it enables the study of the chemical state of active sites on lithium-ion battery electrode surfaces, the valence and coordination structure of surface modification layers in thin film materials, and the interactions of functional groups on biological membranes. These insights provide essential guidance for optimizing surface properties and suppressing undesirable interfacial side reactions.

In complex systems such as multilayer films, heterojunctions, and electrode-electrolyte interfaces, sXAS allows for the selective characterization of the interfacial region. This is achieved by controlling the probing depth through adjustments in the X-ray incident angle. By analyzing the absorption edge features of specific elements at the interface, one can elucidate the atomic arrangement, such as reconstructions induced by lattice matching or mismatch, the nature of interfacial chemical bonding, and charge-transfer behavior across the interface. Furthermore, leveraging the element-specific detection capability of XAS, the diffusion range and concentration profile of elements across the interface can be mapped with precision. This provides direct structural and chemical evidence crucial for understanding fundamental interfacial phenomena and for designing strategies to enhance material performance.

6.2.6 In-situ and Operando sXAS

In-situ and operando sXAS is an advanced extension of traditional static characterization, with its core lying in real time and dynamic tracking of the elemental valence states, electronic structures, and local structural evolution laws of materials under simulated real working environments or the samples' actual operating conditions [15]. The key distinction between the two lies in that in-situ characterization focuses on the sample being in an environment close to real working conditions, while operando characterization places greater emphasis on the sample being in the actual working process. The latter can more directly establish a dynamic correlation among structure, performance, and operating conditions, thereby serving as a crucial technique for revealing the intrinsic working mechanisms of materials.

Compared with static characterization, the core value of this technology is reflected in three aspects. Firstly, it enables dynamic tracking of evolutionary processes by breaking through the limitations of traditional "snapshot" characterization to capture the real time trajectories of elemental valence migration, coordination environment reconstruction, and electronic spin state transitions of materials during operating condition cycles, thereby revealing the complete life cycle of "active species formation-action-attenuation." Secondly, it facilitates the establishment of structure-activity relationships by avoiding structural "artifacts" caused by sample pretreatment, directly acquiring structural information under real operating conditions to accurately correlate changes in the microscopic structure of materials with their macroscopic performance. Thirdly, it possesses the capability for multifield coupling characterization, allowing the integration of multiexternal field regulation modules such as temperature, pressure, atmosphere, electric field, and magnetic field to simulate complex service environments while exploring the synergistic response mechanisms of material structures under multifield interactions.

The realization of in-situ/operando sXAS relies on the precise coupling of spectral detection systems and operating condition simulation modules with core design requirements focusing on four key aspects. First, operating condition-compatible sample cells must fulfill the dual requirements of soft X-ray penetration and environmental tightness. They utilize ultrathin window materials such as 100-nm thick Si_3N_4 membranes to ensure low-attenuation transmission of soft X-rays in the 200–2000 eV range. These cells separate electrolytes from the ultrahigh vacuum environment required for soft X-ray experiments while achieving reliable sealing for solid, liquid, and gas samples. For electrochemical studies, cells are typically integrated with a three-electrode system to enable simultaneous collection of XAS spectra and voltammetric responses, facilitating the capture of transient interfacial processes. Second, online operating condition regulation modules need to integrate dedicated control units tailored to specific application scenarios. Examples include electrochemical workstations for battery charge–discharge cycling, gas flow controllers for catalytic reaction atmospheres and temperature/pressure regulators for extreme-condition studies. These modules adapt to the operating condition simulation needs of diverse research fields including beyond lithium-ion batteries (Figure 6.5). Third, anti-interference and rapid detection design must counteract various signal interferences under operating conditions. High signal-to-noise ratio data acquisition is achieved through optimizing the optical path of sample cells such as oblique incidence to reduce scattering, adopting anti-interference detection modes including TEY for surface-sensitive analysis and total FY for bulk characterization, and enhancing the temporal resolution of beamlines with fast-scanning monochromators. Fourth, given the high absorption of soft X-rays by air, the sample

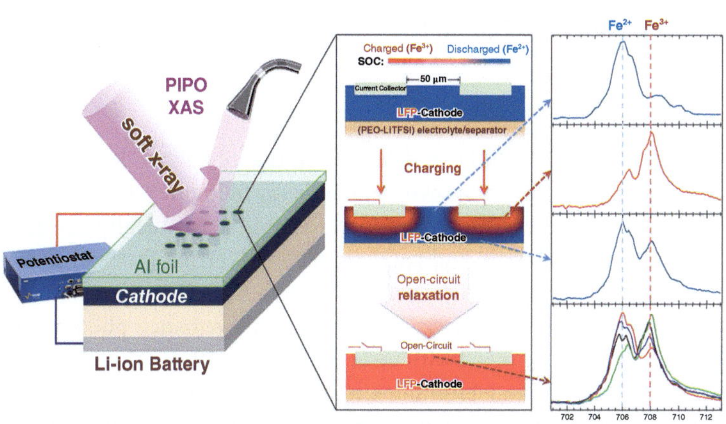

Figure 6.5 A schematic of in-situ cell for sXAS measurement and corresponding sXAS spectra at different state-of-charge [16] / with permission of John Wiley & Sons.

cells and beamline optical paths must be maintained under local vacuum or inert gas protection. This prevents photon attenuation and sample surface oxidation, which is critical for preserving the integrity of spectral data and avoiding artificial surface reactions.

This technology has become a core characterization tool in cutting-edge interdisciplinary research, with typical applications spanning multiple fields. In the field of catalysis, it can track the valence state changes and coordination environment reconstruction of active sites in heterogeneous catalytic reactions, revealing reaction pathways and deactivation mechanisms; in the field of energy materials, it can monitor the metal ion valence migration, chemical bond length changes of electrode materials, as well as the elemental composition and chemical state evolution of interface films, providing a basis for material modification and system optimization; in the field of magnetic materials, it can real-time detect the spin state transitions and magnetic moment coupling changes of TM elements during the application of external fields or temperature cycles, clarifying the microscopic mechanism of magnetic regulation; in the field of thin film and interface engineering, it can in-situ characterize element incorporation, valence state distribution, and interface chemical bond formation during thin film growth, or the dynamic process of interface charge-transfer during the operation of heterojunction devices.

Currently, in-situ and operando sXAS still faces several technical challenges, with corresponding solutions gradually maturing. To address the need for balancing temporal and spatial resolution, temporal resolution for dynamic processes can be enhanced through the pulse mode of synchrotron radiation sources combined with rapid detection technologies. For spatial resolution requirements, three-dimensional characterization of "space-time-structure" can be achieved by integrating focused beams with scanning imaging techniques. Regarding the challenge of analyzing complex environmental signals, data quality optimization can be accomplished via background subtraction algorithms and reference spectrum calibration methods. For matters related to sample stability control, the structural stability of samples during characterization can be maintained by optimizing sample fixation approaches and regulating the intensity of operating conditions appropriately.

6.2.7 Example of a sXAS Beamlines

As an example, the BL02B02 PIPO endstation at the Shanghai Synchrotron Radiation Facility (SSRF) platform is shown schematically in Figure 6.6 [17]. As the first bending magnet-based soft X-ray beamline at SSRF, it is specifically designed for the research of energy materials and catalytic materials. High-flux soft X-rays generated by the bending magnet are first collected by the first elliptical cylindrical mirror (ECM1) installed inside a lead wall to significantly enhance photon flux, then enter a monochromator equipped with three gratings of different line densities (400 lines/mm, 800 lines/mm, and 1100 lines/mm) for wavelength selection. For measuring *K*-edges of low-energy elements such as C, N, and O, the 400 lines/mm grating is selected; for energies above 600 eV (e.g., *L*-edges of TMs), the 800 lines/mm or 1100 lines/mm gratings are used to balance resolution and flux.

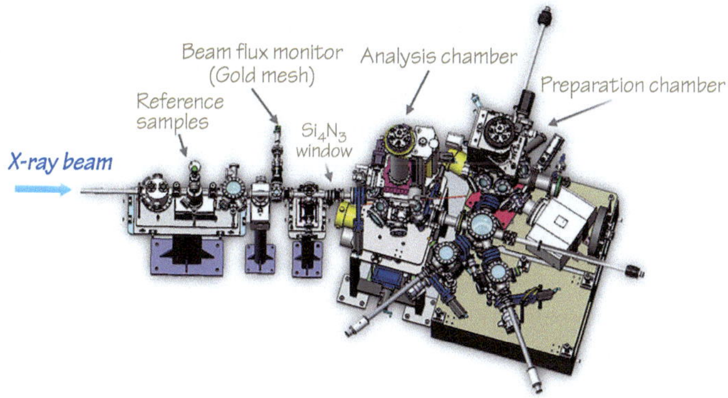

Figure 6.6 Arrangement of the SSRF BL02B02 beamline optics.

The monochromatized beam sequentially passes through a reference chamber (housing $SrTiO_3$, CoO, MnO, NiO single-crystal standards for energy calibration), a gold mesh chamber (for real time monitoring of incident photon intensity), and a slotted gate valve with a central Si_3N_4 window (to protect the beamline during in-situ experiments) before entering the ultrahigh vacuum (UHV) analysis chamber. Inside the analysis chamber, the sample mounted on a temperature-controlled stage can be measured in three complementary modes: TEY (probing depth ~10 nm), TFY (probing depth ~100 nm), and PFY (probing depth ~100 nm). To address the air-sensitivity of most energy and catalytic materials, the beamline is integrated with an inert-atmosphere glovebox for sample transfer to avoid degradation caused by air exposure.

Key performance parameters of the BL02B02 beamline are as follows: energy range of 60–2000 eV, photon flux of ~8×10^{10} photons/s at the O *K*-edge and ~3×10^{10} photons/s at transition metal *L*-edges (up to 2.2×10^{11} photons/s at 250 eV), energy resolution of 12,842 at 224 eV (routine resolution set between 3000–6000), and a focused spot size of ~200 μm × 40 μm. The analysis chamber maintains a UHV better than 5×10^{-9} mbar, equipped with a femtoampere current collector (background noise A) for TEY measurements and a fast silicon drift detector (energy resolution ~100 eV) for PFY/TFY measurements. A laser heater enables sample temperature control from room temperature to 800 °C, and a chopper-lock-in amplifier system filters electrochemical background signals to realize in-situ TEY measurements under electrochemical conditions.

The beamline supports a wide range of sample types, including powders, polycrystals, thin films, and battery electrodes. Its core applications include investigating the evolution of electronic structures of elements during battery charge–discharge cycles, tracking valence state changes of transition metals in catalytic reactions, and serving as a surface characterization tool complementary to EXAFS in single-atom catalysis research. It covers *K*-edges of elements from B to Si (e.g., C, N, O, Na, Mg, Al) and *L*-edges of 4d transition metals. In addition to ex-situ measurements, the beamline is actively developing in-situ capabilities, including in-situ characterization of all-solid-state batteries under operating conditions and in-situ reaction setups for gas/solid and liquid/solid catalytic reactions. With its high photon flux, tunable resolution, multimode detection, and integrated in-situ and air-sensitive sample handling, BL02B02 offers superior sensitivity in probing electronic structures and chemical states compared to conventional laboratory-based XAS setups. In energy and catalysis research, this enables real time, high-precision tracking of dynamic processes (e.g., phase transitions in batteries, catalytic active site evolution), capabilities that exceed the reach of traditional X-ray spectroscopy techniques.

6.3 Application Cases

The preceding sections have established the fundamental principles, technical configurations, and analytical capabilities of sXAS. To concretely illustrate its pivotal role in cutting-edge materials science, particularly in energy storage research, this section presents a series of representative application cases. These examples are meticulously selected to demonstrate how sXAS, with its unique elemental and orbital sensitivity, addresses core scientific challenges that are often intractable by conventional characterization methods. The cases progress in a logical sequence: from probing intrinsic charge compensation mechanisms within bulk electrode materials, to investigating the formation and properties of critical interfaces, further to real time monitoring of dynamic processes under operating conditions, and finally to diagnosing failure mechanisms with spatial resolution. Collectively, these applications underscore the transformative power of sXAS in establishing unambiguous correlations between microscopic electronic structure, local chemistry, and macroscopic functional performance, thereby guiding the rational design of advanced materials and devices.

6.3.1 Application of sXAS in Probing the Charge Compensation Mechanism of Electrode Materials During Electrochemical Cycling

sXAS is an indispensable electronic structure characterization technique that utilizes its elemental specificity and orbital selectivity to probe the electronic properties of battery materials, enabling detailed investigation of redox behaviors during electrochemical processes. This technique is particularly effective in deciphering the charge compensation mechanism, a core issue that is difficult to characterize comprehensively by conventional methods due to the complex coexistence of cationic and anionic redox activities. By monitoring the evolution of the absorption edges of both TMs (e.g., Ni, Co, Mn *L*-edges) and light elements (e.g., O *K*-edge), sXAS provides direct, quantitative evidence to distinguish between cationic redox and anionic redox activity, offering comprehensive insights into the capacity origin and degradation pathways in advanced cathodes.

In lithium-ion battery (LIB) research, sXAS has been widely applied to unravel the electronic structure and charge compensation mechanisms of high-performance cathodes. For instance, studies on Nd-doped $LiNi_{n\cdot 9n}Co_{0.05}Mn_{0.05}O_2$(NCM9055),

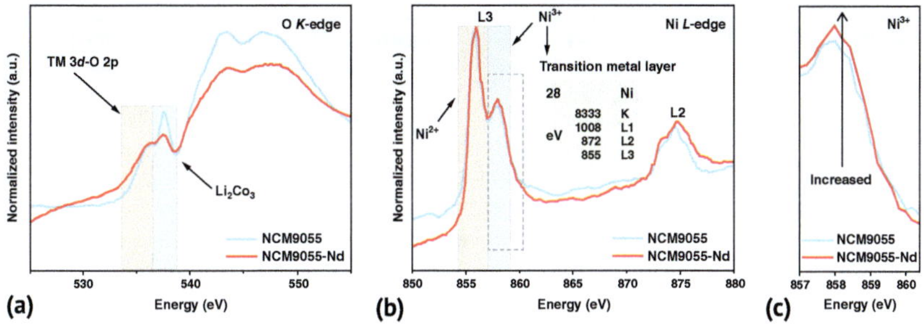

Figure 6.7 O *K*-edge (a) and Ni *L*-edge (b) and its enlarged part (c) of sXAS spectra of cycled NCM9055 and NCM9055-Nd [18] / with permission of John Wiley & Sons.

ex-situ Ni *L*-edge and O *K*-edge sXAS analyses revealed that the doped cathode maintained a higher concentration of Ni^{3+} and a stronger O *K*-edge pre-peak intensity after cycling (Figure 6.7) [18]. Such findings help clarify the synergistic effect of cationic (Ni) and anionic (O) redox in charge compensation, as well as the mechanism by which Nd doping inhibits lattice oxygen loss and improves high-voltage durability.

Beyond LIBs, sXAS is equally pivotal for elucidating complex redox behaviors in sodium-ion battery (SIB) research. For example, in the study of a high-entropy O3-type cathode $Na_{0.9}Ca_{0.05}Cu_{0.15}Ni_{0.25}Mn_{0.4}Ti_{0.1}Sn_{0.05}O_2$ (HEO-NM), ex-situ Ni, Cu, and Mn *L*-edge sXAS analyses were conducted to quantify the valence states of TMs at various states of charge. The results directly revealed that upon charging, Ni^{2+} and Cu^{2+} undergo oxidation to Ni^{3+} (partially Ni^{4+}) and Cu^{3+}, respectively. In contrast, Mn predominantly remains in the +4 oxidation state throughout the entire electrochemical process, with minimal redox participation. Furthermore, O *K*-edge sXAS showed strengthened TM 3d-O 2p hybridization. These findings provided direct electronic structure evidence that the site-specific high-entropy design, combining rigid Ca^{2+} pillars and a flexible multication TM layer, facilitates homogeneous charge compensation, suppresses oxygen loss, and stabilizes the layered structure during deep desodiation, which is crucial for clarifying the synergistic charge compensation mechanism and achieving high cycling stability [19].

Similarly, in the study of the asymmetric-orbital-hybridization designed cathode $NaNi_{0.5}Mn_{0.35}Ti_{0.15}O_2$ (NNMTO), bulk-sensitive PFY mode of Ni *L*-edge sXAS was employed to quantitatively deconvolute the Ni^{2+}, Ni^{3+}, and Ni^{4+} contributions during cycling [20]. The sXAS analysis revealed that in the $NaNi_{0.5}Mn_{0.35}O_2$ (NNMO), the charge process led to a high concentration of Jahn-Teller active Ni^{3+} (44.57%) which was not fully reduced upon discharge, indicating irreversible redox activity (Figure 6.8a, c). In stark contrast, NNMTO exhibited a reversible Ni^{2+}/Ni^{4+} redox couple with a significantly suppressed Ni^{3+} population, directly evidencing that the introduced Ti^{4+} via the Ni-O-Ti pathway successfully modified the nickel redox mechanism (Figure 6.8b, d). Furthermore, to accurately assess the Mn redox activity free from self-absorption artifacts, iPFY mode of Mn *L*-edge sXAS was measured. The spectra confirmed that the Mn oxidation state remained invariant throughout (de)sodiation in both NNMO and NNMTO (Figure 6.8e). This critical

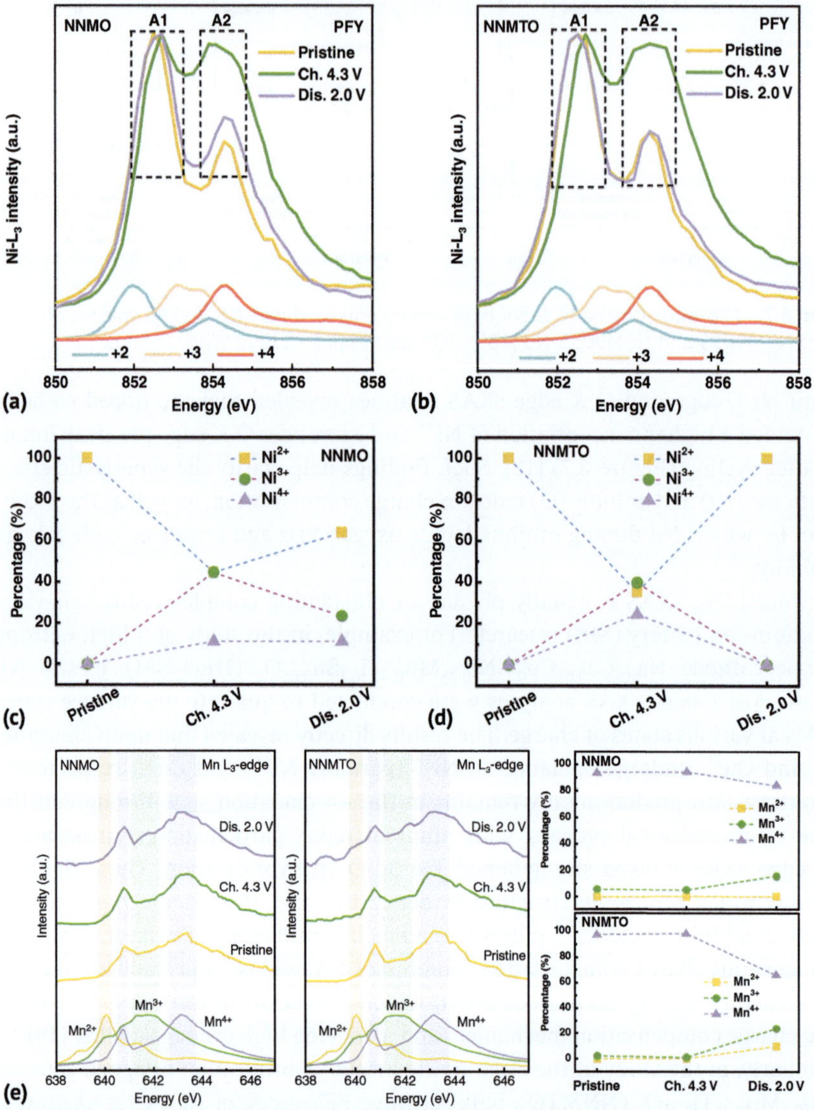

Figure 6.8 (a–d) Ni and (e) Mn L_3-edge XANES spectra with quantitative fitting results for NNMO and NNMTO at various states of charge [20] / with permission of John Wiley & Sons.

finding pinpointed that the electron redistribution and enhanced reversibility originated specifically from the modulated Ni-O orbital hybridization within the asymmetric Ni-O-Ti backbone, rather than from the manganese sites. Such sXAS-derived insights into TM-specific redox activity are indispensable. They provide direct electronic structure-level validation for the proposed orbital hybridization mechanism, explaining how the tailored local chemical environment suppresses oxygen overoxidation and structural distortion, thereby leading to the observed high reversible capacity and cycling stability.

Overall, synchrotron-based soft sXAS provides direct, element-specific insights into the evolution of oxidation states and orbital hybridization during battery operation. By quantitatively probing the transition metal *L*-edges and oxygen *K*-edge, this technique is essential for deconvoluting the complex charge compensation mechanisms involving both cationic and anionic redox. As demonstrated in LIB studies on doped NCM cathodes and in SIB research on high-entropy and orbital-hybridized layered oxides, sXAS offers critical electronic structure evidence that clarifies how specific material designs suppress detrimental phase transitions, stabilize lattice oxygen, and enhance structural integrity. Thus, sXAS proves to be an indispensable diagnostic tool, whose direct electronic structure insights are crucial for deconvoluting degradation pathways and ultimately enabling the rational development of stable, high-capacity cathode materials.

6.3.2 Application of sXAS in Probing Battery Solid Electrolyte Interphase (SEI)

sXAS has emerged as an indispensable technique for probing the formation, composition, and evolution of the SEI in batteries, owing to its exceptional element specificity and tunable probing depth. Its utility extends across a broad spectrum of electrode materials and battery chemistries, yielding direct insights into electronic structures that bridge interfacial chemistry with electrochemical performance.

A fundamental application of sXAS lies in elucidating how the intrinsic properties of active materials govern SEI formation and characteristics. A pertinent example is its application in benchmarking silicon anodes with particle sizes ranging from 50 nm to 5 μm [21]. In this study, ex-situ O *K*-edge and F *K*-edge sXAS analyses were conducted in both surface-sensitive (TEY) and bulk-sensitive (TFY) modes. The spectroscopic data revealed a stark contrast in SEI characteristics. For nanosilicon, the O *K*-edge spectra indicated a thick SEI layer rich in inorganic Li_2CO_3 and organic carbonates (e.g., LEMC, LEDC). Corresponding F *K*-edge spectra showed a strong, persistent LiF signal. This evidences severe and continuous electrolyte decomposition driven by the high surface area. Conversely, microsilicon formed a markedly thinner SEI. Its O *K*-edge spectra were dominated by features of Li_2SiO_3 with weak carbonate signals, while the F *K*-edge showed minimal LiF formation. This direct spectroscopic evidence establishes a clear link: particle size dictates SEI composition and thickness, which in turn governs degradation pathways. The thick, mixed organic-inorganic SEI on nanosilicon contributes to rapid capacity fade, whereas the thin, inorganic-rich SEI on microsilicon facilitates better cycling stability. The case underscores how sXAS provides critical insights for designing tailored mitigation strategies, such as surface engineering for nanosilicon or binder optimization for microsilicon, to manipulate SEI properties for enhanced battery performance.

Beyond its role in characterizing intrinsic material properties, sXAS serves as a decisive tool for verifying advanced interfacial engineering strategies. In a study aimed at constructing an ultra-dense, inorganic-rich fluorinated SEI (F-SEI) on silicon nanoparticles via a surface-selective electrocatalytic MoS_2 coating (Si@MS),

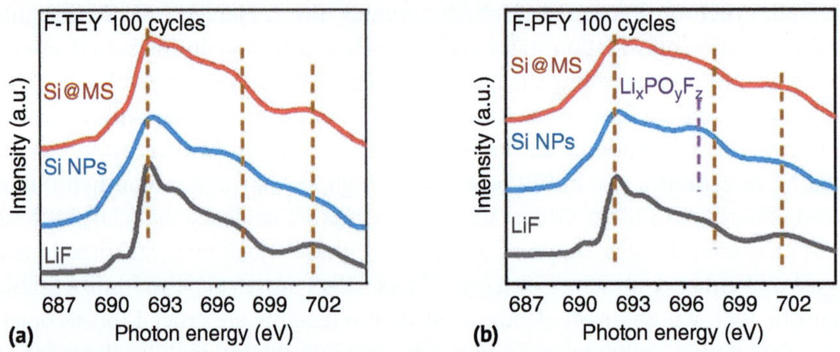

Figure 6.9 F *K*-edge sXAS spectra recorded in (a) TEY and (b) PFY modes for Si@MS versus Si NPs after 100 cycles [22] / with permission of Elsevier.

F *K*-edge sXAS analyses in both surface-sensitive (TEY, ~5 nm) and bulk-sensitive (PFY, ~100 nm) modes provided definitive evidence of the SEI's composition and distribution [22]. The spectra revealed a striking contrast between the engineered Si@MS and bare Si nanoparticles (Si NPs). For Si@MS, the F *K*-edge sXAS in TEY mode showed a strong and persistent LiF signal after 100 cycles, indicating a LiF-enriched surface interphase. More importantly, the bulk-sensitive PFY spectra demonstrated that LiF remained the dominant fluoride component throughout the deeper SEI layers, with minimal signals from undesired by-products such as $Li_xPO_yF_z$ (Figure 6.9). In contrast, the SEI on bare Si NPs exhibited a mixed inorganic-organic mosaic structure, containing both LiF and substantial $Li_xPO_yF_z$, the latter being associated with electrolyte decomposition and interfacial impedance growth. This direct spectroscopic proof linked the selective electrocatalytic coating to the formation of a superior, LiF-dominant interphase, explaining the electrode's exceptional coulombic efficiency and long-term cycling stability.

Furthermore, sXAS is pivotal in understanding and mitigating interfacial challenges in alternative battery systems, such as lithium-sulfur (Li-S) batteries. In a comparative study of ionic-liquid (IL) and conventional organic electrolytes, ex-situ S *K*-edge sXAS (TEY mode) was employed to directly track the sulfur species accumulating on both lithium-metal anodes and sulfur cathodes over extended cycling [23]. The analysis revealed a striking influence of the $LiNO_3$ additive. In IL-based electrolytes without $LiNO_3$, a distinct S^0 signal (~2475 eV) appeared on the anode surface after cycling, visually confirming the deleterious deposition of elemental sulfur/polysulfides via the shuttle effect. This signal was absent when $LiNO_3$ was present, providing direct spectroscopic evidence that $LiNO_3$ promotes a protective SEI that effectively blocks polysulfide reduction on the lithium surface.

In summary, by offering direct, element-specific fingerprints of chemical states and their spatial distribution, sXAS decodes the complex chemistry of battery interfaces. It connects material properties, processing methods, and engineering strategies directly to SEI architecture and stability, providing an indispensable molecular-level guide for designing durable electrodes across multiple battery chemistries.

6.3.3 Application of Operando sXAS in Monitoring Dynamic Interfacial Processes of Batteries During Operation

Operando sXAS has emerged as a powerful and unique tool for monitoring dynamic chemical and structural changes at battery interfaces under realistic operating conditions. Unlike ex-situ or postmortem analyses, operando sXAS provides time-resolved, element-resolved observation of reaction pathways, intermediate species formation, and degradation processes as they occur.

This attribute is particularly crucial for understanding complex interfacial phenomena, such as the polysulfide shuttle in lithium-sulfur (Li-S) batteries or the potential-dependent evolution of interphases. A compelling demonstration of this power is found in the study of the notorious polysulfide shuttle effect in Li-S batteries. Researchers designed a dedicated operando cell compatible with tender X-ray S *K*-edge sXAS, enabling them to probe the sulfur chemistry simultaneously at two critical yet distinct locations: the anode-electrolyte interface (AEI) and the cathode-electrolyte-separator interface (CESI) (Figure 6.10) [24]. During the discharge of a conventional Li-S cell, operando S *K*-edge spectra captured at the AEI revealed the dynamic accumulation of Li_2S and Li_2SO_4 species from the very beginning of the process. This real time spectroscopic evidence directly visualized the shuttling of polysulfides from the cathode, their reduction on the lithium metal anode, and the

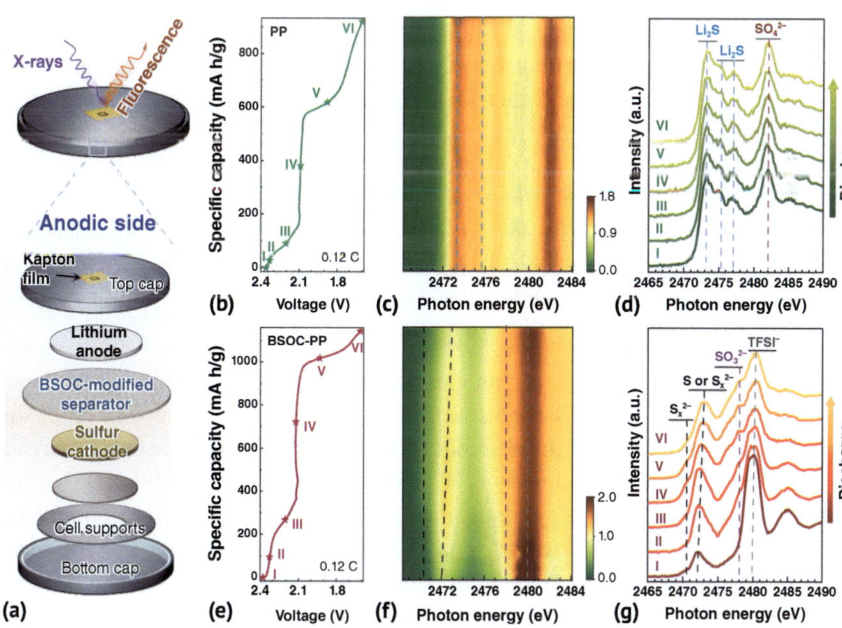

Figure 6.10 In-situ/operando S *K*-edge XAS analysis comparing Li/S cells with bare PP versus BSOC-PP separators. (a) Schematic of the modified coin cell for anode-side detection, featuring a Kapton-sealed X-ray window. (b, e) First discharge voltage profiles at 0.12 C for cells with PP and BSOC-PP separators, respectively. (c, f) Corresponding S *K*-edge XAS maps during discharge. (d, g) Representative spectra extracted at key potentials (marked I–VI in (b) and (e)) [24] / with permission of John Wiley & Sons.

consequent formation of an insulating passivation layer a process directly linked to capacity loss and anode corrosion. In a parallel experiment on the CESI, operando sXAS tracked the deposition and isolation of elemental sulfur (S^0) on the bare separator, identifying another pathway for active material loss. This operando diagnostic capability was then employed to validate a suppression strategy: a functional separator coated with bismuth sulfide/oxide nanoclusters (BSOC). The operando S *K*-edge spectra revealed a dramatically different situation. At the AEI, the characteristic Li_2S signals were absent, indicating effective blockage of polysulfide migration. Meanwhile, at the CESI, persistent polysulfide signals suggested their effective trapping by the BSOC layer. Crucially, the spectral evolution indicated not just physical adsorption but also catalytic conversion of the trapped species. By providing time-resolved and spatially resolved chemical snapshots, operando sXAS deciphered the complete dynamic interplay: the BSOC layer functions as a bi-functional mediator along the polysulfide migration path, simultaneously arresting (trapping) and rehabilitating (catalyzing) the shuttle species in real time, thereby protecting the anode and improving sulfur utilization. This work exemplifies how operando sXAS moves beyond postmortem analysis to become an indispensable tool for the real time visualization of interfacial reactions. It thereby provides unparalleled insight into dynamic failure mechanisms and the real time efficacy of mitigation strategies.

This foundational insight is critically extended to elucidate more complex interfacial dynamics, such as those governing multivalent metal deposition. A seminal investigation into magnesium deposition from borohydride-based electrolytes exemplifies the unique capacity of operando sXAS to identify active interfacial species and deconvolute concomitant chemical transformations [25]. Through the acquisition of potential-dependent Mg *K*-edge spectra, a distinct spectral feature at approximately 1307 eV was resolved under cathodic polarization, energetically situated between the characteristic absorption of metallic Mg (~1303 eV) and solvated Mg^{2+} species from the bulk electrolyte. This feature was conclusively assigned to an interfacial magnesium species (denoted Mg_{SEI}). Its electronic structure, particularly the edge position and fine structure, exhibited compelling alignment with the theoretically predicted contact-ion pair $\left[Mg(\mu\text{-}H)_2BH_2\right]^+$, thereby providing direct spectroscopic evidence for its role as the dominant electroactive precursor facilitating Mg^0 deposition. Complementarily, B *K*-edge spectra captured the dynamic evolution of the borohydride anion (BH^{4-}) at the electrode interface, revealing its transformation into higher-order boron-hydrogen clusters. This in-situ observation established that the formation of the SEI in this system is intrinsically linked to the consumption and reconfiguration of the anionic component. By employing operando sXAS to track the fate of both the cation and anion in tandem, this study established a direct correlation between the macroscopic electrochemical signature of reversible Mg deposition/stripping and the microscopic molecular-scale processes at the interface. It definitively demonstrated that a functional, ion-conducting SEI, derived from the electrolyte itself, can coexist with reversible multivalent metal plating. This work underscores the unparalleled capability of operando sXAS to fingerprint transient intermediates and unravel complex reaction pathways, cementing its role as an indispensable tool for mechanistic interrogation at electrified interfaces.

In summary, operando sXAS decodes the dynamic chemistry of battery interfaces under operating conditions by providing real-time, element-specific spectroscopic fingerprints with spatial resolution. It directly visualizes transient intermediates, reaction pathways, and failure mechanisms, including polysulfide shuttling in Li-S batteries and the formation of active interfacial species in multivalent deposition, thereby linking macroscopic electrochemical behavior to microscopic molecular-scale processes. This capability not only reveals degradation dynamics but also validates interfacial engineering strategies in real time, offering an indispensable analytical perspective for guiding the rational design of stable battery interfaces.

6.3.4 Application of sXAS in Failure Analysis and Heterogeneity Mapping of Battery Materials via Spectromicroscopy

sXAS proves indispensable for directly probing the interfacial chemistry critical to battery failure, particularly in systems where buried interfaces are inaccessible to many other techniques. A focused investigation into the degradation of $LiNi_{0.8}Mn_{0.1}Co_{0.2}O_2$ (NMC811) cathodes in sulfide-based all-solid-state batteries leveraged S *K*-edge sXAS to decode the chemical evolution at the cathode-solid electrolyte interface, a primary suspected failure point [26]. The study specifically analyzed cycled composite electrodes comprising NMC811 and the Li_6PS_5Cl (LPSCl) electrolyte. The S *K*-edge spectra provided direct, element-specific fingerprints of the sulfur species within the LPSCl. For the uncoated NMC811 cathode after cycling, the sXAS data revealed a marked increase in the intensity of a specific peak near 2472.0 eV, which is attributed to the formation of higher-order bridging sulfur units (P-S_x-P). More conclusively, the characteristic spectral signature of the pristine PS_4^{3-} tetrahedral units was significantly diminished, especially in electrodes analyzed at a charged state. This loss of the PS_4^{3-} fingerprint served as direct spectroscopic evidence for the severe and irreversible chemical degradation (oxidation) of the sulfide solid electrolyte at the interface with the high-voltage cathode. In sharp contrast, identical sXAS analysis performed on cathodes protected by a $LiNbO_3$ (LNO) coating showed spectra that closely resembled the pristine electrolyte, with the PS_4^{3-} features largely preserved. This sXAS evidence unambiguously identified the parasitic interfacial reaction as a key chemical driver of failure and quantitatively demonstrated the coating's efficacy in suppressing it. Thus, by isolating and diagnosing the interfacial sulfur chemistry, S *K*-edge sXAS moved beyond correlative inference to provide a causal chemical link between interfacial decomposition, increased impedance, and the overall capacity fade, solidifying its role as a critical tool for definitive failure analysis.

Beyond interfacial failure diagnosis, sXAS is also instrumental in deconvoluting the spatially heterogeneous chemical degradation that underlies battery failure in real-world operational scenarios. A salient example is the postmortem analysis of commercial $LiCoO_2$ cathodes extracted from malfunctioning wireless earbuds, where failure was intrinsically linked to the device's specific configuration and operating microenvironment [27]. Specifically, Co L_3-edge sXAS analyses, performed in both surface-sensitive (TEY) and bulk-sensitive (TFY) modes, revealed a distinct position-dependent chemical signature in the failing cell. The sXAS spectra revealed a striking positional dependence: cathode materials from the top and bottom regions of the cylindrical cell exhibited significant reduction of cobalt (lower oxidation state) at and near the surface, a signature of detrimental phase transitions (e.g., to spinel/rock-salt structures). In contrast, materials from the middle region showed Co L_3-edge spectra similar to those from a healthy control cell. This sXAS-mapped chemical gradient provides direct evidence that the cathode's degradation is nonuniform and spatially correlated with the device's specific configuration and operating environment. By fingerprinting the localized chemical deactivation, sXAS directly links microscopic material changes to the macroscopic performance fade, underscoring its critical role in pinpointing failure origins in real-world applications.

In summary, by delivering element-specific spectroscopic fingerprints with both interface sensitivity and spatial resolution, sXAS decodes the critical chemical evolution driving battery failure. Specifically, it directly probes buried interfacial reactions such as sulfide electrolyte oxidation at high-voltage cathodes, and maps spatially heterogeneous degradation like localized cathode reduction in real-world cells. This ability to chemically fingerprint both interfacial decomposition and non-uniform material deactivation provides a causal link between microscopic chemical changes and macroscopic performance fade, moving beyond correlation to definitive mechanism diagnosis. Thus, sXAS serves as an indispensable analytical tool that pinpoints chemical origins of failure and quantitatively validates interfacial protection strategies, guiding the rational design of durable battery systems.

References

1 F.d. Groot, Multiplet effects in X-ray spectroscopy, Coordination Chemistry Reviews 249(1) (2005) 31–63. https://doi.org/10.1016/j.ccr.2004.03.018.

2 W. Kossel, Zum Bau der Röntgenspektren, Zeitschrift für Physik 1(1) (1920) 119–134. https://doi.org/10.1007/BF01881031.

3 R. Smoluchowski, Über die Feinstruktur der Röntgenabsorptionskanten von Legierungen, Zeitschrift für Physik 95(9) (1935) 588–598. https://doi.org/10.1007/BF01331299.

4 C. Kurylenko, Franges au voisinage de la discontinuité K des rayons X, Le Journal de Physique et le Radium 1(4) (1940) 133–145. https://doi.org/10.1051/jphysrad:0194000104013300.

5 F. Lin, Y. Liu, X. Yu, L. Cheng, A. Singer, O.G. Shpyrko, H.L. Xin, N. Tamura, C. Tian, T.C. Weng, X.Q. Yang, Y.S. Meng, D. Nordlund, W. Yang, M.M. Doeff, Synchrotron X-ray analytical techniques for studying materials electrochemistry in rechargeable batteries, Chemical Reviews 117(21) (2017) 13123–13186. https://doi.org/10.1021/acs.chemrev.7b00007.

6 F. Frati, M. Hunault, F.M.F. de Groot, Oxygen *K*-edge X-ray absorption spectra, Chemical Reviews 120(9) (2020) 4056–4110. https://doi.org/10.1021/acs.chemrev.9b00439.

7 Q. Li, R. Qiao, L.A. Wray, J. Chen, Z. Zhuo, Y. Chen, S. Yan, F. Pan, Z. Hussain, W. Yang, Quantitative probe of the transition metal redox in battery electrodes

through soft X-ray absorption spectroscopy, Journal of Physics D: Applied Physics 49(41) (2016) 413003. https://doi.org/10.1088/0022-3727/49/41/413003.

8 X. Liu, J. Liu, R. Qiao, Y. Yu, H. Li, L. Suo, Y.-s. Hu, Y.-D. Chuang, G. Shu, F. Chou, T.-C. Weng, D. Nordlund, D. Sokaras, Y.J. Wang, H. Lin, B. Barbiellini, A. Bansil, X. Song, Z. Liu, S. Yan, G. Liu, S. Qiao, T.J. Richardson, D. Prendergast, Z. Hussain, F.M.F. de Groot, W. Yang, Phase transformation and lithiation effect on electronic structure of LixFePO4: An in-depth study by soft X-ray and simulations, Journal of the American Chemical Society 134(33) (2012) 13708–13715. https://doi.org/10.1021/ja303225e.

9 W. Yang, X. Liu, R. Qiao, P. Olalde-Velasco, J.D. Spear, L. Roseguo, J.X. Pepper, Y.-d. Chuang, J.D. Denlinger, Z. Hussain, Key electronic states in lithium battery materials probed by soft X-ray spectroscopy, Journal of Electron Spectroscopy and Related Phenomena 190 (2013) 64–74. https://doi.org/10.1016/j.elspec.2013.03.008.

10 L. Tröger, D. Arvanitis, K. Baberschke, H. Michaelis, U. Grimm, E. Zschech, Full correction of the self-absorption in soft-fluorescence extended X-ray-absorption fine structure, Physical Review B 46(6) (1992) 3283–3289. https://doi.org/10.1103/PhysRevB.46.3283.

11 S.L.M. Schroeder, G.D. Moggridge, R.M. Ormerod, T. Rayment, R.M. Lambert, What determines the probing depth of electron yield XAS?, Surface Science 324(2) (1995) L371–L377. https://doi.org/10.1016/0039-6028(94)00779-9.

12 A.J. Achkar, T.Z. Regier, H. Wadati, Y.J. Kim, H. Zhang, D.G. Hawthorn, Bulk sensitive X-ray absorption spectroscopy free of self-absorption effects, Physical Review B 83(8) (2011) 081106. https://doi.org/10.1103/PhysRevB.83.081106.

13 W. Yang, T.P. Devereaux, Anionic and cationic redox and interfaces in batteries: Advances from soft X-ray absorption spectroscopy to resonant inelastic scattering, Journal of Power Sources 389 (2018) 188–197. https://doi.org/10.1016/j.jpowsour.2018.04.018.

14 A. Iglesias-Juez, G.L. Chiarello, G.S. Patience, M.O. Guerrero-Pérez, Experimental methods in chemical engineering: X-ray absorption spectroscopy—XAS, XANES, EXAFS, 100(1) (2022) 3–22. https://doi.org/10.1002/cjce.24291.

15 E.J. Crumlin, Z. Liu, H. Bluhm, W. Yang, J. Guo, Z. Hussain, X-ray spectroscopy of energy materials under in-situ/operando conditions, Journal of Electron Spectroscopy and Related Phenomena 200 (2015) 264–273. https://doi.org/10.1016/j.elspec.2015.06.008.

16 F. Yang, X. Feng, Y.-S. Liu, L.C. Kao, P.-A. Glans, W. Yang, J. Guo, In Situ/Operando (soft) X-ray Spectroscopy Study of Beyond Lithium-ion Batteries, Energy & Environmental Materials 4(2) (2021) 139–157. https://doi.org/10.1002/eem2.12172.

17 X. Meng, Z. Guo, Y. Wang, H. Zhang, Y. Han, G. Zhao, Z. Liu, R. Tai, Design and performance of bending-magnet beamline BL02B at the SSRF, Journal of Synchrotron Radiation 26(2) (2019) 543–550. https://doi.org/10.1107/S1600577518018179.

18 A. Jiang, H.G. Xie, Q.L. Lin, Z. Xiao, F. Wang, X.M. Zhang, W.W. Zhang, W. He, Z.F. Qian, X.Y. Dai, R.H. Wang, Nd doping inhibits lattice oxygen loss to enhance

high-voltage durability of Ni-rich cathodes, Advanced Functional Materials (2025) e19006. https://doi.org/10.1002/adfm.202519006.

19 W.D. Xu, C. Cheng, L. Wang, T. Chen, Z. Zhou, T.R. Yan, S.Q. Shen, P. Zeng, L. Zhang, Synergistic rigidity–flexibility engineering of O3-type sodium layered oxide cathodes through site-specific high-entropy regulation, Advanced Energy Materials 16 (2025) e04557. https://doi.org/10.1002/aenm.202504557.

20 C. Cheng, Y. Shen, C. Chen, S. Tang, Z.Q. Zhuo, Q.J. Niu, C. Yuan, T. Chen, L. Wang, J.H. Guo, D. Sun, L. Zhang, Asymmetric-orbital-hybridization induced electron redistribution enabling stable sodium layered oxides, Advanced Energy Materials 15 (2025) e04261. https://doi.org/10.1002/aenm.202504261.

21 F. Wu, Y. Dong, Y.F. Su, C.X. Wei, T.R. Chen, W.A. Yan, S.Y. Ma, L. Ma, B. Wang, L. Chen, Q. Huang, D.Y. Cao, Y. Lu, M. Wang, L. Wang, G.Q. Tan, J.H. Wang, N. Li, Benchmarking the effect of particle size on silicon anode materials for lithium-ion batteries, Small 19(42) (2023) 2301301. https://doi.org/10.1002/smll.202301301.

22 W.G. Yan, F. Wu, S.Y. Ma, Y.F. Su, B. Wang, Y.B. Guan, T.L. Song, Q. Huang, L. Chen, M. Wang, N. Li, Surface selective electrocatalysis-derived ultra-dense and mechanically robust fluorinated solid electrolyte interface for high-capacity and long-cycling silicon electrode, Materials Today 88 (2025) 284–292. https://doi.org/10.1016/j.mattod.2025.06.037.

23 M. Barghamadi, L. Djuandhi, N. Sharma, A.S. Best, A.F. Hollenkamp, P.J. Mahon, M. Musameh, T. Rüther, In situ synchrotron XRD and sXAS studies on Li-S batteries with ionic-liquid and organic electrolytes, Journal of The Electrochemical Society 167(10) (2020) 100526. https://doi.org/10.1149/1945-7111/ab98a8.

24 L.J. Jia, J. Wang, S.Y. Ren, G.X. Ren, X. Jin, L.C. Kao, X.F. Feng, F.P. Yang, Q. Wang, L.D. Pan, Q.T. Li, Y.S. Liu, Y. Wu, G. Liu, J. Feng, S.S. Fan, Y.F. Ye, J.H. Guo, Y.G. Zhang, Unraveling shuttle effect and suppression strategy in lithium/sulfur cells by in situ/operando X-ray absorption spectroscopic characterization, Energy & Environmental Materials 4(2) (2020) 222–228. https://doi.org/10.1002/eem2.12152.

25 T.S. Arthur, P.A. Glans, N. Singh, O. Tutusaus, K. Nie, Y.-S. Liu, F. Mizuno, J. Guo, D.H. Alsem, N.J. Salmon, R. Mohtadi, Interfacial insight from operando XAS/TEM for magnesium metal deposition with borohydride electrolytes, Chemistry of Materials 29(17) (2017) 7183–7188. https://doi.org/10.1021/acs.chemmater.7b01189.

26 X.L. Zheng, Z.C. Xue, H.C. Hao, Y.K. Cho, Y.S. Li, C. Kim, P. Czaja, S.S. Lee, S. Bone, E. Spielman-Sun, Z.L. Jiang, X.W. Gu, J.N. Weker, G. Yang, J. Nanda, Unravelling electro-chemo-mechanical interplay in layered oxide cathode degradation in solid-state batteries, Science Advances 11(41) (2025) 7189. https://doi.org/10.1126/sciadv.ady7189.

27 G.N. Qian, G.B. Zan, J.Z. Li, D.C. Meng, T.X. Sun, V. Thampy, A.M. Yanyachi, X.J. Huang, H.F. Yan, Y.S. Chu, S. Gul, J.J. Huang, S.D. Kelly, S.J. Lee, J.S. Lee, W.B. Yun, P. Cloetens, P. Pianetta, K.j. Zhao, O.A. Ezekoye, Y.J. Liu, In-device battery failure analysis, Advanced Materials 37(10) (2025) 2416915. https://doi.org/10.1002/adma.202416915.

7

Resonant Inelastic X-ray Scattering for Probing Electronic Excitations in Battery Materials

7.1 Fundamentals and Techniques of Resonant Inelastic X-ray Scattering

The theoretical foundations of resonant inelastic X-ray scattering (RIXS) trace back to the framework of X-ray emission spectroscopy established in 1934, which already encompassed both resonant and non-resonant processes. Over the following decades, the theory continued to be refined and deepened. At the same time, advances in synchrotron radiation sources and innovations in high-resolution optical devices jointly drove significant progress in experimental spectroscopic techniques.

A key breakthrough came in 1991, when researchers used a spectrometer to obtain, for the first time, X-ray absorption near-edge structure measurements in partial fluorescence yield (PFY) mode. The strength of this approach lies in its ability to effectively suppress broadening effects caused by core hole lifetime through high-resolution analysis of the emission spectrum, thereby yielding more accurate absorption coefficients. By 1994, the world's first third-generation synchrotron radiation facility—the European Synchrotron Radiation Facility (ESRF)—began operations. Its soft X-ray brightness achieved an orders-of-magnitude increase, laying an essential source foundation for subsequent high-precision RIXS experiments.

Entering the 21st century, breakthroughs in high-resolution RIXS measurements were achieved successively in both the soft X-ray (e.g., Cu *L*-edge) and hard X-ray (e.g., Cu *K*-edge) regimes, with energy resolutions typically better than 0.5 eV. Over the following two decades, the technique evolved into a key experimental tool for probing collective excitations in quantum materials, significantly advancing the understanding of strongly correlated electronic systems. Since the 2010s, the application scope of RIXS has further expanded into the field of energy materials, contributing notably to the understanding of electrochemical reaction pathways, structural evolution, and degradation mechanisms in high-capacity battery electrode materials.

7.1.1 Fundamental Principles of RIXS

The double differential cross-section, denoted as $\frac{\partial^2\sigma}{\partial\Omega\partial\omega}$, quantifies RIXS spectra. It measures the probability of photon scattering into a solid angle Ω within an energy interval ω to $\omega + \partial\omega$. This scattering process is theoretically framed by the Kramers–Heisenberg formula [1]:

$$\frac{\partial^2\sigma}{\partial\Omega\partial\omega} \propto \sum_f \left| \sum_j \frac{\langle f|T^\dagger_{k_o,\varepsilon_o}|j\rangle \langle j|T_{k_i,\varepsilon_i}|i\rangle}{\omega_i + E_\mathrm{i} + \mathrm{i}\frac{\Gamma_j}{2} - E_j} \right|^2 \delta(E_\mathrm{i} + \omega_i - E_f - \omega_o) \tag{7.1}$$

Within this formula, the incident photon is characterized by its energy ω_i, wave vector k_i, and polarization state ε_i, while the scattered photon is correspondingly described by ω_o, k_o, and ε_o. The initial and final states of the system possess energies E_i and E_f, respectively. A summation over all accessible intermediate states j, each with energy E_j, is required. The resulting RIXS spectrum arises from the coherent superposition of photon absorption $(\langle j|T_{k_i\varepsilon_i}|i\rangle)$ and subsequent emission $(\langle f|T^\dagger_{k_o,\varepsilon_o}|j\rangle)$ processes. Consequently, as delineated by the Kramers–Heisenberg expression, the scattering cross-section exhibits a functional dependence on the energy (ω), direction (k), and polarization (ε) of both the incoming and outgoing photons, with the transition operators themselves being functions of these same variables.

7.1.2 The RIXS Process

The RIXS process follows a photon-in–photon-out scheme, which proceeds through two distinct stages: photon absorption and subsequent photon emission [2]. When the incident photon energy is tuned near the absorption edge of a specific element, it resonantly excites a core electron from that element. As illustrated in Figure 7.1, the electron may be excited either into the valence band (left panel) or into unoccupied states above the Fermi level (right panel).

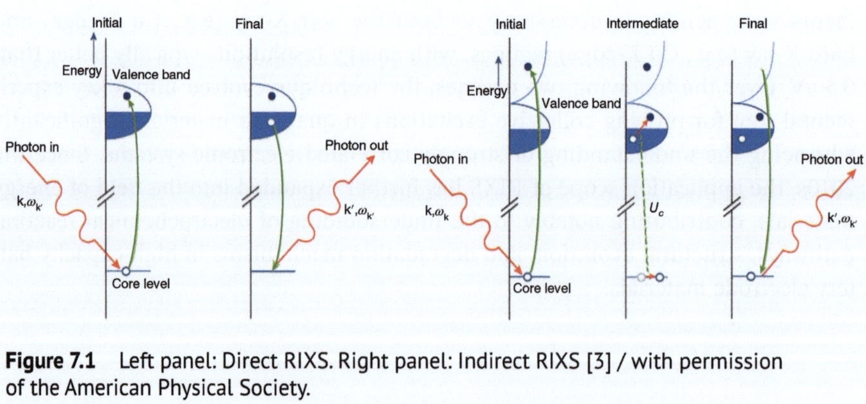

Figure 7.1 Left panel: Direct RIXS. Right panel: Indirect RIXS [3] / with permission of the American Physical Society.

In the first scenario, referred to as direct RIXS, involves an incident photon exciting a core electron into an empty valence state. Subsequently, an electron from a different state within the valence band decays to annihilate the core hole. The final result is a state with an electron–hole excitation, as an electron is created in a previously empty valence state and a hole is left in a filled valence state.

Direct RIXS proceeds via two sequential dipole transitions: an allowed excitation from the core to the valence state, followed by an allowed decay that fills the resulting core hole. For instance, in wide-bandgap insulators, this could involve an initial dipole-allowed 1s → 2p transition, followed by the decay of another electron from the 2p band via a 2p → 1s transition. At the transition metal *L*-edge, it is specifically a 2p → 3d absorption and a subsequent 3d → 2p emission. This mechanism enables direct probing of the valence and conduction states. The resulting spectral line shape is dominated by the direct transitions, though the spectral weight can be redistributed by intermediate-state interactions, including the influence of a strong core hole potential.

In the second scenario, commonly referred to as indirect RIXS, an incoming photon excites a core-level electron into a high-lying unoccupied state located several electron volts above the Fermi energy. The excited electron then relaxes by refilling the core hole, a mechanism that is characteristic of transition metal *K*-edge RIXS processes, such as the 1s → 4p excitation.

For pure indirect RIXS to occur, the dipole-allowed transition from the core state to the conduction band must be weak. In the absence of additional interactions, no inelastic scattering would be observed. However, the intermediate state contains a core hole, which exerts a strong potential on the valence electrons and tends to screen the core hole itself. This core hole potential scatters the valence electrons, thereby creating electron–hole excitations within the valence band. After the 4p → 1s decay, the system is left with these electron–hole excitations, which constitute the detected inelastic signal.

7.1.3 The Experimental Methods and Basis of RIXS

When preparing for a RIXS experiment, the first step is to select the absorption edge at which to work. For transition metal compounds, the primary absorption edges employed are the transition metal *L*-edge and *K*-edge. Although one might initially expect the underlying physics to be similar across edges, experiments reveal that the same valence excitation often carries markedly different spectral weights at different absorption edges. This makes edge selection a key experimental variable for tuning sensitivity and obtaining complementary information. The *L*-edge (2p → 3d) and *M*-edge (3p → 3d) of transition metals (TMs) are comparable in several key physical aspects. Both involve dipole-allowed transitions directly into the 3d orbitals, enabling a complete probe of all 3d orbital degrees of freedom. This provides a unique window for experimentally measuring d–d excitations and uncovering orbital order/orbital excitations. Furthermore, since spin ceases to be a well-defined quantum number in the intermediate state, these processes can also give rise to single-magnon excitations. During the scattering process, the electron excited into the 3d orbital effectively screens the core hole potential. As a result, aside from the dipole-dominated excitations, the spectra show almost no additional complex many-body satellite features, leading to clearer and more direct spectral interpretation.

At the *K*-edge of TMs (1s → 3d/4p), the direct transition into 3d orbitals is quadrupole-allowed and thus has a relatively weak cross-section. The dominant scattering intensity arises from the dipole-allowed 1s → 4p transition. Due to the poor screening of the core hole by the excited 4p electron, the instantaneous Coulomb potential between the 1s hole and the 4p electron induces significant many-body shake-up processes, which can include excitations across the charge transfer gap.

Moreover, the X-rays used in the *K*-edge region typically carry substantial momentum, enabling these charge excitations to be studied as a function of both energy and momentum over a wide range. Because spin–orbit coupling is negligible in the intermediate state, spin is approximately a good quantum number; hence, spin-selection rules prohibit the generation of single-magnon excitations, whereas spin-preserving processes, such as bimagnon excitations, are still allowed.

Following the selection of an absorption edge, the choice of incident energy becomes a critical step in optimizing a RIXS experiment. In practice, a systematic scan over a range of incident energies across the edge is typically performed, as the spectral weight of different features is highly sensitive to the incident photon energy.

At the *L*-edge, tuning the incident energy to the main absorption line resonantly enhances metal-centered orbital transitions. Setting the energy to satellite regions above the main line selectively emphasizes charge transfer excitations. For the *K*-edge, several regimes can be distinguished based on the incident energy: In the pre-edge region, direct excitation into the 3d orbitals makes the process comparable to *L*- and *M*-edge RIXS, primarily probing d–d excitations. The main-edge spectral response is, in contrast, governed by charge transfer excitations. The most efficient screening processes—such as excitations across the charge transfer gap in cuprates—resonate at the lowest energies. Moving to energies well above the edge, the core hole becomes poorly screened, leading to an enhancement of spectral weight for unscreened final states at higher energies.

The relative intensity of the elastic line varies markedly depending on the absorption edge (*K*-, *L*-, or *M*-edge), a key factor that must be considered in both experimental design and data analysis. At the *K*- and *M*-edges, elastic scattering is typically very strong. The tails of the elastic peak often determine whether specific low-energy features can be resolved; here, the dominant constraint is not the full width at half maximum of the energy resolution, even though improved resolution does lead to a narrower elastic line. By contrast, in *L*-edge measurements the elastic contribution is typically much weaker than the inelastic scattering. It is therefore essential to understand and minimize elastic scattering as much as possible in order to successfully extract and interpret the inelastic signals.

When assessing the relative strength of elastic and inelastic contributions, the magnitude of resonance enhancement for the inelastic channel must be considered.

For *L*-edge experiments, the process is typically direct RIXS, which inherently yields stronger inelastic scattering. Moreover, the resonance enhancement at the *L*-edge is substantial. This is evidenced by the intense "white-line" feature in the *L*-edge X-ray absorption spectrum, which contrasts sharply with the weak or absent peaks in *M*- and *K*-edge XAS spectra. Consequently, resonance enhancement is markedly greater for *L*-edge experiments. The combined effect of these factors results in inelastic scattering at the *L*-edge being significantly stronger than that at the *M*- or *K*-edge.

In RIXS experiments, both incident and emitted photons are absorbed by the sample through Auger processes and core-level fluorescence decay. This attenuation, commonly referred to as self-absorption, depends strongly on sample composition, thickness, density, and the scattering geometry employed, directly influencing the probed effective sample volume and the absolute signal intensity. Therefore, to reliably compare spectra collected under different geometries or from different samples, quantitative correction for self-absorption is essential. This consideration is particularly critical for *L*-edge and *M*-edge experiments, where photo absorption cross-sections are generally large, and experimenters must systematically account for self-absorption during data normalization.

7.2 Instrumentation and Analysis of RIXS

The reliable implementation and interpretation of RIXS experiments depend on two core pillars: advanced instrumental design and rigorous data processing. This section focuses on the hardware configuration, experimental geometry, data collection protocols, and analytical methodologies that underpin high-quality RIXS measurements. It first details the structural composition of synchrotron-based RIXS setups, highlighting the distinct optical architectures of hard and soft X-ray systems and the key factors governing their geometric optimization. Subsequently, it addresses critical considerations in data collection, including energy calibration, signal artifacts (e.g., beam damage, self-absorption), and the reliance on relative intensity analysis. Finally, it delves into the theoretical frameworks for data interpretation, such as the connection between RIXS cross-sections and dynamical structure factors, and the practical fitting models used to extract intrinsic material properties from experimental spectra. Together, these components form a comprehensive workflow from experiment design to physical insight extraction.

7.2.1 Instrumentation and Geometries

A synchrotron-based RIXS instrumentation is composed of two main functional units. The beamline is responsible for producing a highly collimated and monochromatic X-ray beam, often with additional focusing at the sample position. The spectrometer, on the other hand, defines the experimental environment at the sample, enables precise control of the scattering geometry to tune the momentum transfer, collects the emitted photons within a specified solid angle, and carries out

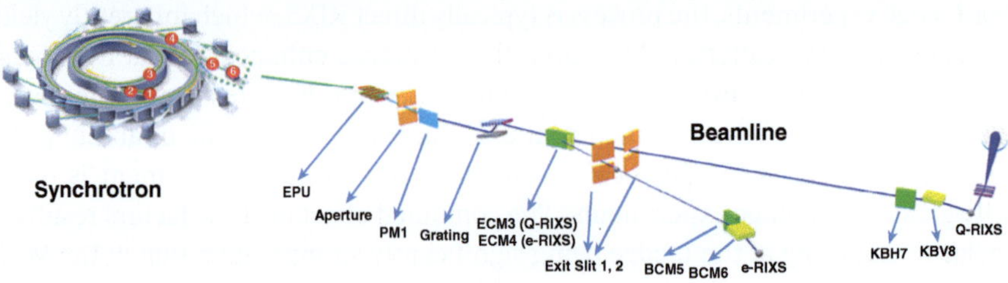

Figure 7.2 Typical RIXS beamline at synchrotron source [5, 6].

high-resolution energy discrimination of the scattered radiation. Figure 7.2 shows a schematic layout of a RIXS beamline.

The advancement of RIXS has been fundamentally coupled to progress in synchrotron source technology. The continual increase in brightness within relevant energy ranges at third- and fourth-generation synchrotrons has enabled a transformative improvement in momentum-resolved RIXS instrumentation [4]. Crucially, dramatic gains in energy resolution have been achieved without compromising flux, allowing experiments to be conducted with good counting statistics within practical timeframes.

The optical architectures of RIXS spectrometers differ in a fundamental way between the hard and soft X-ray regimes, reflecting the distinct diffraction mechanisms available in each case. In the hard X-ray range, spectrometers are almost exclusively based on Bragg diffraction from perfect crystals. The outstanding energy resolution attainable with these instruments ultimately stems from the long-range periodic order of crystal lattices such as silicon or germanium, whose lattice spacing is well matched to the short X-ray wavelength. This approach is not viable for soft X-rays, where the longer wavelengths prohibit the use of crystalline optics that could satisfy the Bragg condition. Instead, soft X-ray systems utilize artificially periodic structures, such as diffraction gratings. This core distinction means that hard and soft X-ray RIXS spectrometers differ essentially in their optical architecture, their pathways to achieving resolution, and their integration with sample environments.

From the perspective of synchrotron radiation experimental physics, RIXS measurements require high photon flux and are therefore primarily conducted at synchrotron facilities. In the soft X-ray regime (300–1500 eV), RIXS beamlines and spectrometers commonly rely on grating-based monochromators to perform energy selection of the undulator radiation incident on the sample, as well as of the photons scattered from it. Since the initial designs, the energy resolution of soft X-ray RIXS spectrometers has been continuously improved. These monochromators operate on a grazing-incidence geometry. Consequently, for a fixed combination of focusing optics, grating, and detector, a longer spectrometer arm delivers better energy resolution.

In the hard X-ray regime (>4000 eV), RIXS relies on perfect crystal Bragg optics. To achieve high energy resolution, a dedicated crystal capable of operating near backscattering geometry is required for each core resonance. Hard X-ray emission spectrometers are primarily classified into two categories: scanning and dispersive. Scanning spectrometers measure the accessible solid angle for individual emission energies within the instrument's energy bandwidth by rotating the crystal analyzer. Dispersive spectrometers use a position-sensitive photon detector to record the full emission spectrum simultaneously. While dispersive instruments offer the advantages of mechanical simplicity and stability, whereas scanning instruments provide greater flexibility in energy coverage and resolution tuning.

The optimal geometrical configuration of a RIXS instrument results from a careful balance of multiple factors: the properties of the X-ray, spatial limitations, the desired energy resolution, and the characteristics of the scattering process under investigation. Regardless of the chosen geometry, the analyzer crystal must possess exceptional quality to guarantee and sustain high energy resolution.

7.2.2 Data Collection of RIXS

In synchrotron-based RIXS experiments, both the beamline monochromator (grating or crystal-based) and the spectrometer's energy scale require calibration. Although absolute energy calibration of the monochromator and detector is possible, it is often omitted in routine practice for reasons of efficiency and necessity. The incident X-ray energy is typically referenced to known absorption edges or elastic scattering features. For RIXS measurements, the exact incident energy is usually of secondary importance. The fundamental calibration is provided by the elastic scattering peak, which establishes the zero energy-loss point and thereby sets the reference for the entire energy-loss spectrum. The absolute scattering intensity in a RIXS experiment is usually not measured. It is influenced by a complex combination of factors including the transmission efficiency of beamline optics, detector active area and efficiency, making precise absolute calibration challenging. Consequently, the focus of RIXS analysis lies in the relative intensity variations within the RIXS map. As shown in Figure 7.3, the RIXS plane displays spectral data as an "excitation-loss map," where different excitonic modes can be identified from their dispersion, spectral weight, and energy levels. The spectral weight allows quantitative comparison of mode intensities, helping to determine the stage of electrochemical reactions in energy storage materials. The energy levels provide key insights into the activation energy required for charge transfer processes.

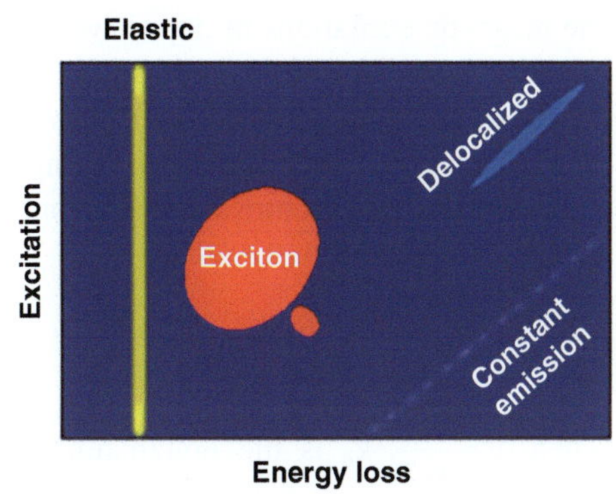

Figure 7.3 RIXS plane: An intensity plot with excitation energy on the vertical axis and energy loss on the horizontal axis [6] / with permission of John Wiley & Sons.

The high-brilliance X-ray beam may induce beam-induced sample damage or alterations, making it essential to systematically evaluate such potential effects in all measurements. However, because RIXS is a bulk-sensitive photon-in-photon-out technique with a relatively large probing depth, beam damage effects are generally less severe than in surface-sensitive electron yield detection. RIXS signals themselves can be influenced by saturation and self-absorption effects. Saturation arises from the energy-dependent penetration depth of X-rays, which alters the effective excited volume as the incident energy is tuned. Self-absorption occurs when emitted X-rays are re-absorbed by the sample itself before escaping, an effect especially pronounced for near-elastic emission close to an absorption edge. For dilute samples, both effects are typically negligible.

7.2.3 Data Analysis and Fitting of RIXS

The microscopic interpretation of the RIXS scattering cross-section remains a central challenge in both experimental and theoretical research. Given the extremely short core hole lifetime (corresponding to an energy broadening Γ typically around 1 eV and a lifetime of ~4 fs), a perturbative expansion can be applied to the denominator of the cross-section formula. This treatment approximates the scattering as the response of the system to a transient core hole perturbation over this short time scale. If one further assumes that the intermediate-state processes are local—that is, confined to a single lattice site—the RIXS cross-section reduces to a form proportional to the spin dynamical structure factor $S(q, \omega)$ [7], thereby linking it directly to magnetic excitations of the system. In practice, however, RIXS can probe a variety of excitations with energies well below 1 eV (and thus lifetimes much longer than 4 fs). This capability stems from the fact that the intermediate state involves more complex many-body correlations and nonlocal processes, enabling access to a rich spectrum of low-energy collective excitations.

RIXS studies on copper oxides have revealed that the line shape of magnetic excitations in doped systems closely resembles that of their parent compounds. Theoretical calculations further confirm that for spin-flip excitations, the dominant intermediate-state processes are confined to a single lattice site [8], allowing the magnetic excitations in doped systems to be described by the spin dynamical structure factor $S(q, \omega)$. In contrast, for charge excitations, the intermediate-state processes are inherently nonlocal, involving correlations across multiple lattice sites, which makes their line shapes complex and difficult to determine definitively.

The dynamical structure factor is theoretically linked to the imaginary part of the dynamical magnetic susceptibility.

$$S(q, \omega) = \frac{Im\chi(q, \omega)}{1 - e^{-\beta\hbar\omega}} \tag{7.2}$$

where $\beta = \frac{1}{k_B T}$, k_B is the Boltzmann constant and T is the temperature. The elementary excitations of the system correspond to the poles of $\chi(q, \omega)$.

If a RIXS excitation at a given momentum can be described by a damped harmonic oscillator model, we can then use this approach to study the RIXS line shape. The motion of a one-dimensional harmonic oscillator under an external force can be written as

$$\frac{d^2x}{dt^2} + \gamma\frac{dx}{dt} + \omega_0^2 x = F(t) \tag{7.3}$$

where γ is the damping coefficient, $F(t)$ is the external force, and ω_0 is the natural oscillation frequency. The response function, also called the generalized susceptibility $\chi(t-t')$, is defined as,

$$x(t) = \int_{-\infty}^{\infty} dt'\chi(t-t')F(t') \tag{7.4}$$

Performing a Fourier transform on the response function,

$$\chi(t) = \int \frac{d\omega}{2\pi} e^{-i\omega t}\chi(\omega) \tag{7.5}$$

Substituting Equation (7.4) into Equation (7.3),

$$\int_{-\infty}^{\infty} \frac{d\omega}{2\pi} \int_{-\infty}^{\infty} dt'\left[-\omega^2 - i\gamma\omega + \omega_0^2\right] e^{-i\omega(t-t')}\chi(\omega)F(t') = F(t) \tag{7.6}$$

Considering $2\pi\delta(t) = \int d\omega\, e^{-i\omega t}$,

$$\chi(\omega) = \frac{1}{-\omega^2 - i\gamma\omega + \omega_0^2} \tag{7.7}$$

Solving for the poles of the above equation yields,

$$\omega_p = -i\gamma/2 \pm \sqrt{\omega_0^2 - \gamma^2/4} \tag{7.8}$$

Therefore, two distinct cases can be identified:

Underdamped $(\omega_0^2 > \gamma^2/4)$: The propagation frequency ω_p possesses both a real and an imaginary part, corresponding to the model of a damped propagating oscillator.

Overdamped $(\omega_0^2 < \gamma^2/4)$: The propagation frequency ω_p becomes purely imaginary. In this regime, the oscillator does not propagate but only decays.

Proceeding to analyze the response function, its real and imaginary parts are respectively:

$$\begin{aligned} \mathrm{Re}\chi(\omega) &= \frac{\omega_0^2 - \omega^2}{(\omega_0^2 - \omega^2)^2 + \gamma^2\omega^2} \\ \mathrm{Im}\chi(\omega) &= \frac{\omega\gamma}{(\omega_0^2 - \omega^2)^2 + \gamma^2\omega^2} \end{aligned} \tag{7.9}$$

The real part of the response function reflects the strength of the system's response to external excitation—the larger its value, the stronger the response at that frequency. The imaginary part is proportional to $\omega\gamma$ (the product of frequency and damping). Under underdamped oscillatory conditions, the peak of the imaginary part generally occurs near the system's natural frequency; however, as $\gamma/2$ approaches ω_0, the deviation between the two gradually increases. Therefore, it is incorrect to determine the eigenenergy of the oscillator simply based on the energy at the peak position.

Currently, most studies employ an asymmetric Lorentzian line shape to fit the RIXS spectra.

$$\begin{aligned} f(\omega) &= \frac{\gamma/2}{(\omega-\omega_1)^2+(\gamma/2)^2} - \frac{\gamma/2}{(\omega+\omega_1)^2+(\gamma/2)^2} \\ &= 2\omega_1 Im\chi(\omega) \end{aligned} \tag{7.10}$$

where ω_1 satisfies $\omega_1^2 + \gamma^2/4 = \omega_0^2$. In the underdamped case, this lineshape agrees with the imaginary part of the response function. However, in the overdamped regime, ω_1 does not exist, so this lineshape cannot be used for fitting; instead, Eq. (7.9) must be employed. More importantly, ω_1 is not the system's natural frequency but rather the oscillation frequency of the damped oscillator. Although ω_1 approaches the natural frequency when the damping γ is small, the two are conceptually distinct.

The dynamic structure factor is given by Eq. (7.2), and the Bose–Einstein distribution must also be included. Finally, to fit the RIXS spectral peaks, the resulting expression must be convoluted with a Gaussian function representing the instrumental resolution, which yields the complete theoretical function used in the actual fitting.

7.3 Application Cases

RIXS, particularly mapping RIXS (mRIXS), has become an indispensable tool in energy materials research, especially for deciphering complex mechanisms in battery cathodes. Leveraging its superior chemical sensitivity and bulk-probing capability, RIXS overcomes limitations of conventional techniques like sXAS and XPS, enabling unambiguous identification and quantitative analysis of redox processes, valence states, and interfacial chemistries. This section focuses on three key application scenarios: unravelling anionic redox mechanisms to break capacity bottlenecks, characterizing TM redox and spin states for performance optimization, and probing solid electrolyte interphase (SEI)/cathode electrolyte interphase (CEI) properties to enhance battery stability. These cases demonstrate how RIXS bridges experimental observations and theoretical insights, guiding the rational design of next-generation high-performance battery materials.

7.3.1 RIXS Unravelling Anionic Redox Mechanisms

The advancement of high-energy-density energy storage systems is fundamentally constrained by the theoretical capacity limits of conventional TM cationic redox, establishing a significant bottleneck in the field. To transcend these limits, triggering anionic redox activity—specifically lattice oxygen redox—has emerged as a vital strategy for next-generation lithium-ion and sodium-ion batteries. Materials such as Li-rich manganese-based oxides and Na-deficient layered oxides operate on the premise that holes can be reversibly stored in oxygen 2p orbitals, unlocking substantial extra capacity.

However, the practical deployment of anionic redox is frequently hampered by severe electrochemical penalties, including large voltage hysteresis, sluggish kinetics, and structural degradation accompanied by irreversible oxygen release. Consequently, understanding the fundamental nature of the oxidized oxygen species is paramount. For years, a central and intense controversy has persisted regarding the fundamental origin of this capacity, specifically centering on distinguishing whether it intrinsically stems from a reversible solid-state lattice redox mechanism or is merely a manifestation of irreversible oxygen gas evolution. Conventional characterization tools fell short of providing a conclusive answer. While O *K*-edge soft X-ray absorption spectroscopy (sXAS) is widely utilized to probe the unoccupied O 2p states, its reliability is compromised by intrinsic physical restrictions. In the pre-edge region of the O *K*-edge (typically 528–533 eV), signals arising from intrinsic oxidized oxygen states (oxygen holes) are often heavily entangled with the broad, dominant spectral features of TM–O hybridization. In standard sXAS, which integrates all emitted photons or electrons, these subtle hole states are often indistinguishable from changes in metal-ligand covalency or surface artifacts, making it arduous to claim the existence of "true" lattice redox.

To address this fundamental ambiguity, mapping of RIXS (mRIXS) has been introduced as a transformative solution. Unlike sXAS, which measures the probability of absorption by counting the total decay products (Total Electron Yield or Total Fluorescence Yield), RIXS is a photon-in-photon-out technique that measures the energy of the emitted photon with high precision. By scanning the incident X-ray energy (excitation energy) across the absorption edge and simultaneously resolving the energy of the emitted X-rays (emission energy), mRIXS generates a two-dimensional intensity map. This capability effectively deciphers a single absorption data point into a full energy-loss spectrum. Physically, this allows researchers to distinguish between different decay channels. For example, elastic scattering (where emission energy equals excitation energy) can be separated from inelastic scattering processes such as d-d excitations, charge transfer excitations, and fluorescence features. It has been established that a distinct spectral feature emerging at an excitation energy of ≈531.0 eV and a specific emission energy of ≈523.7 eV serves as the definitive "fingerprint" for bulk lattice oxygen redox behavior [4, 6]. This feature is distinct from the conventional TM-O hybridization signals, offering a clear spectroscopic indicator of oxygen hole states.

Leveraging this unique fingerprinting capability, researchers have re-examined the universality of anionic redox across various cathode systems. By applying O-*K* mRIXS to a series of classic materials, including $Li_{1.17}Ni_{0.21}Co_{0.08}Mn_{0.54}O_2$ (Li-rich), $LiNi_{1/3}Co_{1/3}Mn_{1/3}O_2$ (NCM111), $LiCoO_2$, $LiNiO_2$, and Li_2MnO_3, an extensive investigation into their charged states was conducted [9]. The experimental approach centered on identifying the established spectroscopic fingerprints of lattice oxygen redox within the mRIXS maps. This involved targeting two key signatures: (i) a distinct, sharp peak appearing at approximately 531 eV excitation and 524 eV emission, and (ii) an associated inelastic signal with minimal energy loss, located near the elastic scattering line. The energy positions of these features align well with the known spectral characteristics of oxidized oxygen species in molecular references such as Li_2O_2 and O_2.

The experimental findings were significant: oxidized oxygen features were unambiguously detected in the charged states of Li-rich, NCM111, $LiCoO_2$, and $LiNiO_2$ (Figure 7.4). However, and crucially, a thorough investigation revealed that these characteristic lattice oxidized oxygen features were entirely absent in Li_2MnO_3 throughout its electrochemical cycling, even when prepared under various synthesis conditions or with different particle sizes. This stark contrast, with Li_2MnO_3 traditionally considered a model system for oxygen redox, provided a critical control reference. These findings fundamentally shifted the community's understanding, indicating that reversible oxygen redox is not an anomaly restricted solely to Li-rich materials but rather a generic property of most TMs oxide cathodes operating at high potentials. Instead, for Li_2MnO_3, Mn^{3+}/Mn^{4+} redox was found to dominate its reversible bulk cycling, with no lattice oxygen redox involved. This case study demonstrates the power of mRIXS to correct historical misconceptions and promote further investigations into the underlying correlations between the primitive TM-O electronic structure and the true nature of oxygen redox.

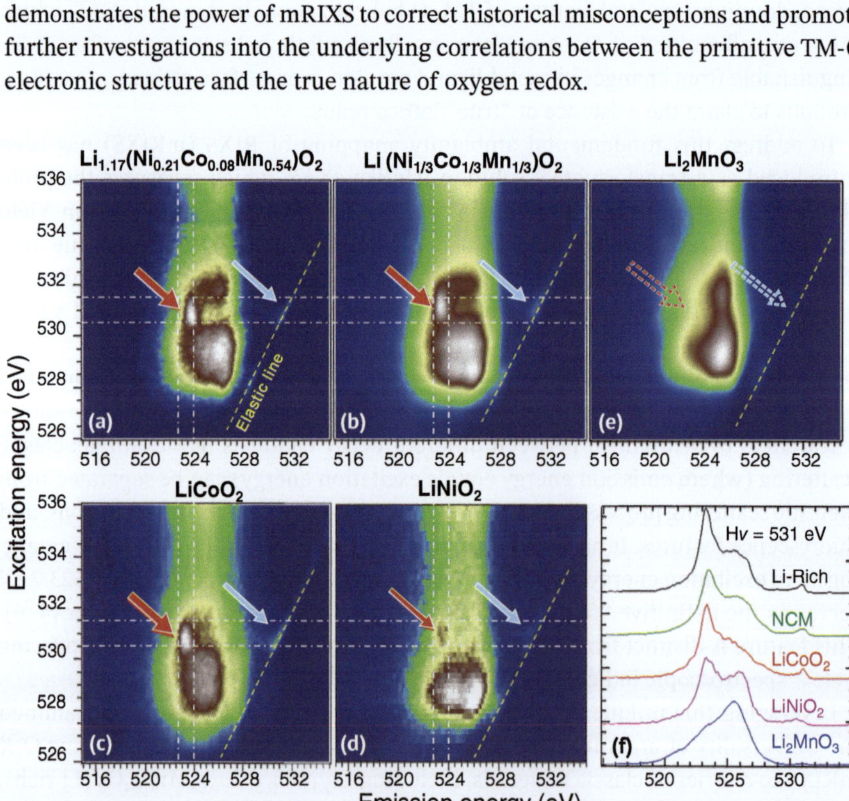

Figure 7.4 O-*K* mRIXS spectra for various cathode materials in their fully charged state Adapted from [10] / with permission of Elsevier.

With the deepening of research, the application of mRIXS has evolved from qualitative identification to rigorous quantitative assessment, which is essential for distinguishing useful reversible redox from harmful gas evolution. Dai et al. established a quantitative framework for evaluating this reversibility. By combining high-efficiency mRIXS with super-PFY (sPFY) analysis—which mitigates self-absorption effects—they investigated two model systems: a P2-type $Na_{2/3}Mg_{1/3}Mn_{2/3}O_2$ and a Li-rich $Li_{1.17}Ni_{0.21}Co_{0.08}Mn_{0.54}O_2$ [10]. By integrating the intensity of the characteristic O-*K* mRIXS feature at 523.7 eV emission energy across charged and discharged states, they successfully quantified the efficiency of the anionic reaction. Their results revealed a high initial-cycle reversibility (79% for the Na-based and 76% for the Li-based material), with substantial capacity retention after long-term cycling. This work highlighted the capability of mRIXS to decouple lattice oxygen activity from other irreversible processes, providing a reliable quantitative basis for evaluating electrode stability.

Ultimately, the power of advanced characterization lies in its ability to guide and validate material design strategies. In the context of sodium-ion batteries, which are plagued by phase transitions and stability issues, Cheng et al. utilized mRIXS to verify that the reversibility of lattice oxygen redox can be significantly improved by modulating the covalency between TMs and oxygen [11]. Using $Na_{0.6}Mg_{0.3}Mn_{0.7}O_2$ (NNMO) as baseline, they introduced highly electronegative Cu^{2+} to substitute part of the electropositive Mg^{2+}, constructing a novel $Na_{0.6}Mg_{0.15}Mn_{0.7}Cu_{0.15}O_2$ (NMMCO) cathode. To explicitly verify the efficacy of this strategy, quantitative mRIXS mapping was employed to track the oxidation state of lattice oxygen during cycling. The results were striking: the Cu-doped NMMCO exhibited a remarkable improvement in lattice oxygen redox reversibility from 73% to 95%, alongside the suppression of detrimental phase transitions (Figure 7.5). This case study exemplifies the indispensability of mRIXS, not just as an analytical tool, but as a critical feedback mechanism for designing high-energy-density electrodes with highly reversible oxygen participation.

In another compelling application, Chen et al. systematically investigated the role of redox-inactive d^0 TMs in the behavior of Li-rich cation-disordered rock salt cathodes (LRCDRS) [12]. The researchers utilized RIXS, specifically O-*K* mRIXS, as a direct and reliable spectroscopic tool to fingerprint oxygen redox reactions in oxide electrodes. This technique was crucial for evaluating reversible oxygen redox and detecting lattice oxygen redox in their model compounds: $Li_{1.3}Nb_{0.3}Mn_{0.4}O_2$ (LNMO) and $Li_{1.2}Ti_{0.4}Mn_{0.4}O_2$ (LTMO).

A clear performance divergence was observed: LNMO provided a higher initial charge storage, but LTMO demonstrated markedly superior cycling durability. RIXS analysis was pivotal in uncovering the mechanistic basis for this disparity. By tracking the well-established spectroscopic fingerprint for oxygen hole states—a distinct feature at an emission energy of approximately 523.7 eV—the measurements confirmed that the anionic redox process was substantially more reversible within the Ti-doped material. This enhanced stability was attributed to the role of Ti^{4+}, which provided a more stable coordination environment for the oxidized oxygen anions, thereby inhibiting their detrimental conversion into O_2 gas. This feature,

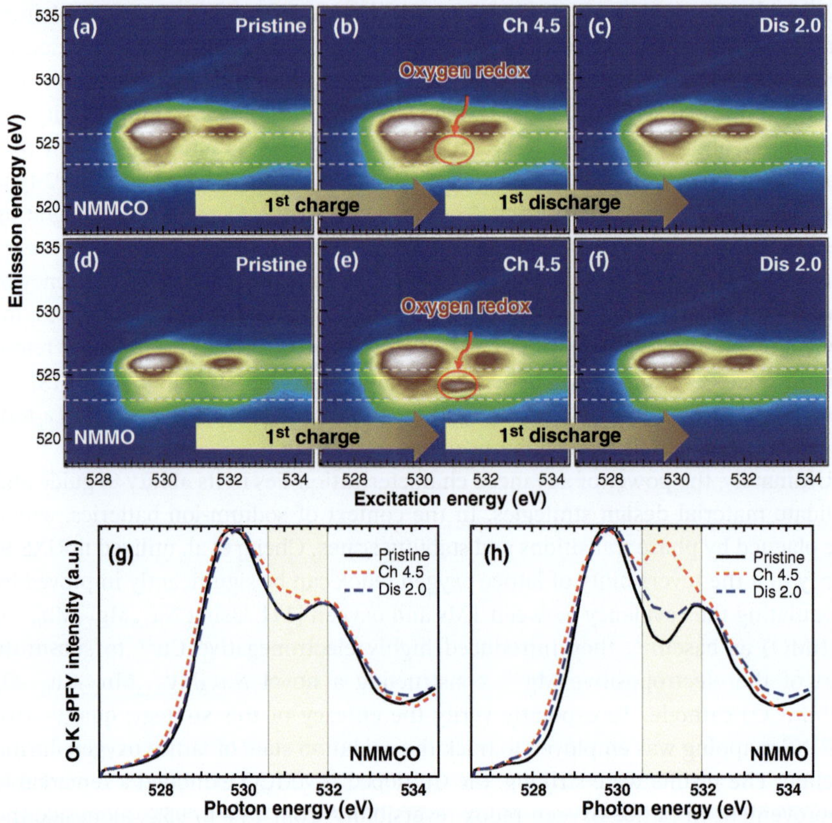

Figure 7.5 O *K*-edge mRIXS maps revealing the distinct oxygen redox behavior in NMMCO and NMMO electrodes at various stages of their initial electrochemical cycle [11] / with permission of John Wiley & Sons.

appearing upon charging and disappearing upon discharging, indicated lattice oxygen redox in both materials. However, in LNMO, this reversibility significantly decayed after only five cycles, whereas in the LTMO electrode, the peak feature was largely maintained even after 50 cycles. Further analysis of RIXS cut spectra showed that after 50 cycles, LTMO retained about 52% of its initial O^{n-} feature intensity, which was nearly three times higher than LNMO's 18%. This distinct contrast confirmed that oxygen redox is considerably more reversible in LTMO, highlighting Ti's effectiveness as a d^0 TM in improving O^{n-} cyclability. This case study demonstrates how mRIXS is an indispensable tool for understanding the intricate redox chemistry and validating strategies for enhancing the stability and performance of high-energy-density cathode materials.

7.3.2 Application of RIXS on TM Redox and Spin States

In lithium- and sodium-ion battery materials, 3d TM (TM = Mn, Fe, Co, Ni) cations serve as the primary centers for charge compensation. Consequently, the redox evolution of these cations dictates critical battery performance metrics, including specific capacity, operating voltage, and structural stability. While conventional sXAS has long been the standard tool for probing TM *L*-edges (2p → 3d transitions), it faces intrinsic physical limitations. Due to the limited soft X-ray penetration depth, the fluorescence yield mode is usually adopted to detect the bulk information of investigated materials, which yield fluorescence signal may be absorbed by material itself and lead to a distorted spectral shape (especially for TM *L*-edge). This spectral lineshape distortion hinders further quantification and restricts the investigation of the redox mechanism of battery cathodes. Compared to conventional sXAS, mRIXS introduces a critical new dimension of information by resolving emitted photons along their emission energies. This added dimensionality naturally disentangles electronic states that are otherwise merged in sXAS spectra, providing an incisive method to fingerprint bulk redox processes. Consequently, mRIXS offers superior chemical sensitivity, which is particularly essential for identifying novel oxidation couples and resolving the complex redox behaviors of both TMs and oxygen that remain obscure in standard spectroscopy. With its low cost, high abundance and low toxicity, Mn is one of the most important TM elements used in battery materials. Defining and characterizing its valence state is of vital importance for the study of cathode materials. Researchers have suggested that Mn^{1+} might be present in coordination compounds, but XPS and XAS have not been included until now. Through investigating Prussian blue analog $Na_{1.24}Mn[Mn(CN)_6]\cdot 2.1H_2O$, the presence of Mn^{1+} was successfully verified by Mn-*L* mRIXS [13]. Only a slight difference can be observed in the XAS spectra of Mn^{2+} and Mn^{+} at around 643.5 eV. In sharp contrast, there are significant differences in the mRIXS spectra of Mn^{2+} and Mn^{1+} (Figure 7.6). During the mRIXS measurement process, in the low-spin Mn^{+} system, there exists d-d excitation that does not exist in the normal Mn^{2+} system, thus providing the ultimate experimental verification of the Mn^{1+}/Mn^{2+} redox in the anode. The example of Mn^{1+} not only demonstrates the prominent role of mRIXS spectroscopy in characterizing abnormal TM oxidation states, but also shows that the complexity of actual electrochemical systems is far beyond imagination.

On a deeper level, the unique advantage of mRIXS lies in its ability to go beyond qualitative identification and achieve quantitative analysis of TM bulk redox processes. While traditional sXAS faces limitations in quantitatively analyzing TM *L*-edges due to lineshape distortions caused by self-absorption effects, which severely restricts accurate quantification of valence states, Mn-*L* mRIXS experiments can provide more accurate bulk Mn redox information by avoiding these distortions through the inverse PFY (iPFY) results. For instance, in a study investigating Li_2MnO_3 materials at different states of charge including LMO100, LMO60, and LMO20, researchers employed mRIXS to quantitatively fit the Mn-*L* iPFY spectra, enabling precise evaluation of bulk Mn valence states [14]. The findings indicated that, during charging, the bulk Mn valence state was predominantly +4, very similar to its pristine state; however, during discharge, Mn in the bulk existed as Mn^{2+} and Mn^{3+} (Figure 7.7). Through this quantitative analysis, the study further revealed that the Mn valence state in LMO100 undergoes more pronounced changes during charge and discharge, with a greater amount

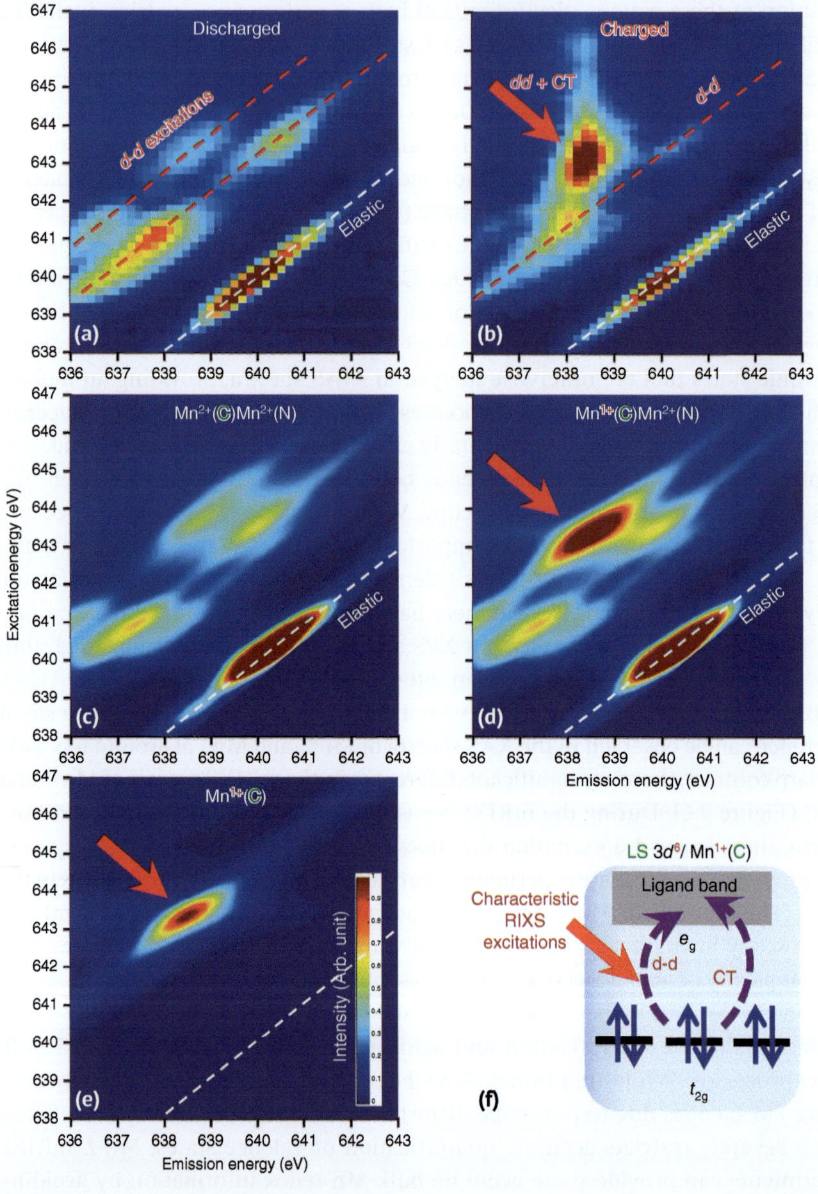

Figure 7.6 Experimental verification of the Mn^{1+} oxidation state using Mn *L*-edge RIXS maps [13] / Springer Nature / CC BY 4.0.

of Mn^{2+} formed during discharge. This not only quantifies the evolution of Mn valence during electrochemical cycling but also highlights the powerful capability of mRIXS in deeply understanding how different material characteristics, such as stacking faults, influence TM redox activity. This mRIXS-based quantitative approach lays a solid foundation for accurately elucidating the complex redox mechanisms of TMs in battery materials.

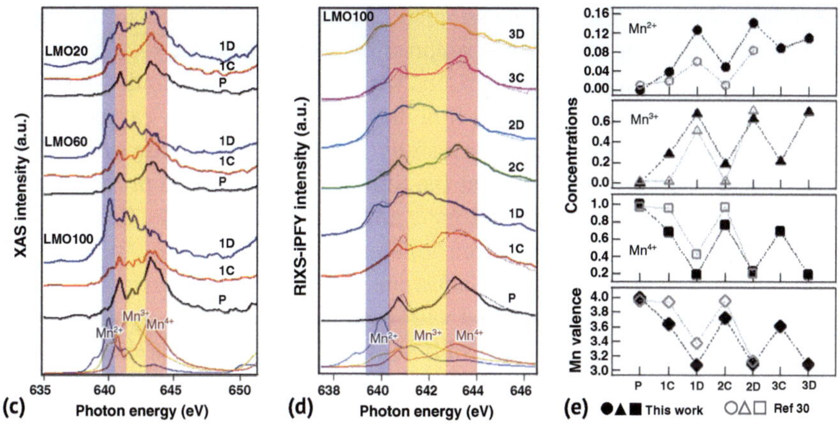

Figure 7.7 Mn-*L* TEY and iPFY spectra of Li_2MnO_3, and corresponding quantification results [14] / with permission of John Wiley & Sons.

7.3.3 RIXS Characterization of Battery Interfaces (SEI/CEI)

The operational performance and long-term stability of rechargeable batteries, including lithium- and sodium-ion systems, are fundamentally dictated by the properties of the passivating layers formed at the electrode-electrolyte boundary—namely the SEI on the anode and its cathode counterpart (CEI) [15]. However, the chemical reactions at these interfaces are notoriously complex and dynamic: the anode side is characterized by continuous electrolyte decomposition and SEI formation, while the cathode interface often suffers from intrinsic fatigue driven by structural collapse and gas release. Although conventional sXAS possesses inherent surface sensitivity for detecting reaction products, it frequently lacks the necessary chemical sensitivity to resolve subtle interfacial effects, particularly when the signals are dominated by the bulk materials in the vicinity of the interface.

The introduction of RIXS offers a transformative solution to these characterization limits. Unlike conventional XAS, which maps unoccupied states via photon counts or current, RIXS records the energy distribution of emitted photons, thereby resolving a single absorption data point into a full energy distribution curve. This capability provides a new dimension of information along the emission energy scale, granting RIXS significantly improved chemical sensitivity. Consequently, RIXS serves as a powerful "fingerprinting" tool capable of disentangling complex signals that are indistinguishable in standard spectroscopy.

This unique capability to resolve complex interfacial chemistries is vividly exemplified by the characterization of the SEI on graphite anodes in Li-ion batteries. In a representative study, researchers utilized soft X-ray emission and absorption spectroscopy near the C 1s edge to probe the electronic structure and chemical composition of SEI layers formed on graphite electrodes cycled in EC:DMC and PC:EC electrolytes [16]. The power of RIXS lies in its element-specific and chemically sensitive nature through dipole-selective excitations. By tuning the incident photon energy to characteristic absorption features (e.g., 285 eV, 288.6 eV, and 290.3 eV), the researchers obtained resonant emission spectra that selectively enhanced signals from specific SEI components while suppressing the background.

Through rigorous spectral comparisons with reference compounds—including lithium oxalate, lithium succinate, lithium methoxide, lithium carbonate, and ethylene carbonate—the study successfully identified the precise constituents of the SEI. The results revealed that the SEI on cycled graphite primarily comprises lithium oxalate, lithium succinate, and lithium methoxide. Crucially, the RIXS approach allowed for the clean separation of contributions from the SEI layer and the underlying graphite substrate via spectral subtraction, revealing distinct emission features associated with each compound. This application not only highlights the ability of RIXS to differentiate between various carbon species with both elemental and orbital selectivity but also demonstrates its practical advantage in avoiding sample-charging issues common in surface-sensitive techniques like XPS. Thus, RIXS has proven to be an indispensable tool for offering deep insights into interfacial reactions, stability, and composition under realistic cycling conditions.

7.4 Summary and Perspectives

mRIXS, commonly known as RIXS, has emerged as a transformative and powerful characterization technique for advancing the understanding of battery cathode materials, particularly in elucidating cationic and anionic redox mechanisms.

RIXS offers a new dimension of information along the emission energy, providing significantly improved chemical sensitivity compared to conventional sXAS and XPS. This 2D spectroscopic capability allows it to overcome several limitations inherent in traditional techniques. For instance, RIXS effectively disentangles specific chemical states that are often merged in standard sXAS spectra, which can suffer from entangled TM-O hybridization signals and shallow probing depths. It provides non-distorted TM *L*-edge spectra, overcoming lineshape distortions and self-absorption effects inherent in sXAS, enabling quantitative analysis of bulk TM redox processes using iPFY. Additionally, it avoids sample-charging issues common in XPS.

Despite its advanced capabilities, RIXS faces several limitations and presents significant opportunities for future development. A primary challenge is that RIXS is a photon-hungry technique due to the low branching ratio of specific core-hole decay channels, often requiring hours or even days for reliable measurements. Future developments, such as ultra-high-efficiency spectrometers and optical designs enabling simultaneous measurement of both incident and emission energies (2D scheme), promise to drastically reduce data acquisition times, potentially by tens or hundreds of times.

The enhanced X-ray brightness from fourth-generation synchrotron sources increases the risk of radiation damage, especially for fragile energy materials, which can lead to misinterpretation of results (e.g., beam-induced molecular O_2 instead of intrinsic electrochemical products). More detailed studies on radiation damage effects are urgently needed, alongside strategies like using itinerant samples or imaging modes that do not require significant beam focusing.

Optimizing energy resolution remains a continuous challenge, and balancing multiple resolutions (energy, momentum, spatial) for specific problems is complex. The fundamental form of oxidized oxygen species and their correlation with kinetic issues like voltage hysteresis also remain subjects of intense debate. Future prospects include further optimization of high-quality grating optics and elongated instrument configurations for superior energy resolution. Deeper theoretical development is needed to fully comprehend the precise oxidized states during OR processes and to clarify the fundamental connection between OR reaction and kinetic issues.

For in-situ/operando measurements, soft X-ray RIXS can be challenging in liquid electrolytes due to low X-ray transmissivity through optical windows and electrolyte evaporation in ultra-high vacuum (UHV) environments. However, the technique is highly promising for solid-state batteries, which are compatible with UHV environments, allowing for in-situ studies of electrode-electrolyte interfaces during cycling. Developments in operando hard X-ray RIXS could also enable measurements in ambient conditions.

The spatial distribution of anionic redox, including local active sites or defects, is still unclear. The development of 2D RIXS microscopy, combining RIXS with scanning microscopy, could provide spatial maps of redox activity, addressing this gap.

Finally, current theoretical simulation methods like Dynamical Mean Field Theory have limitations, and the large datasets generated from RIXS experiments demand advanced analysis tools. Future efforts will focus on improving theoretical simulations (e.g., cluster approximations, *ab initio* calculations) for accurate interpretation. Furthermore, the integration of artificial intelligence and machine learning tools is expected to play a significant role in accelerating RIXS data analysis, facilitating spectral classification, redox-state tracking, and anomaly detection.

In conclusion, RIXS is an indispensable tool that has profoundly advanced our understanding of complex redox mechanisms in battery materials. Addressing its current limitations through ongoing instrumental advancements, theoretical refinements, and innovative data analysis approaches will further unlock its potential, guiding the rational design of next-generation high-performance battery cathodes.

References

1 H.A. Kramers, W. Heisenberg, Über die Streuung von Strahlung durch Atome, Zeitschrift für Physik 31(1) (1925) 681–708. https://doi.org/10.1007/BF02980624.

2 X. Liu, W. Yang, Z. Liu, Recent progress on synchrotron-based in-situ soft X-ray spectroscopy for energy materials, Advanced Materials 26(46) (2014) 7710–7729. https://doi.org/10.1002/adma.201304676.

3 L.J.P. Ament, M. van Veenendaal, T.P. Devereaux, J.P. Hill, J. van den Brink, Resonant inelastic X-ray scattering studies of elementary excitations, Reviews of Modern Physics 83(2) (2011) 705–767. https://doi.org/10.1103/RevModPhys.83.705.

4 C. Cheng, Z. Zhuo, S. Chen, X. Zhou, C. Yuan, P. Zeng, J. Guo, L. Zhang, Cationic and anionic redox of battery cathodes investigated by advanced synchrotron-based mapping of resonant inelastic X-ray scattering, Advanced Functional Materials 34(39) (2024) 2403442. https://doi.org/10.1002/adfm.202403442.

5 Z. Zhang, J. Liu, Y. Lei, D. Wang, Y. Cui, D. Sun, Design of RIXS beamline at Shenzhen innovation light facility, Nuclear Instruments and Methods in Physics Research Section A: Accelerators, Spectrometers, Detectors and Associated Equipment 1064 (2024) 169439. https://doi.org/10.1016/j.nima.2024.169439.

6 D. Zhou, J. Yu, S. Liu, D. Wong, J. Wang, Q. Wang, Z. Zhang, X. Yao, H. Yu, Exploring anionic redox chemistry of battery cathodes with resonant inelastic X-ray scattering, Advanced Materials 37(40) (2025) 2508246. https://doi.org/10.1002/adma.202508246.

7 Y. Lu, M.W. Haverkort, Nonperturbative series expansion of green's functions: The anatomy of resonant inelastic X-ray scattering in the doped Hubbard model, Physical Review Letters 119(25) (2017) 256401. https://doi.org/10.1103/PhysRevLett.119.256401.

8 C.J. Jia, E.A. Nowadnick, K. Wohlfeld, Y.F. Kung, C.C. Chen, S. Johnston, T. Tohyama, B. Moritz, T.P. Devereaux, Persistent spin excitations in doped antiferromagnets revealed by resonant inelastic light scattering, Nature Communications 5(1) (2014) 3314. https://doi.org/10.1038/ncomms4314.

9 Z. Zhuo, C.D. Pemmaraju, J. Vinson, C. Jia, B. Moritz, I. Lee, S. Sallies, Q. Li, J. Wu, K. Dai, Y.-d. Chuang, Z. Hussain, F. Pan, T.P. Devereaux, W. Yang, Spectroscopic signature of oxidized oxygen states in peroxides, The Journal of Physical Chemistry Letters 9(21) (2018) 6378–6384. https://doi.org/10.1021/acs.jpclett.8b02757.

10 K. Dai, J. Wu, Z. Zhuo, Q. Li, S. Sallis, J. Mao, G. Ai, C. Sun, Z. Li, W.E. Gent, W.C. Chueh, Y.-d. Chuang, R. Zeng, Z.-x. Shen, F. Pan, S. Yan, L.F.J. Piper, Z. Hussain, G. Liu, W. Yang, High reversibility of lattice oxygen redox quantified by direct bulk probes of both anionic and cationic redox reactions, Joule 3(2) (2019) 518–541. https://doi.org/10.1016/j.joule.2018.11.014.

11 C. Cheng, C. Chen, S. Chu, H. Hu, T. Yan, X. Xia, X. Feng, J. Guo, D. Sun, J. Wu, S. Guo, L. Zhang, Enhancing the reversibility of lattice oxygen redox through modulated transition metal–oxygen covalency for layered battery electrodes, Advanced Materials 34(20) (2022) 2201152. https://doi.org/10.1002/adma.202201152.

12 D. Chen, J. Wu, J.K. Papp, B.D. McCloskey, W. Yang, G. Chen, Role of Redox-inactive transition-metals in the behavior of cation-disordered Rocksalt cathodes, Small 16(22) (2020) e2000656. https://doi.org/10.1002/smll.202000656.

13 A. Firouzi, R. Qiao, S. Motallebi, C.W. Valencia, H.S. Israel, M. Fujimoto, L.A. Wray, Y.-D. Chuang, W. Yang, C.D. Wessells, Monovalent manganese based anodes and co-solvent electrolyte for stable low-cost high-rate sodium-ion batteries, Nature Communications 9(1) (2018) 861. https://doi.org/10.1038/s41467-018-03257-1.

14 B. Wang, Z. Zhuo, H. Li, S. Liu, S. Zhao, X. Zhang, J. Liu, D. Xiao, W. Yang, H. Yu, Stacking faults inducing oxygen anion activities in Li_2MnO_3, Advanced Materials 35(22) (2023) 2207904. https://doi.org/10.1002/adma.202207904.

15 W. Yang, T.P. Devereaux, Anionic and cationic redox and interfaces in batteries: Advances from soft X-ray absorption spectroscopy to resonant inelastic scattering, Journal of Power Sources 389 (2018) 188–197. https://doi.org/10.1016/j.jpowsour.2018.04.018.

16 A. Augustsson, M. Herstedt, J.H. Guo, K. Edström, G.V. Zhuang, J.P.N. Ross, J.E. Rubensson, J. Nordgren, Solid electrolyte interphase on graphite Li-ion battery anodes studied by soft X-ray spectroscopy, Physical Chemistry Chemical Physics 6(16) (2004) 4185–4189. https://doi.org/10.1039/b313434b.

8

Synchrotron-based Hard X-ray Absorption Spectroscopy for Local Structure and Redox Chemistry in Battery Materials

8.1 Fundamentals of Hard X-ray Absorption Spectroscopy

X-ray absorption spectroscopy evolved from early discoveries of element-specific absorption edges to modern X-ray absorption near edge structure (XANES) and extended X-ray absorption fine structure (EXAFS) techniques. Enabled by quantum theory and synchrotron radiation, these methods provide detailed, element-resolved insights into local atomic structure, coordination, and electronic states.

8.1.1 XANES

The history of X-ray absorption spectroscopy (XAS) is intimately connected to the early developments in X-ray physics and atomic spectroscopy. Following Wilhelm Conrad Röntgen's discovery of X-rays in 1895, researchers quickly recognized that X-rays were absorbed differently by various materials. However, systematic studies of absorption edges-sharp increases in absorption at element-specific energies-began in the early 20th century. In 1909, Charles Glover Barkla, while investigating secondary X-ray emission, observed that each element exhibits characteristic "absorption limits" (later termed absorption edges) that correspond to the binding energies of inner-shell electrons. This work, for which Barkla received the 1917 Nobel Prize in Physics, established the element-specific nature of X-ray absorption and laid the groundwork for understanding atomic energy levels.

A crucial advance came in the 1920s with the work of Dirk Coster and George de Hevesy, who systematically measured absorption edges across the periodic table and identified the fine structure just above the edge—the region later called XANES. They noted that the absorption coefficient did not simply jump at the edge but displayed oscillations extending several hundred electronvolts beyond it. These oscillations, initially referred to as "Kossel-Kronig structure," were recognized as containing information about the local atomic environment, although a full theoretical explanation would not emerge until decades later.

The development of quantum mechanics in the 1920s and 1930s provided the theoretical framework to interpret absorption edges. Hans Bethe's 1930 paper on the theory of X-ray absorption edges, together with contributions by John H. Van Vleck, Rudolf Peierls, and others, related the edge position to the binding energy of core electrons and the fine structure to transitions into unoccupied electronic states. However, the detailed analysis of the near edge region as a probe of local coordination and oxidation state only became routine with the advent of synchrotron radiation sources in the 1970s and 1980s.

The modern era of XANES began with the availability of high-brilliance, tunable X-ray beams from synchrotron storage rings. These sources enabled precise, energy-resolved measurements of absorption edges with high signal-to-noise ratios, making it possible to extract subtle chemical and structural information. Pioneering work by Farrel Lytle, Dale Sayers, Edward Stern, and others in the 1970s established the theoretical formalism for interpreting XANES and its extended counterpart—EXAFS. Today, XANES is a standard tool in materials characterization, complementing diffraction and scattering techniques by providing element-specific insights into oxidation states, coordination symmetry, and electronic structure.

1. Fundamental Principles of XANES
   Drawing upon the core physical interactions between photons and matter, XAS is fundamentally grounded in the exponential attenuation of an X-ray beam propagating through a material. This behavior is quantitatively captured by the Lambert–Beer law. For a monochromatic incident X-ray beam with initial intensity I_0 passing through a sample of thickness d, the transmitted intensity I_T can be expressed as:

$$I_T = I_0 \exp(-\mu d) \tag{8.1}$$

Here, μ denotes the linear absorption coefficient, which depends on both the material composition and the incident photon energy [1].

The absorption coefficient exhibits a sharp rise when the incident photon energy coincides with the binding energy of a core electron. At this critical energy, the photon is absorbed via the photoelectric effect, resulting in the ejection of the core electron and the formation of a core hole (Figure 8.1). This results in a distinct "absorption edge" in the XAS spectrum, appearing as a step-like rise in the absorption coefficient. Because each element has a unique set of core-level binding energies governed by its nuclear charge, XAS provides unambiguous elemental specificity.

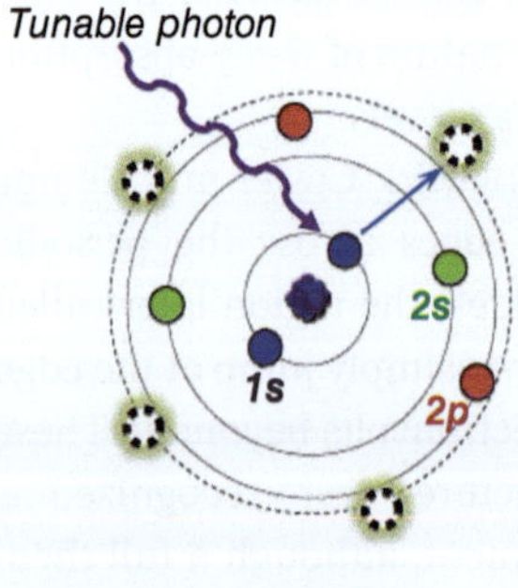

Figure 8.1 An atomic schematic of core-shell electron activated process [2] / with permission of Elsevier.

Moreover, even for the same element, changes in oxidation state alter the effective nuclear charge felt by core electrons, leading to measurable shifts in the absorption edge energy. Thus, XAS can also resolve the oxidation state and local chemical environment of the absorbing element.

A typical XAS spectrum is divided into two regions: XANES, spanning from roughly 20 eV below to 50 eV above the edge, and EXAFS, extending from about

50 eV to ~1000 eV beyond the edge. The XANES region contains rich information encoded in the edge position, spectral shape, and characteristic pre-edge or post-edge features, which reflect the electronic structure, oxidation state, and local coordination geometry. This chapter focuses specifically on the physical origin, interpretation, and applications of XANES.

2. Absorption and Electronic Transitions
Upon X-ray absorption, a core-level electron is excited from its initial bound state. It may either transition to an unoccupied bound orbital at a higher energy or be ejected into the continuum of free states, becoming a photoelectron. The spectral features in the XANES region can be categorized according to the final state of this excited electron:

Pre-edge region: Characterized by relatively sharp, discrete peaks. These originate from bound-to-bound transitions, where the core electron is excited to localized, unoccupied molecular or crystal orbitals near the Fermi level.

Edge region: Dominated by the ionization threshold. The absorption rises steeply as the photon energy becomes sufficient to excite the electron into the continuum of free states (bound-to-continuum transitions).

Post-edge region: Exhibits broad oscillations and features resulting from the interference of the scattered photoelectron wave with itself after interacting with neighboring atoms.

These electronic transitions are governed by quantum mechanical selection rules, which imprint detailed information about the electronic structure and local symmetry of the absorbing atom onto the XANES spectrum [3].

3. Dipole and Quadrupole Transitions
The primary selection rule for X-ray absorption is the dipole selection rule: $\Delta l = \pm 1$, where l is the orbital angular momentum quantum number. This dictates the allowed final states for a given initial core orbital. In condensed matter-such as molecules, liquids, and solids-atomic orbitals hybridize to form molecular orbitals or bands. Consequently, the final states are hybridized orbitals, and their interpretation involves concepts from molecular orbital theory, crystal field theory, multiple scattering theory, and band structure.

Quadrupole transitions, governed by $\Delta l = \pm 2$, are formally forbidden in the dipole approximation. However, they gain weak intensity in non-centrosymmetric environments through mixing (hybridization) of orbitals. A classic example is the pre-edge feature in transition metal *K*-edge spectra (1s → 3d), which is a quadrupole transition. Its intensity is significantly enhanced by metal 3d-ligand p orbital hybridization in low-symmetry coordination geometries (e.g., tetrahedral or distorted octahedral sites). Therefore, the intensity of the pre-edge peak serves as a sensitive probe of local geometric symmetry around the absorbing atom.

4. *K*-Edge Features and Their Structural Significance
In transition metal compounds, the *K*-edge pre-edge feature provides direct insight into the metal atom's coordination geometry. In general, quadrupole

transitions (e.g., 1s → 3d) are more strictly forbidden in highly symmetric environments. Consequently, the pre-edge peak is relatively weak in perfectly octahedral coordination (Oh symmetry), as observed in many compounds. In contrast, the lower symmetry of a tetrahedral field (Td) relaxes the selection rules, allowing the quadrupole transition to gain significant intensity, resulting in a distinct and stronger pre-edge peak. Therefore, the intensity of the pre-edge feature can be used to diagnose changes in local geometric symmetry within a material (Figure 8.2).

Immediately following the absorption edge, a prominent maximum often occurs, known as the "white line." This feature primarily reflects the transition to unoccupied p-states (np character for a *K*-edge). The intensity of the white line is inversely proportional to the electron density in these np states. For instance, in single-atom catalysts, the white line intensity typically lies between that of the corresponding bulk metal (low intensity, filled d-band) and its oxide (high-intensity, empty/partially empty d-band). Furthermore, based on crystal field theory, the different splitting patterns of d-orbitals in tetrahedral versus octahedral fields also influence the white line intensity. Specifically, the white line is generally weaker in tetrahedral coordination than in octahedral coordination-a trend opposite to that observed for the pre-edge feature.

For single-atom catalysts with square-planar geometry (D_4h symmetry), an additional characteristic "fingerprint peak" appears at an energy slightly above the main

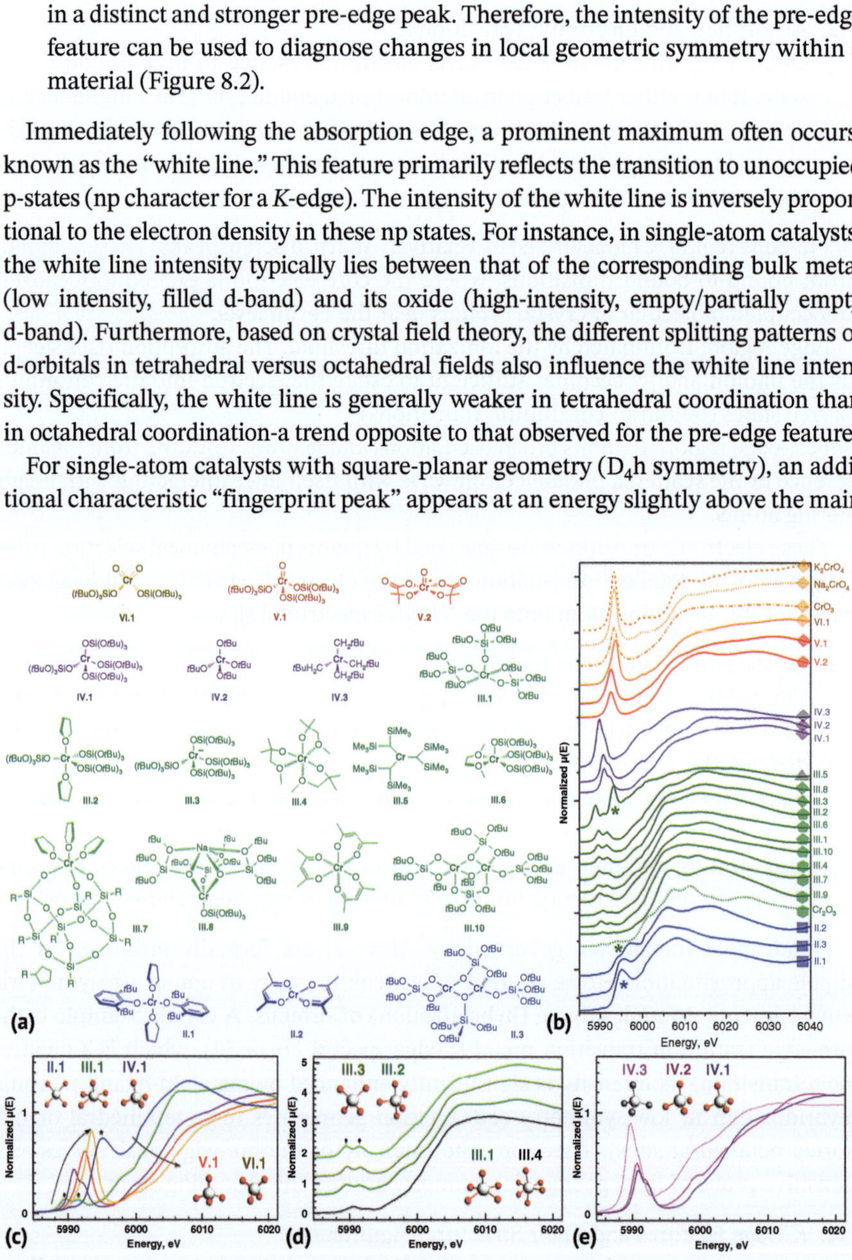

Figure 8.2 The Cr *K*-edge XANES of tetrahedron and octahedron configuration [4] / with permission of American Chemical Society.

pre-edge. This feature corresponds to the 1s → $4p_z$ transition, which is symmetry-allowed in this configuration. This distinct signature is highly valuable for identifying square-planar geometries and can be used to monitor dynamic changes in the bonding plane of the central atom during catalytic reactions.

5. *L*-Edge Features and Their Structural Significance
   L-edge spectroscopy is the most powerful XAS method for directly probing the electronic structure of d-orbitals. When the oxidation state of an element increases, the number of d-electrons decreases, leading to a higher density of unoccupied d-states. This is reflected in the L_3-edge spectrum as a strengthening of the white line intensity, often accompanied by a slight blue shift (higher energy). Conversely, reduction (gain of d-electrons) leads to decreased white line intensity.

Furthermore, the relative magnitudes of the crystal field splitting energy and the electron pairing energy determine whether a transition metal ion stabilizes in a high-spin or a low-spin electronic configuration. This spin state is exquisitely probed by *L*-edge XAS, as the intensity and detailed profile of the white line are directly correlated with the number and specific energetics of unoccupied d-orbitals.

Due to the inherent splitting of d-orbital energy levels in a ligand field, the L_3-edge often exhibits clear multiplet splitting into distinct sub-peaks. The pattern and energy separation of these components serve as a fingerprint, allowing researchers to distinguish between different coordination geometries (e.g., octahedral vs. tetrahedral) and to quantify ligand field strengths (Figure 8.3) [5].

8.1.2 EXAFS

The observation of oscillatory structures extending hundreds of electronvolts beyond X-ray absorption edges dates back to the early days of X-ray spectroscopy. In the 1920s, theoretical physicists including R. de L. Kronig provided the first explanation for these oscillations, attributing them to the scattering of photoelectrons by neighboring atoms. This became known as the Kronig structure,

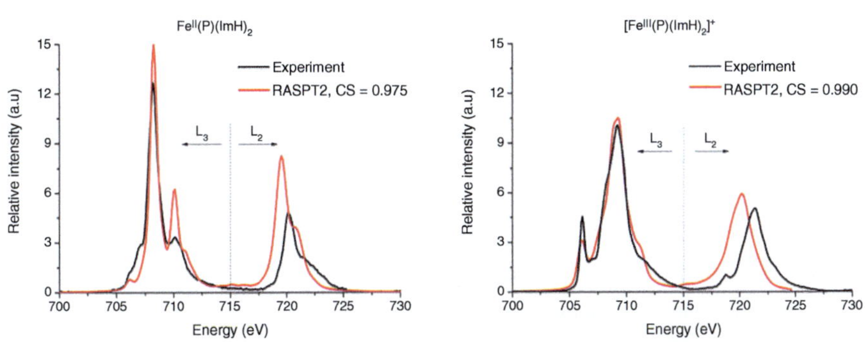

Figure 8.3 The L_3/L_2-edge of Fe^{II}/Fe^{III} [5] / with permission of American Chemical Society.

recognized as a potential source of structural information, but its interpretation remained largely qualitative for decades due to experimental and theoretical limitations.

The modern era of EXAFS began in the early 1970s with the pioneering work of Dale E. Sayers, Edward A. Stern, and Farrel W. Lytle. Their critical insight was to express the EXAFS oscillations in terms of the photoelectron wave vector k and apply a Fourier transform to convert the oscillatory signal from k-space to a real-space radial distribution function. This breakthrough, published in 1971, established EXAFS as a quantitative, short-range structural probe. The development of the theoretical framework for single-scattering and multiple scattering contributions by Stern, Sayers, and Lytle, coupled with the formulation of the EXAFS equation, transformed the technique from a spectroscopic curiosity into a powerful analytical tool.

The concurrent rise of synchrotron radiation sources in the 1970s and 1980s provided the high-intensity, tunable X-ray beams necessary for routine EXAFS measurements. This synergy between theory and instrumentation propelled EXAFS into mainstream materials science, chemistry, and biology, enabling the determination of local atomic environments-bond distances, coordination numbers, and disorder in crystalline, amorphous, and liquid systems alike.

1. Fundamental Principles of EXAFS
   According to the classical theoretical framework, the oscillatory component of the X-ray absorption spectrum, denoted as $\chi(k)$ and observed beyond the absorption edge, originates from the scattering of the emitted photoelectron by neighboring atoms. This model treats the photoelectron in its final state as an outgoing spherical wave that is subsequently scattered back towards the absorbing atom. The standard EXAFS equation for a single scattering shell is:

   The standard EXAFS equation for a single scattering shell is:

$$\chi(k)=\sum_j \frac{N_j S_0^2 F_j(k)}{kR_j^2} \exp\left(-2\sigma_j^2 k^2\right) \exp\left(-\frac{2R_j}{\lambda(k)}\right) \sin\left[2kR_j + \phi_{ij}(k)\right] \quad (8.2)$$

where:

- k: Photoelectron wave number, $k = \sqrt{\frac{2m(E-E_0)}{\hbar^2}}$
- j: Index over coordination shells
- N_j: Coordination number of the j-th shell
- S_0^2: Amplitude reduction factor (accounts for many-body effects)
- $F_j(k)$: Backscattering amplitude from atoms in the j-th shell
- R_j: Average distance to the j-th shell
- σ_j^2: Debye–Waller factor (mean-square disorder in R_j)
- $\lambda(k)$: Photoelectron mean free path
- $\phi_{ij}(k)$: Total phase shift (from both central and backscattering atoms)

This equation forms the basis for EXAFS fitting, where theoretical or empirical standards are used to extract structural parameters (R, N, σ^2) from experimental data.

a. Photoelectron Scattering Process
Upon X-ray absorption, a core electron is ejected with kinetic energy $E_k = E - E_0$.
This electron propagates as a spherical wave with wavelength $\lambda_e = 2\pi/k$.
Neighboring atoms scatter this electron wave back towards the absorber.
The constructive or destructive interference between the outgoing and backscattered waves modulates the probability of the absorption event, causing the observed oscillations in $\mu(E)$.

b. Fourier Transform and Radial Distribution
The raw EXAFS oscillation $\chi(k)$ is weighted by k^n (typically $n = 1, 2, 3$) to amplify high-k data.
A Fourier transform of $k^n\chi(k)$ yields a pseudo-radial distribution function $\tilde{\chi}(R)$:

$$\tilde{\chi}(R) = \frac{1}{\sqrt{2\pi}} \int_{k_{\min}}^{k_{\max}} k^n \chi(k) e^{-2ikR} W(k)\, dk \tag{8.3}$$

Peaks in $\tilde{\chi}(R)$ correspond to coordination shells, but are shifted by ~0.3–0.5 Å from the true distances due to the phase shift $\phi(k)$.
Window functions $W(k)$ (e.g., Hanning, Kaiser-Bessel) are applied to minimize termination artifacts [6].

c. Information Content of EXAFS
EXAFS is sensitive to local structural parameters within a radius of ~3–6 Å around the absorbing atom:

- Interatomic distances (R_j) with precision of ±0.01–0.02 Å.
- Coordination numbers (N_j) with typical accuracy of ±10–20%.
- Disorder (σ_j^2), including thermal and static disorder.
- Elemental identity of neighboring atoms (via $F_j(k)$ dependence on atomic number Z).

2. Key Phenomenon: The Origin of Extended Oscillations and Multiple Scattering
The extended fine structure is a direct consequence of the wave nature of the photoelectron and its interaction with the local atomic environment. Two primary phenomena govern the EXAFS signal: single scattering and multiple scattering.
Single Scattering: The photoelectron scatters once from a neighboring atom before returning to the absorber. This dominates the EXAFS signal for linear or near-linear paths (absorber–scatterer–absorber angle ~180°) and is the basis of the standard EXAFS equation.
Multiple Scattering: The photoelectron undergoes two or more scattering events before returning. This becomes significant for:

Linear or collinear atomic arrangements (e.g., M-X-M in linear molecules, metal oxides with linear O-M-O paths).

High-Z scatterers (e.g., Pt, Au) where scattering amplitudes are large.
Ordered systems with well-defined geometry (e.g., crystals, clusters).
Multiple scattering contributions provide information beyond pair distances, such as bond angles and three-body correlations.

3. Amplitude and Phase Transferability
   A cornerstone of EXAFS analysis is the concept of transferability. The scattering amplitude $F_j(k)$ and phase shift $\phi_{ij}(k)$ depend primarily on the atomic numbers of the absorber and scatterer, and their separation. These functions can be calculated ab initio (e.g., using FEFF, GNXAS) or extracted from well-characterized reference compounds (experimental standards). This allows for the fitting of unknown structures using theoretical or empirical scattering profiles.

8.1.3 Wavelet Transform Method for XAS

1. Principle and Fundamental Theory of Wavelet Transform
   In synchrotron radiation characterization techniques, the acquired experimental data-such as diffraction patterns, absorption spectra, or scattering images-often contain multidimensional and non-stationary information spanning from atomic to macroscopic scales. Conventional data analysis tools, like the Fourier transform, excel at revealing the global frequency composition of a signal but fall short in locating the specific distribution of these frequency components in time or space. To overcome this limitation, wavelet transform (WT) has emerged as a powerful mathematical framework that provides simultaneous insight into both the time-frequency domain characteristics of a signal. It is particularly well-suited for analyzing complex data exhibiting transient variations, localized features, or fractal structures.

As a method for digital signal processing, the core idea of WT is to decompose a signal in a multiscale, multiresolution manner using a localized "wavelet" function as the basis. This wavelet basis function is localized in both time (or space) and frequency. By applying scaling (to change frequency) and translation (to shift time position) operations, the signal can be analyzed at different resolutions. This fundamentally differs from the Fourier transform, which uses infinitely extended sinusoidal basis functions. While Fourier analysis effectively describes the global frequency content, it completely loses time-localization information.

a. Wavelet Basis Function
   The basis function for WT is a finite-energy oscillatory function with zero mean, concentrated within a limited time interval. It is governed by a scale parameter a and a translation parameter b, and is defined as:

$$\psi_{a,b}(t) = \frac{1}{\sqrt{|a|}}\psi\left(\frac{t-b}{a}\right) \tag{8.4}$$

Where $\psi(t)$ is the mother wavelet [7].

b. Scaling and Translation Operations
A family of wavelet basis functions is generated by varying the parameters a and b. The factor $1/\sqrt{|a|}$ ensures energy normalization across different scales.
Scale Parameter a: Controls the frequency characteristics. A larger a dilates the wavelet in time, corresponding to the analysis of lower frequency components. A smaller a compress the wavelet, analyzing higher frequency components.
Translation Parameter b: Determines the temporal position of the wavelet, allowing it to locate local features within the signal.

c. Inner Product Operation
The essence of the WT is to compute the inner product (similarity) between the signal and scaled/translated versions of the mother wavelet. For a continuous signal $f(t)$, its Continuous Wavelet Transform is:

$$W_f(a,b)=\langle f,\psi_{a,b}\rangle=\frac{1}{\sqrt{|a|}}\int_{-\infty}^{\infty}f(t)\psi^*\left(\frac{t-b}{a}\right)dt \tag{8.2}$$

Where ψ^* denotes the complex conjugate. The resulting wavelet coefficients $W_f(a,b)$ quantify the similarity between the signal and the wavelet at scale a and position b. By computing these coefficients across scales and positions, a two-dimensional time-frequency energy distribution map (scalogram) is obtained, revealing localized energy concentrations in the signal [8].

d. Multiresolution Analysis
WT decomposes a signal through a multiresolution framework. It projects the signal space onto a sequence of nested subspaces, each associated with a specific scale.

At coarse scales (low frequencies), the subspaces have high frequency resolution, capturing the global trend of the signal.

At fine scales (high frequencies), the subspaces have high time resolution, highlighting local details, transients, or abrupt changes.

This allows WT to adaptively analyze signal features across different time and frequency scales.

In practice, Discrete WT (DWT) is often used, which discretizes the scale and translation parameters using dyadic (powers of two) sampling: $a = 2^j$, $b = k\cdot 2^j$. The basis functions become:

$$\psi_{j,k}(t) = 2^{-j/2}\psi(2^{-j}t - k) \tag{8.6}$$

and the discrete wavelet coefficients are:

$$d_{j,k}=\langle f,\psi_{j,k}\rangle \tag{8.7}$$

DWT is intrinsically linked to Multiresolution Analysis (MRA) theory. MRA represents the signal space as nested approximation subspaces V_j, where each V_j is spanned by scaled and translated versions of a scaling function $\phi(t)$, representing the "approximation" of the signal at resolution 2^{-j}. The detail lost when moving from a finer subspace V_{j-1} to a coarser one V_j is captured in the wavelet subspace W_j, spanned by the wavelet functions $\psi_{j,k}(t)$.

e. Orthogonality and Completeness
Most practical WTs employ orthogonal or biorthogonal wavelet bases. Orthogonality ensures that the basis functions at different scales and translations are independent, preventing information redundancy. Furthermore, these wavelet bases form a complete set, meaning the original signal can be perfectly reconstructed from its wavelet coefficients without loss of information, providing a stable and unique representation.

f. Common Wavelet Basis Functions
The choice of the mother wavelet influences the analysis results and should align with the signal characteristics and application goals. Commonly used wavelet families include:

g. Haar Wavelet
The simplest discontinuous, orthogonal wavelet. It has high computational efficiency and is suitable for detecting step changes or edges in signals.

h. Daubechies Wavelets (dbN)
A family of orthogonal wavelets with compact support and varying degrees of smoothness (e.g., db4, db8). Widely used for signal denoising, compression, and feature extraction due to their good balance between localization and smoothness.

i. Morlet Wavelet
A complex-valued wavelet formed by modulating a complex exponential with a Gaussian window. It offers excellent time-frequency localization and is particularly useful for time-frequency analysis and identifying oscillatory modes in signals.

j. Mexican Hat Wavelet (Ricker Wavelet)
Proportional to the second derivative of a Gaussian function. Its shape resembles a sombrero, making it effective for peak detection and edge analysis in continuous WTs.

k. Symlets and Coiflets
These are modifications of Daubechies wavelets optimized for higher symmetry. Increased symmetry helps reduce phase distortion during signal decomposition and reconstruction, which is beneficial for certain applications.

2. Advantages of WT Analysis
As a time-frequency analysis technique, WT decomposes a complex signal into a series of simple, orthogonal basis functions. This decomposition provides a clear view of the signal's characteristics across different times and frequencies. In practice, WT processes a given signal in a "layered" or hierarchical manner, breaking it down into components of varying scales to reveal its internal structure.

Typically, the two-scale DWT approach is used, decomposing the signal into high-frequency and low-frequency components at each level.

- The high-frequency components correspond to rapidly changing, detailed parts of the signal (e.g., edges, noise, transients).

- The low-frequency components represent the slowly varying, smoother, or global trend of the signal.

This approach preserves crucial information about subtle changes while potentially discarding less significant coarse variations. By effectively mining the complexity of signals, WT plays a vital role in numerous scientific fields, including noise suppression, image reconstruction, and the analysis of microstructural features in materials science.

3. Energy Compaction
   In synchrotron-based imaging, WT exhibits a powerful energy compaction property. This means it can concentrate the essential information (energy) of an image into a relatively small number of significant wavelet coefficients. This enables the removal of a large amount of redundant data with a high compression ratio while simultaneously preserving the image's critical features, making it highly effective for image compression.

Experimental results demonstrate that wavelet-based compression methods can maintain high reconstructed image quality even at high compression ratios. Under identical compression ratios, wavelet transform significantly outperforms Fourier transform, offering superior image quality and a notable resistance to "blocking artifacts." This implies that during image transmission and storage, wavelet compression can save substantial bandwidth and storage space. Upon decompression, the images retain high clarity and visual fidelity, effectively avoiding common issues like blurring and distortion.

a. Noise Control and Denoising
   During the processing of synchrotron data, the multiscale decomposition capability of wavelet transform plays a crucial role in noise control. It separates the signal and noise by concentrating their energy into different sets of wavelet coefficients. By applying thresholding techniques to these coefficients-effectively removing or shrinking those primarily associated with noise-the meaningful signal can be recovered. This time-frequency localized analysis is exceptionally well-suited for handling non-stationary signals (common in experimental data), thereby enhancing the accuracy and reliability of signal processing and reducing errors in subsequent analysis.
b. Logarithmic Congruence with Human Vision
   The mathematical properties of wavelet decomposition align well with the logarithmic sensitivity of the human visual system. This congruence means that images processed or compressed using wavelets often appear more natural and perceptually clear to an observer. This characteristic is particularly beneficial in fields like medical imaging and satellite imagery, where researchers rely on visual interpretation to extract critical information. By maintaining visual fidelity, wavelet-based processing aids in improving the accuracy and reliability of data analysis derived from such images.

8.2 Equipment and Analysis

This section outlines hard XAFS experimental setups, measurement modes, data preprocessing, and analysis workflows. Advances in synchrotron beamlines, detectors, and modeling enable precise atomic-scale structural characterization across diverse materials, including thin films and energy-related systems.

8.2.1 Standard Experimental Setup

Although the setup of synchrotron beamlines is mostly assisted by specialist beamline scientists, it remains highly useful to understand the systematic principles behind the measurements. Hard XAS (hXAS) beamlines are tailored to hard X-rays with an energy range from a few keV up to approximately 100 keV, and are mainly used for probing the *K*-edges of heavy elements. The main components of such a beamline (Figure 8.4) are as follow: (1) Source and beam control: High-brilliance hard X-rays are generated by undulators or wigglers in an electron storage ring, with collimating optics to control beam shape and divergence, suitable for the range of a few KeV to ~100 KeV. (2) Harmonic rejection mirror: Coated with high-reflectivity materials (e.g., Pt/Ni) optimized for hard X-ray energies, it not only reduces harmonic content but also maintains beam integrity across the 10-100 KeV range, filtering stray high-order radiation critical for hard XAFS high-energy resolution. The reflective mirror is positioned at a glancing angle within the X-ray beam, ensuring that X-rays of fundamental energy are directed toward the sample via reflection. (3) Monochromator (often Si(111)/Si(311) double crystals): The commonly used double-crystal monochromator consists of two parallel crystals, while the other type is the channel-cut monochromator, which is a single crystal with a slot nearly cut through it. By rotating the crystals within the white X-ray beam, the monochromator precisely tunes the energy of hard X-rays, achieving a resolution of ~0.1 eV to match the fine energy shifts of hard X-ray absorption edges [9]. (4) Slits define the X-ray profile: For hard X-rays, their micrometer-to-millimeter adjustments not only balance the beam flux and spatial resolution required for this high-energy radiation but also enable a key trade-off that involves enhancing the energy resolution of incident X-rays at the cost of a slight loss in X-ray intensity. Available in fixed and adjustable designs, fixed slits have pre-cut openings with heights ranging from 0.2 to 1.0 mm and widths of several centimeters, while adjustable slits utilize independently movable metal plates to precisely define each edge of the hard X-ray beam. (5) Detectors: two common detectors for XAFS are transmission-type ionization

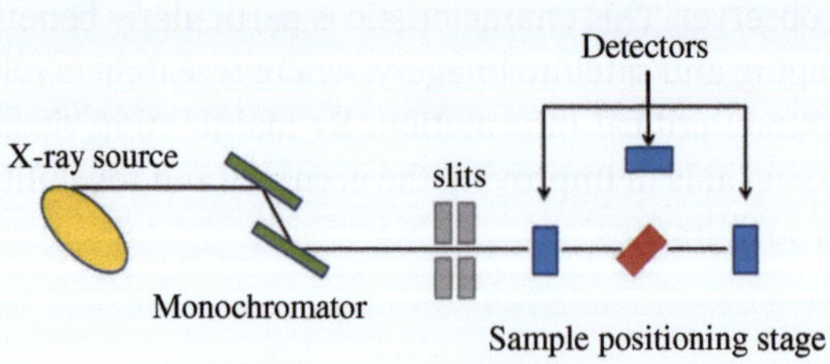

Figure 8.4 Schematic of the basic components of a XAFS spectroscopy beamline [10] / Rice University / CC BY 4.0.

chambers and fluorescence-type silicon drift detectors. The former are ideal for high-concentration samples, as they ionize gas within their chambers using X-rays, generate current through applied voltage bias, and provide a wide linear response range plus high stability. The latter suit low-concentration samples, as they produce electron-hole pairs in silicon via sample-emitted fluorescent photons, collect carriers via drift fields, and offer high-energy resolution and count rate.

XAFS measurements on homogeneous samples are routinely carried out via either the transmission mode or 45° fluorescence detection mode. However, thin-film samples are mostly substrate-supported thin-film structures. When precise probing of the local structure of thin films at the atomic scale is required, it not only demands the adoption of high-sensitivity XAFS techniques but also necessitates minimizing signal interference from the substrate as much as possible. To address this challenge the grazing-incidence fluorescence XAFS (GIXAFS) technique has been developed. It operates by making the incident X-rays strike the sample surface at a near-parallel angle which significantly reduces the X-ray penetration depth into the sample. Only the surface-layer atoms of the sample are excited to generate fluorescent signals thus effectively eliminating signal interference from the substrate. Since this technique detects hard X-ray signals tests can be performed under conventional experimental conditions using common detectors in conjunction with GIXAFS instrumentation. This system is mainly composed of three parts (Figure 8.5) [11]: (1) The front slit is positioned behind the I_0 detector and serves to optimize the X-ray beam incident on the sample. (2) The sample stage resides downstream of the front slit, affixed to a vertical translation stage as well as an angular goniometer. A Lytle detector is positioned at the location where the vertically incident X-ray beam is aligned with the sample to collect fluorescent signals. A soller slit is installed between the detector and the sample to improve the measurement signal-to-noise ratio. Additionally, the combination of a tantalum shielding can a soller slit and a filter is used to reduce the scattered radiation generated by the sample. Concurrently, photographic film sheets and thin, small-sized lead foils are incorporated into the setup to mitigate diffractive interference arising from the thin-film substrate. (3) Rear slits are positioned downstream of the sample and upstream of the I1 photodiode-based detector, while all such slits are securely affixed to precision vertical translation stages.

The approaches discussed earlier all rely on continuously variable monochromatic light sources. Their XAFS spectra are captured in the course of wavelength scanning and this process usually leads to relatively long experimental durations. By contrast the energy-dispersive method can obtain a full XAFS spectrum within a time frame on the millisecond scale. This method directly makes use of a polychromatic X-ray beam. A bent crystal is applied to disperse the polychromatic beam which is then focused onto the sample surface. After the beam passes through the sample it undergoes a second dispersion process. X-rays with varying wavelengths are separated from each other in the

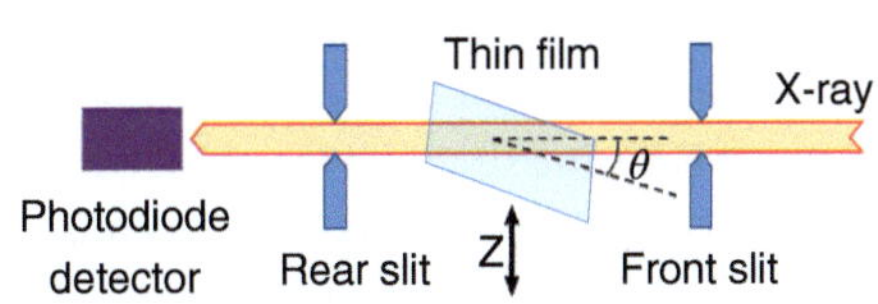

Figure 8.5 Diagram for the alignment procedure of the GIXAS setup [11] / with permission of Springer Nature.

spatial dimension during this step. A linear position-sensitive detector is employed to record these X-rays at different spatial locations. Each location corresponds to a specific X-ray energy and this spatial separation allows the entire absorption spectrum to be obtained efficiently.

The development of hard X-ray absorption spectrometers progresses in tandem with advances in light source technology, detection techniques and data analysis methods. During the early exploration phase from the 1920s to 1960s, laboratory X-ray sources enabled qualitative elemental analysis through the observation of absorption edges, though their limited energy and intensity restricted its application. Subsequent improvements in X-ray tube technology extended the energy range to the hard X-ray region (>5 keV) yet yielded experimental data of unsatisfactory quality. The 1970s marked a critical theoretical breakthrough as Sayers et al. proposed the Fourier transform analysis method in 1971, which uncovered the physical essence of EXAFS and laid the foundation for quantitative structural analysis. In the maturation stage spanning the 1980s to 1990s, mid-generation synchrotron facilities spurred the construction of dedicated XAFS beamlines. Optimized silicon/germanium monochromators and ionization chamber detectors significantly enhanced energy resolution, expanding applications to catalyst and battery research while enabling valence state and coordination number analysis. Since the 21st century, ultra-bright synchrotrons have driven the development of microbeam XAFS, rapid scanning and multidetection modes. Silicon drift detector arrays and advanced optical designs have further boosted resolution, and integration with machine learning has advanced data processing capabilities, making XAFS an indispensable tool for atomic-scale dynamic studies such as battery cycling.

8.2.2 Measurement Modes and Sample Preparation

Transmission mode and fluorescence mode are the two most widely used measurement approaches in XAFS experiments, with significant differences in their operational logic, applicable scenarios, and technical requirements. Transmission mode is the most frequently employed method in XAS experiments. Its core operational logic involves analyzing the absorption properties by quantifying the difference in photon flux of the X-ray beam before (I_0) and after (I_1) it penetrates the sample. This mode serves as the standard technical solution for XAS measurements in the hard X-ray regime, and is particularly suitable for sample systems with high concentrations of the target element. To ensure the reliability of measurement data, samples must meet three key requirements. First, they need to have a thickness compatible with experimental requirements to avoid excessive attenuation of X-rays due to being too thick or insufficient signal-to-noise ratio (SNR) due to being too thin. Second, they must ensure spatial uniformity to prevent measurement deviations caused by local concentration variations. Third, they should be free from structural defects such as pinholes and cracks, which could allow X-rays to penetrate directly through defective regions and interfere with the absorption signal.

The measurement logic of fluorescence mode differs from that of transmission mode. It requires simultaneously acquiring the incident X-ray photon flux (I_0) and the fluorescent X-ray photon flux (I_f) emitted by the sample after X-ray absorption occurs. The absorption properties of the sample are then calculated through the correlation between these two fluxes. This mode offers greater advantages in terms of sample type compatibility. On one hand, it can be used for thick samples (no strict thickness control is needed to prevent complete attenuation of X-rays). On the other hand, it enables effective analysis of sample systems with low concentrations of the target element—even when the concentration of the target element in the sample is as low as the ppm level or even lower, accurate measurements can still be achieved through the amplification effect of fluorescent signals. It should be noted that if this mode is applied to high-concentration samples, the fluorescent X-rays will undergo secondary absorption by absorbing atoms in the sample during their propagation inside the sample, leading to attenuation of the fluorescent signal intensity. This phenomenon is defined as "self-absorption" in the field of XAS, and it is a key interference factor that requires correction when measuring high-concentration samples using fluorescence mode.

From the perspective of different measurement modes, the transmission mode is typically used for high-concentration samples, where thickness is the core parameter determining measurement quality. On one hand, the total linear absorption of all atoms, which is characterized by the product of the total linear absorption coefficient $\mu(E)$ and thickness t, must be controlled within 2.5 absorption lengths to prevent excessive attenuation of incident X-rays. On the other hand, the product of the difference in linear absorption coefficients for the target absorbing atoms and thickness t should be close to 1 absorption length. This value corresponds exactly to the ideal intensity of the absorption edge step; if precise control of $\Delta\mu(E)\,t = 1$ is required, the corresponding sample thickness can be derived using specific formulas [10].

$$t = \frac{1}{\Delta\mu} = \frac{1.66\sum_i n_i M_i}{\rho\sum_i n_i\left[\sigma_i(E_a) - \sigma_i(E_b)\right]} \tag{8.8}$$

ρ is the compound density, n is the elemental stoichiometry, M is the atomic mass, σE is the adsorption cross-section in barns/atom. E_a and E_b are the just above and below the energy edge.

For low-concentration samples, the total X-ray absorption properties become the focus of regulation. In this case, emphasis is placed on the areal density of sample. The intensity of the absorption edge step is determined by the product of the areal density and the difference in mass absorption coefficients, where the mass absorption coefficient difference $\Delta(\mu(E)/\rho)$ refers to the difference between the total mass absorption coefficients before and after the absorption edge. To ensure a sufficient SNR for the absorption edge signal, the ideal value of the absorption edge step should be controlled to 1, expressed mathematically as $(\rho t) \times \Delta(\mu(E)/\rho) = 1$. The specific value of $\Delta(\mu(E)/\rho)$ is calculated using the mass fraction of each element in the sample and the mass absorption coefficients of the corresponding elements before and after the absorption edge.

$$\frac{\Delta\mu(E)}{\rho} = \sum f_i \left[\left(\frac{\Delta\mu(E)}{\rho} \right)_{i,(E_a)} - \left(\frac{\Delta\mu(E)}{\rho} \right)_{i,(E_b)} \right] \tag{8.9}$$

$(\mu(E)/\rho)_i$ is the mass adsorption coefficient just above (E_a) and below (E_b) of the energy and f_i is the mass fraction of the element i.

8.2.3 Data Preprocessing and Analysis

8.2.3.1 Raw Data Preprocessing

Raw data preprocessing aims to eliminate experimental interference and lay the foundation for data reliability, with its core centered on energy calibration and noise suppression. Energy calibration is the primary step. Due to potential minor drifts in synchrotron radiation beam energy, calibration using standard reference materials is required. During the experiment, the absorption edge energy of the reference material must be measured simultaneously to align the absorption edge energy of the sample spectrum with the standard value of the reference material. The calibration accuracy must be controlled within ± 0.1 eV to ensure the accuracy of subsequent bond length calculations. Additionally, format conversion and data extraction need to be performed on the raw data. The absorption coefficient $\mu(E)$ is calculated from the experimentally recorded light intensity signals (where I_0 denotes the incident light intensity and I represent the transmitted light intensity), ensuring the data is stored as a two-dimensional "energy-absorption coefficient" array to facilitate subsequent analysis.

8.2.3.2 Background Subtraction

The core of background subtraction is to separate the contribution from the local atomic structure from the atomic inner-shell absorption and background absorption arising from long-range structures, with the key lying in the rational construction of the background absorption function $\mu_0(E)$ (Figure 8.6a). Background absorption mainly comprises two components: one is the absorption of atomic inner-shell electrons (atomic background), and the other is the absorption generated by other atoms in the sample that are distant from the absorbing atoms (structural background). The commonly used subtraction method is the combined "Victoreen equation + cubic spline function" approach: First, the Victoreen equation is employed to fit the atomic background before the absorption edge (ranging from −200 eV to −50 eV relative to the absorption edge energy E_0); subsequently, the cubic spline function is used to fit the structural background after the absorption edge (ranging from +50 eV to +800 eV relative to E_0). The spline node spacing is generally set to 50–100 eV to ensure the background curve is smooth and can accurately conform to the baseline trend of the absorption spectrum (Figure 8.6b, c).

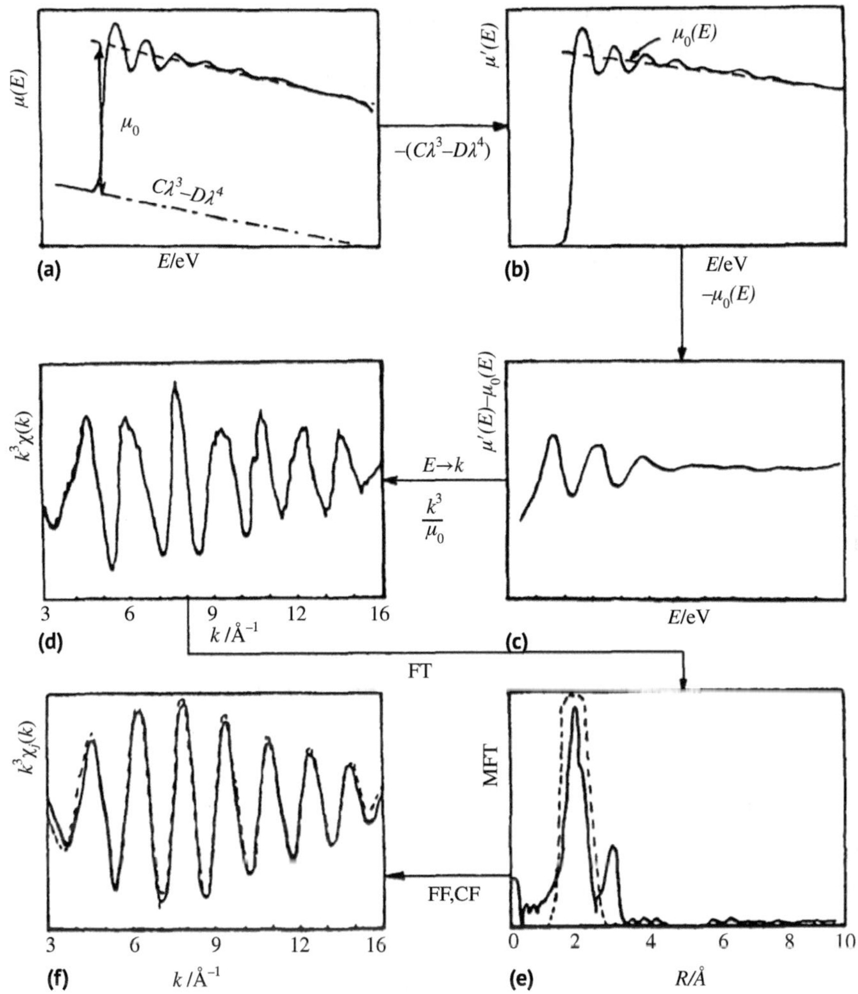

Figure 8.6 EXAFS data processing procedure.

8.2.3.3 Normalization

The purpose of normalization is to eliminate the effects of sample thickness, concentration variations, and fluctuations in absorption edge intensity, thereby enabling the comparability of EXAFS signals between different samples or experiments. Its core lies in standardizing the net absorption coefficient to a uniform scale. The normalization process consists of two steps: first, the absorption edge step value $\Delta\mu_0 = \mu_{\text{max}} - \mu_{\text{min}}$ is calculated, where μ_{max} denotes the maximum absorption coefficient after the absorption edge (ranging from + 100 eV to + 200 eV) and μ_{min} represents the limit value of the background fit before the absorption edge. Second, the net absorption coefficient is divided by $\Delta\mu_0$ to obtain the normalized absorption coefficient $\chi_{\text{norm}}(E)$, which standardizes the absorption edge step to 1. Consequently, the EXAFS signal manifests as oscillations of $\chi_{\text{norm}}(E)$ around 1.

8.2.3.4 Conversion to *k*-space and Weighting

The essence of the EXAFS signal lies in the interference effect of neighboring atoms around the absorbing atom on scattered electrons. This effect is more readily manifested in *k*-space (wave vector space), requiring the conversion of the normalized signal in energy space to *k*-space followed by weighting processing. In the course of conversion, E_0 requires explicit definition with its typical definitions including the energy associated with the midpoint of the absorption step region of edge and the energy corresponding to the peak of the first derivative of absorption coefficient. This approach ensures the consistency of k-value computations. Weighting processing is intended to enhance the EXAFS signals in the high-*k* region (high-*k* regions correspond to long-range scattering and are more sensitive to bond lengths and coordination numbers), facilitating subsequent Fourier transformation (Figure 8.6d). Common weighting methods include k^1, k^2, and k^3 weighting, with the specific choice depending on sample characteristics: for light atom ligands (e.g., O, N), since the scattering amplitude decays rapidly with increasing *k*, k^2, or k^3 weighting is preferred; for heavy atom ligands (e.g., Pt, Au), the scattering amplitude decays slowly, so k^1 weighting can be adopted [12].

8.2.3.5 Fourier Transformation

The core of Fourier transformation lies in converting the oscillatory signal in *k*-space into the radial distribution function in R-space (distance space) (Figure 8.6e), which intuitively presents the shell distribution of neighboring atoms around the absorbing atom. It is a critical step in EXAFS data processing. The parameter selection for Fourier transformation directly affects the resolution of the R-space spectrum: firstly, the integration range in *k*-space is generally set to $k = 2 - 10$ Å^{-1}. Secondly, the choice of window function commonly includes the Hanning window or the Blackman window. The width of the window function needs to balance resolution and sidelobe suppression with the half-width of the Hanning window typically set to 1.0–1.5 Å^{-1} while the Blackman window offers superior sidelobe suppression but slightly reduces resolution. Thirdly, the number of points for Fourier transformation is usually 512 or 1024, ensuring the step size in R-space after transformation is less than to guarantee the accuracy of peak position determination.

After Fourier transformation, the radial structure function $F(R)$ is obtained. Peak locations align with the interatomic separations separating the absorber atom and adjacent atomic species. The peak intensity is related to the coordination number and atomic number of the scattering atoms, while the peak width reflects the bond length disorder. The $F(R)$ spectrum allows preliminary determination of the number of neighboring atom shells, their approximate distances, and relative contents, providing a basis for constructing subsequent fitting models.

8.2.3.6 Fitting Analysis

Fitting analysis is a core step for the quantitative extraction of structural parameters. Based on the single-scattering approximation theory and combined with scattering paths calculated by FEFF, structural parameters are optimized via the least squares method to obtain quantitative results such as coordination number N, bond length R, disorder parameter σ^2, and phase shift correction $\Delta\Phi$ (Figure 8.6f). The construction of the fitting model must adhere to the "principle of parsimony" to avoid overfitting. First, the fitting range is determined according to the Fourier-transformed $F(R)$ spectrum, typically selecting the R range corresponding to major peaks. Subsequently, potential scattering paths are identified based on the chemical composition of sample and known structural information, and the scattering amplitude $f(k)$ and phase shift $\Phi(k)$ for each path are calculated using FEFF software. For samples with unknown structures, the type of scattering atoms can be initially determined by comparison with standard samples before constructing the fitting model.

The constraint and optimization of fitting parameters are critical to ensuring result reliability: the coordination number N must be constrained based on stoichiometry and structural rationality. The initial value of the bond length R is taken from the apparent distance derived from Fourier transformation, with the fitting range generally restricted to within ± 0.1 Å. The disorder parameter σ^2 reflects thermal and structural disorder of the bond length, which must be a positive value. The phase shift correction $\Delta\Phi$ is usually set as a free parameter to compensate for minor discrepancies between FEFF calculations and experimental conditions. The evaluation of fitting results relies on the goodness-of-fit R-factor, which is generally required to be $R < 0.05$. Additionally, the residual curve should be checked for the absence of significant systematic fluctuations to ensure the fitting model accurately reflects the experimental signal.

8.2.4 Beamline Overview

XAFS spectroscopy serves as a core technique for probing the atomic-scale structure of materials. Its performance is inherently dependent on the development level of synchrotron radiation beamlines. Globally, a suite of high-performance beamlines has been established, underpinning cutting-edge research in fields such as energy science, materials engineering, and high-pressure physics [13–15]. For instance, the BL11B beamline at the Shanghai Synchrotron Radiation Facility (SSRF) is a hard X-ray general-purpose spectroscopic beamline, covering an energy range of 5–30 keV with a photon flux exceeding 5×10^{11} photons per second (phs/s). In contrast, the SSRF BL14W1 beamline is dedicated to XAFS spectroscopy measurements featuring high-energy resolution, high spectral purity, and high SNR. The 4B7A beamline at the Beijing Synchrotron Radiation Facility focuses on the medium-energy region (2.1–5.7 keV), enabling XAFS studies of low-Z elements (e.g., P and S) by covering their absorption edges. Although the newly constructed High-Energy Photon Source is centered on imaging applications, its energy range of 5–15 keV allows for extended XAFS-related research. The ID22 beamline at the European Synchrotron Radiation Facility integrates multiple techniques, including X-ray fluorescence, XAS, XRD, and 2D/3D X-ray imaging [16]. This versatility enables comprehensive characterization of materials in terms of compositional heterogeneity, local atomic structure, and chemical state. The 3ID and 13BM beamlines at the Advanced Photon Source (APS,

USA) are renowned for their exceptional versatility, supporting in-situ characterization across diverse research fields. Meanwhile, the BL04B1 and BL10XU beamlines at the SPring-8 facility (Japan) are distinguished by their ultra-high brightness, making them ideal for analyzing low-concentration samples.

8.3 Application Cases

This section highlights hard XAS applications in battery research, including element-specific redox tracking, EXAFS-based local structure probing, in-situ time-resolved studies, and surface-sensitive reflection modes, enabling comprehensive insight into bulk and interfacial electrochemical processes.

8.3.1 Monitoring Redox Reactions

XAS offers a significant advantage through its element-specific selectivity, enabling precise analysis of the local chemical environment and oxidation state of a specific element within a multielement system. This capability is crucial for identifying which components actively participate in redox reactions at varying states of charge (SOCs), particularly in electrodes composed of multiple transition metals [17]. For instance, Cr *K*-edge XANES spectroscopy was employed to investigate the redox behavior of chromium in $Li_{1.2}Cr_{0.4}Mn_{0.4}O_2$ (L-LCMO) [18]. The study demonstrated that introducing cation disorder in this layered, lithium-rich material effectively reduces voltage hysteresis and enhances the reversibility of the Cr^{3+}/Cr^{6+} redox couple (shown in Figure 8.7). Notably, a variant with partial cation disorder (PD-LCMO) exhibited a substantial reversible capacity exceeding 200 mAh/g over 100 cycles at 100 mA/g, attributed to suppressed collective Cr migration. This strategic manipulation of cation disorder in redox-active elements not only modulates multielectron redox processes but also controls transition metal TM migration dynamics. Such insights open new avenues for engineering high-performance cathode materials in Li-ion batteries by leveraging partial disorder as a design tool.

In addition, in the region of all-solid-state batteries (ASSBs), XAS have been demonstrated as an effective technique to understand the reaction mechanisms and structural evolution at solid-state electrolyte (SSE) interfaces. For instance, Li

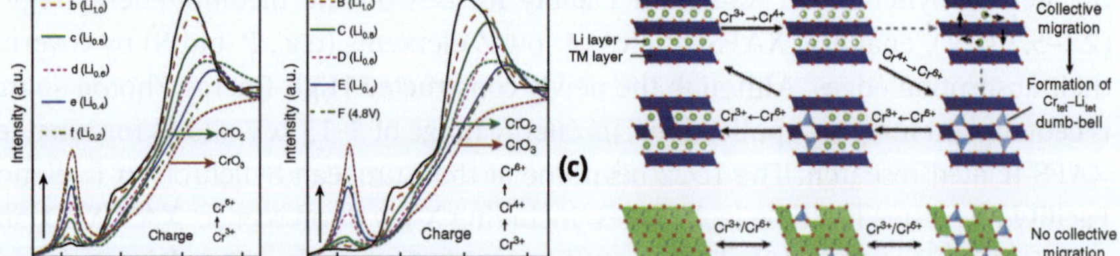

Figure 8.7 (a, b) Cr *K*-edge XANES spectra at different state of charge. Schematic illustration of Cr redox and migration in (c) L-LCMO and (d) PD-LCMO Adapted from [18].

et al. [19] employed XANES in conjunction with complementary characterization techniques to study the interfacial evolution between $Li_{10}GeP_2S_{12}$ (LGPS) SSE and $LiNi_{0.8}Mn_{0.1}Co_{0.1}O_2$ (NCM811) cathodes, with and without $LiNbO_x$ (LNO) coating. Analysis of the S *K*-edge spectra revealed that uncoated NCM811-LGPS cells experienced SSE decomposition during cycling, evidenced by peak shifts in the 2470~2471 eV range. In contrast, LNO-coated cathodes exhibited spectral stability, indicating interface preservation. Both systems displayed comparable Ni *K*-edge evolution, but distinct Mn/Co *K*-edge behaviors highlighted the intrinsic instability of Mn during delithiation/lithiation. In uncoated cathodes, parasitic interfacial reactions caused structural degradation and impurity formation, which diminished ionic conductivity. The LNO coating effectively suppressed sulfur diffusion, stabilized the cathode-electrolyte interface, and enhanced ASSB performance.

Another notable application of XAS involves investigating the operational mechanism of the $S_{9.3}I$ cathode in ASSBs via S *K*-edge analysis [20]. During discharge to 1.7 V, the S–S π^* feature in S_8 shifted to lower energy levels, corresponding to the formation of short-chain LiS_x species. Further discharge to 1.3 V induced additional spectral shifts linked to Li_2S and long-chain LiS_x. Remarkably, these structural transformations were fully reversible, with the S_8 feature restored upon charging back to 3.0 V. This study underscores the utility of XAS in capturing reversible redox chemistry and interface stability in ASSBs.

8.3.2 Probing Local Structure

The EXAFS technique provides a unique capability to probe the local structural characteristics of individual elements within electrode materials, offering critical insights that conventional XRD cannot achieve. In case of NMC ($LiNi_{1/3}Co_{1/3}Mn_{1/3}O_2$) systems, XRD struggles to distinguish Ni, Co, and Mn due to their nearly identical X-ray scattering cross-sections. While XRD confirms that $LiNi_{1/3}Co_{1/3}Mn_{1/3}O_2$ shares a crystallographic structure with $LiCoO_2$, it lacks the resolution to differentiate the atomic environments of these transition metals. Such limitation can be overcome by EXAFS, which reveals the existence of element-specific short-range ordering in the transition metal layers of NMC materials. Tsai et al. conducted a quantitative analysis of EXAFS spectra for Ni, Co, and Mn in a charged NMC sample. Their findings demonstrated that Mn and Ni exhibit similar trends in bond length variations and Debye–Waller factors (which quantify local structural disorder), whereas Co shows markedly distinct behavior. This observation led to the conclusion that Ni and Mn form a short-range ordered arrangement, a phenomenon undetectable through XRD alone. Such element-level resolution highlights the critical role of EXAFS in uncovering hidden structural correlations in complex multielement electrode systems.

The Jahn-Teller (J-T) distortions in octahedral transition metal layers of layered oxides are closely linked to structural degradation mechanisms. Specifically, these distortions can induce an unfavorable phase transition from the layered structure to the spinel configuration during cycling, which accelerates capacity fading. In a notable example, recent investigations have demonstrated pronounced local structural reconfigurations around Cu atoms, primarily driven by J-T distortions [21].

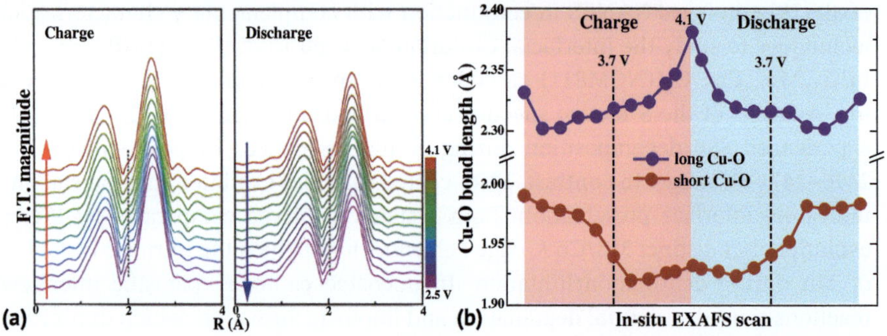

Figure 8.8 (a) In-situ Cu *K*-edge FT-EXAFS spectra during the first electrochemical cycle, and (b) corresponding fitting results of Cu-O bond length evolution [21] / Springer Nature / CC BB 4.0.

Ex-situ nPDF analysis has shown that the pristine material features a distorted CuO_6 octahedron, attributed to imbalanced electron occupation in the e_g orbitals of Cu^{2+} within a six-fold O coordination environment (Oh symmetry). To further dissect these dynamics, time-resolved in-situ Cu *K*-edge EXAFS measurements were employed during the first electrochemical cycle (Figure 8.8). At the initial voltage plateau of 3.66 V, the long Cu–O bond (associated with the dz^2 orbital) remains nearly constant (<0.01 Å variation), whereas the short Cu–O bond (linked to the $dx^2{-}y^2$ orbital) undergoes significant contraction, reducing from ~1.98 Å to ~1.93 Å. This discrepancy exacerbates the J-T distortion in the CuO_6 octahedron. Post-3.66 V plateau, as the material transitions to the P3' phase, the short bond shortens further to ~1.92 Å, and the long bond slightly elongates to >2.31 Å. At the subsequent higher voltage plateau of 3.99 V, the trend inverts: the short bond stabilizes, while the long bond rapidly extends to >2.35 Å. This reversal starkly contrasts with the initial desodiation behavior, suggesting distinct charge compensation mechanisms governed by electron redistribution rather than the conventional $Cu^{2+} \rightarrow Cu^{3+}$ oxidation model. During the subsequent discharge process, the bond lengths reverse their trends observed during charging, nearly recovering to the initial values of ~2.33 Å (long bond) and ~1.98 Å (short bond) by the end of the first cycle. The high reversibility of these local structural changes correlates with the material's minimal voltage hysteresis and exceptional cycling stability.

WT analysis serves as a complementary tool to FT-EXAFS, offering unique advantages in spectral interpretation. While FT-EXAFS primarily identifies interatomic distances, WT-EXAFS provides dual insights: it not only reveals the specific distances present in the EXAFS spectrum but also pinpoints the exact atomic coordination environments contributing to each distance. This two-dimensional visualization (simultaneously in R-space and k-space) enables precise discrimination of overlapping scattering paths, particularly valuable for resolving contributions from distinct structural motifs with similar interatomic distances. A case in point is a Cu-based composite sample, where WT-EXAFS clearly distinguishes Cu–Cu scattering contributions from metallic Cu and Cu oxides, delivering a direct, neighbor-specific view of local atomic arrangements.

Overall, XANES is widely employed to track the valence states of targeted elements as a function of SOC, offering essential information for understanding charge compensation mechanisms and the reversibility of redox reactions. When combined with EXAFS, this dual approach becomes indispensable: EXAFS excels at probing local structures in electrode materials with disordered or amorphous configurations, whether in their as-synthesized form or after extended cycling, while XANES focuses on electronic structure evolution. These complementary capabilities are vital for establishing structure-property relationships, diagnosing degradation pathways, and informing the rational design of advanced Li-ion batteries.

8.3.3 Time-resolved Hard XAS and Its Application to Dynamic Studies

In-situ XAS is a powerful technique for investigating structural evolution and reaction kinetics in battery materials during electrochemical cycling. By acquiring XAS spectra in real time during charge and discharge processes, it becomes possible to monitor dynamic electrochemical phenomena, such as local structural modifications around specific element sites and the rate capabilities of distinct redox centers.

To implement in-situ XAS experiments, it is essential to develop specialized electrochemical cells equipped with X-ray-transparent windows. The absorption of X-rays is approximately proportional to the atomic number (Z), so window materials composed solely of low-Z elements are preferred. Beryllium ($Z = 4$) is an intuitive choice because of its low Z, but its handling during cell assembly is challenging, and oxidation can render it highly toxic. As a practical alternative, polymer films like Kapton are widely adopted in such experiments. The electrochemical half-cell setup typically comprises a current collector, a Li metal counter electrode, and a separator, all submerged in an electrolyte solution composed of $LiPF_6$ salt dissolved in carbonate-based solvents. For transmission detection mode, the X-ray beam must transmit through all these components to reach the detector with adequate photon flux. However, for low-Z elements such as sulfur (S), phosphorus (P), and silicon (Si), fluorescence detection mode is the only viable option, as the organic electrolyte in the cell absorbs most of the photons. The first in-situ XAS investigation targeting battery systems was conducted at Brookhaven National Laboratory, focusing on the nickel oxidation state in nickel metal hydride batteries within a plate-like cell configuration. Since then ,in-situ techniques have been adapted to various rechargeable battery chemistries. Modern XAS studies often employ modified in-situ coin or pouch cells tailored to specific X-ray beamline setups. These advancements have enabled the characterization of diverse lithium-ion battery materials, including layered transition metal oxides, spinel structures, polyanionic frameworks, conversion-type electrodes, and intercalation anodes, significantly expanding the scope of in-situ XAS in battery research.

The time-resolved XAS technique is a critical approach for unraveling structural dynamics linked to rate capability in electrode materials containing multiple redox-active centers. An illustrative case study involves its application to Li-rich layered mixed-phase $Li_{1.2}Ni_{0.15}Co_{0.1}Mn_{0.55}O_2$ (LMR-NMC), specifically targeting the unresolved question of rate-limiting factors in this class of materials. LMR-NMC cathodes are historically recognized for their suboptimal rate performance, with theoretical models frequently attributing this to delayed lithium extraction within Li_2MnO_3 nanodomains. Recently, direct experimental evidence supporting this hypothesis had remained scarce. To address this, researchers applied a constant voltage of 5 V to the LMR-NMC cathode to simultaneously oxidize Ni^{2+}, Co^{3+}, and Mn^{2+} ions. During this process, Ni, Co, and Mn *K*-edge XAS spectra were collected in real time, enabling precise tracking of redox reaction kinetics. The comparative analysis of FT-EXAFS magnitude for these elements reveals distinct kinetic profiles. Ni exhibits rapid shifts in both position and intensity of its first coordination peak (Ni–O) within the initial 100 seconds, corresponding to the sequential oxidation of $Ni^{2+} \rightarrow Ni^{3+} \rightarrow Ni^{4+}$. Notably, the EXAFS features stabilize after 160 seconds, suggesting the $Ni^{2+} \rightarrow Ni^{4+}$ transformation is nearly complete within the first 3 minutes. Contrastingly, Mn shows a progressive decline in Mn–O coordination shell intensity throughout the 900-second experiment, directly highlighting its markedly slower delithiation kinetics. Co occupies an intermediate position, with its Co–O peak changes ceasing after 200 seconds. These findings definitively identify Mn oxidation as the primary kinetic bottleneck, providing the first direct experimental evidence that Li_2MnO_3 nanodomains are indeed the limiting factor in LMR-NMC systems. This insight not only clarifies the mechanistic basis for poor rate capability but also offers actionable guidance for designing engineering strategies to optimize the performance of lithium-rich layered materials.

In ASSBs research, Li et al. [22] employed operando XANES spectroscopy to probe the chemical stability of a vacancy-rich $Li_9N_2Cl_3$ SSE against Li metal during plating and stripping cycles. The experiments were conducted in a vacuum-compatible XAS chamber using an identical operando cell configuration as previously described (Figure 8.9). Owing to the shallow X-ray incidence angle, the detected fluorescence signals originated predominantly from the interfacial regions of interest. Throughout the electrochemical cycling, XANES spectra were acquired in real time. A first derivative analysis of the spectral evolution over the entire process revealed no significant changes in the XANES features, demonstrating the intrinsic stability of the vacancy-rich $Li_9N_2Cl_3$ SSE toward metallic lithium.

In addition, researchers also conducted in-situ XANES studies to identify the atmospheric components responsible for reactions with Li_3InCl_6 SSE through a dedicated reaction chamber [23]. The chamber was sealed with a Be window and Mylar film to ensure compatibility with the synchrotron X-ray chamber and fluorescence detector. Li_3InCl_6 samples were exposed to three distinct gas environments: dry air, a mixture of argon Ar and moisture, and a mixture of dry air with moisture. Following the purge process, Cl *K*-edge and In L_3-edge XANES spectra were collected to monitor the evolution of Li_3InCl_6 along with the exposure time. It revealed that the reason for the degradation of the ionic conductivity of Li_3InCl_6 in ambient air is moisture, which reacts with Li_3InCl_6 upon exposure. Of particular relevance is the finding that Li_3InCl_6 exhibits significantly greater stability in

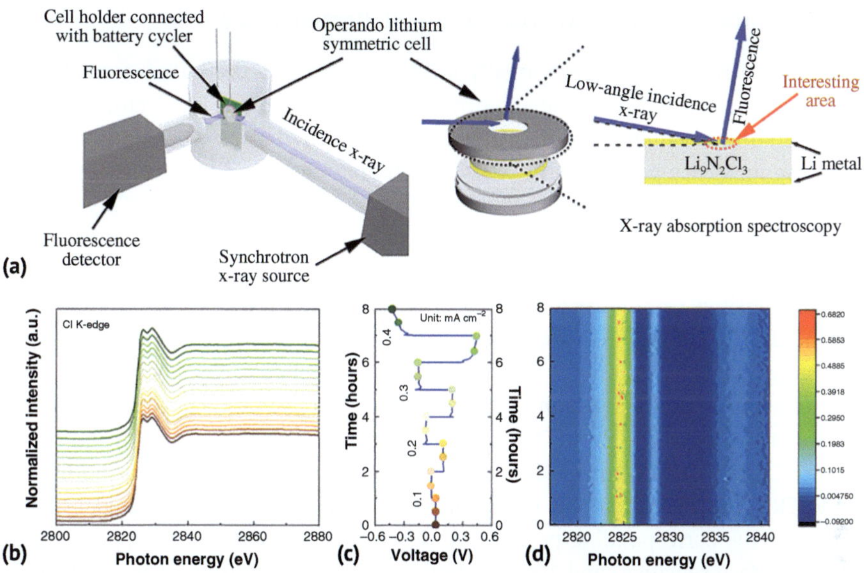

Figure 8.9 (a) Schematic illustration of the operando XANES studies of chemical stability of $Li_9N_2Cl_3$ SSE. (b) Operando Cl *K*-edge XANES spectra, (c) the corresponding discharge/charge voltage profiles, and (d) the first derivative mapping of Cl *K*-edge XANES spectra [22] / with permission of American Association for the Advancement of Science.

low-humidity air. While a portion decomposes into non-ionically conductive In_2O_3 and LiCl, releasing HCl gas, the majority of the material forms the corresponding hydrate. This enhanced stability suggests that the practical industrial application of this electrolyte is feasible within a commercial dry-room environment for battery manufacturing.

Moreover, dispersive XAS instrumentation enables acquisition of high-quality time-resolved XAS spectra at temporal resolutions as brief as milliseconds, facilitating the study of dynamic structural processes in batteries under high-rate operation. This groundbreaking application was first demonstrated by Yu et al. [24], who utilized the technique to elucidate the structural evolution of $LiFePO_4$ during chemical and electrochemical delithiation. Their work centered on quantifying the phase fractions of $LiFePO_4$ (or $FePO_4$) under varying charging conditions. The findings indicated that the average structural transformations in $FePO_4$ upon Li extraction predominantly follow a $LiFePO_4/FePO_4$ two-phase reaction pathway, with consistent behavior observed across low (1 C) and high (30 C) charging scenarios. This observation highlights the resilience of the two-phase reaction mechanism in $LiFePO_4$ systems, even under extreme rate conditions where rapid ion dynamics critically influence battery performance.

Overall, the in-situ XANES characterization plays an irreplaceable role in probing the actual reaction process within batteries. Moreover, it causes almost no damage to the samples and is of great significance for in-depth analysis of the charging and discharging mechanisms and failure mechanisms of batteries.

8.3.4 Surface Probing Tool: Hard XAS in Reflection Mode

Although hard XAS is conventionally utilized for probing the bulk information of materials, it can also be adapted as a surface-sensitive technique for electrode studies. By exploiting the highly collimated synchrotron X-ray beams in conjunction with glancing incidence geometry, the X-ray penetration depth can be restricted to a few nanometers, enabling surface-specific probing of electrode materials. This surface-sensitive approach is especially critical for in operando investigations of electrode surface dynamics under battery operation conditions. In contrast to most other surface-sensitive techniques like XPS and soft XAS, which necessitate vacuum environments for sample analysis, hard XAS eliminates such constraints, making it uniquely suited for real-time or real-condition measurements. Notably, this technique was first employed by Takamatsu et al. [25], who pioneered its application as total reflection fluorescence XAS (TRF-XAS) to investigate the surface structural evolution of $LiCoO_2$ during charge-discharge cycles. The electrochemical setup for operando TRF-XAS measurements is depicted in Figure 8.10. At incident angles of 0.2° and 2.2°, the corresponding penetration depths are estimated to be approximately 3 nm (surface-sensitive) and 100 nm (bulk-sensitive), respectively. The results indicate that the $LiCoO_2$ thin-film electrode undergoes progressive surface degradation

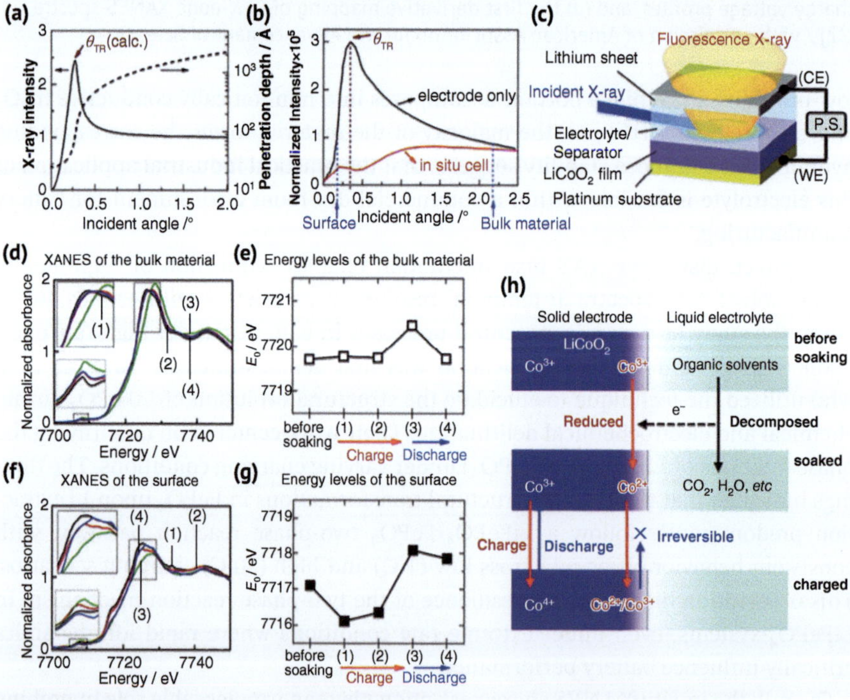

Figure 8.10 (a, b) Incident-angle-dependent X-ray intensity and penetration depth. (c) Schematic illustration of the cell used for in-situ TRF-XAS measurements. (d–g) Co *K*-edge XANES spectra at different incident angles, and the corresponding absorption energy levels. (h) Schematic illustration of the degradation mechanism at the interface [25] / with permission of John Wiley & Sons.

during electrochemical cycling. Notably, Co would be reduced immediately upon immersion in the organic electrolyte, even prior to current application in the cell. This phenomenon is in line with previous report, which found the analogous behavior in NMC materials, where both Mn and Co would be reduced after exposure to electrolytic solutions. As a consequence, irreversible electronic structural changes occurred at the electrode surface during subsequent charge-discharge cycles—a stark contrast to the reversible redox reactions observed in the bulk material. Subsequently, Yamamoto et al. applied this technique to $LiFePO_4$ thin-film electrodes. Their findings revealed that the electrode/electrolyte interface of $LiFePO_4$ remained free of reduction-induced damage, unlike the $LiCoO_2$ case. These studies underscore the critical role of TRF-XAS as versatile and reliable technique for analyzing electrode surfaces under in-situ/operando conditions and establishing a direct correlation between the stability of electrode/electrolyte interfaces and electrochemical performances.

A notable application involves studying the relationship between electronic structural stability at the electrode/electrolyte interface and the cyclic performance of lithium-ion batteries, specifically for $LiFePO_4$ and $LiCoO_2$ systems [26]. In-situ TRF-XAS was employed to directly monitor the electronic structure at this interface during electrolyte immersion and Li+ ion extraction/insertion processes. While the $LiFePO_4$ surface remains stable upon electrolyte immersion, the $LiCoO_2$ surface undergoes reduction. Furthermore, the electronic structural changes on the $LiFePO_4$ surface exhibit high reversibility during Li^+ ion cycling, unlike the irreversible alterations observed on the $LiCoO_2$ surface. The stability of $LiFePO_4$ is attributed to the absence of significant potential fluctuations in the surface space charge layer. The pronounced reversible behavior of the $LiFePO_4$ interface electronic structure is directly linked to its stable cycling capacity. These findings demonstrate the critical role of interfacial electronic structure in determining the long-term performance of Li-ion batteries.

References

1 R.W. Ricci, M.A. Ditzler, L.P. Nestor, Discovering the beer-lambert law, Journal of Chemical Education 71(11) (1994) 983–985. https://doi.org/10.1021/ed071p983.

2 W. Yang, X. Liu, R. Qiao, P. Olalde-Velasco, J.D. Spear, L. Roseguo, J.X. Pepper, Y.-d. Chuang, J.D. Denlinger, Z. Hussain, Key electronic states in lithium battery materials probed by soft X-ray spectroscopy, Journal of Electron Spectroscopy and Related Phenomena 190 (2013) 64–74. https://doi.org/10.1016/j.elspec.2013.03.008.

3 A.L. Ankudinov, J.J. Rehr, Development of XAFS theory, Journal of Synchrotron Radiation 10 (2004) 366–368. https://doi.org/10.1107/S0909049503009130.

4 D. Trummer, K. Searles, A. Algasov, S.A. Guda, A.V. Soldatov, H. Ramanantoanina, O.V. Safonova, A.A. Guda, C. Copéret, Deciphering the Phillips catalyst by orbital analysis and supervised machine learning from Cr pre-edge XANES of molecular libraries, Journal of the American Chemical Society 143(19) (2021) 7326–7341. https://doi.org/10.1021/jacs.0c10791.

5 M. Guo, E. Källman, R.V. Pinjari, R.C. Couto, L. Kragh Sørensen, R. Lindh, K. Pierloot, M. Lundberg, Fingerprinting electronic structure of heme iron by Ab initio modeling of metal *L*-edge X-ray absorption spectra, Journal of Chemical Theory and Computation 15(1) (2018) 477–489. https://doi.org/10.1021/acs.jctc.8b00658.

6 E.A. Stern, Theory of the extended X-ray-absorption fine structure, Physical Review B 10(8) (1974) 3027–3037. https://doi.org/10.1103/PhysRevB.10.3027.

7 S. Chang, L. Hu, A novel 3D entangled wavelet transformation and its properties, Journal of Mathematical Physics 60(10) (2019). https://doi.org/10.1063/1.5102060.

8 Z. Xia, H. Zhang, K. Shen, Y. Qu, Z. Jiang, Wavelet analysis of extended X-ray absorption fine structure data: Theory, application, Physica B: Condensed Matter 542 (2018) 12–19. https://doi.org/10.1016/j.physb.2018.04.039.

9 F. De Groot, A. Kotani, Core Level Spectroscopy of Solids, CRC press 2008.

10 P.M.V. Raja, A.R. Barron, Physical methods in chemistry and nano science, Rice University 2019. https://hdl.handle.net/1911/112293.

11 Z. Yin, G. Zhang, Y. Xie, Y. Chen, S. Chu, C. Shao, D. Song, L. Zheng, P. An, J. Zhang, A new mobile grazing-incidence X-ray absorption fine spectroscopy endstation at Beijing synchrotron radiation facility, Radiation Detection Technology and Methods 6(2) (2022) 194–200. https://doi.org/10.1007/s41605-022-00317-5.

12 S. Calvin, XAFS for Everyone, CRC press 2024.

13 L. Sun, X. Pan, Y.n. Xie, J. Zheng, S. Xu, L. Li, G. Zhao, Accelerated dynamic reconstruction in metal–organic frameworks with ligand defects for selective electrooxidation of amines to azos coupling with hydrogen production, Angewandte Chemie 136(21) (2024) e202402176. https://doi.org/10.1002/anie.202402176.

14 Y. Li, Y. Wu, T. Li, Y. Yao, H. Cai, J. Gao, G. Qian, Amorphous engineering of scalable metal-organic framework-derived electrocatalyst for highly efficient oxygen evolution reaction, Small 20(28) (2024) 2311356. https://doi.org/10.1002/smll.202311356.

15 Y. Pan, Y. Li, A. Nairan, U. Khan, Y. Hu, B. Wu, L. Sun, L. Zeng, J. Gao, Constructing FeNiPt@ C trifunctional catalyst by high spin-induced water oxidation activity for Zn-air battery and anion exchange membrane water electrolyzer, Advanced Science 11(19) (2024) 2308205. https://doi.org/10.1002/advs.202308205.

16 G. Martínez-Criado, R. Tucoulou, P. Cloetens, P. Bleuet, S. Bohic, J. Cauzid, I. Kieffer, E. Kosior, S. Labouré, S. Petitgirard, Status of the hard X-ray microprobe beamline ID22 of the European synchrotron radiation facility, Synchrotron Radiation 19(1) (2012) 10–18. https://doi.org/10.1107/S090904951104249X.

17 F. Lin, Y. Liu, X. Yu, L. Cheng, A. Singer, O.G. Shpyrko, H.L. Xin, N. Tamura, C. Tian, T.C. Weng, X.Q. Yang, Y.S. Meng, D. Nordlund, W. Yang, M.M. Doeff, Synchrotron X-ray analytical techniques for studying materials electrochemistry in rechargeable batteries, Chemical Reviews 117(21) (2017) 13123–13186. https://doi.org/10.1021/acs.chemrev.7b00007.

18 J. Huang, B. Ouyang, Y. Zhang, L. Yin, D.H. Kwon, Z. Cai, Z. Lun, G. Zeng, M. Balasubramanian, G. Ceder, Inhibiting collective cation migration in Li-rich cathode materials as a strategy to mitigate voltage hysteresis, Nature Materials 22(3) (2023) 353–361. https://doi.org/10.1038/s41563-022-01467-z.

19 X. Li, Z. Ren, M. Norouzi Banis, S. Deng, Y. Zhao, Q. Sun, C. Wang, X. Yang, W. Li, J. Liang, X. Li, Y. Sun, K. Adair, R. Li, Y. Hu, T.-K. Sham, H. Huang, L. Zhang, S. Lu, J. Luo, X. Sun, Unravelling the chemistry and microstructure evolution of a cathodic interface in sulfide-based all-solid-state li-ion batteries, ACS Energy Letters 4(10) (2019) 2480–2488. https://doi.org/10.1021/acsenergylett.9b01676.

20 J. Zhou, M.L. Holekevi Chandrappa, S. Tan, S. Wang, C. Wu, H. Nguyen, C. Wang, H. Liu, S. Yu, Q.R.S. Miller, G. Hyun, J. Holoubek, J. Hong, Y. Xiao, C. Soulen, Z. Fan, E.E. Fullerton, C.J. Brooks, C. Wang, R.J. Clement, Y. Yao, E. Hu, S.P. Ong, P. Liu, Healable and conductive sulfur iodide for solid-state Li–S batteries, Nature 627(8003) (2024) 301–305. https://doi.org/10.1038/s41586-024-07101-z.

21 X. Wang, L. Yin, A. Ronne, Y. Zhang, Z. Hu, S. Tan, Q. Wang, B. Song, M. Li, X. Rong, S. Lapidus, S. Yang, E. Hu, J. Liu, Stabilizing lattice oxygen redox in layered sodium transition metal oxide through spin singlet state, Nature Communications 14(1) (2023) 7665. https://doi.org/10.1038/s41467-023-43031-6.

22 W. Li, M. Li, P.-H. Chien, S. Wang, C. Yu, G. King, Y. Hu, Q. Xiao, M. Shakouri, R. Feng, B. Fu, H. Abdolvand, A. Fraser, R. Li, Y. Huang, J. Liu, Y. Mo, T.-K. Sham, X. Sun, Lithium-compatible and air-stable vacancy-rich $Li_9N_2Cl_3$ for high–areal capacity, long-cycling all–solid-state lithium metal batteries, Science Advances 9 (2023) eadh4626. https://doi.org/10.1126/sciadv.adh4626.

23 W. Li, J. Liang, M. Li, K.R. Adair, X. Li, Y. Hu, Q. Xiao, R. Feng, R. Li, L. Zhang, S. Lu, H. Huang, S. Zhao, T.-K. Sham, X. Sun, Unraveling the origin of moisture stability of halide solid-state electrolytes by in situ and operando synchrotron X-ray analytical techniques, Chemistry of Materials 32(16) (2020) 7019–7027. https://doi.org/10.1021/acs.chemmater.0c02419.

24 X. Yu, Q. Wang, Y. Zhou, H. Li, X.Q. Yang, K.W. Nam, S.N. Ehrlich, S. Khalid, Y.S. Meng, High rate delithiation behaviour of $LiFePO_4$ studied by quick X-ray absorption spectroscopy, Chemical Communications 48(94) (2012) 11537–11539. https://doi.org/10.1039/c2cc36382h.

25 D. Takamatsu, Y. Koyama, Y. Orikasa, S. Mori, T. Nakatsutsumi, T. Hirano, H. Tanida, H. Arai, Y. Uchimoto, Z. Ogumi, First in situ observation of the $LiCoO_2$ electrode/electrolyte interface by total-reflection X-ray absorption spectroscopy, Angewandte Chemie International Edition 51(46) (2012) 11597–11601. https://doi.org/10.1002/anie.201203910.

26 K. Yamamoto, T. Minato, S. Mori, D. Takamatsu, Y. Orikasa, H. Tanida, K. Nakanishi, H. Murayama, T. Masese, T. Mori, H. Arai, Y. Koyama, Z. Ogumi, Y. Uchimoto, Improved cyclic performance of lithium-ion batteries: An investigation of cathode/electrolyte interface via *in situ* total-reflection fluorescence X-ray absorption spectroscopy, Journal of Physical Chemistry C 118(18) (2014) 9538–9543. https://doi.org/10.1021/jp5011132.

9

Electron Paramagnetic Resonance Spectroscopy for Defect and Redox Characterization in Battery Materials

9.1 Basic Principles

Electron paramagnetic resonance (EPR) technology is a spectroscopic detection approach that utilizes unpaired electrons to analyze the chemical structure information of paramagnetic substances [1–3]. Originating from physics research, EPR technology was initially employed to investigate aspects such as the electronic structure of specific complex atoms [4, 5], atomic dipole moments, molecular structures, and crystal structures. Subsequently, it was extended to the analysis and characterization of organic compounds in the field of biochemistry. Currently, the EPR testing technique has been extensively applied in numerous fields, including physics, chemistry, biology, medicine, and energy science [6–8]. Notably, EPR can acquire the structural and dynamic information [9] of substances during physical processes and chemical reactions without interfering with the reactions, making it an effective in-situ mechanism analysis method [6–8].

9.1.1 Principles of Magnetic Resonance

In an atom, electrons are capable of executing two distinct types of motion. The first type of motion is orbital, occurring around the atomic nucleus, whereas the second is spin motion, which involves rotation about an axis passing through the electron's center. These two forms of motion give rise to an orbital magnetic moment and a spin magnetic moment, respectively. When a free electron is placed in an external magnetic field B_0, the interaction between the electron and the field results in the splitting of its spin energy levels. Specifically, the spin energy levels transition from a degenerate state to two separate energy levels. This phenomenon is known as Zeeman splitting, which refers to the energy level separation associated with the electron spin magnetic moment in an external magnetic field [10]. The splitting of the electron spin energy levels and the corresponding resonance condition are schematically illustrated in Figure 9.1.

The higher energy level, designated as the α state, corresponds to the magnetic quantum number

$$E_+ = m_{S\alpha} = +1/2 \tag{9.1}$$

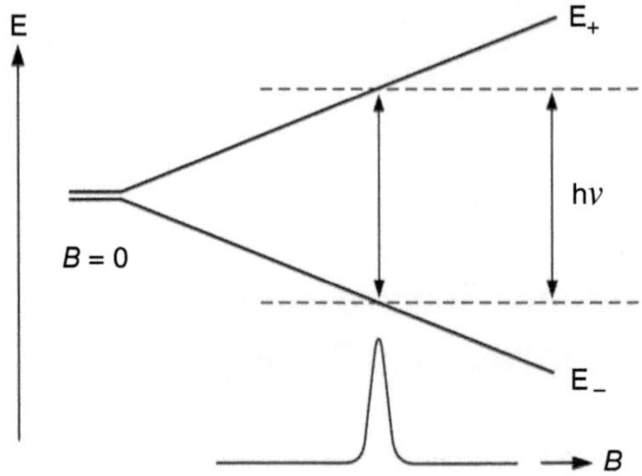

Figure 9.1 Electron Zeeman splitting as a function of magnetic field strength, *B*, and its corresponding resonance condition [10] / with permission of John Wiley & Sons.

and has an energy

$$E = +1/2g_e\beta_e B \tag{9.2}$$

The lower energy level, designated as the β state, corresponds to

$$E_- = m_{S\beta} = -1/2 \tag{9.3}$$

and has an energy

$$E = -1/2g_e\beta_e B \tag{9.4}$$

The energy separation between these two spin states is proportional to the strength of the external magnetic field B_0:

$$h\nu = \Delta E = g_e\beta_e B_0 \tag{9.5}$$

where m_S is the magnetic quantum number, g_e is a dimensionless g-factor, and

$$\beta_e = \frac{he}{2m} \tag{9.6}$$

denotes the Bohr magneton.

Based on the generation of spin energy level splitting, when an electromagnetic wave with a central frequency of ν is applied perpendicular to the external magnetic field B_0, and when the energy of this electromagnetic wave coincides with the Zeeman energy level spacing, that is, Equation 9.5 when is fulfilled, Some electrons in the lower energy level can absorb energy from the electromagnetic wave and transition to the higher energy level. This is the resonance absorption phenomenon, also referred to as the EPR phenomenon. The absorption signal generated by the stimulated transition serves as the origin of the EPR electron signal. After being processed by an electronic system, the EPR spectral line can be acquired. The absorption signal recorded by the EPR spectrometer is the first derivative linear, also known as the first derivative spectral line.

9.1.2 Absorption-relaxation-absorption Process

The unpaired electrons that give rise to EPR are not all in the low-energy state. From the Maxwell-Boltzmann distribution, the distribution law of the number of particles under thermodynamic equilibrium can be described as follows:

$$\frac{n_u}{n_d} = \exp\left(-\frac{hv}{kT}\right) \tag{9.7}$$

In this context, n_u represents the number of particles occupying the high-energy state, k denotes the Boltzmann constant, and T stands for the thermodynamic temperature. From the equation, it is evident that the number of particles in the lower energy level exceeds that in the upper energy level, thereby leading to a net energy absorption.

Under magnetic resonance conditions, when particles in the lower energy level absorb energy from electromagnetic waves and transition to the higher level, some in the higher level return via relaxation and release energy to the lattice through radiation. This process is referred to as paramagnetic relaxation. It is a process in which unpaired electrons exchange energy with the surrounding environment through non-radiative transitions. Moreover, it is a process by which the spin system returns from a non-equilibrium state to a thermal equilibrium state after being perturbed by electromagnetic waves.

The absorption-relaxation-absorption process essentially constitutes a process of energy transfer among electromagnetic waves, the sample, and the environment. The occurrence of the relaxation phenomenon represents the step of energy dissipation, which guarantees the continuous occurrence of the resonance absorption phenomenon.

9.1.3 Field Modulation

The outcomes of the most extensively utilized continuous wave-EPR (CW-EPR) spectra typically consist of spectral lines that represent the first derivative of the measured absorption signal. This procedure depends on an external high-frequency modulation field to resolve the absorption signal. The modulation coil is wound outside the resonant cavity such that the direction of the generated modulation magnetic field aligns with that of the external magnetic field. A high-frequency and low-amplitude modulation magnetic field is superposed on the slowly scanned main magnetic field. The modulation frequency is generally set at 100 kHz. By detecting the peak-to-peak amplitude using a phase-sensitive detector, the first derivative signal of the absorption signal can be acquired. Given that this field modulation method utilizes a phase-sensitive detector, only signals modulated at the same frequency (100 kHz) can be detected. This detection approach, which employs high-frequency modulation and phase-sensitive detection, can significantly enhance the signal-to-noise ratio. To further improve the resolution of the instrument, modern spectrometers are also capable of observing the measurement results in the form of second derivative curves. The dependence of EPR signal amplitude on the microwave power under different detection conditions is shown in Figure 9.2.

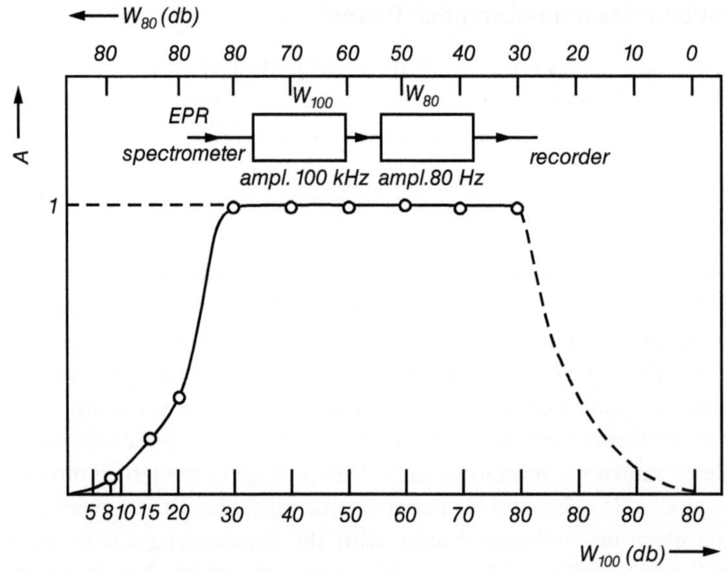

Figure 9.2 Amplitude of the EPR signal vs. again W level in two channels [11] / with permission of John Wiley & Sons.

9.1.4 Phenomena/Effects

The Zeeman effect refers to the phenomenon in which the spectral lines of atoms emit light and split and polarize in an external magnetic field. What Zeeman first observed and Lorentz gave a theoretical explanation for was the phenomenon where the spectral lines split into three. Later, more complex and difficult-to-explain situations than the three-split phenomenon were discovered. Thus, the former is termed the normal Zeeman effect and the latter the anomalous Zeeman effect, both of which are fully explained by quantum mechanics.

Around 1895, the Dutch physicist Zeeman (Zeeman) continued Faraday's unfinished exploration and began to study the influence of magnetic fields on the spectra of light sources. Through precise experimental observations, he found that under the action of a strong magnetic field, the D spectral line of the sodium light source showed a broadening phenomenon. Zeeman ingeniously improved the experimental device and excluded interference factors such as temperature. Finally, he confirmed that this "broadening" of the spectral line was actually the splitting of the spectral line, proving that the magnetic field can indeed change the frequency and period of light. This phenomenon was later called the "Zeeman effect."

To account for this phenomenon, Zeeman resorted to Lorentz's electron theory. This theory posits that the vibration of charged "ions" (i.e., electrons) within matter gives rise to light. Lorentz provided guidance to Zeeman, suggesting that if this theory were correct, then when observed from different directions, the edges of the split spectral lines should display specific polarization states (circularly polarized light along the magnetic field direction and plane-polarized light perpendicular to this direction). Zeeman's experimental findings precisely corroborated these predictions. More significantly, he estimated the charge-to-mass ratio (e/m) of charged particles based on the width of the spectral line splitting. The obtained value was comparable to J. J. Thomson's electron charge-to-mass ratio measured from cathode rays, providing strong independent evidence for the existence of electrons.

Nonetheless, further in-depth investigations uncovered new issues. Zeeman and other scientists (such as Preston, Runge, Paschen, etc.) discovered that the number and spacing of the spectral lines of many atoms splitting in the magnetic field did not align with Lorentz's predictions derived from the theory. This phenomenon was termed the "anomalous Zeeman effect," whereas the originally theory-consistent phenomenon was designated as the "normal Zeeman effect." Lorentz's classical theory was unable to explain this, thereby becoming a problem that perplexed the field of physics for nearly three decades.

The definitive solution to this problem hinged upon the advancement of quantum mechanics. In 1925, Uhlenbeck and Goudsmit put forward the hypothesis that electrons possess "spin." Subsequently, in 1926, Heisenberg and Jordan integrated electron spin with orbital magnetic moments and successfully accounted for the anomalous Zeeman effect from the vantage point of quantum mechanics. Consequently, the comprehensive panorama of the Zeeman effect was unveiled: it served not only as an early confirmation of the existence of electrons but also as a crucial experimental foundation that propelled the development of physics and quantum theory.

9.2 Equipment and Analysis

The EPR spectrometer is primarily composed of a microwave bridge, an electromagnet system, a transmission system, a microwave resonant cavity, a detection system, and a modulation system. The EPR spectrometer is primarily composed of a microwave bridge, an electromagnet system, a transmission system, a microwave resonant cavity, a detection system, and a modulation system, as schematically illustrated in Figure 9.3 [12].

1. Microwave Bridge: A microwave bridge comprises components dedicated to the generation, control, and detection of microwave radiation, which typically occupies the frequency range of 1–100 GHz. Commonly employed microwave sources include klystrons and Gunn diodes. Klystrons are widely utilized in spectrometer systems owing to their stability, high power output, and low noise characteristics, thereby contributing to enhanced instrumental resolution. An isolation modulator is positioned directly after the microwave source to attenuate reflected signals and preserve frequency stability.
2. Electromagnet System: In an EPR spectrometer, the electromagnet system plays a critical role in generating a magnetic field that meets stringent requirements for intensity, stability, and homogeneity. The magnetic assembly consists of both magnets and magnetic field sensors. Two primary magnet types are

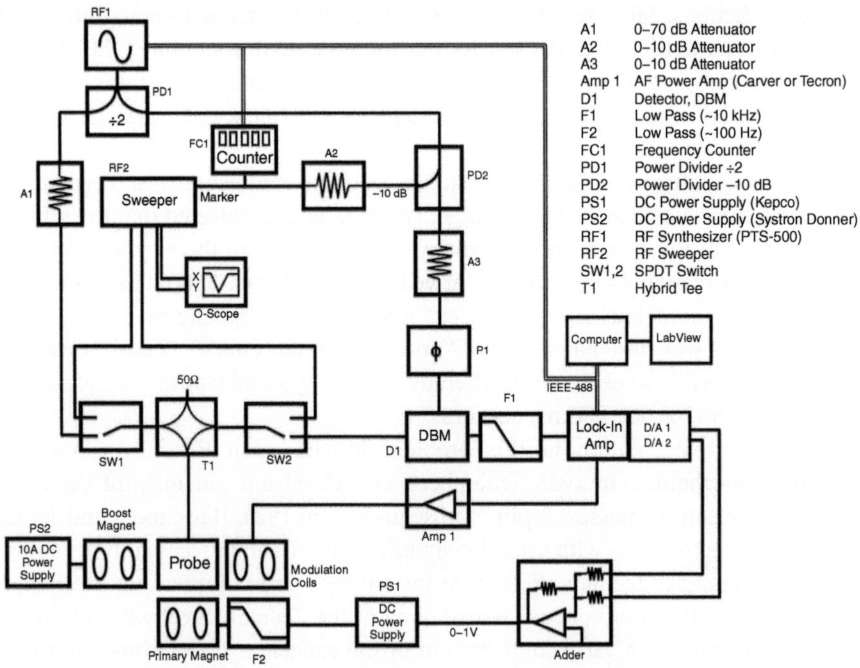

Figure 9.3 Block diagram of the EPR [12] / with permission of John Wiley & Sons.

employed in EPR instrumentation: electromagnets, which can produce fields of up to 1.5 T and are particularly suited for Q-band frequency measurements, and superconducting magnets, which are optimized for operation at W-band frequencies and beyond. The choice of magnet is carefully determined by the required magnetic field range to ensure optimal correspondence with the microwave operating frequency.

3. Transmission System: The transmission system is essential for delivering the generated microwaves to the target location and comprises interconnected components such as isolators, attenuators, circulators, and waveguides. A directional coupler divides the electromagnetic waves from the microwave source into two separate pathways. One pathway directs the signal toward the resonant cavity, whereas the other extends to the reference arm. Variable attenuators are integrated into both arms to allow fine adjustment of microwave power. Along the cavity pathway, the microwaves interact with the sample, producing a resonance absorption signal that facilitates detailed analysis. In the reference arm, a phase shifter is positioned after the variable attenuator to adjust the phase relationship between the reference and reflected signals.
4. Resonant Cavity: The resonant cavity constitutes the central component of an EPR spectrometer. This metallic enclosure—typically square or cylindrical in geometry—serves as the dedicated site for sample placement and plays an essential role in signal generation, amplification, and detection. Within the cavity, weak signals emitted by the sample are enhanced via resonance

phenomena. When the energy level splitting closely matches the frequency of incident microwave photons, the resulting absorption line can be detected using a solid-state diode. The capacity of the resonant cavity to store microwave energy is quantified by the quality factor-Q. A higher Q value corresponds to increased spectrometer sensitivity, reflecting an enhanced ability to detect subtle changes in the sample's magnetic properties.

5. Detection System: The microwave signal is transmitted from the resonant cavity via a circulator, after which it undergoes detection before entering a 100 kHz narrow-band amplifier. It is then sent to a phase-sensitive detector that selects signals matching the reference signal in frequency and phase. The output from the phase-sensitive detector is subsequently amplified and recorded by a computer.
6. Modulation System: Most external components, such as microwave source generators and detectors, are integrated into the microwave bridge control unit. The input and output microwave frequencies are controlled via a modulation system, which functions to reduce noise and improve the sensitivity of the spectrometer.

In the EPR instrument, the magnetic field produced by the electromagnet system is oriented perpendicularly to both the electromagnetic wave delivered to the resonant cavity via the waveguide and the scanning coil. When the frequency ν_0 of the electromagnetic wave and the magnetic field strength H_0 satisfy the resonance condition, the sample inside the resonant cavity undergoes resonance and absorbs energy. The resulting absorption signal is amplified within the cavity, detected, and further processed by the phase-sensitive detector before being displayed on an oscilloscope and automatically recorded.

9.2.1 Influencing Factors

1. Temperature Effect: The impact of temperature on the signal is predominantly manifested in two aspects. Firstly, thermal fluctuations interfere with the test outcomes. Secondly, the survival duration of high-temperature signals is brief and their capture is challenging. Consequently, for certain measurements, such as the detection of solid surface pores and the examination of the unstable valence states of transition metals, signals can be more effectively collected in a low-temperature environment of liquid nitrogen.
2. Sample Concentration: When the sample concentration is too high, the distance between molecules is relatively close, which will make the electron spin-spin interaction stronger and affect the instrument resolution; when it is too low, it will be close to the detection limit of the instrument and affect the sensitivity of the results. Therefore, a dispersant is needed to appropriately adjust the sample concentration.
3. Magnetic Field Modulation: When the amplitude of the magnetic field modulation is excessively large or the frequency is too close to the spectral line, it can have an impact on the output signal of the detector, leading to spectral line distortion and deformation.

4. Sample Type: Ascertaining the sample type and the nature of the paramagnetic center holds significant reference value for optimizing the microwave frequency and magnetic field strength. For instance, samples in aqueous solutions, particularly biological specimens, predominantly absorb microwaves via the dielectric process, and the dielectric loss is profoundly influenced by the frequency. Consequently, to ensure the effective observation of the EPR signal, a lower microwave frequency must be chosen.

9.2.2 Work Pattern

CW-EPR represents the traditional operating mode of EPR. In this mode, the resonance signal is detected by keeping the microwave frequency fixed while scanning the magnetic field. The sample is positioned within a resonant cavity, where the microwave interacts with the sample. When the resonance condition is satisfied with the Equation 9.5, the unpaired electrons absorb microwave energy and undergo a transition, thereby generating a first derivative spectral line. This mode is well-suited for the detection of most conventional samples and is distinguished by its simple operation and stable signals.

9.2.3 Quantitative Analysis

EPR spectroscopy currently represents the only methodology capable of directly detecting unpaired electrons in a given sample. Among its various applications, quantitative EPR is particularly important, as it enables precise determination of the number of unpaired electron spins present. This information is critically valuable across multiple disciplines, including detailed investigations of reaction kinetics, accurate elucidation of reaction mechanisms, and a range of commercial applications. As a result, the quantification of unpaired electron spins via EPR has remained a persistent focus of scientific research. Presently, two principal approaches to quantitative EPR are employed: relative quantification and absolute quantification.

In the relative quantitative EPR method, quantification is achieved by comparing the integrated area of the EPR absorption spectrum of an unknown sample with that of a standard sample containing a known spin concentration. However, the integrated intensity of the EPR signal depends not only on the number of unpaired electron spins but also on several experimental factors. These include instrumental parameter settings, the dielectric properties of the sample, as well as its dimensions, geometry, and precise positioning within the resonance cavity. Therefore, to obtain reasonably accurate quantitative results, it is essential that the standard and unknown samples exhibit similar characteristics, possess comparable size and shape, and are measured in identical positions within the cavity.

Absolute quantitative EPR directly determines the number of unpaired electron spins in a sample without a reference standard. This approach requires accurate knowledge of the second integral area of the EPR spectrum—typically obtained from the first derivative spectrum—along with relevant experimental parameters, sample volume, the cavity distribution function, and appropriate correction factors. The procedure involves first acquiring the EPR spectrum of the sample, followed by computation of the second integral area of the first derivative signal. By integrating this value with the experimental conditions and instrumental factors, the absolute concentration of unpaired spins can be directly calculated.

9.3 Introduction of New Technologies

9.3.1 In-situ Electrochemistry

One of the key characteristics of EPR spectroscopy technology lies in its ability to characterize samples without interfering with physical, chemical, and biological reaction processes. Consequently, it holds great potential for in-situ characterization. The EPR signal intensity and double integral value increase during charging and reversibly recover during discharging.

Furthermore, through a comprehensive analysis of the EPR spectral line shape and its temperature dependence, it has been revealed that there are two spin behaviors within the double-capacitance behavior of activated carbon: spins at defects, which generate narrow signals; and surface aromatic spins, which give rise to wide signals. The narrow signal exhibits a striking potential-dependent response across all electrolyte solutions, while the broad signal reveals its transformation only under more concentrated conditions, adding a layer of nuance to the overall spectral behavior.

These results suggest that the increase in the narrow signal is attributed to the formation of free radicals through the reduction of functional groups, thereby leading to an elevation in the unpaired electron density. The potential dependence of the wide signal may reflect additional ion adsorption in the deep porous structure of activated carbon.

9.3.2 Combination

When used in combination with traditional EPR instruments, the method of fixing the microwave frequency and scanning the magnetic field is adopted, which is called a CW-EPR instrument. In recent years, with the development of additional devices, pulse-EPR and rapid scanning-EPR technologies have emerged, enabling the EPR technology to enter a diverse field, obtain more diverse information, and further expand the sample measurement range.

9.3.3 Pulse EPR

The pulse-wave excited EPR technology is actually a relaxation measurement tool. Under the excitation of the pulse wave, the relaxation phenomenon of the spin

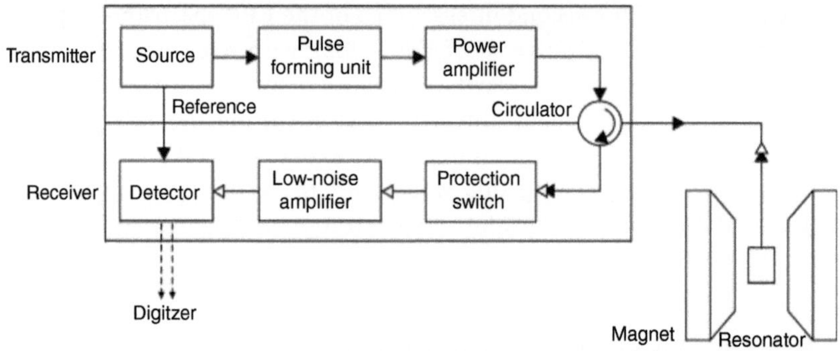

Figure 9.4 Block diagram of a typical pulse EPR spectrometer [13] / with permission of Taylor & Francis Group.

system is amplified, providing information contrasting with that of continuous wave excited resonance. The implementation of the pulse technology only requires simple attachment addition and adjustment to the microwave source device. By setting a pulse program control switch in the microwave source, pulsed microwaves with a certain interval can be output. Pulse EPR has multi-frequency, multi-resonance, and arbitrary waveform modes. A typical configuration of a pulse EPR spectrometer is shown in Figure 9.4.

The advantages of pulse EPR are as follows: it can improve the g-factor resolution; amplify the microwave power; the resonance transition signal excited by pulsed microwaves can be detected without magnetic field modulation; by changing the pulse sequence of the exciting electromagnetic wave, more extensive information about paramagnetic compounds can be obtained. However, compared with CW-EPR, pulse EPR cannot excite to obtain the entire spectral information, and its sensitivity is limited due to a larger bandwidth.

9.3.4 Rapid Scanning-EPR

The rapid scanning-EPR (RS-EPR) technology combines the generation principles of CW-EPR and pulse EPR, and is more suitable for studying fast reaction kinetics. During rapid scanning-EPR, the magnetic field resonantly scans the sample in a time shorter than the electron spin relaxation time. The basic configuration of a rapid-scan EPR spectrometer is schematically illustrated in Figure 9.5.

The advantages of the rapid scanning signal are that it can provide full-spectrum detection and is suitable for providing higher microwave power under unsaturated conditions, especially suitable for low-frequency EPR imaging; the rapid scanning-EPR spectrum can simultaneously measure the absorption and dispersion components of the spin system response, which can elevate dramatically the signal-to-noise ratio. Compared with pulse EPR, rapid scanning-EPR has higher sensitivity, better spatial resolution, shorter acquisition time, and can perform more preclinical imaging measurements per unit time. This technology is particularly important for conducting in vivo clinical research.

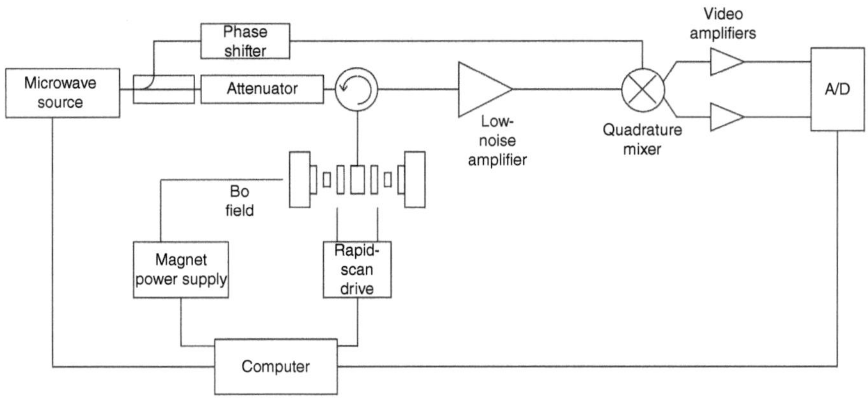

Figure 9.5 Block diagram of a rapid-scan spectrometer [14] / with permission of John Wiley & Sons.

9.4 Application Cases

EPR spectroscopy, particularly in its operando or in-situ configuration, has become a pivotal tool for real-time probing of paramagnetic species and dynamic processes in working batteries. By directly detecting unpaired electrons associated with transition metal ions, oxygen radicals, metallic lithium, and defect states, operando EPR not only reveals redox mechanisms and degradation pathways but also enables spatially resolved imaging of reactive intermediates—capabilities that are otherwise inaccessible to conventional electrochemical or structural techniques. The following sections highlight representative applications of EPR across cathode, anode, and solid-state systems, with an emphasis on insights gained under realistic operating conditions.

9.4.1 In-situ EPR Characterization of Lithium Deposition and Dendrite Evolution Under Fast Charging

Kang et al. [15] explored lithium deposition and dendrite growth in NCM811|| graphite full cells under 4 C extreme-fast charging/discharging long-term cycling via operando EPR spectroscopy and EPRI, quantifying lithium deposition, visualizing dendrite morphology/distribution, and analyzing lithium stripping/ reintercalation kinetics to support optimizing charging and suppressing dead lithium (Figure 9.6a). EPR (X-band, ~9.4 GHz, 0.5 G modulation, 2.206 mW power) used Dysonian fitting for graphite/lithium/Super P signals (Figure 9.6b), with double integral (DI) quantifying deposition (spin-proportional) and A/B characterizing deposit size (linked to 1.3 μm Li skin depth); EPRI achieved YX-plane (front-view) imaging, overcoming traditional side-view (YZ/XZ) limits. Under 4.3 V CCCV charging, slow DI rise (loose, divergent dendrites), A/B ~1.6 (1.45 μm deposits), EPRI showed edge-distributed dead lithium (Figure 9.6c), and intact electrodes post-disassembly (Figure 9.6d); under 4.6 V CC overcharging (Figure 9.6e), nearly

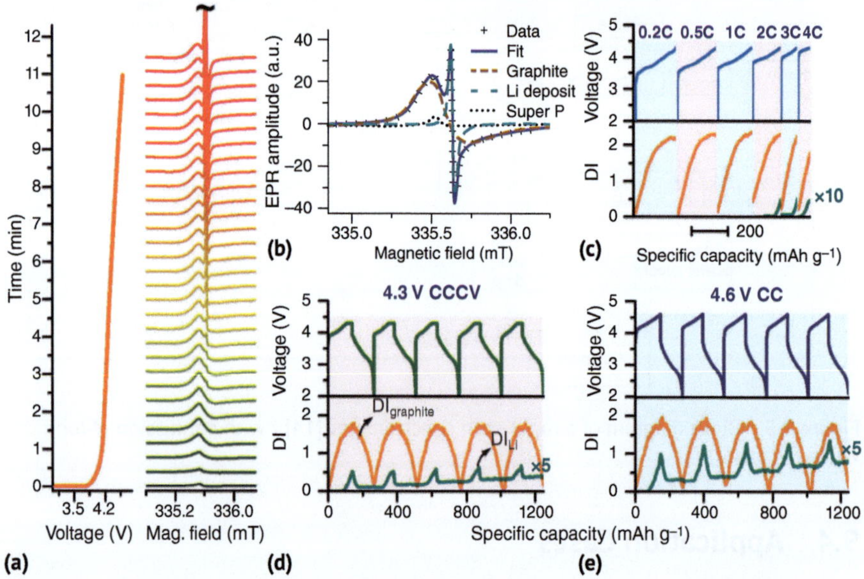

Figure 9.6 (a) AVoltage profile and operando EPR spectra recorded on an NCM811|| graphite cell during 4 C charging to 4.6 V. (b) Three-component fitting results of one slice of the operando data. (c) Detection of Li deposition at different charge rates with a cutoff voltage of 4.3 V. The voltage profiles, DI of graphite, and DI of Li metal are shown by blue, orange, and turquoise lines, respectively. (d, e) Operando EPR experiments of the cells which undergo (d) 4.3 V CCCV charge and (e) 4.6 V CC charge at a 4 C rate. Note that DI is proportional to the detected spin amount in graphite or Li metal. Density of states at the Fermi level must be considered when comparing the mole amounts of the substances [15] / with permission of American Chemical Society.

doubled deposition (faster DI), A/B ~1.6, but EPRI showed sharp peaks (long, vertical dendrites, Figure 9.6e), exposed anode copper foil and 7Li NMR confirmed morphology differences. Long-term EPR showed dead lithium stabilized (DI ~1.5) post-dozens of cycles (saturation at ~85/55 cycles for 4.3 V/4.6 V) (Figure 9.7), with EPRI revealing first-cycle-determined dead lithium distribution (Figure 9.8, unchanged post-50 cycles).

Liu et al. [16] characterized lithium dendrites in Li/LLZO/Li all-solid-state batteries using 240 GHz continuous wave-EPR spectroscopy (tested at 290 K, magnetic field scanning range of 8.53–8.585 T), with relevant results corresponding to Figure 9.9. The pristine LLZO showed no EPR signal, the moderately cycled sample (cycled for 2 days) exhibited a weak signal (mossy amorphous dendrites), and the short-circuited failed sample presented a strong and narrow signal (dendritic crystalline dendrites). The signal intensity reflects dendrite accumulation, and the linewidth change reveals the transformation of dendrites from amorphous to crystalline, providing a basis for distinguishing the two dendrite formation mechanisms: non-uniform plating at the interface and lithium-ion reduction at grain boundaries.

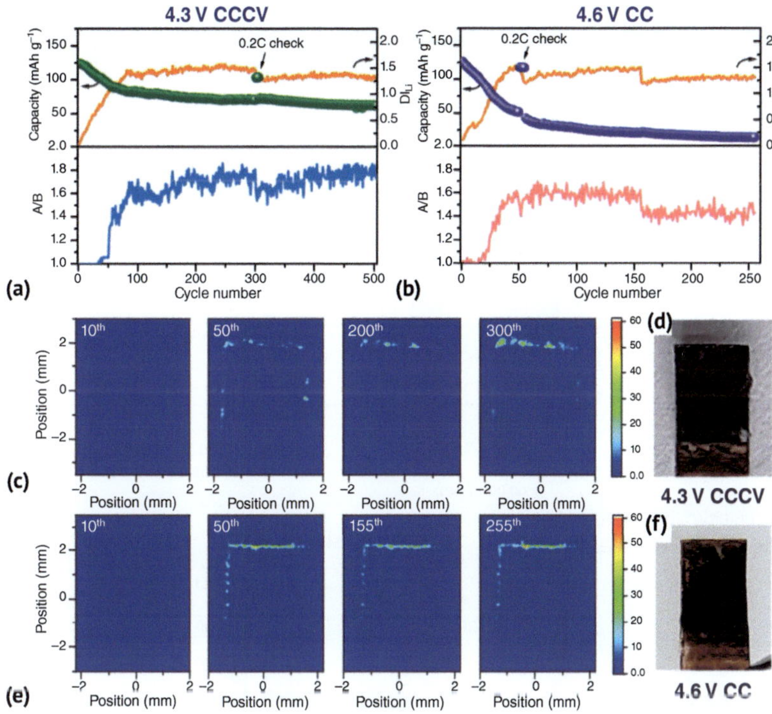

Figure 9.7 Operando EPR and EPRI results of the long-term cycling tests, showing the accumulation processes of dead Li in full cells. (a, b) Cycling performance and EPR DI results of Li deposits for the cells using (a) 4.3 V CCCV and (b) 4.6 V CC cycling protocols. (c, e) Spatial–spatial EPR images of the dead Li in the cells at different cycles using (c) 4.3 V CCCV and (e) 4.6 V CC cycling protocols. The color scale in the figure caption refers to lithium signal intensity, and it gets stronger from blue to red. Some vertical stripes are artifacts introduced during image reconstruction. (d, f) Optical photos of the anodes after long-term cycling corresponding to (c) and (e), respectively. The color dots are used to verify the decayed capacity with a 0.2 C rate [15] / with permission of American Chemical Society.

9.4.2 Analysis of Active Species and Redox Mechanisms in Cathode Materials

Zuo et al. [17] investigated the relationship between voltage hysteresis and the electronic structure of Li-rich cathode materials (LRMs), using EPR spectroscopy as a core characterization tool to distinguish the electron spin states of transition metal ions in two Li_2RuO_3 model materials with C2/c and $P2_1/m$ symmetries, as well as practical Mn-based LRMs, with the relevant results corresponding to Figure 9.10. Specifically, EPR measurements were performed at 5 K to enhance signal sensitivity, focusing on detecting unpaired electrons associated with the spin configurations of Ru^{4+} and Ni^{3+} ions. For $P2_1/m$-Li_2RuO_3, the EPR spectrum showed no detectable signal (EPR-silent), which is consistent with the low-spin (LS) configuration of Ru^{4+} $\left(t_{2g}^4\right)$ where electrons are fully paired in the t_{2g} orbitals, resulting in no net

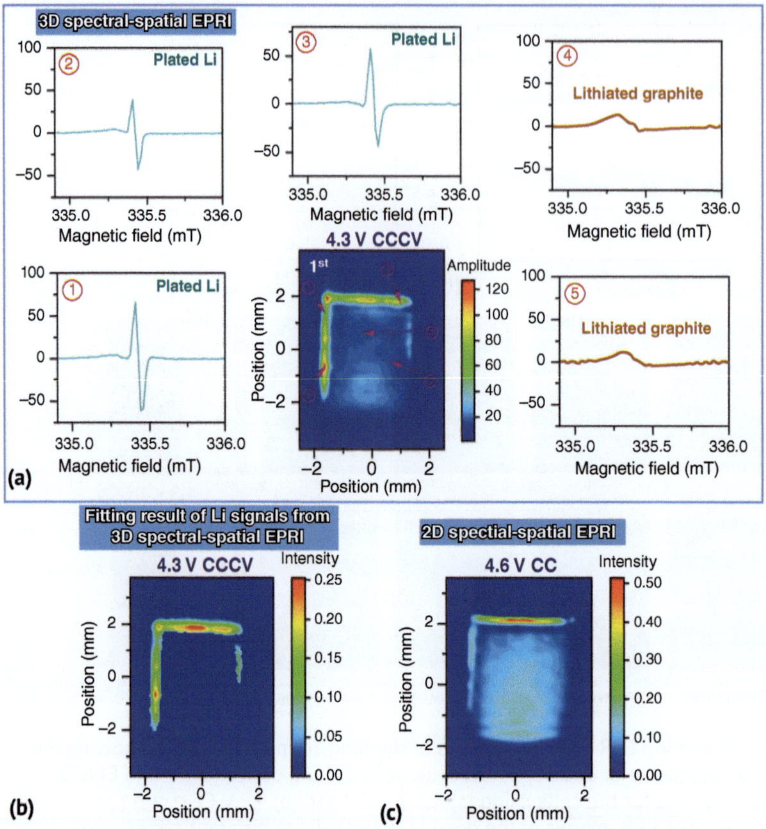

Figure 9.8 EPRI results at the first discharge for the two protocols. (a) 3D spectral–spatial EPRI for the 4.3 CCCV protocol. The EPR spectra on the electrode plane are resolved. The EPR spectra at five selected points demonstrate that the graphite signals are distributed on the electrode plane, while the plated Li signals concentrate at the electrode edges. (b) Fitting result of the Li signals from the 3D spectral–spatial image. The Li profile is in line with the spatial–spatial images shown in the main text. (c) 2D spatial–spatial EPRI for the 4.6 CC protocol. The signals from plated Li and lithiated graphite cannot be resolved. However, by comparing the image to those shown in the main text, the Li profile also concentrates at the electrode edges [15] / with permission of American Chemical Society.

unpaired spin density. In contrast, C2/c-Li_2RuO_3 exhibited a sharp isotropic EPR peak, a key signature of the high-spin (HS) configuration of Ru^{4+} $\left(t_{2g}^3\ e_g^1\right)$ with an isolated unpaired electron in the e_g orbital. This distinct EPR response directly confirmed the difference in electron spin states between the two Li_2RuO_3 polymorphs, which was further corroborated by macro magnetic measurements (Curie-Weiss fitting yielding effective paramagnetic moments of 0.588μ_B/Ru for P2_1/m-Li_2RuO_3 and 2.49μ_B/Ru for C2/c-Li_2RuO_3. Extending to Mn-based LRMs, the EPR principle was applied to rationalize the spin states of Ni^{3+}: $Li_{1.2}Ni_{0.36}Mn_{0.44}O_2$ with two e_g electrons (HS Ni^{3+}, $t_{2g}^5\ e_g^2$) showed a higher effective paramagnetic moment (3.26μ_B/Ni), while $Li_{1.15}Ni_{0.377}Mn_{0.473}O_2$ with an isolated e_g electron (LS Ni^{3+}, $t_{2g}^6\ e_g^1$)

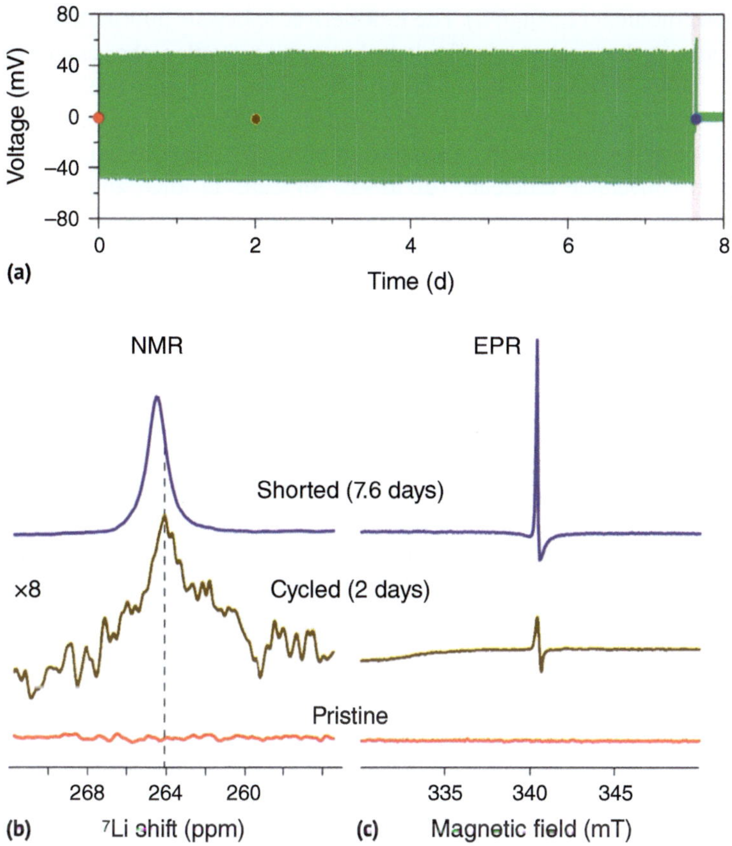

Figure 9.9 Li microstructure formation in cubic LLZO. (a) Voltage profile of the Li/LLZO/Li cell polarized with a current density of 0.1 mA cm^{-2}. A short circuit occurs in the shaded region. (b, c) ^{7}Li MAS (25 kHz) NMR (c) and EPR (d) spectra of Li dendrite formed at three stages—pristine, cycled for two days and shorted—with colors corresponding to the points marked with color-coded dots in a [16] / with permission of Springer Nature.

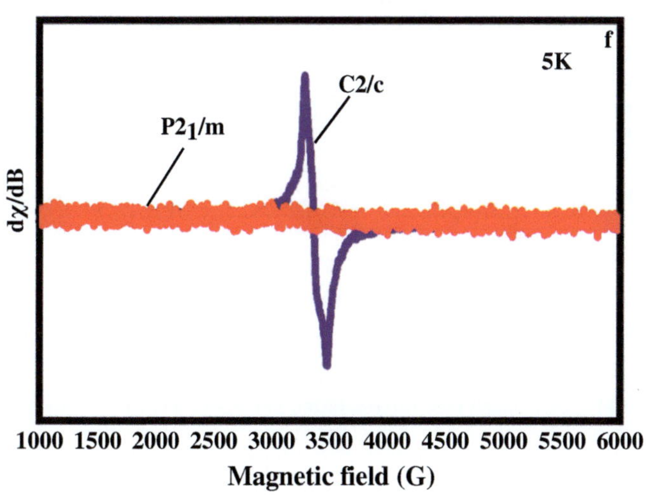

Figure 9.10 EPR measurements of $P2_1/m$– and $C2/c$-Li_2RuO_3 under 5 K [17] / with permission of American Chemical Society.

exhibited a lower moment (2.51μ_B/Ni). The EPR-derived spin state information was critical to elucidating the mechanism of voltage hysteresis reduction: the isolated eg electron filling in HS C2/c-Li_2RuO_3 and LS $Li_{1.15}Ni_{0.377}Mn_{0.473}O_2$ enhanced metal-oxygen covalency, reduced the charge transfer band gap, and lowered the oxidation potential of anion redox, ultimately leading to minimized voltage hysteresis and improved energy conversion efficiency (up to 95% for C2/c- Li_2RuO_3 and 93% for the Mn-based LRM).

To investigate the sodium-ion (Na^+) storage mechanism of hard carbon (HC) in different electrolytes for sodium-ion batteries, Liu et al. [18] employed ex-situ EPR spectroscopy as a core characterization technique, combined with spectral deconvolution, to successfully distinguish two key Na species in HC: adsorbed-Na and quasi-metallic-Na (Figures 9.11 and 9.12). Specifically, in the EPR spectra, the broad peak (ΔHpp ≈ 27 G) was attributed to adsorbed-Na on the carbon surface (derived from charge transfer involving delocalized π-electrons of the carbon skeleton during electrochemical sodiation), while the narrow peak (ΔHpp ≈ 3 G) corresponded to quasi-metallic-Na in the HC anode. By monitoring the evolution of these EPR signals during the charge-discharge process, the study revealed distinct Na storage behaviors in ether-based (e.g., DEGDME) and ester-based (e.g., EC/DEC) electrolytes: in the ether-based electrolyte, the initially formed adsorbed-Na in the discharge slope region (1.0–0.1 V) was efficiently converted into quasi-metallic-Na

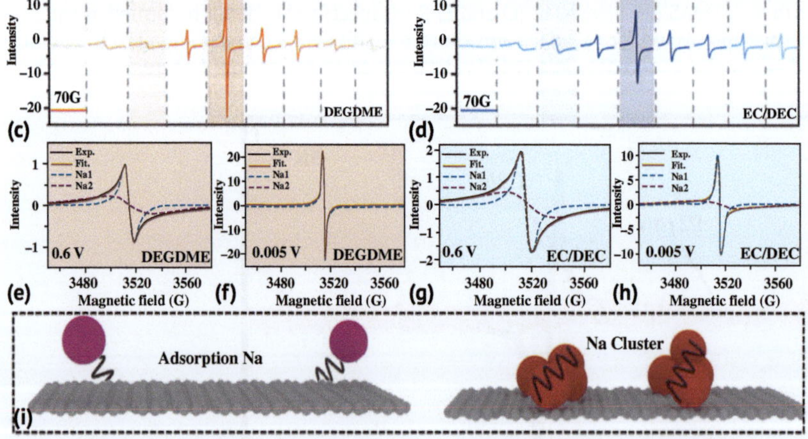

Figure 9.11 Discharge/charge profiles and experimental EPR spectra of HC in (a, c) DEGDME- and (b, d) EC/DEC-based electrolytes. Fitted EPR spectra with Lorentzian line shape at (e, g) slope region and (f, h) platform region. (i) Schematic illustration of adsorption-Na and quasi-metallic-Na [18] / with permission of John Wiley & Sons.

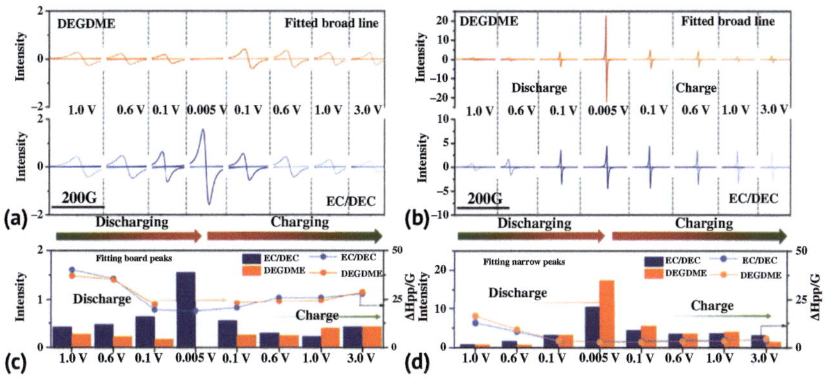

Figure 9.12 The (a) fitted board Lorentzian line shape and (b) fitted narrow Lorentzian line shape corresponding to the experimental spectra of HC at different potentials in both electrolyte systems with normalized mass loading of 1 mg. Intensity and line width variation trend of (c) fitted board EPR peaks and (d) fitted narrow EPR peaks [18] / with permission of John Wiley & Sons.

(intercalated into graphite interlayers) in the discharge plateau region (0.1–0.005 V), and this quasi-metallic-Na could be effectively desodiated during charging (minimizing "dead sodium" formation); in contrast, the ester-based electrolyte hindered this conversion, leading to the persistent coexistence of adsorbed-Na and quasi-metallic-Na throughout charge-discharge cycles, with quasi-metallic-Na prone to irreversible retention as dead sodium after charging. Additionally, the EPR signal intensity in the low-voltage region was significantly higher for HC in the ether-based electrolyte, indicating more unpaired electrons (and thus more activated electrochemical active sites) in HC, which further validated the superior Na^+ storage and conversion kinetics of HC in ether-based electrolytes. Overall, ex-situ EPR spectroscopy provided direct qualitative and quantitative evidence for deciphering the electrolyte-dominated Na^+ storage mechanism in HC, laying a critical foundation for understanding the enhanced electrochemical performance of HC in ether-based electrolytes.

Hu et al. [19] studied the enhancement mechanism of Ti substitution on the oxygen redox reversibility and cycling stability of P2-type sodium battery cathode material ($Na_{0.72}Li_{0.24}Mn_{0.76}O_2$, referred to as NLMO). Using dual-mode (perpendicular ⊥/parallel || polarization) EPR technology (X-band, ⊥ mode tested at 90 K, || mode tested at 2 K), they systematically characterized the evolution of manganese ion valence states, oxidized oxygen species, and ion coupling effects in the Ti-substituted material ($Na_{0.72}Li_{0.24}Ti_{0.10}Mn_{0.66}O_2$, referred to as NLMTO-0.1) during charge-discharge cycles, providing direct evidence for the mechanism of Ti substitution regulating oxygen redox reversibility. Specifically, under ⊥ mode (Figure 9.13a), both NLMO and NLMTO-0.1 exhibited a broad Lorentzian line shape signal with $g \approx 2.00$, attributed to antiferromagnetically coupled Mn^{4+}-O clusters. Under || mode (Figure 9.13b), NLMO showed an asymmetric signal with $g \approx 4.01$, corresponding to residual Mn^{3+} ($3d^4$, $S = 1$ spin state) from the solid-state reaction, while the signal of NLMTO-0.1 was significantly weakened, confirming that 10% Ti^{4+} substitution can

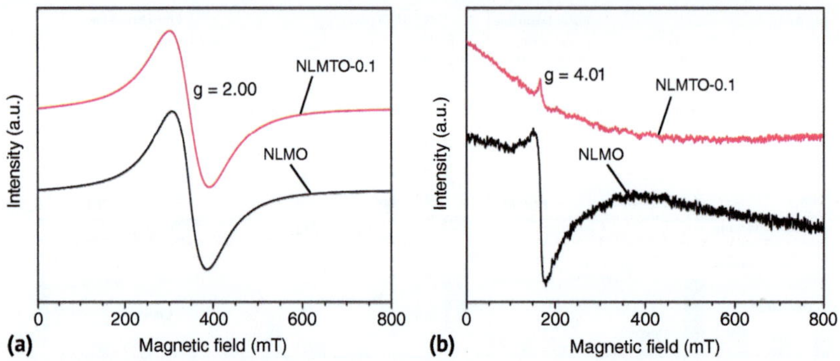

Figure 9.13 (a) ⊥-mode EPR spectra under 90 K and (b) ||-mode EPR spectra under 2 K of pristine NLMO and NLMTO-0.1 Adapted from [19].

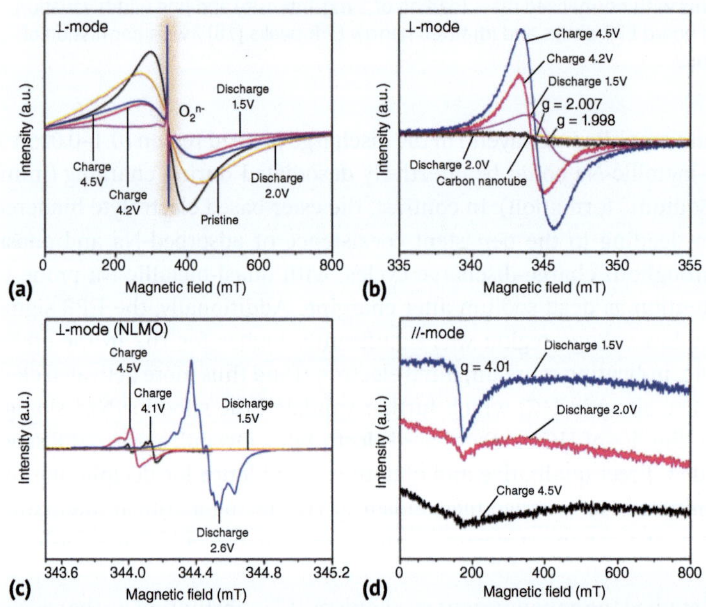

Figure 9.14 (a) ⊥-mode EPR spectra of cycled NLMTO-0.1 under the representative SoC; spectra are normalized based on the mass of cathode materials and measured Q values. The yellow-shaded region stands for oxidized oxygen species O^{2n-}. (b) Fine scanning ⊥-mode EPR spectra of cycled NLMTO-0.1 in the region of 335–355 mT. The O^{2n-} EPR line shape of charged NLMTO-0.1 is symmetric without a hyperfine pattern. (c) Fine scanning ⊥-mode EPR spectra of cycled NLMO in the region of 343.5–345.2 mT. The O^{2n-} EPR signals of charged NLMO possess various hyperfine patterns. (d) ||-mode EPR spectra of cycled NLMTO-0.1 under the representative SoC [19] / with permission of American Chemical Society.

effectively reduce the Mn^{3+} content and suppress Jahn-Teller distortion and oxygen vacancy formation. ⊥ mode EPR characterization of NLMTO-0.1 at different states of charge (SoC) showed (Figure 9.14a) that the intensity of the broad Mn^{4+} signal decreased significantly after cycling, because O^{2-} was oxidized to O_2^{n-} ($1 \leq n \leq 3$) during charging, weakening the coupling between Mn^{4+} and oxygen species. In the

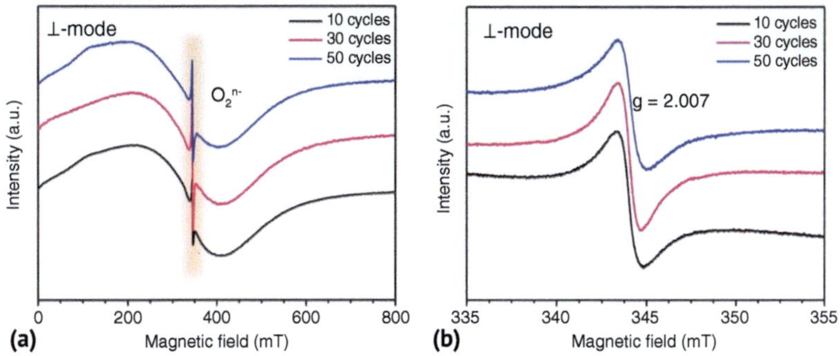

Figure 9.15 (a) ⊥-mode EPR spectra of fully charged NLMTO-0.1 (4.5 V) under various cycles. The yellow-shaded region stands for O_2^{n-}. (b) Fine scanning ⊥-mode EPR spectra of fully charged NLMTO-0.1 (4.5 V) in the region of 335–355 mT. The O_2^{n-} EPR signal is still detected for the fully charged sample after 50 cycles, demonstrating the high reversibility of oxygen redox chemistry for NLMTO-0.1 [19] / with permission of American Chemical Society.

fine scanning spectrum in the range of 335–355 mT (Figure 9.14b), a sharp signal with an effective g-value of 2.007 appeared when charged to 4.2 V and 4.5 V, clearly attributed to EPR-active oxidized oxygen species (O_2^{3-} or O_2^-, S = 1/2), confirming the occurrence of oxygen redox. This signal had no hyperfine splitting structure, in sharp contrast to the multiple hyperfine splitting of the O_2^{n-} signal of charged NLMO (Figure 9.14c), indicating that Ti^{4+} substitution effectively alleviated the anisotropic coupling between Mn^{4+} and O_2^{n-}. During discharge, the O_2^{n-} signal basically disappeared at 2.0 V, reflecting the reversibility of oxygen redox; a new signal with g ≈ 1.998 appeared when discharged to 1.5 V, corresponding to Ti^{3+} ($3d^1$, S = 1/2) reduced from doped Ti^{4+} ions. || mode EPR further confirmed that no Mn^{3+} signal was detected in NLMTO-0.1 before 2.0 V, and the conversion between Mn^{4+} and Mn^{3+} mainly occurred in the range of 2.0–1.5 V, clearly separated from the oxygen redox range (4.5–2.0 V) (Figure 9.14d). In addition, EPR tracking of cycled NLMTO-0.1 showed (Figure 9.15) that even after 50 cycles, a clear O_2^{n-} signal (g = 2.007) could still be detected when charged to 4.5 V, with no significant attenuation in intensity. This is the first time EPR has confirmed that oxygen redox in sodium battery cathode materials can achieve high reversibility over 50 cycles, which is also one of the core reasons why Ti^{4+} substitution improves the cycling stability of the material (85.1% capacity retention after 80 cycles).

Hu et al. [20] used EPR technology to accurately capture the evolution of unpaired electron signals of anions (oxygen) and cations (manganese) in P2-type fluorine-substituted sodium battery cathode material (P2-$Na_{0.65}Li_{0.22}Mn_{0.78}O_{1.99}F_{0.01}$, referred to as NLMOF) during charge-discharge cycles, providing direct evidence for the oxygen redox reversibility and structural stability mechanism, as shown in Figure 9.16. Experiments were performed using an X-band EPR spectrometer to conduct in-situ characterization of NLMOF electrodes at different SoCs at 90 K: the pristine NLMOF only exhibited a broad anisotropic signal with g ≈ 2.00, corresponding to antiferromagnetically coupled Mn^{4+} ions in the material, and its signal characteristics were consistent with the electronic spin interaction of Mn^{4+}.

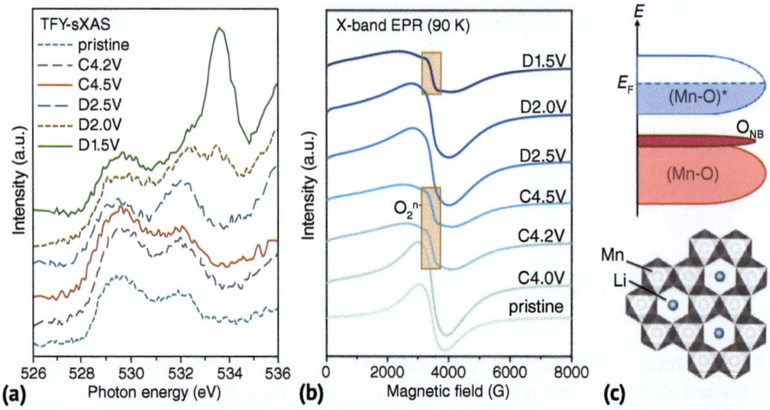

Figure 9.16 (a) O *K*-edge sXAS spectra under TFY mode and (b) X-band EPR spectra under 90 K of NLMOF at various SoC during the first charge and discharge processes. (c) Schematic band structure showing the origin of oxygen redox activity in Mn^{4+}-based Li-substituted layered compounds. "ONB" denotes O 2p nonbonding state [20] / with permission of American Chemical Society.

When charged to 4.2 V, the EPR signal of Mn^{4+} further broadened, and a new signal with g = 2.007 appeared, which was attributed to oxidized oxygen species O_2^{n-} (n = 1,2,3), confirming that oxygen anions underwent oxidation during charging and charge compensation was mainly derived from oxygen redox. At the end of charging to 4.5 V, the broadening of the Mn^{4+} signal intensified because the electron-electron dipolar interaction between oxidized O_2^{n-} and Mn^{4+} was stronger than that between unoxidized O^{2-}, leading to further changes in the spin environment of Mn^{4+}. During discharge, the EPR signal of Mn^{4+} gradually recovered to near the pristine state in the range of 2.5–2.0 V, and the characteristic signal of O_2^{n-} tended to disappear, indicating that oxygen species underwent reversible reduction and the oxygen redox process had good reversibility. At the end of discharge to 1.5 V, the Mn^{4+} signal broadened again, and a weak O_2^{n-} signal could still be observed. It is speculated that this was due to the reduction reaction of molecular oxygen trapped inside the material and sealed coin cells, and this process may be accompanied by side reactions between the electrolyte and oxidized oxygen to form carbonate species (corroborated by the O *K*-edge sXAS results in Figure 9.16a). Notably, compared with the EPR results of the unsubstituted material P2-$Na_{0.66}Li_{0.22}Mn_{0.78}O_2$ (referred to as NLMO) (Figure 9.16c), the EPR signal of oxidized O_2^{n-} in NLMOF had no hyperfine splitting structure, indicating that trace fluorine substitution (0.5%) effectively alleviated the anisotropic coupling between Mn^{4+} and O_2^{n-}, reducing local structural distortion caused by irreversible oxygen loss. Combined with the operando DEMS results showing that the O_2 release amount of NLMOF was significantly lower than that of NLMO, it further confirmed that the reversible O_2^{n-} signal captured by EPR was directly related to lattice oxygen stability. Fluorine substitution enhanced the flexibility of Mn-O bonds and increased the oxygen vacancy formation energy, suppressing irreversible oxygen precipitation and Jahn-Teller distortion, ultimately ensuring the excellent cycling stability of the material (87.8% capacity retention after 50 cycles).

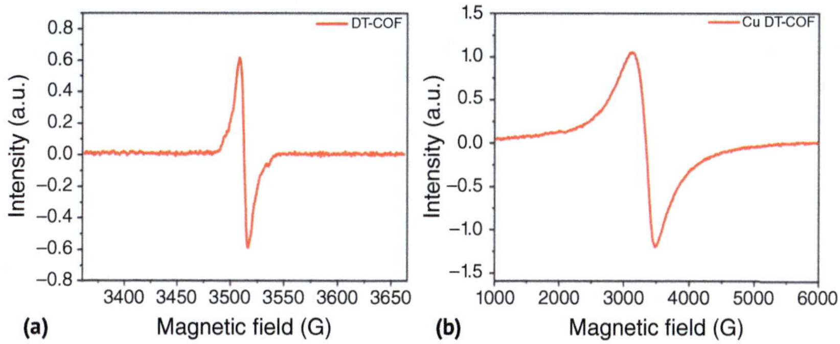

Figure 9.17 EPR spectra of (a) DT-COF and (b) Cu-DT COF Adapted from [21].

9.4.3 Evaluation of Storage Mechanism and Defect Characteristics in Anode Materials

To gain a deeper understanding of the lithium storage mechanism of Cu-DT COF, EPR measurements were conducted on both DT-COF and Cu-DT COF [21] (Figure 9.17). After coordination with copper ions, the EPR signal of Cu-DT COF exhibited a stronger intensity. This enhancement can be attributed to two main factors: first, the increase in radicals generated by the conversion of the C=O double bond in the keto-enamine structure of DT-COF to the C-O single bond in the enol-imine structure of Cu-DT COF; second, the introduction of copper ions themselves, which also contributes to the strengthening of the EPR signal. Furthermore, electrostatic potential (ESP) images of DT-COF and Cu-DT COF were analyzed to identify potential Li^+ storage sites. The results showed that electron-rich regions were primarily concentrated around N and O atoms, indicating these atoms serve as the primary Li^+ storage sites. Notably, the ESP equivalent minima of Cu-DT COF were significantly lower than those of DT-COF, suggesting that the Li^+ storage performance around N and O atoms was further improved after copper ion coordination. This finding was well-consistent with the electrochemical data, providing crucial insights into the enhanced lithium storage capability of Cu-DT COF.

To verify the existence and content difference of oxygen vacancies, EPR spectroscopy was conducted on black TiO_2 yolk-shell spheres (B-YST) prepared by calcination in a nitrogen atmosphere and white TiO_2 yolk-shell spheres (W-YST) calcined in air (shown in Figure 9.18) [22]. The results showed that W-YST exhibited no obvious EPR signal response, while B-YST displayed a strong EPR signal at a g-value of 2.002—a characteristic signal that directly confirmed the presence of abundant oxygen vacancies in B-YST. Combined with the XPS test results, where the proportion of oxygen vacancies in B-YST (approximately 33.4%) was higher than that in W-YST (22.5%), it was further confirmed that the oxygen-deficient

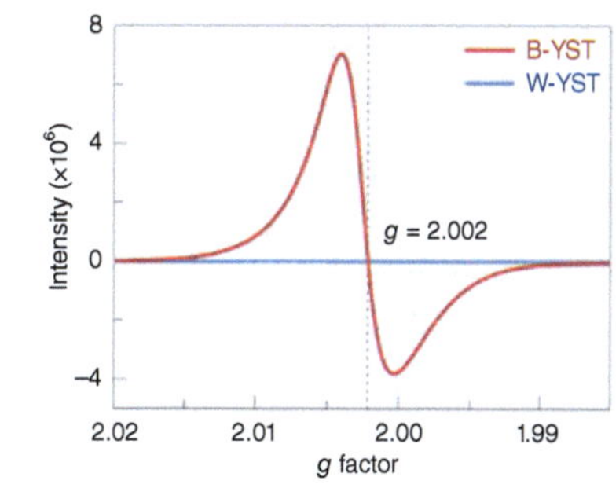

Figure 9.18 EPR absorbance spectra of B-YST and W-YST [22] / with permission of John Wiley & Sons.

environment created by the nitrogen atmosphere successfully introduced a large number of oxygen vacancies in B-YST. These oxygen vacancies played a crucial role in improving the electronic conductivity of the material, reducing charge transfer resistance, and enhancing lithium-ion storage performance.

9.5 Conclusion and Prospective

EPR spectroscopy has emerged as a powerful and indispensable tool for probing the electronic and structural dynamics of energy storage materials under operational conditions. As highlighted throughout this review, EPR's unique sensitivity to unpaired electrons enables direct, quantitative insights into redox processes involving both anionic (e.g., lattice oxygen) and cationic (e.g., transition metals such as Mn and V) species, thereby offering critical evidence for mechanisms like reversible oxygen redox—a cornerstone of high-capacity cathode design. Furthermore, EPR effectively identifies and characterizes intrinsic point defects, such as oxygen vacancies, elucidating their role in enhancing electronic conductivity and ion storage kinetics.

Perhaps most significantly, recent advances in operando EPR combined with EPR imaging (EPRI) have revolutionized the study of lithium metal anodes. This approach overcomes long-standing limitations of conventional characterization techniques by enabling real-time quantification of metallic lithium deposition, spatially resolved visualization of dendrite morphology and distribution, and kinetic analysis of lithium stripping and reintercalation processes. These capabilities have not only clarified the distinct growth modes and failure pathways associated with different fast charging protocols but have also directly informed practical strategies—such as post-charge relaxation—to mitigate dead lithium formation and improve cycling stability.

In summary, EPR has evolved from a niche spectroscopic method into a central analytical platform that bridges atomic-scale electronic structure, defect chemistry, and macroscopic electrochemical performance. Its continued integration with complementary techniques and its extension to emerging battery chemistries (e.g., Na^+, solid-state systems) promise to further accelerate the rational design of next-generation batteries with enhanced energy density, safety, and longevity.

References

1 B. Wang, L.W. Le Fevre, A. Brookfield, E.J.L. McInnes, R.A.W. Dryfe, Resolution of lithium deposition versus intercalation of graphite anodes in lithium ion batteries: An *in situ* electron paramagnetic resonance study, Angewandte Chemie International Edition 60(40) (2021) 21860–21867. https://doi.org/10.1002/anie.202106178.

2 Y. Bai, Z. Wang, N. Qin, D. Ma, W. Fu, Z. Lu, X. Pan, Two-step redox in polyimide: Witness by in situ electron paramagnetic resonance in lithium-ion batteries, Angewandte Chemie International Edition 62(18) (2023). https://doi.org/10.1002/anie.202303162.

3 C. Szczuka, P. Jakes, R.A. Eichel, J. Granwehr, Identification of LiPF6 decomposition products in li-ion batteries with endogenous vanadyl sensors using pulse electron paramagnetic resonance and density functional theory, Advanced Energy and Sustainability Research 2(12) (2021) 2100121. https://doi.org/10.1002/aesr.202100121.

4 C. Wang, Y. Xie, Y. Huang, S. Zhou, H. Xie, H. Jin, H. Ji, Li3PO4-enriched SEI on graphite anode boosts Li+ De-solvation enabling fast-charging and low-temperature lithium-ion batteries, Angewandte Chemie International Edition 63(21) (2024). https://doi.org/10.1002/anie.202402301.

5 J. Wang, T. He, P. Li, L. Meng, Y. Ding, H. Tian, H. Bai, X. Lou, H. Zhang, An *in situ* thermal cross-linking binder for silicon-based lithium ion battery, Journal of Colloid and Interface Science 649 (2023) 795–803. https://doi.org/10.1016/j.jcis.2023.06.158.

6 F. Schneider, K. Mobius, M. Plato, The use of electron paramagnetic resonance in organic chemistry, Angewandte Chemie International Edition in English 4(10) (2003) 856–867. https://doi.org/10.1002/anie.196508561.

7 M. Ju, J. Kim, J. Shin, EPR spectroscopy: A versatile tool for exploring transition metal complexes in organometallic and bioinorganic chemistry, Bulletin of the Korean Chemical Society 45(10) (2024) 835–862. https://doi.org/10.1002/bkcs.12899.

8 Y. Zhang, Y. Wan, D.Y. Kong, Y.J. Xu, J.H. Pan, Q. Li, B. Wang, M.B. Wu, H. Hu, Electron paramagnetic resonance spectroscopy in carbon materials for energy storage: A review, Small Methods 9(11) (2025) e01362. https://doi.org/10.1002/smtd.202501362.

9 T. Moragues, M. Agrachev, S. Mitchell, G. Jeschke, J. Pérez-Ramírez, A.J. deMello, Droplet-based EPR spectroscopy for real-time monitoring of liquid-phase catalytic reactions, Small Methods 9(7) (2025) 2401771. https://doi.org/10.1002/smtd.202401771.

10 F. Gerson, W. Huber, Physical fundamentals of electron spin resonances, Electron Spin Resonance Spectroscopy of Organic Radicals 2003, pp. 3–9. https://doi.org/10.1002/3527601627.ch1.

11 J. Stankowski, S. Waplak, V.A. Yurin, Electric field modulation of the EPR spectrum of Cr3+ in ferroelectric TGS monocrystals, Physica Status Solidi (a) 22(1) (1974) K41–K43. https://doi.org/10.1002/pssa.2210220153.

12 L.E. Switala, W.J. Ryan, M. Hoffman, S. Javier, B.E. Black, J.P. Hornak, A wide-line low frequency electron paramagnetic resonance spectrometer, Concepts in Magnetic Resonance Part B: Magnetic Resonance Engineering 47B(2) (2017). https://doi.org/10.1002/cmr.b.21355.

13 S. Stoll, Pulse EPR, eMagRes 2017, pp. 23–38. https://doi.org/10.1002/9780470034590.emrstm1510.

14 G.R. Eaton, S.S. Eaton, Rapid-scan electron paramagnetic resonance, eMagRes 2016, pp. 1529–1542. https://doi.org/10.1002/9780470034590.emrstm1522.
15 S. Kang, X. Lou, M. Shen, F. Geng, B. Hu, Visualizing lithium deposition and identifying two types of dendrites in extreme-fast-charging full cells across the entire lifespan by operando EPR and EPR imaging, The Journal of Physical Chemistry Letters 15(50) (2024) 12248–12256. https://doi.org/10.1021/acs.jpclett.4c03022.
16 H. Liu, Y. Chen, P.-H. Chien, G. Amouzandeh, D. Hou, E. Truong, I.P. Oyekunle, J. Bhagu, S.W. Holder, H. Xiong, P.L. Gor'kov, J.T. Rosenberg, S.C. Grant, Y.-Y. Hu, Dendrite formation in solid-state batteries arising from lithium plating and electrolyte reduction, Nature Materials 24(4) (2025) 581–588. https://doi.org/10.1038/s41563-024-02094-6.
17 Y. Zuo, H. Shang, J. Hao, J. Song, F. Ning, K. Zhang, L. He, D. Xia, Regulating the potential of anion redox to reduce the voltage hysteresis of li-rich cathode materials, Journal of the American Chemical Society 145(9) (2023) 5174–5182. https://doi.org/10.1021/jacs.2c11640.
18 G. Liu, Z. Wang, H. Yuan, C. Yan, R. Hao, F. Zhang, W. Luo, H. Wang, Y. Cao, S. Gu, C. Zeng, Y. Li, Z. Wang, N. Qin, G. Luo, Z. Lu, Deciphering electrolyte dominated Na+ storage mechanisms in hard carbon anodes for sodium-ion batteries, Advanced Science 10(36) (2023) 2305414. https://doi.org/10.1002/advs.202305414.
19 C. Li, C. Zhao, B. Hu, W. Tong, M. Shen, B. Hu, Unraveling the critical role of Ti substitution in P2-NaxLiyMn1–yO2 cathodes for highly reversible oxygen redox chemistry, Chemistry of Materials 32(3) (2020) 1054–1063. https://doi.org/10.1021/acs.chemmater.9b03765.
20 C. Zhao, Q. Yang, F. Geng, C. Li, N. Zhang, J. Ma, W. Tong, B. Hu, Restraining oxygen loss and boosting reversible oxygen redox in a P2-type oxide cathode by trace anion substitution, ACS Applied Materials & Interfaces 13(1) (2020) 360–369. https://doi.org/10.1021/acsami.0c16236.
21 D. Luo, H. Zhao, F. Liu, H. Xu, X. Dong, B. Ding, H. Dou, X. Zhang, Outstanding lithium storage performance of a copper-coordinated metal-covalent organic framework as anode material for lithium-ion batteries, Energy & Environmental Materials 7(5) (2024). https://doi.org/10.1002/eem2.12732.
22 Z. Li, Y. Peng, C. Liu, X. Zhang, X. Li, Y. Huang, Y. Ren, D. Ji, G. Cao, Oxygen-deficient TiO2 Yolk-shell spheres for enhanced lithium storage properties, Energy & Environmental Materials 5(1) (2021) 238–244. https://doi.org/10.1002/eem2.12156.

10

Raman Spectroscopic Characterization of Structural and Interfacial Evolution in Battery Materials

10.1 Fundamental Principles and Theory

While the practical utility of Raman spectroscopy in battery materials and industrial applications—driven by its nondestructive nature and minimal sample preparation—is well-established, a profound grasp of its underlying physical principles is essential for fully exploiting its analytical capabilities. The transition of this technique from a complex laboratory curiosity to a robust, mainstream tool hinges on navigating the inherent physical challenges of light-matter interaction, specifically the vast disparity in intensity between elastic Rayleigh scattering and the faint, inelastic Raman signals. Therefore, elucidating the classical and quantum mechanical frameworks governing these scattering phenomena is critical for understanding both the historical constraints of the method and the specific instrumental breakthroughs, such as the advent of lasers and charge-coupled device (CCD) detectors, that have unlocked its high-resolution potential.

10.1.1 The Development of Classical Theory and Quantum Mechanics

Raman spectroscopy is a nondestructive, rapid, and high-resolution analytical technique that provides information on molecular vibrations and rotations by analyzing the inelastically scattered light resulting from the interaction between light and matter. Although prior to the 1980s, Raman spectroscopy was confined to laboratories and utilized by high-level experts, with tests being time-consuming and often unsuccessful, breakthrough advancements in instrumentation over recent decades—particularly the development of lasers and CCD detectors—have transformed it into a widely applicable analytical technique. In the field of battery materials research, the advantages of Raman spectroscopy lie in its typically non-contact nature, requiring no sample modification, and its ability to collect Raman information through transparent containers or windows made of glass, sapphire, or plastic. In industrial production, the lack of a need for prior sample preparation is a primary reason many users choose Raman spectroscopy over other established analytical techniques (Figure 10.1).

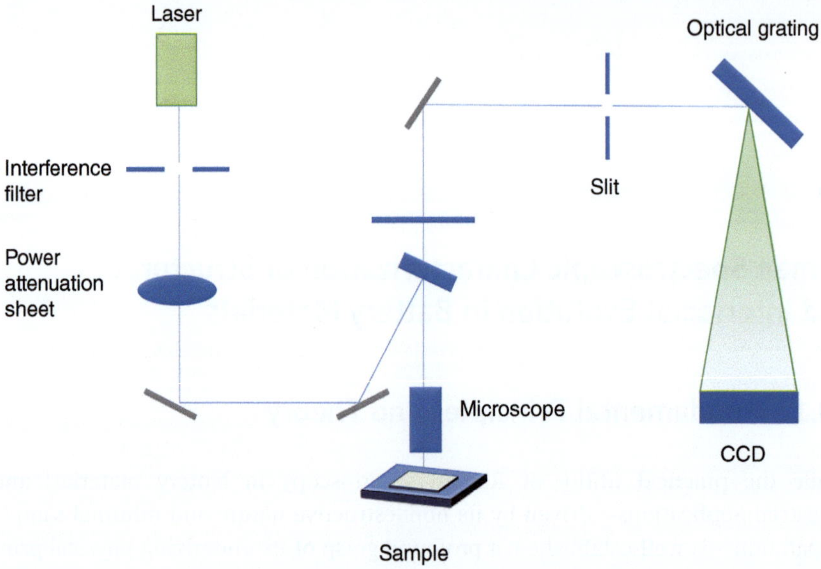

Figure 10.1 Schematic diagram of the Raman testing device.

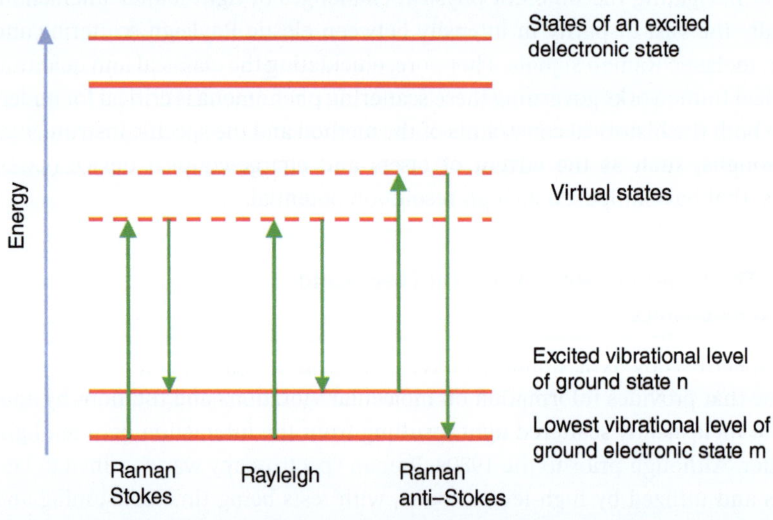

Figure 10.2 Diagram of the Rayleigh and Raman scattering processes.

When a beam of light is incident on a sample, it can be transmitted, absorbed, or scattered. Most of the scattered light has the same wavelength as the incident light; this is known as elastic scattering or Rayleigh scattering. Rayleigh scattering is intense, typically 10^3 to 10^5 times stronger than Raman scattering. Only a small fraction of the scattered light experiences a shift in wavelength due to interactions with molecular vibrations and rotations within the sample. This phenomenon is known as inelastic scattering, or Raman scattering (Figure 10.2).

The classical theory of Raman scattering is described based on the change in molecular polarizability (denoted as α). Light, as an oscillating electromagnetic field, can induce oscillations in the electron cloud of a molecule, leading to the emission of light, i.e., scattering. Atoms are held together by chemical bonds, and changes in their positions alter the polarizability of the electron cloud. Consequently, molecular vibrations lead to a periodic modulation of the scattered light intensity. The intensity of the scattered light is the sum of three different frequency components: the original light frequency, the Stokes frequency (original frequency minus the vibrational frequency), and the anti-Stokes frequency (original frequency plus the vibrational frequency).

However, the classical theory incorrectly predicts that the relative intensities of Stokes and anti-Stokes Raman scattering are equal. A detailed physical understanding of the Raman effect requires a quantum mechanical explanation using energy level diagrams. At room temperature, most molecules reside in the lowest vibrational energy state (ground state, *m*). Incident photons excite the molecule from the ground state to a higher energy virtual state. A virtual state is an unstable energy level whose energy depends on the frequency of the incident light.

If the molecule returns from the virtual state to a higher vibrational level (*n*), the emitted photon has less energy and a longer wavelength than the incident photon. This is known as Stokes Raman scattering. If a molecule is already in an excited vibrational state (*n*) and absorbs a photon to reach a virtual state, subsequently returning to the ground state (*m*), the emitted photon has more energy and a shorter wavelength than the incident photon. This is called anti-Stokes Raman scattering. According to the Boltzmann distribution, the population of molecules in the ground state is far greater than that in excited states, so the intensity of Stokes Raman scattering is always greater than that of anti-Stokes Raman scattering.

The Raman scattering intensity is directly proportional to the incident light intensity and to the fourth power of the scattered light frequency $(\nu_0 - \nu)^4$. This relationship indicates that increasing the intensity of the incident light or using incident light of a higher frequency (shorter wavelength) can significantly enhance the Raman scattering intensity. A Raman spectrum is typically plotted as Raman shift, with units of wavenumbers (cm^{-1}), which represents the energy difference between the scattered light and the incident light.

10.1.2 Vibration Modes, Selection Rules, and Polarizability Tensor

The number and types of molecular vibrations are determined by the number of atoms. For a molecule with N atoms, the number of vibrational degrees of freedom (i.e., the total number of vibrations that can appear in a spectrum) is 3N − 6 for nonlinear molecules and 3N − 5 for linear molecules. For example, a nonlinear triatomic molecule has three vibrational modes: symmetric stretching, bending or deformation, and asymmetric stretching. Molecular vibrations can be visualized using a "ball-and-spring" model, where stronger bonds correspond to higher frequencies and heavier atoms to lower frequencies.

A vibrational mode is Raman-active if it induces a change in the molecule's polarizability during the vibration. The selection rules are as follows: A vibration is infrared-active if it causes a change in the dipole moment. A vibration is Raman-active if it causes a change in the molecule's polarizability.

For instance, the stretching vibration of oxygen (O_2) changes its polarizability but not its dipole moment, so it exhibits a peak in the Raman spectrum but not in the infrared spectrum. In contrast, for carbon dioxide (CO_2), the symmetric stretch causes a large change in polarizability, resulting in strong Raman scattering. However, its deformation (bending) mode causes a change in the dipole moment with almost no change in polarizability, thus showing strong infrared absorption but very weak or no Raman scattering. In the context of group theory, a vibrational mode is Raman-active if and only if the irreducible representation of that vibration has the same symmetry as one of the components of the polarizability tensor (which transform as quadratic functions like x^2, y^2, z^2, xy, xz, yz).

The polarization of the scattered Raman light depends on the direction of the change in the electron cloud's polarizability induced by the vibration. For single-crystal samples, the Raman spectrum is dependent on the orientation of the crystal axes relative to the polarization of the incident and detected light. This relationship is crucial for analyzing crystal and molecular structures. Mathematically, the Raman-active vibrational modes of a crystal are described by the Raman tensor. The change in polarization is described by a physical quantity called the depolarization ratio (ρ).

10.1.3 Enhanced Raman Effect (Surface-Enhanced Raman Spectroscopy (SERS)/Resonance Raman Spectroscopy (RRS))

The Raman scattering effect is inherently very weak; only one in every 10^6 to 10^8 photons may undergo Raman scattering. This is a major intrinsic drawback of conventional Raman spectroscopy. Enhanced Raman spectroscopy techniques can effectively overcome this limitation.

SERS is one of the most prominent enhancement techniques. When molecules are in proximity to nanoparticles or rough surfaces of certain metals (e.g., silver (Ag), gold (Au), copper (Cu)), the Raman scattering signal is dramatically enhanced. This effect can increase the Raman cross-section by up to 10^7 times. When SERS is combined with **RRS**, the enhancement can be as high as 10^{14} to 10^{15}, making single-molecule Raman spectroscopy possible. To achieve significant enhancement, the metal surface must have an appropriate degree of roughness.

The enhancement mechanisms of SERS are broadly categorized into two types: the electromagnetic mechanism and the chemical mechanism. The electromagnetic mechanism arises from the interaction of the excitation light with the metal surface, leading to an amplified electric field experienced by the analyte. The chemical mechanism results from the overlap of wavefunctions between the analyte and the metal.

The principle of **RRS** is that when the wavelength of the excitation light matches an electronic transition of the molecule, resonance occurs, increasing the Raman scattering intensity by a factor of up to 10^6 compared to conventional scattering.

10.1.4 Structural Information of Signal Strength and Frequency Shift

The height, width, area, position (frequency shift), and shape of Raman peaks all contain valuable information about the sample, such as its composition, molecular orientation, crystalline form, and stress/strain state.

Although symmetry theory cannot predict the intensity of Raman-active vibrations, estimations can be made based on empirical rules. For example, vibrations of polar chemical bonds tend to produce weak Raman intensities; stretching vibrations usually scatter more strongly than bending vibrations; symmetric vibrations generally produce stronger Raman scattering than antisymmetric vibrations; and crystalline materials typically exhibit stronger and more numerous Raman peaks than amorphous materials.

In spectral analysis, strain (or stress) can induce a shift in Raman peaks, enabling Raman spectroscopy to be used for quantitative strain measurement. Furthermore, the full width at half maximum of Raman peaks often increases with structural disorder in carbon materials, a property used to characterize the degree of structural order in battery materials like carbon fibers or graphene.

10.2 Instrumentation and Analysis

The practical realization of Raman spectroscopy as a robust analytical tool is contingent upon the sophistication of its instrumentation, which must address the stringent requirement of isolating weak inelastic scattering signals from intense Rayleigh background noise. The evolution of this technique has been driven by critical advancements in hardware components—specifically, the transition to high-brightness laser excitation sources, the adoption of multichannel CCD detectors, and the development of specialized optical configurations such as Fourier-Transform and dispersive systems. These instrumental innovations have not only optimized signal acquisition speed and fluorescence suppression but have also enabled a diverse range of operational modes. From high-resolution confocal microscopy and spatial chemical imaging to remote fiber-optic monitoring, the following section delineates the essential equipment and analytical methodologies that define modern Raman spectroscopy.

10.2.1 Typical Equipment and Light Sources

A Raman spectrometer is designed to achieve two primary goals: first, to block Rayleigh scattering and other stray light from entering the detector, and second, to disperse the Raman scattered light into its constituent frequencies (wavelengths) and direct them onto the detector. To meet these objectives, the instrument must be equipped with an excitation source, an optical system, a spectrometer (or spectrograph), a detector, and a computer processing system.

The laser, as an excitation source, has been a key factor in the rapid development of Raman spectroscopy. Its main advantages are extremely high brightness, strong directionality, and excellent monochromaticity. A wide variety of lasers are available, including gas lasers (e.g., argon-ion lasers), solid-state lasers (e.g., Nd:YAG lasers), and semiconductor lasers. Selecting the appropriate wavelength is crucial to avoid fluorescence interference.

Fourier transform Raman spectrometers, based on the Michelson interferometer, represent one of the most significant advancements in modern Raman spectroscopy. They typically use an Nd:YAG laser to provide near-infrared excitation (e.g., 1064 nm), which can significantly reduce or eliminate the fluorescence background from most samples.

Dispersive Raman spectrometers separate light of different wavelengths and use a multielement detector placed in the focal plane to measure the intensity of different wavelength beams simultaneously. This multichannel capability significantly reduces detection time.

The introduction of **CCD detectors** is another major breakthrough in modern Raman spectroscopy. CCD detectors offer the multichannel advantage, transforming Raman spectroscopy into a rapid testing technique capable of acquiring a complete Raman spectrum in seconds or even less. This has vastly improved efficiency, replacing the traditional single-channel photomultiplier tube (PMT).

10.2.2 Typical Working Mode

Micro-Raman Spectroscopy combines a Raman spectrometer with an optical microscope, allowing the study of very small sample areas by focusing the laser beam to a small spot (a few micrometers in diameter).

Confocal Micro-Raman Spectroscopy enhances axial (depth) resolution by incorporating a confocal pinhole. This technique can typically achieve an axial resolution of 1 to 2 μm. With confocal microscopy, it is possible to perform point-by-point analysis along the axial direction of a transparent sample, such as measuring the stress/strain distribution along a fiber axis.

Raman Imaging (or Mapping) involves creating an image of a sample where each pixel contains spectral information. By analyzing the intensity of characteristic peaks related to composition or structure, two-dimensional or three-dimensional spatial distribution maps can be generated. Imaging modes include point-mapping (the most time-consuming), line-mapping, and area-mapping.

The introduction of **Fiber-Optic Probe Raman Spectroscopy** has greatly expanded the application range of Raman analysis. Fiber probes simplify the alignment between the sample and the spectrometer, enabling measurements on samples located far from the instrument. This is highly significant for online monitoring on industrial production lines or for analysis in hazardous environments.

Nonimaging probes have very high collection efficiency but were historically limited by "contamination" from the Raman scattering and fluorescence background of the fiber itself. Modern probes often incorporate filters to solve this issue.

Focusing probes have confocal properties that reject light originating from outside the focal point, making them particularly suitable for analyzing the contents of thick-walled or transparent containers. Focusing probes are highly effective at rejecting the Raman and fluorescence background from silica optical fibers.

10.2.3 Qualitative Analysis

Raman spectroscopy is a powerful tool for qualitative analysis. A Raman spectrum often contains many well-defined, resolvable peaks, making it possible in principle to distinguish between a wide variety of samples.

The reference spectrum method involves comparing the spectrum measured from a sample with a database of known spectra to identify the substance. The spectral subtraction method is used to identify components in a multicomponent mixture. By subtracting the spectra of known components, the remaining spectrum corresponds to the solute.

The combined use of Raman and Infrared (IR) spectroscopy is an important strategy for qualitative analysis. IR spectroscopy is sensitive to changes in dipole moment, whereas Raman spectroscopy is sensitive to changes in polarizability, making the two techniques physically complementary. Their joint application significantly enhances the reliability of identification. For example, in the CO_2 molecule, the symmetric stretching vibration yields strong Raman scattering but very weak or no IR absorption, while the deformation mode shows strong IR absorption but very weak or no Raman scattering.

10.2.4 Quantitative Analysis

The basis for quantitative analysis using Raman spectroscopy is the linear proportional relationship between the measured Raman peak intensity of an analyte and its concentration. A plot of the analyte's Raman peak area (integrated intensity) versus its concentration is known as a **calibration curve**.

To correct for the influence of factors such as sample transparency and laser power on peak area or height, nearly all quantitative Raman analysis methods employ some form of **internal standard** when constructing the calibration curve.

In cases where the analyte's Raman peak overlaps with other peaks, peak height measurement may be preferred, as it is less sensitive to partial peak overlap than peak area measurement. For multicomponent materials where unique characteristic peaks are difficult to find or where peaks severely overlap, advanced **chemometric methods**, such as partial least squares, can be applied to simultaneously consider the area, shape, or frequency shift of multiple peaks.

10.2.5 Enhancing Analytical Capabilities Through Combined Techniques

Integrating different characterization methods into a single instrument to obtain multiple types of information simultaneously is a major advancement in Raman spectroscopy. This approach overcomes the positioning difficulties and time consumption associated with using separate instruments sequentially.

The combination of **Raman spectroscopy with atomic force microscopy (AFM)** allows for the simultaneous observation of sample morphology and the detection of chemical composition, molecular structure, aggregate state, and micro-regional strain in a given area, while also providing high-resolution data on morphology, local elasticity, and electrostatic forces. This hyphenated technique promises to elevate research to a new level, especially for nanomaterial design and thin-film studies.

The combination of **Raman spectroscopy with scanning electron microscopes (SEM)** utilizes high-resolution secondary electron images from the SEM to display the sample's morphological structure and identify micro-regions of interest. Raman spectroscopy can then be used to rapidly and definitively analyze the chemical composition and physical structure (e.g., crystallinity, stress/strain) of that specific region. SEMs are often equipped with an Energy-Dispersive X-ray Spectroscopy (EDX) accessory for elemental analysis of elements with atomic numbers greater than 5.

10.3 Advanced Techniques and In-situ Applications

The evolution of Raman spectroscopy from a static characterization method to a dynamic, in-situ diagnostic tool represents a paradigm shift in materials science, particularly within the domain of energy storage and nanotechnology. While standard Raman analysis provides essential molecular fingerprints, the integration of advanced environmental controls and near-field optics enables the real-time monitoring of complex physicochemical processes—ranging from electrochemical phase transitions and Solid Electrolyte Interphase (SEI) formation to stress distribution under mechanical load. By overcoming traditional limitations such as the optical diffraction limit through Tip-Enhanced Raman Spectroscopy (TERS) and synergizing with complementary techniques like AFM and SEM, modern Raman platforms offer a multidimensional perspective that allows for the simultaneous correlation of chemical, structural, and topographical properties at the nanometer scale.

10.3.1 Electrochemical In-situ Raman Testing Technique

Raman spectroscopy offers unique advantages for studying electrochemical systems, making it an ideal tool for battery materials research.

First, water exhibits a very weak scattering signal in Raman spectroscopy, causing minimal interference with the analyte's spectrum. This makes Raman spectroscopy an ideal tool for analyzing aqueous electrolytes, in stark contrast to infrared spectroscopy, where water strongly absorbs mid-infrared light.

Second, the noncontact and nondestructive nature of Raman analysis allows for the collection of information through transparent containers (e.g., glass, quartz) or windows, making it suitable for in vivo samples or real-time diagnostics on production lines.

In the construction of in-situ test environments, the use of fiber-optic probes enables the laser beam to be focused inside an electrochemical cell located far from the

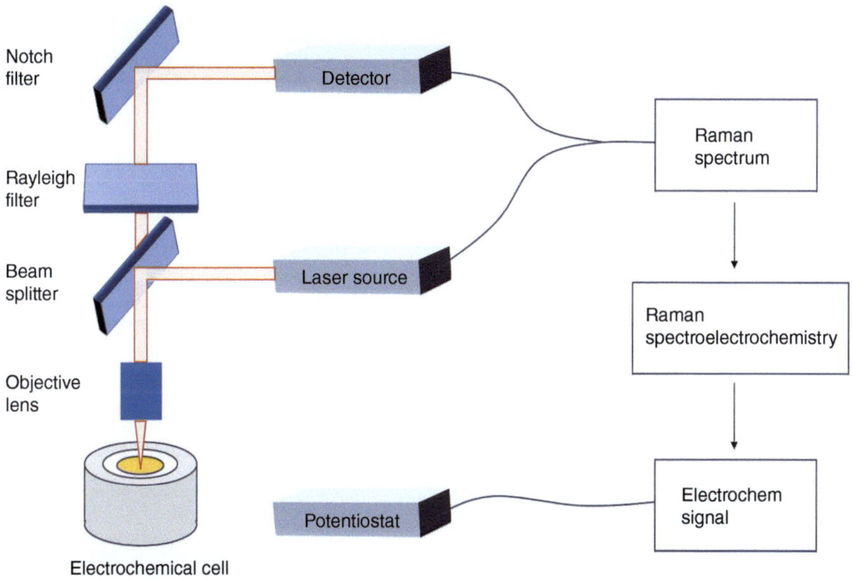

Figure 10.3 Schematic diagram of in-situ Raman testing process.

spectrometer and to collect the resulting signal. With confocal micro-Raman spectroscopy, point-by-point analysis along the axial (depth) direction of electrode materials, electrolytes, or interfaces can be achieved. This is crucial for investigating complex electrode/electrolyte interfaces, such as the SEI layer in batteries (Figure 10.3).

Examples of Raman spectroscopy applications in battery materials include: Monitoring Phase Transitions in Electrode Materials: Raman spectroscopy is highly sensitive to changes in a material's crystal and molecular structure. It can monitor the evolution of structural order/disorder in electrode materials during lithium insertion/extraction. For example, the Raman spectral behavior of carbon materials corresponds to changes in the order/disorder of the graphite lattice, which can be quantified by monitoring the D- and G-bands. Studying the SEI Layer: Enhanced techniques like SERS and TERS can be used to analyze weak molecular signals on electrode surfaces and at interfaces. Combined with nanoscale spatial resolution, these methods can overcome the optical diffraction limit to perform high-resolution analysis of the local chemical composition and microstructure of the SEI film (Figures 10.4 and 10.5).

10.3.2 In-situ Technologies for Thermal and Strain Environments

Raman spectroscopy is a powerful tool for studying the micromechanics of materials. During charging and discharging, battery electrode materials undergo volume changes that can generate significant mechanical stress/strain, affecting their cycling stability.

For many materials, including graphene, carbon nanotubes, and silicon, the frequency shift of one or more characteristic peaks is sensitive to strain, often exhibiting a simple linear relationship. This physical phenomenon allows Raman spectroscopy

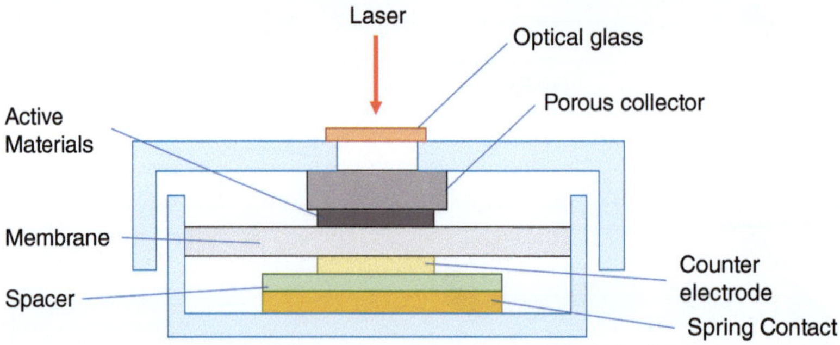

Figure 10.4 Schematic diagram of in-situ Raman testing devices.

Figure 10.5 Optical image of in-situ Raman testing devices.

to be used for the quantitative measurement of strain distribution and interfacial stress transfer in composite materials or electrodes under external loads.

For example, in silicon carbide fibers, the two peaks from free carbon shift to lower frequencies with increasing tensile strain, while the peak from SiC grains shifts to higher frequencies; both show a good linear relationship with fiber strain. The G- and 2D-bands of graphene are also highly sensitive to strain.

Furthermore, Raman and fluorescence spectroscopy have made outstanding contributions to the measurement of residual stress/strain in materials. For ceramic or metal matrix composites used in high-temperature environments, the thermal residual stress caused by mismatched coefficients of thermal expansion can be mapped using Raman or fluorescence spectroscopy.

Temperature and pressure also affect the position and intensity of Raman peaks. Increasing the temperature populates higher energy levels, which can enhance the intensity of anti-Stokes scattering. Additionally, higher temperatures or pressures increase molecular collision rates, leading to the broadening of Raman peaks. To avoid sample damage from laser-induced heating, techniques such as using a sample rotation cell or continuous sample movement can be employed to dissipate the heat.

10.3.3 Near-field Optics and Nano-enhanced In-situ Technology

Far-field optics are constrained by the diffraction limit (approximately half the wavelength of light), which restricts the minimum resolvable feature size. Near-field optics, which studies the optical field within one wavelength of a source or object, can in principle break this diffraction limit.

Tip-Enhanced Raman Spectroscopy (TERS) combines AFM with near-field optics. The principle of TERS is that when a very sharp tip is brought close to a sample surface, the Raman effect in the near-field region around the tip is greatly enhanced. The tip is typically made from or coated with a material that supports surface plasmons, such as an isolated noble metal nanoparticle. Experimental studies have demonstrated that TERS can achieve spatial resolution on the nanometer scale. This high resolution makes TERS a powerful tool for investigating micro-domains and internal structures of nanomaterials, such as carbon nanotubes.

TERS can be considered a derivative of SERS, with its enhancement mechanism related to the amplified electric field caused by surface plasmons on the metal tip.

10.3.4 Multifunctional Combined System

Integrating Raman spectroscopy with other analytical techniques into a single instrument is a key strategy for improving analytical efficiency and obtaining more comprehensive information.

The combination of Raman spectroscopy and AFM, especially when incorporating TERS mode, is a significant advanced hyphenated technique. In an AFM-Raman system, the AFM probe can serve as the TERS tip, dramatically increasing the Raman intensity and enabling simultaneous nanoscale analysis of a sample's (e.g., graphene) topography and chemical properties.

The combination of Raman and Infrared (IR) spectroscopy can provide complementary Raman and IR spectral information from the same micro-region. Since IR spectroscopy is strongly affected by water absorption, this combination is particularly useful for more reliable identification of aqueous materials (such as biological samples).

Raman spectrometers can also be coupled with other instruments, such as SEM, Transmission Electron Microscopes, Kelvin Probe Force Microscopes (KPFM), and

Electrostatic Force Microscopes (EFM). For instance, in an AFM-Raman system, it is possible to simultaneously acquire a graphene sample's topography image, G-band intensity map, surface potential (KPFM image), and local elastic properties (from Peak Force QNM or force-distance curves).

10.4 Application of Raman Spectroscopy in the Study of Battery Materials

Raman spectroscopy, as a nondestructive analytical technique with high spatial resolution (typically submicron level), operates on the principle of inelastic scattering phenomena during light-matter interactions. Incident photons exchange energy with molecular vibrations and lattice vibrations (phonons), generating characteristic scattered spectra with frequency shifts [1]. Raman shifts directly correspond to vibrational modes of chemical bonds or lattice structures within materials, providing comprehensive information about chemical bonds, crystal structures, molecular configurations, stress states, and local environments. In complex electrochemical systems like secondary batteries (e.g., lithium-ion, sodium-ion, and solid-state batteries), materials undergo dynamic processes during charge-discharge cycles-including reversible/irreversible structural transformations, phase transitions, interfacial side reactions, and component migration [2, 3].

These microscopic changes directly determine macroscopic electrochemical performance metrics such as capacity, rate performance, cycle life, and safety. Traditional bulk characterization techniques often struggle to capture these localized, dynamic, and sometimes weak signals. Leveraging its high sensitivity to chemical bond/lattice vibrations, excellent spatial resolution, and compatibility with diverse testing environments (e.g., in-situ electrochemical cells, high/low temperatures, and atmosphere control), Raman spectroscopy has become an indispensable "probe" for revealing multiscale structural and chemical evolution in battery materials during synthesis, storage, and electrochemical cycling processes [4, 5].

10.4.1 Crystal Structure Analysis and Phase Transition Tracking of Active Materials

Raman spectroscopy, with its highly specific response to chemical bonds and lattice vibrations, has become a key technique for analyzing and tracking the fine crystal structures, local phase compositions, and phase evolution behaviors of battery electrode materials during lithium/sodium intercalation/deintercalation processes. This is due to the fact that specific crystal symmetry, atomic coordination environments, and bonding states in materials excite unique photonic vibration modes, which manifest as distinctive Raman "fingerprint" spectra.

1. **Laminated oxide cathode materials**
   The Raman spectrum's exceptional sensitivity to local atomic arrangements and chemical bond vibrations makes it an indispensable tool for analyzing the

fine structure of layered transition metal oxides ($LiMO_2$, M=Ni, Co, Mn, Al, and their combinations), identifying defects, tracking dynamic phase transitions, and revealing decay mechanisms [6, 7]. The Raman-active vibration modes in these materials primarily originate from the vibrations of the MO_6 octahedral unit, with the most distinctive being a pair of strong peaks in the ~480–500/cm and ~580–600/cm ranges, typically attributed to E_g and A_{1g} vibration modes respectively. The E_g mode is mainly associated with the symmetric bending vibrations of transition metal-oxygen (M-O) bonds within the ab plane, showing sensitivity to the layered order of crystal structures. The A_{1g} mode corresponds to the asymmetric stretching vibrations of M-O bonds along the c-axis, demonstrating extreme sensitivity to interlayer stacking and bond length variations along the c-axis. Consequently, the precise positions of these two characteristic peaks, their relative intensity ratio (I_{A1g}/I_{Eg}), half-width at half maximum, and symmetry collectively form a "fingerprint" set of information for evaluating the local structural state of materials.

In nickel-rich NCM (e.g., NCM811) or NCA materials, the similar ionic radii of Ni^{2+} (0.69 Å) and Li^+ (0.76 Å) during synthesis or cycling lead to Li/Ni mixing, where some Ni^{2+} ions migrate to the Li layer. This phenomenon disrupts the ideal layered order and introduces local lattice distortion. Raman spectroscopy provides clear evidence: as mixing increases, the A_{1g} peak broadens significantly and shifts toward lower wave numbers, while its intensity ratio with the E_g peak changes. For instance, M. Zuba et al. found through systematic studies of NCM materials with varying Ni content that the half-width of the A_{1g} peak strongly correlates with Li/Ni mixing degree determined by X-ray diffraction refinement, offering a key method for rapid, nondestructive evaluation of structural order. This mixing not only hinders lithium-ion diffusion but also serves as a precursor to surface structural degradation.

Layered oxide cathodes, particularly high-nickel materials, undergo irreversible structural transformation from layered to spinel phases (e.g., $LiNi_2O_4$) or even insulating rock salt phases (e.g., NiO) during high-voltage cycling (>4.3 V vs. Li^+/Li). This degradation process involves oxygen loss and migration/reduction of transition metal ions. Raman spectroscopy sensitively detects these phase transitions: the broad peak around 650/cm, typically associated with spinel formation, while the rock salt phase may cause significant weakening or disappearance of characteristic peaks (E_g and A_{1g}) in the original layered structure. In-situ Raman studies by K. Marker et al. clearly demonstrate that during high-voltage cycling of NCM811 cathodes, the A_{1g} peak gradually weakens while a new peak emerges and intensifies at 670/cm, visually confirming the surface transition to spinel structure. This surface restructuring is the primary cause of voltage decay and impedance increase.

During lithium-ion intercalation/deintercalation, the material undergoes a series of reversible phase transitions (e.g., H1→M→H2 phase transitions). These transitions are accompanied by continuous or abrupt changes in lattice parameters, particularly along the c-axis, which affect the vibration frequency of M-O bonds. In-situ/operational Raman spectroscopy enables real-time monitoring of characteristic peak positions as they evolve with potential/capacity. The systematic "hardening" (blue shift) of the A_{1g} peak toward higher wavenumbers with increasing lithium removal depth directly reflects the increased stretching force constant of M-O bonds caused by c-axis contraction. During phase transitions, spectral features such as splitting, disappearance, or appearance of new peaks may be observed. This provides experimental evidence for validating theoretical phase transition pathways and evaluating the reversibility differences between materials like NCM and NCA.

This technical framework is equally applicable to layered oxide cathodes in sodium-ion batteries (e.g., Na_xTMO_2, where TM denotes transition metals). For instance, both P2-type and O3-type structures exhibit distinguishable characteristic peaks in their Raman spectra. Raman spectroscopy can be utilized to investigate phase evolution during sodium-ion insertion/extraction, as well as complex behaviors such as transition metal layer sliding and Na^+/vacancy order-disorder transitions. By analyzing changes in vibration modes, correlations between structural stability and electrochemical performance can be established, such as identifying the impact of Jahn-Teller distortion in manganese-based/copper-based layered oxides.

In summary, Raman spectroscopy has evolved beyond basic phase identification to serve as a multidimensional probe for investigating the microstructural orderliness, surface chemical stability, electrochemical phase transition mechanisms, and degradation mechanisms of layered oxide cathode materials. This technique directly correlates macroscopic electrochemical performance (e.g., capacity fade and voltage drop) with microscopic structural changes at the atomic level (e.g., cation mixing, oxygen loss, and phase reorganization), providing critical scientific insights for the rational design and failure analysis of high-performance, high-stability cathode materials.

2. **Polyanionic cathode similar to Prussian blue analogs**

Polyanionic materials (e.g., phosphates, silicates, sulfates) and Prussian blue analogs (PBAs) serve as crucial cathode systems, characterized by high structural stability and tunable operating voltages. Raman spectroscopy plays an irreplaceable role in studying these materials, as it is highly sensitive to vibrations of strongly covalently bonded anionic groups with high localization. This technique provides direct evidence of crystal structures, local coordination environments, charge ordering, and structural evolution during electrochemical processes.

Polyanionic materials are characterized by their rigid, strongly covalent anionic frameworks, such as $(PO_4)^{3-}$, $(SiO_4)^{4-}$, or $(SO_4)^{2-}$. These groups generate sharp, high-intensity Raman characteristic peaks, whose precise positions are highly sensitive to variations in crystal field effects, bond lengths, and bond angles. For instance, in the sodium-ion battery cathode material $Na_3V_2(PO_4)_3$ (NVP), peaks around 980/cm and 1010/cm correspond to the vibration of V-O bonds and the symmetric stretching of P-O bonds, respectively. During sodium intercalation/deintercalation, Raman spectra clearly demonstrate continuous shifts in these characteristic peaks rather than simple disappearance and reappearance. This strongly indicates that NVP undergoes a highly reversible single-phase solid solution reaction mechanism, rather than the traditional two-phase reaction. The continuous changes correspond to subtle yet measurable distortions in V-O bond lengths and $[PO_4]$ tetrahedral structures caused by sodium intercalation/deintercalation, providing a structural basis for understanding its excellent rate performance and cycling stability [8].

In the lithium battery field, the classic material $LiFePO_4$ exhibits striking similarities in XRD spectra between its insulating $FePO_4$ phase and ion-conductive $LiFePO_4$ phase (both with olivine structures), yet Raman spectroscopy can clearly distinguish them. The characteristic $[PO_4]$ tetrahedral peaks of $FePO_4$ show a significant shift toward higher wavenumbers (approximately 10–20/cm) compared to $LiFePO_4$. Through in-situ Raman mapping technology, researchers can "film" the real-time movement and morphological evolution of the $LiFePO_4$/$FePO_4$ two-phase interface at individual particles or electrode levels during charge-discharge cycles (Figure 10.6) [9]. This method directly validates the classic core-shell model or domino model, while revealing how current density and particle size influence phase transition kinetics. Additionally, Raman spectroscopy detects Li_3PO_4 by-products and disordered carbon structures (D and G peaks) formed on material surfaces due to side reactions, providing comprehensive information for evaluating material performance.

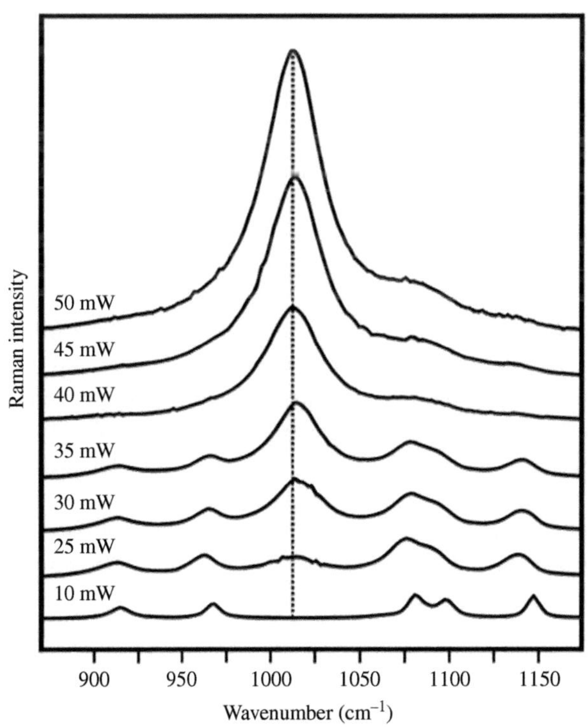

Figure 10.6 Raman spectra of $FePO_4$ collected as a function of power. The wavelength of the excitation laser is 532 nm [9] / with permission of John Wiley & Sons.

PBAs, as promising cathodes for sodium/potassium-ion batteries, exhibit ideal Raman spectroscopic targets in their open framework structure due to the C ≡ N-bridge bonds. The C ≡ N stretching vibration peaks in the ~2100–2200/cm range are highly sensitive to transition metal oxidation states, coordination environments (primarily the types and quantities of cations at interstitial sites), and crystal field strength. For instance, in $Na_xFe[Fe(CN)_6]$, the valence state of Fe shifts between Fe^{2+} and Fe^{3+} as sodium ions (x) intercalate/deintercalate, causing significant near-linear shifts in the C ≡ N peaks (exceeding 30/cm). This shift not only enables precise calibration of the material's state of charge (SOC) but also reveals complex multistep sodium intercalation/deintercalation mechanisms. Studies show that seemingly smooth platforms in some PBAs' charge-discharge curves may indicate two or more discontinuous or continuous phases in the C ≡ N peak through in-situ Raman spectroscopy, revealing intermediate solid solution phases or intricate charge rearrangement processes-details unobservable by conventional XRD. Additionally, Raman spectroscopy effectively detects common $[Fe(CN)_6]$ vacancy defects in PBAs and their effects on structural stability and hydration/deintercalation behavior, as vacancies alter the local environment of C ≡ N bonds, leading to peak broadening or shoulder effects.

3. **Silicon-based and alloy anode materials**

Raman spectroscopy plays an irreplaceable role in the study of silicon-based and alloy anode materials. It can accurately identify the phase transition of materials before and after lithium/sodium intercalation, intuitively reveal the microscopic stress distribution caused by significant volume changes, and dynamically analyze the evolution process of the electrode–electrolyte interface, providing crucial insights for understanding their failure mechanisms and performance optimization.

Silicon undergoes complex multistep phase transitions during (de)lithiation, progressing from crystalline silicon (c-Si) to amorphous lithium-silicon alloys (a-Li_xSi), and finally to crystalline $Li_{15}Si_4$. The sharp first-order optical phonon peak at ~520/cm in c-Si serves as its characteristic fingerprint. During the initial lithiation stage, this peak gradually broadens and weakens indicating that lithium-ion insertion causes lattice expansion and loss of long-range order-essentially the amorphization process. operando Raman studies further reveal that after complete amorphization, continued lithiation causes the broadened spectrum to gradually exhibit the characteristic broad peak of $Li_{15}Si_4$ around ~380/cm and ~490/cm, signifying the formation of the final crystalline phase. This phase transition pathway is crucial for understanding the voltage plateau and capacity release of silicon anodes. More importantly, confocal Raman surface scanning technology enables visualization of spatial nonuniformity in lithiation extent across electrode surfaces and even within individual silicon particles. The study found that during cycling, the lithiation front does not advance uniformly but is influenced by particle size, conductive network connectivity, and local current density, leading to stress concentration and preferential crack initiation [10]. For instance, mapping analysis of silicon-carbon composite anodes revealed significant lithiation lag in incomplete carbon-coated regions or large silicon particles, where Raman spectra still retain strong c-Si characteristic peaks. These regions often serve as crack initiation sites post-cycling. This non-uniform lithiation is one of the key factors contributing to silicon particle pulverization and rapid capacity decay.

The Raman peak shift ($\Delta\omega$) exhibits a linear relationship with the material's stress (σ), expressed as $\Delta\omega = \Pi\sigma$, where Π denotes the phonon deformation potential. Thus, precise measurement of silicon's characteristic peak shifts before and after cycling enables quantitative analysis of micro-stress generated during lithium intercalation. Research reveals that post-de-lithiation, irreversible volume expansion and plastic deformation leave substantial tensile stress in silicon particles, manifested as permanent low-frequency shifts in Raman peaks relative to the original state. This residual stress accelerates mechanical failure in subsequent cycles.

The substantial volume changes in silicon anodes cause continuous rupture and reconstruction of the SEI, leading to sustained consumption of both electrolyte and active lithium. Raman spectroscopy enables in-situ monitoring of the dynamic evolution of SEI composition on silicon surfaces. For instance, when comparing conventional carbonate electrolytes with those containing fluorinated ethylene carbonate (FEC) additives, Raman spectra clearly demonstrate that the latter forms a more stable SEI layer enriched with LiF (identifiable by its weak characteristic peak) and organic lithium salts earlier during cycling. This stable SEI effectively inhibits continuous electrolyte decomposition, which is crucial for enhancing the cycle life of silicon anodes.

Raman spectroscopy is equally applicable to studying other alloying reaction anode materials. For tin-based anodes, it distinguishes tin oxides (e.g., SnO_2, with characteristic peaks at ~630/cm and ~770/cm) from metallic tin (with broad peaks around ~110 and ~200/cm). During simulation, the transition of tin to amorphous/crystalline Na_xSn alloys can be tracked. For antimony-based anodes, the evolution of Raman characteristic peaks (~110 and ~150/cm) during alloying provides insights into reaction mechanisms. In sodium-ion batteries, operando Raman has successfully monitored the "transformation-alloying" two-step reaction mechanism of antimony anodes, clearly revealing the emergence and disappearance of the Na_xSb intermediate phase.

In conclusion, Raman spectroscopy has become a core characterization tool for in-depth analysis of complex electrochemical behaviors and failure mechanisms in silicon-based and alloy anode materials, by providing multidimensional in-situ and spatially resolved information on phase transitions, stress, and interfacial chemistry.

10.4.2 Study on the Electrode–electrolyte Interface Process and Side Reaction

The degradation of battery performance and safety issues are mainly caused by complex physicochemical processes at the electrode–electrolyte interface. Raman spectroscopy, especially when combined with confocal microscopy, can perform

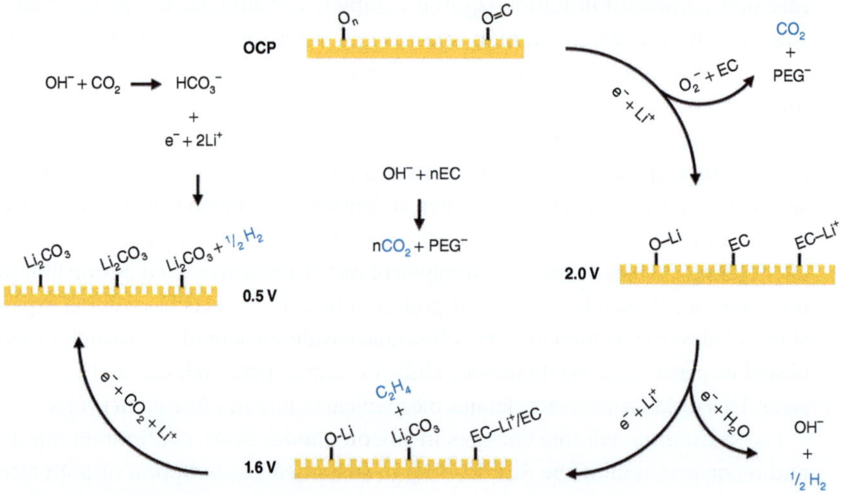

Figure 10.7 Schematic illustration of SEI formation process on silicon anode [11] / with permission of John Wiley & Sons.

micro-scale and in-situ analysis of solid-liquid or solid-solid interfaces, revealing the formation, evolution, and compositional information of SEI/cathode-electrolyte interface (CEI) [11], as shown in Figure 10.7.

1. **Study on the SEI of the anode**
The chemical composition, structural stability, and dynamic evolution of SEI are pivotal interfacial science issues determining the first-cycle performance, cycle life, and safety of batteries-particularly those with highly active anodes. Raman spectroscopy, leveraging its molecular vibration fingerprinting capability and micro-area analysis properties, has become a key technique for in-situ/ex-situ analysis of SEI chemical composition, spatial distribution, and formation mechanisms.

In lithium-ion battery systems, Raman spectroscopy is widely used to investigate the SEI chemistry on various anode materials [12]. In traditional carbonate-based electrolytes (e.g., $LiPF_6$/ethylene carbonate (EC)-DMC), the formation of SEI on graphite anode surfaces is closely related to the solvation sheath structure of lithium ions (Li^+). Using in-situ Raman spectroscopy, researchers can monitor real-time molecular changes between the electrolyte bulk phase and the electrode interface. Near the first lithium intercalation potential (~0.8 V vs. Li^+/Li), the intensity of the characteristic EC molecular peak (~717/cm, cyclic breathing mode) shows significant reduction at the interface region, accompanied by new broad peaks corresponding to lithiated EC decomposition products (e.g., the C=O stretching vibration at ~1850/cm, originating from dicyanurate compounds). Combined with sharp double peaks at ~1090/cm and ~1490/cm (ν_1 and ν_3 vibration modes of Li_2CO_3) and broad bands in the ~300–400/cm range (phonon bands of inorganic materials like Li_2O/LiF), Raman spectroscopy clearly reveals the classic SEI multilayer structure dominated by alkyl lithium carbonate ($ROCO_2Li$) and inorganic lithium salts (Li_2CO_3, LiF) [3]. By comparing different electrolyte formulations (e.g., altering solvent ratios, adding film-forming additives like FEC, VC), Raman spectroscopy can quantitatively evaluate the effects of additives on the organic/inorganic component ratios, crystallinity, and uniformity within the SEI. For instance, the preferential reduction of VC additives generates an elastic organic layer rich in poly(VC), where the intensity of Raman characteristic peaks (such as C=C stretching vibration ~1640/cm) strongly correlates with cycling stability. The significant volume changes in silicon anodes during cycling lead to continuous rupture and regeneration of the SEI, resulting in irreversible consumption of active lithium and electrolyte. Raman spectroscopy mapping technology demonstrates unique value in visualizing the chemical composition distribution and thickness heterogeneity of the SEI on silicon particles after cycling. Research reveals that in conventional electrolytes, the SEI on silicon particles is rich in organic polymer components (broad C—H and C=C stretching vibration regions ~2800–3100/cm and ~1550–1650/cm) with uneven thickness, corresponding to poor cycling performance. When FEC additives are introduced, Raman mapping shows a significant increase in inorganic components (especially LiF, whose phonon modes exhibit broad peaks at ~300–400/cm) with more uniform distribution. More importantly, monitoring the ~520/cm characteristic peak of the silicon bulk phase reveals that FEC-derived LiF-rich SEI more effectively inhibits particle amorphization and mechanical pulverization of silicon particles, as evidenced by higher retention of crystalline silicon peak signals after cycling. The Raman spectroscopy demonstrates exceptional advantages in correlating interface chemistry (SEI composition) with structural evolution (silicon lattice). For lithium metal anodes, Raman spectroscopy faces challenges of weak signals and photothermal effects, yet critical information can still be obtained through carefully designed experiments (e.g., using low-power lasers and enhanced substrates). In-situ Raman spectroscopy enables investigation of cyclic evolution of interface chemical composition during lithium deposition/dissolution. For instance, it can monitor the formation of organic solvent decomposition products (e.g., lithium alkyl carbonates) during initial deposition stages, while higher concentrations of inorganic compounds like Li_2O or LiH may be detected in high-current density or localized hotspots. Raman spectroscopy provides a direct method for identifying "dead lithium" (metallic lithium that has lost electrical contact with current collectors): The plasma exciton edge bands or extremely low-frequency phonon modes of metallic lithium remain extremely weak, whereas the characteristic Raman peaks of inorganic residues (e.g., Li_2CO_3, Li_2O) become prominent when metallic lithium is passivated by insulating SEI layers. Consequently, strong Li_2CO_3 signal regions appearing on lithium anode surfaces after cycling likely correspond to "dead lithium" accumulation zones. Additionally, when combined with spatial resolution techniques, Raman spectroscopy can track the chemical environment at lithium dendrite growth fronts, providing insights into dendrite growth mechanisms and strategies for inhibition development.

In sodium-ion battery systems, the formation mechanism of SEI differs significantly from that in lithium battery systems, with Raman spectroscopy providing crucial insights. For mainstream hard carbon anodes, studies show that SEI formed in carbonate-based electrolytes typically contains abundant organic components derived from solvent decomposition (e.g., sodium alkyl carbonate), exhibiting strong Raman characteristic peaks (~1450 and ~1650/cm). The relative content of inorganic components like Na_2CO_3 (~1085/cm) and NaF (~320/cm) is regulated by salt concentration and additives. In-situ Raman observations reveal that SEI formed on hard carbon surfaces in ether-based electrolytes (e.g., $NaPF_6$ in DGDME) is thinner and richer in inorganic components, which is closely related to its excellent rate performance and cycling stability. For alloy anodes (e.g., tin, antimony), significant volume changes lead to continuous SEI cracking and regeneration. Raman spectroscopy combined with mapping technology can visualize the distribution of post-cycling SEI components (e.g., Na_2CO_3, organic sodium salts) on electrode surfaces relative to active materials (e.g., Sn or Na_xSn alloys), revealing the chemical correlation between "dead zone" formation and active material failure, providing guidance for designing stable interface protective layers.

Furthermore, a spatially resolved confocal resonant Raman microscopic analysis to investigate anode-grown Beobachter biofilms, as shown in Figure 10.8 [13]. The results confirm the presence of an intra-biofilm redox gradient whereby the probability that a heme is in the reduced state increases with increasing distance from the anode surface. Such a gradient is required to drive electron transport toward the anode surface by electron hopping via cytochromes. The results also indicate that at open circuit, when electrons are expected to accumulate in redox cofactors involved in electron transport due to the inability of the anode to accept electrons, nearly all c-type cytochrome hemes detected in the biofilm are oxidized.

In conclusion, Raman spectroscopy, with its unique chemical recognition and spatial resolution capabilities, not only enables static analysis of the

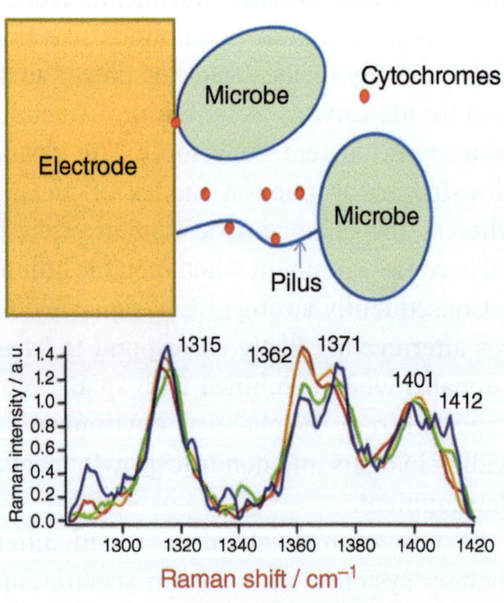

Figure 10.8 Confocal resonance Raman microscopy is applied to study anode-grown Beobachter sulphureousness biofilms with spatial resolution [13] / with permission of John Wiley & Sons.

final chemical composition of SEI but also dynamically reveals its nucleation, growth, and restructuring processes in different systems such as lithium and sodium batteries. This provides an indispensable microscopic perspective and experimental basis for the rational design of high-performance electrolytes and interface modification strategies.

2. **The CEI and the dissolution of transition metals**
In high-voltage, high-energy-density battery systems, the stability of the CEI and the dissolution of transition metal (TM) ions are interdependent, forming the core interface issue that leads to battery performance degradation. Raman spectroscopy, with its molecular vibration "fingerprint" recognition capability and spatial resolution advantages, provides an irreplaceable analytical tool for in-situ/ex-situ analysis of the chemical composition and evolution kinetics of the CEI, as well as for directly detecting TM dissolution and its cascading destructive effects on the electrode structure and anode interface.

In lithium-ion batteries, high-nickel layered oxides (such as $LiNi_{0.8}Co_{0.1}Mn_{0.1}O_2$,NCM811)andlithium-richmanganese-based($Li_{1.2}Ni_{0.13}Co_{0.13}Mn_{0.54}O_2$, LRMO) cathode materials work under high voltage (>4.3 V vs. Li^+/Li), where the oxidative decomposition of the electrolyte and the release of lattice oxygen from the cathode surface synergistically drive the formation and growth of CEI. Through ex-situ Raman mapping techniques, the chemical composition distribution of CEI on the surface of cathode particles after cycling can be visualized. For example, on the surface of NCM811 electrodes subjected to high-voltage cycling, characteristic signals from electrolyte oxidation products can be detected: peaks at ~730/cm and ~1450–1500/cm typically belong to polycarbonate or oligoether organic polymers; strong peaks at ~1090/cm and ~1490/cm indicate the formation of large amounts of Li_2CO_3. The spatially uneven distribution of these species is directly related to differences in local current density and crystal orientation. In-situ Raman further reveals the dynamic formation process of CEI: when initially charged to above 4.5 V, the intensity of characteristic peaks representing solvent molecules (such as EC) (~710/cm) can be observed to weaken, while new peaks representing oxidation products gradually strengthen, indicating the nucleation and growth of CEI at the high-voltage plateau.

The dissolution of TM originates from the instability of the cathode material's surface lattice structure under electrolyte/high-voltage conditions. For manganese-based materials (e.g., LRMO), Raman spectroscopy serves as a critical tool for monitoring Mn^{3+} Jahn-Teller distortion. This distortion causes MnO_6 octahedral distortion, manifested as significant broadening and asymmetry of Mn-O vibration peaks in the Raman spectrum between ~500–650/cm. Post-cycling signal degradation in this region directly reflects reduced surface Mn^{3+} concentration, indicating TM dissolution and structural reorganization of the surface into spinel or rock salt phases. More crucially, Raman spectroscopy enables direct detection of TM ion deposition at the anode side and its modification of the SEI. Raman surface scanning of post-cycling graphite or silicon anodes reveals non-intrinsic, broadened transition metal oxide (e.g., MnO, NiO) vibration signals (typically wide peaks at ~500–700/cm). These deposited TM compounds (particularly Mn species) exhibit catalytic activity that accelerates further decomposition and recombination of organic components in the anode SEI, promoting the formation of thicker, higher-impedance SEI layers. This process can be tracked through the evolution of characteristic vibration peaks such as C—H and C=O in the SEI region.

Cathodes in sodium-ion batteries, such as layered oxides ($NaTMO_2$, TM = Mn, Fe, Ni, Cu, etc.) and polyanionic compounds, also face challenges like interfacial side reactions and TM dissolution, particularly under moisture-sensitive and high-voltage conditions. Taking P2-type $Na_{2/3}Ni_{1/3}Mn_{2/3}O_2$ as an example, it undergoes complex phase transitions and sodium vacancy ordering at high voltages, accompanied by surface side reactions. Raman spectroscopy can distinguish between CEI species formed by electrolyte oxidation (similar to lithium batteries but potentially containing sodium salts like Na_2CO_3, with C-O vibration peaks at ~1065/cm) and residual alkali ($NaOH/Na_2CO_3$) generated from material surface reactions with H_2O/CO_2. More importantly, monitoring the Mn-O vibration mode (~500–600/cm) during cycling helps evaluate Mn's oxidation state and local environmental stability. Research reveals irreversible broadening and shifting of peaks in this region during charge-discharge cycles, correlating with Mn^{3+} dissolution and the collapse of layered structures into disordered rock salt. Dissolved Mn ions also migrate and deposit on the hard carbon anode surface, catalyzing the formation of thicker SEI layers rich in NaF and carbonates, as confirmed by Raman analysis of the anode.

Materials such as sodium vanadate ($Na_3V_2(PO_4)_3$, NVP) exhibit high structural stability and low risk of transition metal leaching. Raman spectroscopy proves valuable in studying material interfaces by precisely monitoring the integrity of surface carbon coatings and their interactions with electrolytes. The carbon layer on NVP particles is typically characterized by the intensity ratio of D peak (~1350/cm, disordered carbon) to G peak (~1580/cm, graphitized carbon, I_D/I_G). Post-cycling, comparing ID/IG values between particle surfaces and interiors helps determine whether the carbon coating has been compromised by electrolyte corrosion or volume changes. When the carbon layer breaks down, active materials become exposed to direct electrolyte reactions, potentially causing localized vanadium leaching (though to a much lesser extent than layered oxides) and abnormal thickening of the CEI layer. This can be inferred by detecting new phosphate or vanadium oxide-related vibration peaks in exposed areas, distinct from those in intact carbon-coated regions.

In summary, Raman spectroscopy provides a comprehensive chemical framework for understanding the twin failure mechanisms of cathode interfacial reactions and transition metal dissolution. Through end-to-end analysis spanning CEI components to cathode structure and cross-cathode analysis, this study establishes direct diagnostic evidence and optimization strategies for suppressing these side reactions via surface modification and electrolyte engineering.

3. **Characterization of electrolyte liquid phase structure and concentration gradient**
Raman spectroscopy offers unique advantages for analyzing the microscopic solvation structure of electrolytes and the ionic concentration distribution at electrode interfaces. By monitoring characteristic vibration peaks of solvents, anions, and their coordination states, it enables noninvasive acquisition of ion-solvent and ion-ion interaction data in electrolytes. The technique also visualizes mass transfer processes at the electrode/electrolyte interface during charge-discharge cycles, providing critical insights into concentration polarization, dendrite growth, and interfacial stability.

In lithium-ion battery systems, the decomposition of electrolytes and ion migration on graphite or lithium metal anode surfaces are closely related to local concentration gradients at the interface. For example, in systems using conventional $LiPF_6$/carbonates electrolytes, the cyclic breathing vibration mode (~712/cm) of EC molecules is highly sensitive to the solvation state of lithium ions. This peak position remains relatively fixed when the battery is at rest. However, during high-current charging (lithiation), rapid reduction and consumption of Li^+ at the anode interface cause instantaneous Li^+ concentration drops, leading to dissociation of solvent molecules from the first solvation shell. This manifests as a shift of the EC cyclic breathing peak toward higher wavenumbers. Through time-resolved in-situ Raman spectroscopy with lateral scanning, we can quantitatively map the Li^+ concentration gradient from the electrode surface to the bulk electrolyte. This gradient is a key factor driving concentration polarization and voltage rise in batteries. Further research reveals that the tip effect during lithium metal deposition exacerbates this concentration gradient, promoting both solvent co-embedded lithium and dendrite growth. Raman spectroscopy validation shows that using high-concentration electrolytes or localized high-concentration electrolytes can significantly increase Li^+ concentration at the interface and alter solvation structures, thereby suppressing concentration fluctuations and promoting uniform deposition. At high-voltage cathodes (e.g., NCM811), the electrolyte not only faces oxidation risks but its solvation structure also affects interface reaction kinetics. Raman spectroscopy can be applied to study electrolyte systems of novel lithium salts such as lithium bis(trifluoromethylsulfonyl) imide (LiTFSI). TFSI-anions exhibit two distinct peaks at approximately 740/cm and 750/cm, corresponding to their cisoid and transgauche conformations respectively. The peak intensity ratio (I740/I750) serves as a direct indicator for assessing the formation of ionic pairs (CIP) and aggregates (AGG). In-situ Raman studies demonstrate that when the cathode is charged to high voltages (>4.3 V vs. Li^+/Li), the interfacial electric field attracts additional anions, leading to an increase in anion aggregation (AGG) and a higher coordination ratio of anions in the solvation sheath. This "anion-enriched" solvation structure has been shown to facilitate the formation of a more stable CEI with abundant inorganic components, though it may concurrently reduce interfacial Li^+ mobility. These findings provide direct spectroscopic evidence for the design of "anion-derived" interfacial chemistry [14].

In sodium-ion battery systems, ether-based electrolytes (e.g., $NaPF_6$ in DIGLYME) have garnered significant attention due to their excellent anode compatibility. The C-O-C stretching vibration peak (~850/cm) and C-H deformation vibration peak of diethylene glycol dimethyl ether (DIGLYME) molecules are highly sensitive to Na^+ coordination states. Raman spectroscopy studies reveal that during initial sodium intercalation in hard carbon anodes, the coordination effect between ether oxygen atoms and Na^+ at the interface intensifies, causing red shifts and broadening of characteristic peaks. Notably, spectral comparisons between bulk and near-surface hard carbon electrodes demonstrate dynamic changes in Na^+ concentration at the interface during sodium intercalation. Under fast charging conditions, rapid consumption of interface leads to reorganization of solvation sheath structures and even decoordination of solvent molecules, which is closely related to the migration energy barrier of sodium ions and initial Coulombic efficiency [15].

For layered oxide or Prussian blue-based cathodes, the antioxidant stability of the electrolyte is critical. Raman spectroscopy can be used to study the dissociation degree and solvation structure of different sodium salts (such as $NaPF_6$, NaTFSI, NaFSI) in carbonate solvents. For example, the solvation sheath of $NaPF_6$ typically coordinates more solvent molecules than $LiPF_6$, which affects its oxidation stability. In-situ Raman testing can monitor changes in electrolyte components near the cathode surface during charging. Research has found that when the charging potential approaches the electrolyte's theoretical oxidation limit, the characteristic peak intensity of solvent molecules (such as EC or PC) begins to show irreversible decay, accompanied by the formation of new peaks (e.g., characteristic peaks of polycarbonate, a product of carbonate oxidation). This intuitively reveals the initial potential and process of electrolyte oxidation decomposition at the cathode interface, providing a key method for evaluating and screening high-voltage sodium-based electrolyte systems.

Raman spectroscopy enables in-situ, micro-area monitoring of characteristic vibration modes in electrolytes, achieving a leap from observing the average solvation structure of the liquid electrolyte to precisely visualizing local concentration gradients at the electrode interface and the dynamic reconfiguration of the solvation sheath. This information is crucial for understanding polarization sources at both electrodes in full cells, designing electrolytes that facilitate uniform ion transport, and constructing stable interfaces. It serves as an indispensable bridge connecting the microscopic ionic environment with macroscopic electrochemical performance.

10.4.3 Transport Mechanism and Interface Stability of Solid Electrolyte

Raman spectroscopy plays a unique role in solid-state battery research, as it can elucidate both the intrinsic ionic conduction mechanism of solid electrolytes (SSEs) and the chemical and physical contact issues at the solid-solid interface.

1. **Structure and ion transport of solid electrolyte**

Raman spectroscopy is a key tool for analyzing the bulk structure and microscopic mechanisms of ion transport in SSEs, as well as for diagnosing their compatibility with electrode interfaces. By detecting the "fingerprint" information of chemical bonds and lattice vibrations in materials, it can nondestructively and in-situ reveal the structural basis of ion conduction pathways, phase transition behaviors, and interfacial side reaction products, the scheme of in-situ Raman spectroelectrochemical cells is shown in Figure 10.9 [16].

For polyethylene oxide (PEO)-based polymer electrolytes, Raman spectroscopy simultaneously reveals the evolution of both the polymer matrix and lithium/sodium salt components. The C-O-C stretching vibration modes (~860/cm and ~1130/cm) of PEO chains are sensitive to their conformation and crystallinity. In the amorphous phase, the highly curled conformation of PEO chains facilitates ion transport, with corresponding Raman peaks differing from those in crystalline phases. Monitoring their relative intensities enables in-situ tracking of reversible crystallization-melting processes during charge-discharge cycles, allowing evaluation of temperature and current effects on transport pathways. Crucially, Raman spectroscopy directly detects the conformation of anion species and ion-pair states. Taking lithium bis(trifluoromethanesulfonyl) imide (LiTFSI) as an example, its anion TFSI exhibits distinct peaks at ~740/cm (S-N-S bending, coordination-sensitive) and ~1240/cm (CF_3 symmetric stretching), clearly distinguishing free anions, contact ion pairs (CIPs), and aggregates (AGGs). Studies show that higher salt concentrations or the introduction of specific additives (e.g., ceramic fillers) alter the relative proportions of these species, promoting lithium salt dissociation and enhancing lithium-ion migration number (tLi^+). For sodium-ion systems like NaTFSI/PEO, the anion

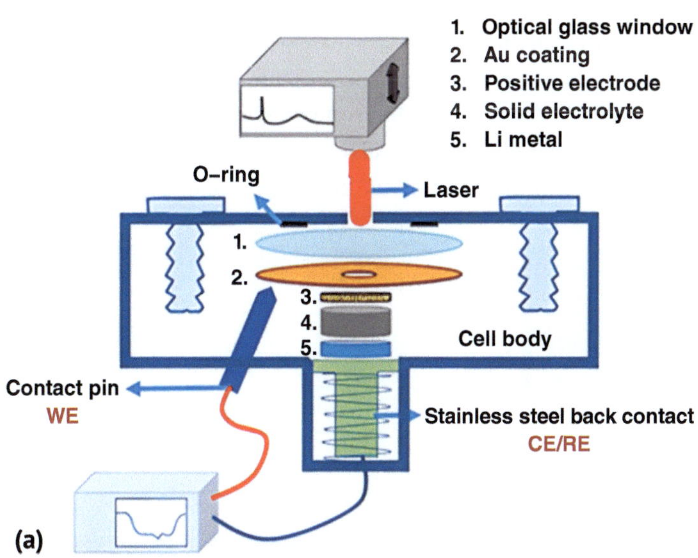

Figure 10.9 The scheme of in-situ Raman spectroelectrochemical cells [16] / with permission of John Wiley & Sons.

spectrum follows similar patterns but shifts peaks due to cation mass differences, providing direct evidence for understanding Na^+ solvation and transport mechanisms.

In inorganic SSEs, Raman spectroscopy serves as a sensitive probe for phase identification and evaluation of local structural disorder. For Li3PS4 glass ceramics, the P-S stretching vibration (~420/cm) and P-S-P bending vibration modes of $[PS_4]^{3-}$ tetrahedrons constitute their characteristic fingerprints. These peaks exhibit significant sharpening and displacement during transitions from glassy to crystalline states (e.g., formation of $Li_7P_3S_{11}$ or $Li_{10}GeP_2S_{12}$ phases). In-situ Raman spectroscopy can track structural evolution during preparation processes such as heat treatment. Crucially, when contacting high-voltage layered oxide cathodes (e.g., $LiCoO_2$, NCM), Raman spectroscopy directly detects interface-generated products from SSE chemical oxidation decomposition, including elemental sulfur (~150–250/cm broad peak), P2S5 (~420–430/cm peak shape change), and oxygen-containing thiophosphate species (e.g., POS_3^{3-}, new peak near ~650/cm), as shown in Figure 10.10 [17]. The formation of these insulating phases is the primary cause of interface impedance surge. The Raman spectra of garnet-type LLZO ($Li_7La_3Zr_2O_{12}$) effectively distinguish high-ion-conductivity cubic phase (c-LLZO) from low-conductivity tetragonal phase (t-LLZO). t-LLZO exhibits more pronounced Raman peaks due to its lower crystal symmetry. By monitoring the appearance or disappearance of specific peaks (e.g., those associated with Zr-O stretching vibrations), researchers can evaluate the effects of dopants (e.g., Al, Ta) on phase stability and detect adverse phase transitions during cycling. At the interface with the lithium metal anode, Raman spectroscopy identifies interface chemical changes caused by reactions between the Li_2CO_3 impurity layer on the LLZO surface and lithium, or by lithium dendrite penetration.

Solid-state sodium-ion battery systems have garnered significant attention due to their cost advantages, with Raman spectroscopy playing a pivotal role in their characterization. Sodium sulfide fast-ion conductors like Na_3PS_4 exhibit tetrahedral vibration modes of $[PS_4]^{3-}$ similar to lithium, but the peak positions shift systematically due to the heavier Na^+ mass and bond length variations.

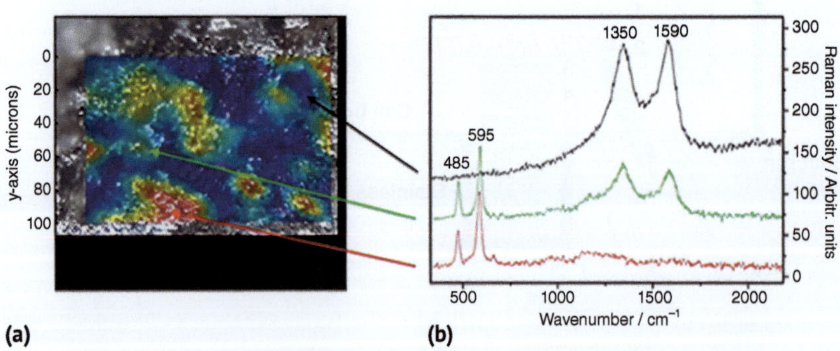

Figure 10.10 The Raman spectrum of $LiCoO_2$ [17] / with permission of Elsevier.

Raman spectroscopy enables the study of glass-ceramic transition and monitoring of interface stability with metallic sodium anodes. Experimental findings reveal that reduction products such as Na_2S (a broad peak around 250/cm) may form at the Na/Na_3PS_4 interface, with their formation kinetics closely related to interface impedance growth. Side reactions at interfaces with high-voltage cathodes (e.g., $NaCrO_2$ or polyanionic materials) can also be visualized through similar in-situ Raman mapping. NASICON-type electrolytes like $Na_3Zr_2Si_2PO_{12}$ (NZSP) primarily show Raman spectra reflecting tetrahedral vibrations of $[SiO_4]$ and $[PO_4]$ groups (~900–1100/cm range). The spectral fine structure (number, position, and width of peaks) is sensitive to the connection modes between ZrO_6 octahedrons and $(Si/P)O_4$ tetrahedrons, Na^+ occupancy sites, and structural order, directly correlating with ionic conductivity. In interface studies, Raman spectroscopy effectively detects potential Na^+/H^+ exchange between NZSP and metallic sodium (leading to alkaline layer formation) or carbonate impurities generated from reactions with CO_2/H_2O in air, all of which degrade interfacial contact.

In summary, Raman spectroscopy, by identifying and analyzing specific vibrational modes, not only elucidates the intrinsic structural factors determining ion transport in various solid electrolyte systems (polymers, sulfides, oxides), but also directly reveals the chemical and electrochemical degradation processes occurring at interfaces when these materials are matched with lithium/sodium-metal anodes and different types of cathodes. This provides indispensable dynamic information and microscopic evidence for the rational design and interface optimization of solid electrolyte materials.

2. **Solid electrolyte and electrode interface**

   The performance bottleneck of solid-state batteries stems from the high impedance and chemical/electrochemical instability at the solid-solid interface. Raman spectroscopy, with its chemical specificity and micro-area analysis capability, has become a key tool for deciphering interface microstructures, identifying reaction products, and revealing failure mechanisms. Its applications have been extensively applied to both cathode and anode interfaces in lithium and sodium battery systems.

   1. Analysis of the interface of lithium battery system

      At the interface between high-voltage oxide cathodes (such as $LiNi_{0.8}Mn_{0.1}Co_{0.1}O_2$, NCM811) and sulfide solid electrolytes (such as Li_6PS_5Cl, LPSCl), there is significant chemical and electrochemical incompatibility. Raman spectroscopy can clearly capture the side reactions at the interface. Studies have shown that when NCM811 is directly contacted with LPSCl, even without electrochemical cycling, characteristic peaks of the LPSCl electrolyte (such as the P-S stretching vibration of $[PS_4]^{3-}$ tetrahedron at ~420/cm) can be detected as weakening or disappearance in the Raman spectrum at the interface region, while new characteristic peaks belonging to P_2S_5 (~390/cm and ~430/cm) and Li_2S (~375/cm) appear simultaneously. This confirms that residual lithium compounds (such as

Li_2CO_3, LiOH) or lattice oxygen on the surface of the cathode material spontaneously react with the sulfide electrolyte, leading to the formation of an insulating interface layer [18]. During electrochemical cycling, in-situ Raman further reveals that when charged to high voltages (>4.2 V vs. Li^+/Li), active oxygen species released from the cathode exacerbate the oxidative decomposition of the sulfide electrolyte, potentially generating higher oxidation state oxygen-containing sulfur-phosphorus compounds (such as $PO_xS_y^{3-}$ species, with characteristic Raman shifts in the range of ~650–850/cm). The ionic conductivity of this interface phase is extremely low, which is the primary cause of the sharp increase in interface impedance and capacity decay.

Raman spectroscopy addresses two critical challenges at the interface between lithium metal anodes and oxide solid-state electrolytes (e.g., garnet-type $Li_7La_3Zr_2O_{12}$, LLZO): identifying interface by-products and evaluating lithium deposition uniformity. LLZO is air-sensitive, forming a Li_2CO_3 passivation layer on its surface (with characteristic peaks at ~1090/cm and ~1490/cm) that severely hinders lithium-ion transport and creates significant interfacial impedance. Raman mapping provides visual insights into the distribution of Li_2CO_3 on LLZO particles, enabling assessment of surface modification processes like acid treatment or heat treatment. Crucially, optically transparent in-situ batteries allow Raman spectroscopy (combined with low-wavenumber signal collection) to monitor lithium deposition on LLZO surfaces. During uniform deposition, the characteristic lithium signal (a broad peak typically spanning ~100–200/cm) shows uniform enhancement at the interface. Conversely, lithium dendrite growth manifests as localized Raman signal intensification or spot-like distribution. Combined with subsequent SEM morphology analysis, this reveals direct correlations between dendrite initiation hotspots and LLZO grain boundaries or surface defects. For polymer electrolyte-lithium metal interfaces, Raman tracking of lithium salt anions (e.g., $TFSI^-$) during surface reduction/decomposition, along with polymer structural stability analysis, further enhances understanding.

2. Interface analysis of sodium electrolyte system
   In sodium-ion solid-state batteries, interface issues are equally prominent. For example, when NASICON-type solid electrolytes (such as $Na_3Zr_2Si_2PO_{12}$, NZSP) are directly compounded with oxide cathodes (such as $Na_{0.7}MnO_2$), the Raman spectrum shows significant changes in the SiO_4/PO_4 characteristic peaks of NZSP (~900–1100/cm) at the interface, along with new broad peaks attributed to amorphous silicon/phosphate glass phases. This indicates that during high-temperature sintering or prolonged contact, element mutual diffusion and interface reactions occur between the cathode and electrolyte, forming a high impedance layer. For sulfide sodium-ion solid electrolytes (such as Na_3PS_4), the interface with high-voltage cathodes (such as $Na_3V_2(PO_4)_3$@C, NVP) is more fragile. Raman mapping reveals a sharp decrease in the intensity of the

$[PS4]^{3-}$ characteristic peak at the contact edge between NVP particles and the Na_3PS_4 electrolyte, accompanied by signals of products like Na_2S, confirming the oxidation of $[PS_4]^{3-}$ by V^{5+}. Through Raman spectroscopy screening, researchers found that oxygen doping or the construction of buffer layers (such as $NaNbO_3$) on the surface of Na_3PS_4 can effectively suppress interface side reactions, with evidence directly coming from the increased retention of the original electrolyte characteristic peaks in the interface region.

The study of sodium-metal anode interfaces with solid electrolytes represents an emerging application field in Raman spectroscopy. Similar to lithium-based systems, sodium-metal interfaces with β-Al_2O_3 or NASICON-type electrolytes may form passive layers containing Na_2CO_3 (Raman peak at ~1085/cm). In-situ Raman observations of sodium deposition reveal that interface contact uniformity directly determines sodium deposition morphology. At poorly wetted interfaces, sodium tends to nucleate locally and form dendrites, exhibiting dramatic spatial fluctuations in Raman signals. Notably, when using alloy anodes such as Sn or Sb, Raman spectroscopy can dynamically track alloying reactions at the solid electrolyte interface. For instance, in-situ experiments demonstrate the gradual disappearance of characteristic peaks (near ~100 and ~200/cm) of metallic Sn as sodiation progresses, while broadened signals of amorphous or crystalline NaxSn alloys emerge. This provides direct spectroscopic evidence for understanding reaction kinetics and volume strain management in solid-state systems with alloy anodes.

In summary, Raman spectroscopy has become an indispensable tool for analyzing complex interface issues in solid-state batteries through its "chemical imaging" capability. Whether in lithium or sodium-based systems, it can qualitatively and semi-quantitatively identify reaction products at the cathode or anode interfaces, visualize the distribution of interfacial phases, evaluate the effectiveness of interface modification, and in-situ reveal the dynamic evolution of interfaces at micron or even nanoscale levels. This provides crucial experimental evidence and deep insights for the rational design of stable, low-impedance solid-solid interfaces.

10.4.4 Integration of Multimodal and In-situ/Condition Characterization Techniques

In order to obtain more comprehensive, dynamic, and real information, Raman spectroscopy is increasingly combined with other characterization techniques, and a variety of advanced in-situ/operational testing methods are developed.

1. **Combined with microscopy**
   1. **Confocal micro-Raman spectroscopy and mapping techniques**
      By combining hyperspectral sensitivity with submicron spatial resolution (typically reaching ~0.5 μm), this technology enables quantitative imaging

of chemical composition and structural distribution in battery electrodes and interfaces [1–11]. In lithium-ion batteries, it is widely used to visualize phase distribution of electrode materials and nonuniform charge state (SOC) during cycling. For instance, cross-sectional Raman mapping of secondary particles in high-nickel layered cathodes (e.g., NCM811) clearly reveals significant gradients in cation mixing (manifested as broadening and shifting of the Alg peak) between particle surfaces and internal regions after long-term cycling, which directly correlates with the origins of crack initiation and capacity decay. In silicon-carbon composite anodes, simultaneous monitoring of Raman signal spatial distribution and evolution in silicon phase (~520/cm peak) and graphite phase (G peak ~1580/cm) allows intuitive analysis of mechanical/electrochemical coupling/decoupling behaviors between silicon and carbon components during cycling, as well as localized stress concentration issues caused by lithium intercalation nonuniformity. This technology is equally critical for sodium-ion batteries. When studying Prussian blue analog cathodes (e.g., $Na_xFe[Fe(CN)_6]$), high-precision mapping of C ≡ N stretching vibration peaks (~2150/cm) can map sodium-ion intercalation/deintercalation kinetics in secondary particles, revealing bottleneck regions for sodium-ion transport. For hard carbon anodes, the intensity ratio (ID/IG) and position of the D peak (~1350/cm, disordered carbon) and G peak (~1580/cm, graphitized carbon) in the Raman spectrum are sensitive to sodium storage sites (e.g., defects, nanopores, interlayer spaces of graphite microcrystals). Mapping techniques can be used to characterize the spatial heterogeneity of the internal structure of hard carbon particles and its correlation with sodium storage performance.

2. **Tip-enhanced Raman spectroscopy (TERS)**

   By integrating Raman spectroscopy with AFM and leveraging the plasmon enhancement effect of metal tips, this technique has achieved spatial resolution surpassing optical diffraction limits, reaching the nanoscale (<10 nm), as shown in Figure 10.11. This breakthrough enables nanoscale chemical imaging of the topmost 1–3 nm layer of the electrode–electrolyte interface (SEI/CEI) film [19, 20]. In lithium metal anode research, tip-enhanced Raman spectroscopy (TERS) provides nanoscale compositional mapping of SEI films, directly distinguishing key

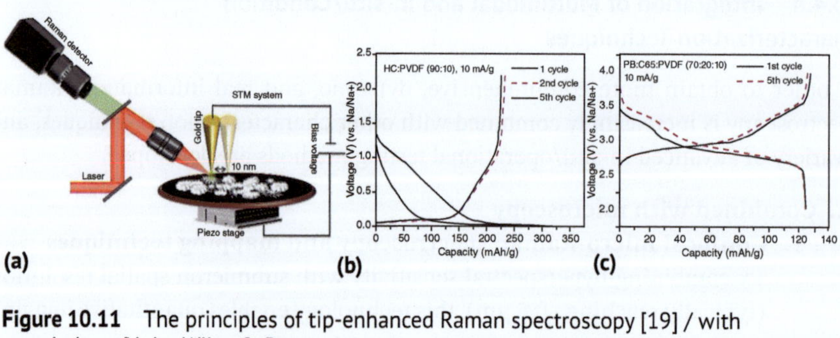

Figure 10.11 The principles of tip-enhanced Raman spectroscopy [19] / with permission of John Wiley & Sons.

components like LiF, Li_2CO_3, and organic materials distributed in nanodomains. It reveals the true nanomorphology of "mosaic" or "layered" structures, offering unprecedented insights into SEI's mechanical properties and ionic transport anisotropy. For cathode materials, TERS can detect chemical compositions on individual primary particles and their dependence on underlying crystal orientations (e.g., exposed (003) and (104) planes in NCM materials), elucidating the mechanisms of interfacial side reactions at the nanoscale.

3. **The coupling system of Raman spectroscopy and scanning electron microscopy**

   The integration of direct correlation between morphology, composition, and crystal structure information within inert atmosphere glove boxes has been established. This technology demonstrates unique advantages in analyzing degradation mechanisms of cyclically used electrodes [3], as shown in Figure 10.12.

   For instance, when observing intergranular cracks in sodium-based layered oxide cathodes (e.g., $NaNi_{1/3}Fe_{1/3}Mn_{1/3}O_2$) under SEM, in-situ Raman point-scan or mapping can immediately be performed on crack edges and bulk particles. This comparative analysis identifies whether structural degradation residues like P2/O3 phases or sodium carbonate accumulate at crack sites, thereby clarifying whether cracks result from mechanical stress or interface side reactions. For silicon-based anodes after cycling, the system enables

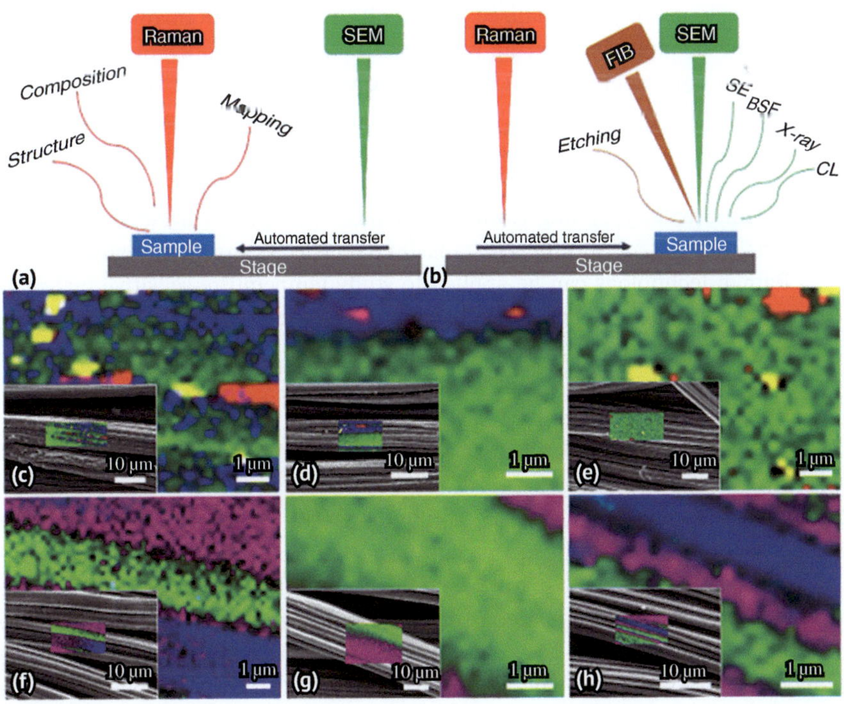

Figure 10.12 Schematic of combined SEM and Raman testing [3] / with permission of John Wiley & Sons.

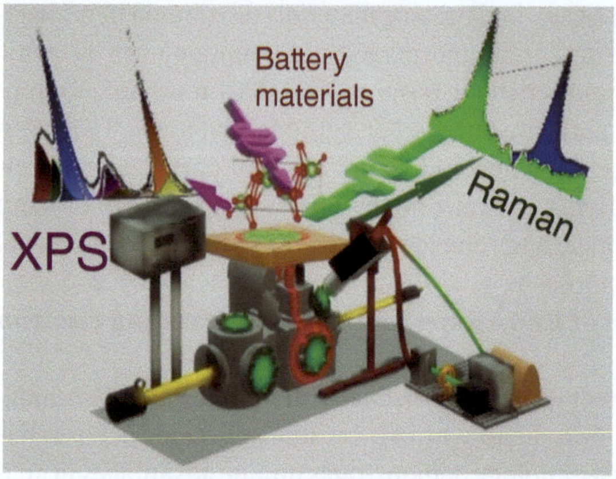

Figure 10.13 Schematic illustration of combining XPS and Raman characterization Adapted from [21].

Raman analysis directly on SEM-observed fracture or expansion regions to confirm irreversible formation of lithium-silicon alloy phases (e.g., $Li_{15}Si_4$) or permanent amorphous structural changes, precisely correlating physical deformation with chemical transformations. This multimodal integrated technique significantly enhances the attribution capability for complex interfacial phenomena and degradation mechanisms.

2. **Advanced in-situ and operating condition characterization**
   The development of specialized in-situ electrochemical Raman cells enables real-time dynamic monitoring under simulated battery operating conditions (constant current charge/discharge, cyclic voltammetry, and varying temperatures). This provides direct evidence for understanding reaction pathways, intermediate products, phase transition kinetics, and degradation mechanisms. In recent years, operando Raman testing-continuous spectral acquisition during actual battery cycling (rather than post-cycling) has avoided relaxation effects, yielding more authentic transient data [21], as shown in Figure 10.13.

   Raman technology can be applied to study temperature-sensitive materials (e.g., lithium metal deposition morphology) or intermediate states of freezing reactions.

10.5 Summary and Outlook

Raman spectroscopy, with its unique ability to identify chemical bonds and lattice vibrations through "fingerprint" recognition, combined with micro-area analysis advantages and excellent in-situ compatibility, has been deeply integrated into every aspect of battery material development from fundamental research to product development. It not only excels in analyzing the static structures of single components like electrodes and electrolytes, but also plays an irreplaceable role in revealing critical scientific issues during dynamic electrochemical processes, including structural evolution, phase transition pathways, interfacial side reactions, and ion transport mechanisms. In the future, as Raman spectroscopy continues to evolve (with ultrafast time resolution, higher spatial resolution, and lower detection limits) and integrates more deeply with advanced characterization techniques, big data analysis, and artificial intelligence, its application boundaries will expand further. For instance, combining machine learning for automated analysis and modeling of massive in-situ Raman spectral data could enable predictive design of battery materials and interface behaviors, accelerating the development of next-generation high-performance, long-life, and high-safety battery systems. Against the backdrop of emerging battery systems like solid-state batteries, lithium-sulfur batteries, and sodium/potassium-ion batteries facing more complex interface and reaction mechanism challenges, Raman spectroscopy, as a powerful "molecular microscope," will demonstrate increasingly significant value.

References

1 Y. Ozaki, S.J.P.A.o.R.S. Šašic, Introduction to Raman Spectroscopy, 2008, pp. 1–28. https://doi.org/10.1002/cmtd.202200042.

2 W.J.C.M. Zheng, Beginner's guide to Raman spectroelectrochemistry for electrocatalysis study, Chemistry Methods 3(2) (2023) e202200042. https://doi.org/10.1002/9780470225882.ch1.

3 Y. Zheng, T. Deng, N. Yue, W. Zhang, X. Zhu, H. Yang, X. Chu, W.J.J.o.R.S. Zheng, Raman spectroscopy and correlative-Raman technology excel as an optimal stage for carbon-based electrode materials in electrochemical energy storage, Journal of Raman Spectroscopy 52(12) (2021) 2119–2130. https://doi.org/10.1002/jrs.6178.

4 S. Guan, E. Liao, S. Sun, Q. Peng, K. Zeng, K. Shin, X. Guo, X. Zhou, Raman spectroscopy, Energy Storage Materials Characterization 2025, pp. 397–417. https://doi.org/10.1002/9783527834679.ch16.

5 X.M. Lin, J.F.J.C. Li, Applications of *in situ* Raman spectroscopy on rechargeable batteries and hydrogen energy systems, ChemElectroChem 10(4) (2023) e202201003. https://doi.org/10.1002/celc.202201003.

6 R. Baddour-Hadjean, J.P.J.L.I.R.B.M. Pereira-Ramos, Technology, N. Applications, Raman investigation of cathode materials for lithium batteries, Lithium Ion Rechargeable Batteries: Materials, Technology, and New Applications 2009, pp. 103–162. https://doi.org/10.1002/9783527629022.

7 J.A. Gilbert, V.A. Maroni, Y. Cui, D.J. Gosztola, D.J. Miller, D.P.J.A.M.I. Abraham, Composition and impedance heterogeneity in oxide electrode cross-sections detected by Raman spectroscopy, Advanced Materials Interfaces 5(9) (2018) 1701447. https://doi.org/10.1002/admi.201701447.

8 C. Fang, Y. Huang, W. Zhang, J. Han, Z. Deng, Y. Cao, H. Yang, Routes to high energy cathodes of sodium-ion batteries, Advanced Energy Materials 6(5) (2016) 1501727. https://doi.org/10.1002/aenm.201501727.

9 C.M. Burba, J.M. Palmer, B.S. Holinsworth, Laser-induced phase changes in olivine FePO4: A warning on characterizing LiFePO4-based cathodes with Raman spectroscopy, Journal of Raman Spectroscopy 40(2) (2009) 225–228. https://doi.org/10.1002/jrs.2114.

10 T. Gruber, D. Thomas, C. Röder, F. Mertens, J.J.J.o.R.S. Kortus, Raman spectroscopic studies of LixSiy compounds, Journal of Raman Spectroscopy 44(6) (2013) 934–938. https://doi.org/10.1002/jrs.4308.

11 N. Gogoi, T. Melin, E.J.J.A.M.I. Berg, Elucidating the step-wise solid electrolyte interphase formation in lithium-ion batteries with operando Raman spectroscopy, Advanced Materials Interfaces 9(22) (2022) 2200945. https://doi.org/10.1002/admi.202200945.

12 D. Ketenoglu, G. Spiekermann, M. Harder, E. Oz, C. Koz, M.C. Yagci, E. Yilmaz, Z. Yin, C.J. Sahle, B.J.S.R. Detlefs, X-ray Raman spectroscopy of lithium-ion battery electrolyte solutions in a flow cell, Synchrotron Radiation 25(2) (2018) 537–542. https://doi.org/10.1107/S1600577518001662.

13 N. Lebedev, S.M. Strycharz-Glaven, L.M. Tender, Spatially resolved confocal resonant Raman microscopic analysis of anode-grown *Geobacter sulfurreducens* biofilms, ChemPhysChem 15(2) (2014) 320–327. https://doi.org/10.1002/cphc.201300984.

14 L. Lin, L. Suo, Y.-s. Hu, H. Li, X. Huang, L. Chen, Epitaxial induced plating current-collector lasting lifespan of anode-free lithium metal battery, Advanced Energy Materials 11(9) (2021) 2003709. https://doi.org/10.1002/aenm.202003709.

15 Y. Luo, F. Zheng, L. Liu, K. Lei, X. Hou, G. Xu, H. Meng, J. Shi, F. Li, A High-power aqueous zinc–organic radical battery with tunable operating voltage triggered by selected anions, ChemSusChem 13(9) (2020) 2239–2244. https://doi.org/10.1002/cssc.201903083.

16 Y. Zhou, C. Doerrer, J. Kasemchainan, P.G. Bruce, M. Pasta, L.J. Hardwick, Observation of Interfacial Degradation of Li6PS5Cl against Lithium Metal and $LiCoO_2$ via In situ Electrochemical Raman Microscopy, Batteries & Supercaps 3(7) (2020) 647–652. https://doi.org/10.1002/batt.201900218.

17 T. Gross, C. Hess, Raman diagnostics of $LiCoO_2$ electrodes for lithium-ion batteries, Journal of Power Sources 256 (2014) 220–225. https://doi.org/10.1016/j.jpowsour.2014.01.084.

18 R. Dubey, J. Sastre, C. Cancellieri, F. Okur, A. Forster, L. Pompizii, A. Priebe, Y.E. Romanyuk, L.P.H. Jeurgens, M.V. Kovalenko, K.V. Kravchyk, Building a better li-garnet solid electrolyte/metallic li interface with antimony, Advanced Energy Materials 11(39) (2021) 2102086. https://doi.org/10.1002/aenm.202102086.

19 S. Dinda, S. Trivedi, A. Roy, F.D. Pammer, M.J.A.E.M. Fichtner, Capturing nano-scale inhomogeneity of the electrode electrolyte interface in sodium-ion batteries through tip-enhanced Raman spectroscopy, Advanced Energy Materials 13(42) (2023) 2302176. https://doi.org/10.1002/aenm.202302176.

20 Z.Q. Tian, X.M.J.D.i.E.S.I.b.M.F. Zhang, Electrochemical surface-enhanced Raman spectroscopy (EC-SERS): Early history, principles, methods, and experiments, Developments in Electrochemistry: Science Inspired by Martin Fleischmann 2014, pp. 113–135. https://doi.org/10.1002/9781118694404.ch7.

21 M. Radtke, K. Kopp, C.J.C.M. Hess, Coupling long-range Raman with X-ray photoelectron spectroscopy for complementary bulk and surface characterization of battery materials, Chemistry Methods 2(1) (2022) e202100058. https://doi.org/10.1002/cmtd.202100058.

11

Gas Evolution and Surface Chemistry Analysis of Battery Materials by Differential Electrochemical Mass Spectrometry (DEMS) and Time-of-Flight Secondary Ion Mass Spectrometry (TOF-SIMS)

Mass spectrometry provides a direct approach to probing the formation mechanisms and evolution of critical interfaces in battery research at the molecular level, leveraging its high sensitivity for molecular identification and dynamic monitoring capabilities. Functional mass spectrometry approaches, such as time-of-flight secondary ion mass spectrometry (TOF-SIMS) and differential electrochemical mass spectrometry (DEMS), enables precise identification of the chemical composition and spatial distribution of electrode/electrolyte interphases (e.g., solid electrolyte interphase (SEI)/cathode electrolyte interphase (CEI)), real time tracking of side reaction pathways and gaseous products during cycling, and quantitative analysis of lithium-ion distribution, phase transitions, and active material loss. This technique transcends the limitations of macro-electrochemical characterization and serves as an indispensable core analytical method for uncovering the fundamental causes of battery performance degradation and guiding the design of advanced materials and interface engineering.

11.1 Principles and Instrumentation of DEMS

11.1.1 Fundamentals of DEMS

In the field of lithium-ion battery research, DEMS technology has garnered significant attention. This is due to its ability to perform in-situ, real time, quantitative detection of gaseous and volatile products. DEMS tightly integrates electrochemical signals, such as voltage, capacity, and current, with the real-time generation of reaction gases. Its key advantage is its ability to detect gas types and concentrations produced by electrode reactions in real time, synchronously and precisely. This provides valuable insights into the internal reaction mechanisms of batteries and offers robust analytical results for performance degradation and safety assessments. Electron transfer between reactant molecules in lithium battery electrochemical redox reactions is often complex. The microscopic electrochemical mechanisms of many redox reactions, particularly at interfaces, remain poorly understood.

In 1971, Bruckenstein and Gad demonstrated the role of mass spectrometry in electrochemical gas analysis [1]. They developed an in-situ DEMS monitoring system based on a porous electrode structure design. This structure allowed gas products generated on

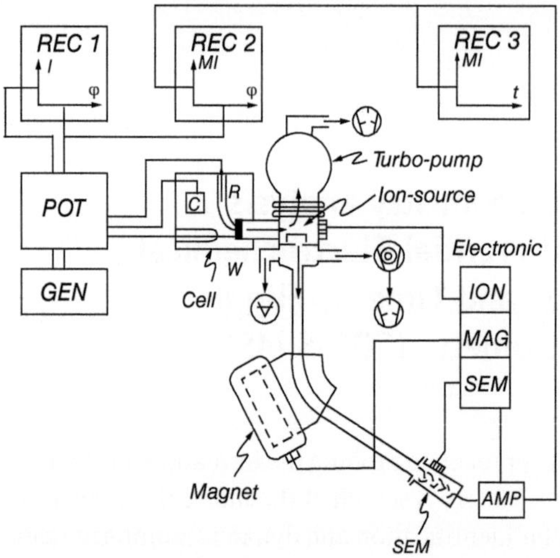

Figure 11.1 View of the complete apparatus showing the electrochemical cell with inlet system, the mass spectrometer arrangement and the electronic equipment [2] / with permission of John Wiley & Sons.

the electrolyte side to diffuse into the vacuum chamber of the mass spectrometer, where they were detected after electron bombardment. Compared to other methods at the time, this approach enabled relatively rapid product detection, thereby demonstrating that in-situ DEMS technology could be used to detect gases produced by reactions within electrodes. In 1984, Wolter and Heitbaum adapted the Teflon membrane inlet system to develop a novel technique capable of quantitatively measuring gaseous and volatile products from electrochemical reactions with high sensitivity. The device they designed is shown in Figure 11.1. This technique achieves in-situ, real time monitoring of electrode reaction products by coupling an electrochemical cell, a membrane inlet system, and a mass spectrometer. Specifically, products generated at the porous working electrode are directly introduced into the ion source of the mass spectrometer via the membrane inlet system. The intensity of the acquired mass signal is proportional to the Faradaic current of the electrode reaction. Due to the system's rapid response (short rise time), the current signal can be correlated with specific mass spectral signals in real time and quantitatively during potential scanning. This configuration enables the system to detect product generation rates as low as 5×10^{-11} mol/s [2].

The technique was thus designated as DEMS, a name chosen to acknowledge its conceptual lineage from Bruckenstein's work while underscoring the methodological advancements that distinguish it. At the time, the DEMS system they studied involved the use of a separation membrane as a physical barrier to isolate the in-situ electrochemical cell from the mass spectrometer, thereby forming the gas inlet port. This membrane must fulfill two critical functional requirements: First, it must exhibit selective permeability toward volatile species, permitting the passage of gaseous reaction products while maintaining impermeability to electrolytes. Second, the response time for gas transport from the electrode interface to the mass spectrometer detector must be sufficiently brief. Only when these conditions are met can the temporal characteristics of electrochemical processes (variations in potential/current) be precisely synchronized with mass spectrometric signals, enabling authentic differential measurements. However, Teflon membrane interfaces are constrained by sluggish gas transport kinetics, resulting in prolonged system response times and diminished gas transfer efficiency, which inevitably introduces quantitative inaccuracies. In 2007, Novak et al. first reported a novel electrochemical cell as shown in Figure 11.2, which was specifically designed for in-situ gas analysis using DEMS [3].

To achieve high-sensitivity online detection, gaseous products generated by electrochemical reactions are extracted in real time through a capillary at the top of the cell and directly introduced into a quadrupole mass spectrometer. This design enables the system to achieve a detection sensitivity approximately 100 times that of conventional methods. Subsequently, capillary-based carrier gas purging configurations have been widely developed and implemented. This sampling methodology now achieves gas transfer efficiencies approaching 100%, significantly enhancing quantitative accuracy. Furthermore, designing an effective gas flow path for the in-situ cell can greatly improve detection efficiency and accuracy.

The vacuum system employed by DEMS utilizes a multistage pumping combination to establish an ultra-high vacuum environment (10–5 to 10–9 mbar), thereby eliminating molecular collision interference during ion transport. Sampling

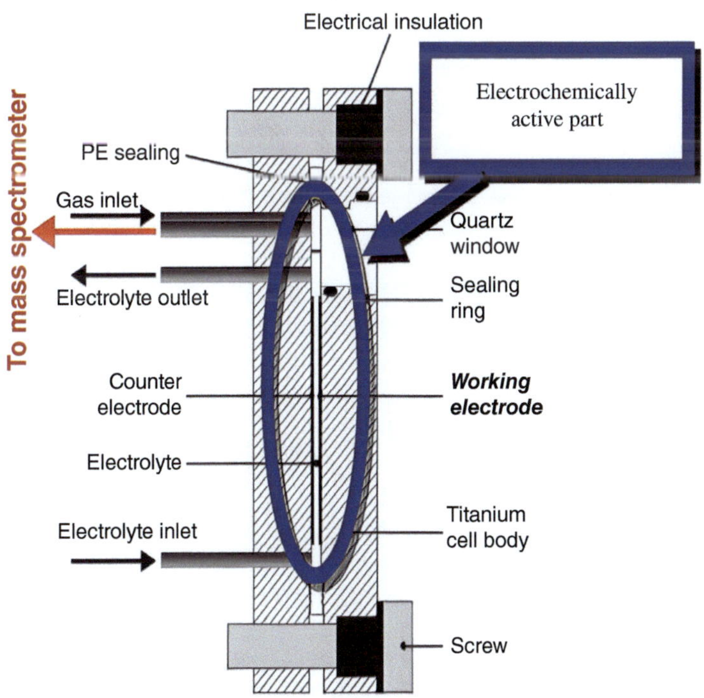

Figure 11.2 Schematic drawing of the improved DEMS measurement cell for headspace gas analysis, for measurements at elevated temperature, heating plates can be attached to either side [3] / with permission of Elsevier.

interfaces are purged with carrier gas. Electrochemical cell products are efficiently transferred via capillary restrictors, while an integrated cryogenic trap selectively condenses solvent vapors. Upon entering the chamber, the ion source ionizes gaseous substances through 70 eV electron impact (EI) ionization, generating characteristic molecular ions. Subsequently, the mass analyzer (typically a quadrupole filter) separates ions based on their mass-to-charge ratio (m/z). The final detector (secondary electron multiplier, SEM) amplifies and records ion intensity, enabling real time reconstruction of time-resolved mass spectra synchronized with electrochemical parameters.

The working principle of DEMS is based on electron impact ionization (EI) technology. In this process, gas molecules are ionized through bombardment by electrons emitted from a high-temperature filament (shown in the Figures 11.3 and 11.4 as 70 eV). The resulting ions are then transported via the extraction optics within the ion source to a quadrupole mass filter for mass separation. Finally, depending on the signal intensity, detection is carried out by either a Faraday cup or an electron multiplier.

Figure 11.3 EI ionizer cut away. [Copied from the Hiden product brochure.]

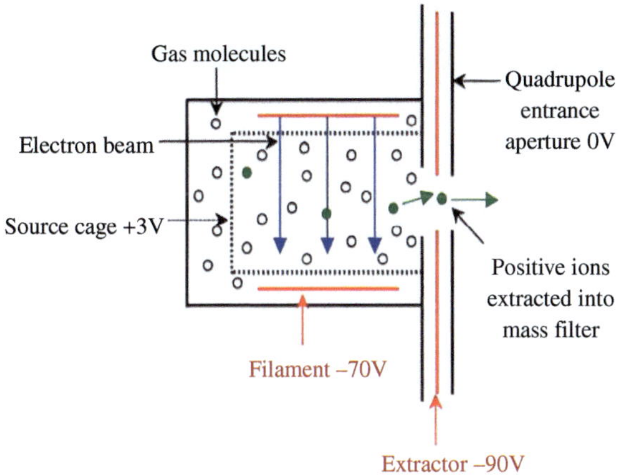

Figure 11.4 Diagram illustrating EI ion source operation. [Copied from the Hiden product brochure.]

11.1.2 Fundamentals of TOF-SIMS

TOF-SIMS is one of the most widely used surface analysis techniques. It is a micro-beam micro-area in-situ analytical technique, which is applicable to the analysis of various solid materials in organic, inorganic, energy, biological, medical, electronic, geological/archaeological, environmental, and other fields. It has high detection sensitivity, which can reach $10^{-6} \sim 10^{-9}$ orders of magnitude (varying from element to element), high mass resolution (mass resolution >10,000) and high spatial resolution (<50 nm), and can not only detect all elements and their isotopes, including H–U, but also can detect monoatomic ions or low mass number fragment ions, as well as high mass number fragment ions or molecular ions, and can also identify different organic compound molecules. The technique can perform mass spectrometry, imaging and depth profiling, and can also characterize the spatial distribution of different elements and compounds in two- and three-dimensional ion images. Compared to other analytical techniques (e.g. Nano-SIMS), TOF-SIMS is able to reduce damage to the sample surface to less than 1% of the single-atom/molecule layer by injecting a lower dose of ions to the sample surface during the test.

TOF-SIMS is a highly sensitive surface analysis technique that provides elemental, isotopic, and molecular information from the outermost few atomic layers of a sample (typically 1–3 nm), with capabilities for 2D chemical mapping and 3D depth profiling. In TOF-SIMS, a pulsed high-energy primary ion beam (e.g., $Bi3^+$ or Arn^+ cluster ions) bombards the sample surface (Figure 11.5). The resulting energy transfer through collisional cascades enables surface atoms and molecules to overcome their binding energy and be sputtered from the material. A small fraction of these ejected particles become ionized via electron exchange or chemical ionization processes, forming secondary ions that carry chemical information from the topmost surface region. These secondary ions are accelerated by a high extraction voltage (typically several kilovolts) to a constant kinetic energy (KE = qV) and directed into

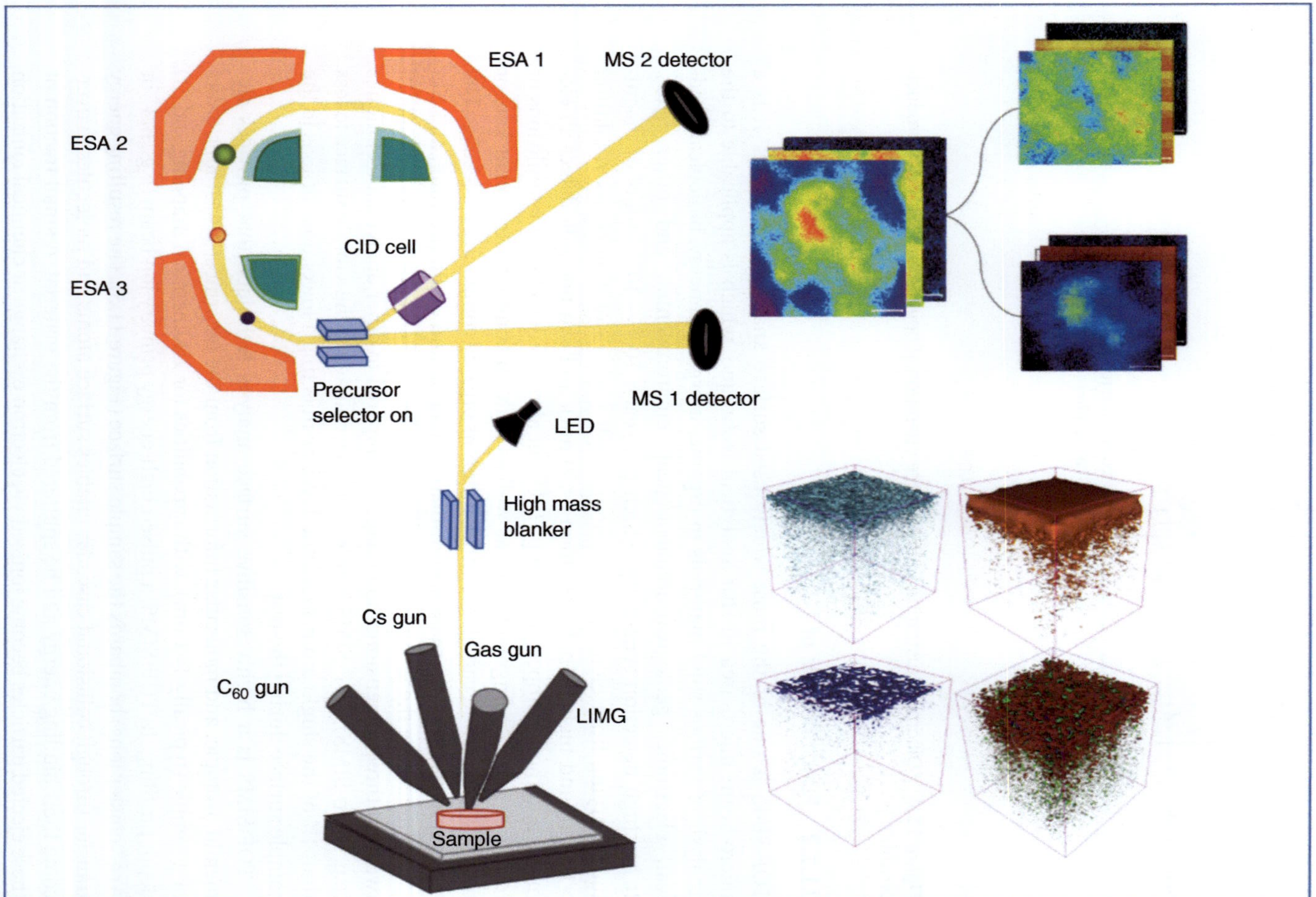

Figure 11.5 Schematic diagram of TOF-SIMS working principle.

a field-free drift tube. According to the relationship KE = 0.5 mv2, their velocity is inversely proportional to the square root of their mass-to-charge ratio (m/z). Thus, lighter ions reach the detector earlier than heavier ones. By precisely measuring the flight time (t) from the emission pulse to detection with a microchannel plate (MCP) assembly, the m/z values are determined, converting time-domain data into a high-resolution mass spectrum (mass resolution >10,000). The pulsed detection mode allows simultaneous acquisition of the full mass range per incident pulse. When combined with primary beam rastering, submicrometer-scale chemical imaging is achieved. Static SIMS (using low ion doses) preserves molecular information, whereas dynamic SIMS (high ion doses) enables depth profiling. Through time-to-mass conversion, TOF-SIMS reveals detailed elemental distributions, molecular composition, and 3D chemical architecture of material surfaces, making it particularly valuable in energy materials research for analyzing electrode interfaces, catalyst surfaces, and SEIs.

TOF-SIMS foundational concept dates back to the mid-20th century. In 1946, American physicist W. E. Stephens first systematically articulated the principle of time-of-flight mass spectrometry (TOF-MS), which determines the mass-to-charge ratio of ions by precisely measuring their flight time through a field-free drift tube. This concept laid the theoretical groundwork for subsequent high-speed, full-spectrum detection techniques. By the late 1970s, a pivotal breakthrough was achieved through the convergence of two technologies. German scientist Alfred Benninghoven and his research team made significant advancements in static secondary ion mass spectrometry (S-SIMS) by successfully integrating a TOF mass analyzer with SIMS. This fundamental integration leveraged the parallel detection capability of TOF-MS, enabling the acquisition of complete mass spectra under extremely low primary ion dose conditions that a key requirement for static surface analysis. Early instruments primarily employed pulsed single-atom primary ion beams (such as Ar^+ or Ga^+) coupled with linear flight tube configurations. The 1980s marked a critical period of instrumental development and initial commercialization for TOF-SIMS. As related technologies matured, the instrument design became standardized in the early part of the decade. The founding of the German company IONTOF in 1983 represented a major milestone in the industrialization of the technique. By the mid-to-late 1980s, the introduction of the first commercial system (e.g., TOF-SIMS I) signified its transition from a laboratory prototype to a practical analytical tool. Since the beginning of the 21st century, continuous improvements in performance and the establishment of robust commercial platforms have led to the widespread adoption of TOF-SIMS in both industrial and academic research. Currently, the high-end TOF-SIMS instrument market is primarily served by two leading manufacturers: Germany's IONTOF (with product series such as TOF.SIMS 5 and TOF.SIMS M6) and Japan's PHI (with representative models including the nanoTOF II and nanoTOF 3+). These companies continue to drive progress in the field, focusing on enhancements in spatial resolution, mass spectral accuracy, and advanced three-dimensional analytical capabilities.

11.2 Applications of DEMS in Battery Research

In the field of lithium battery research, DEMS technology can detect gases that traditional electrochemical methods cannot capture, thereby enabling analysis of the reaction mechanisms responsible for their generation. During in-situ cell operation, DEMS allows both intermediate gases produced during reactions and final products to be detected and quantitatively analyzed [4]. By making invisible gas information visible through DEMS, elucidating reaction mechanisms and subsequently optimizing performance, this technology has become an essential tool in lithium-ion battery research [5, 6].

11.2.1 Elucidation of Electrolyte Decomposition Mechanisms

Currently, the electrolyte in batteries consists mainly of organic components, which leads to severe gas evolution issues. This also causes gas to build up during prolonged battery operation, which poses certain hazards. Therefore, employing DEMS's in-situ quantitative detection mechanism to track gas evolution behavior and correlate it with electrochemical reactions over time enables the gas generation mechanism to be investigated. This approach effectively suppresses electrolyte gas evolution, enhances battery operational safety and provides valuable information for engineering applications.

The flammability of commercial carbonate-based electrolytes is a significant issue, as it increases the risk of thermal runaway in batteries during accidents. In particular at high voltages, the electrolyte decomposes readily to produce CO_2 gas, causing the battery to swell. This further increases safety risks and reduces service life. Although high-concentration electrolytes can improve high-voltage stability and flame retardants can mitigate safety hazards, their limited application and development is hindered by their high viscosity, low ionic conductivity and high cost. Therefore, developing an electrolyte that exhibits both intrinsic nonflammability and high stability at high voltages is essential for achieving high-energy-density, high-safety lithium-ion batteries. Zeng et al. [7] investigated the behaviors of gas generation at the interface between a TFEP-based electrolyte and lithium-rich manganese oxide (LRMO) cathodes, designing a novel fluorophosphate-based electrolyte and employing in-situ techniques, including DEMS. The results contrasted starkly with those from traditional carbonate electrolytes (as shown in Figure 11.6): In conventional electrolytes, significant and persistent carbon dioxide release was observed starting at 4.4 V, indicating severe oxidative decomposition. By contrast, no detectable carbon dioxide signals were observed throughout the high-voltage charge-discharge cycle using the TFEP-based electrolyte, which demonstrates its exceptional oxidative stability.

Leveraging this key attribute, the battery exhibited outstanding electrochemical performance. The LRMO cathode achieved a high initial Coulombic efficiency of 78%, a high reversible capacity of approximately 225 mAh/g, and excellent long-term cycling stability (72% capacity retention after 100 cycles). Simultaneously, the

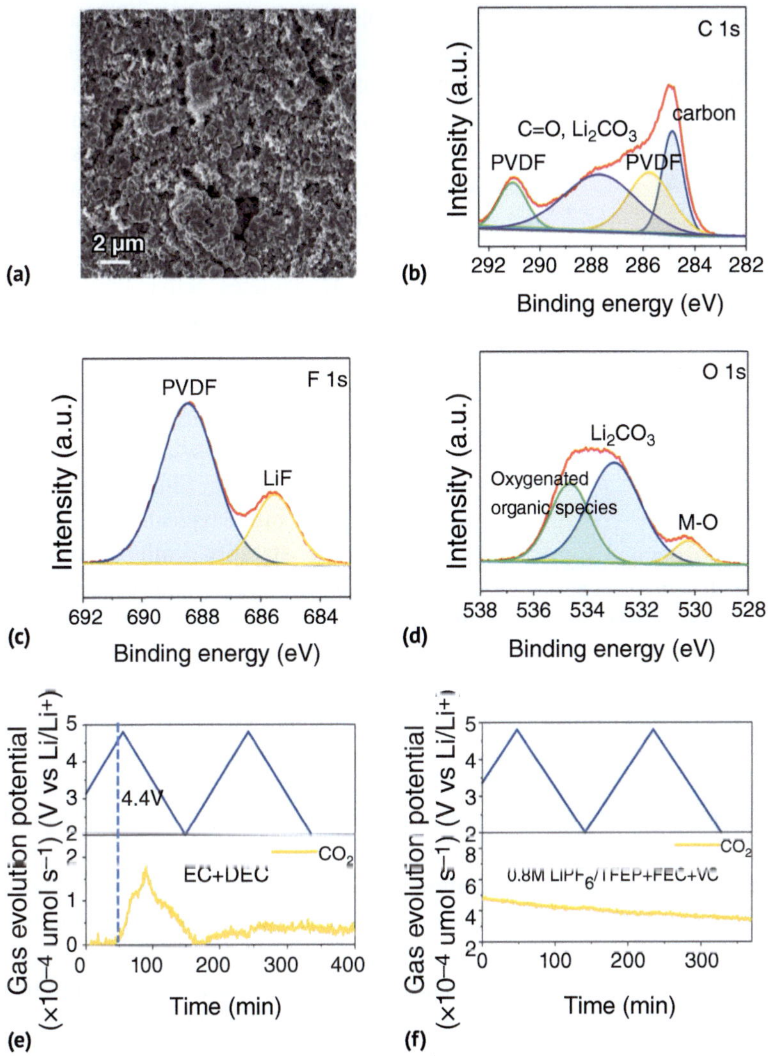

Figure 11.6 Decomposition performance of TFEP-based electrolyte on the LRMO cathode. (a) SEM image and (b–d) XPS spectra of the LRMO electrodes cycled in 0.8 M $LiPF_6$/TFEP + FEC + VC electrolytes for 10 cycles. B, C1s; C, F1s; D, O1s. DEMS data for LRMO electrode in (e) 1.0 M $LiPF_6$/ethylene carbonate (EC): DEC (1:1 by vol) electrolyte (e) and TFEP-based electrolyte (f), respectively, at a scan rate of 0.5 mV s−1. Blue line: scanning potential. Yellow line: evolution rate of CO_2 [7] / with permission of John Wiley & Sons.

synergistic effect of fluoroethylene carbonate (FEC) and vinylene carbonate (VC) additives facilitated the formation of a dense and robust SEI on the silicon-silicon carbide-carbon (SSC) anode, ensuring effective compatibility with the high-capacity anode material. Ultimately, the full cell configured with the SSC anode and LRMO cathode delivered an energy density exceeding 500 Wh/kg, while also achieving a high level of safety inherent to the nonflammable electrolyte.

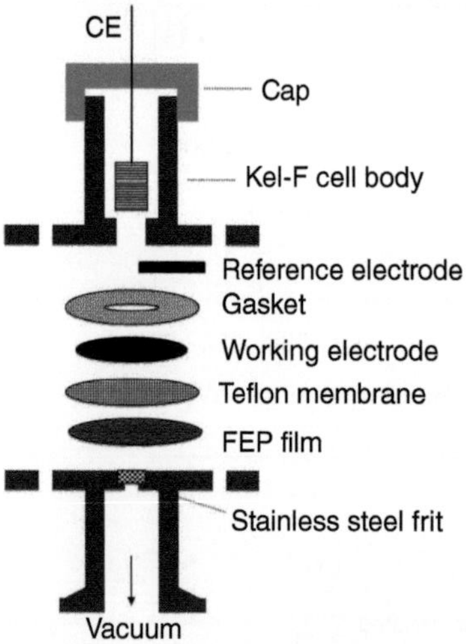

Figure 11.7 Schematic diagram of the DEMS cell for Li-ion battery studies [8] / with permission of American Chemical Society.

High-voltage cathode materials, such as $LiNi_{0.5}Mn_{1.5}O_4$ with an operating voltage of approximately 4.7 V, are pivotal for enhancing the energy density of lithium-ion batteries. However, their operating potentials often exceed the thermodynamic stability window of conventional carbonate-based electrolytes. Electrolyte decomposition at high voltages not only depletes active lithium and causes performance degradation but also leads to gas generation (CO_2 and O_2), resulting in cell swelling and potentially triggering thermal runaway and explosion, thereby posing serious safety risks. When high-voltage cathode materials are employed, their operating voltage range typically exceeds the thermodynamic stability limit of conventional electrolyte solvents. Consequently, the interfacial stability between the electrode and the electrolyte emerges as a critical research focus. Furthermore, in many in-situ DEMS cells that utilize membranes for sampling, issues such as significant solution resistance and electrode resistance are commonly encountered. The accuracy of the collected data can be significantly improved if the in-situ cell configuration better simulates the actual environment of real battery reactions. To address these problems Wang et al. designed a three-electrode in-situ DEMS cell configuration as shown in Figure 11.7 [8]. This setup allows for a closer simulation of the actual operating conditions of lithium-ion batteries. The use of practical, commercially available lithium-ion battery electrode sheets enables the simulation of in-situ electrochemical reactions and facilitates subsequent DEMS investigation. It is asserted that the designed in-situ cell specifically detects gases generated solely at the working electrode. This design effectively minimizes interference from gases produced at the counter electrode, thereby enhancing the accuracy of the measurements.

Based on a systematic comparison of the electrochemical reactions between three electrolytes (propylene carbonate (PC), EC/DMC, and EC/DEC) and different electrodes regarding the resulting gas evolution behavior of CO_2 and O_2, they found that among the three electrolytes, the EC/DMC electrolyte was the most stable, while EC/DEC exhibited the poorest stability. With the LCO electrode, only CO_2 evolution was detected below 5 V in the EC/DEC electrolyte, as shown in Figure 11.8, whereas for $LiMn_2O_4$, the onset potential for CO_2 evolution was the highest (approximately 4.9 V) in the EC/DMC electrolyte. In the PC electrolyte, CO_2 evolution began at around 4.7 V. In the EC/DEC electrolyte, its stability was the worst, showing a high CO_2 production rate and slow kinetics. Furthermore, they analyzed the possible pathways for gas evolution. On $LiMn_2O_4$ and $LiNi_{0.5}Mn_{1.5}O_4$, CO_2 primarily originates from the direct oxidation of the solvent. In the $LiCoO_2$/EC/DEC system, the high-voltage reaction causes the release of lattice oxygen from LCO, which

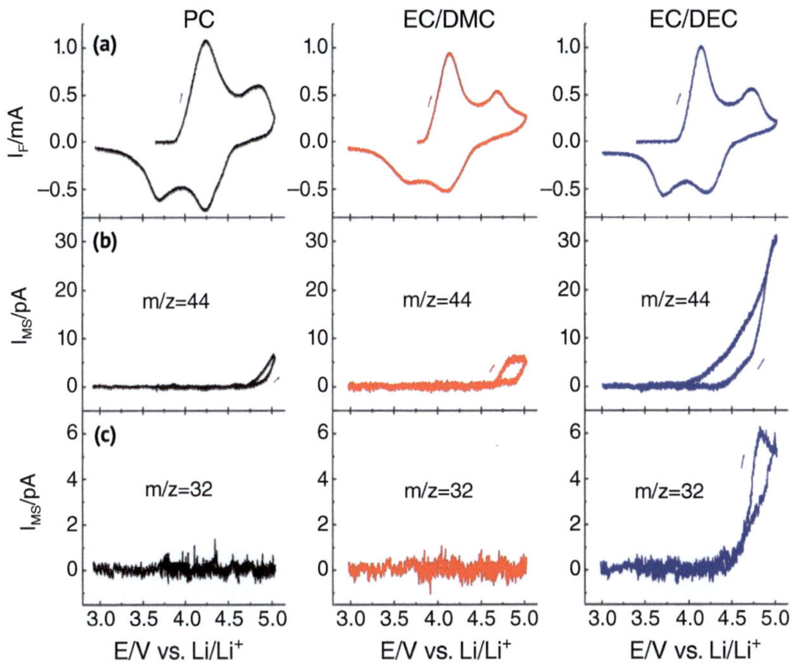

Figure 11.8 DEMS data for $LiCoO_2$ electrode in 1 M $LiPF_6$ + PC (left panel), 1 M $LiPF_6$ + EC/DMC (1:1 wt %) (middle panel), and 1 M $LiPF_6$ + EC/DEC (1:1 wt %) (right panel), respectively, at a scan rate of 0.2 mV/s. (a) Electrochemical data, (b) intensity of mass spectra at m/z = 44 (CO_2^+), (c) intensity of mass spectra at m/z = 32 (O_2^+) [8] / with permission of American Chemical Society.

subsequently forms O_2. This O_2 may then catalytically react with the electrolyte, leading to an increase in CO_2 gas production. Furthermore, compared to $LiCoO_2$, the spinel-structured $LiMn_2O_4$ and $LiNi_{0.5}Mn_{1.5}O_4$ exhibited relatively moderate catalytic activity towards electrolyte oxidation.

Through this innovative DEMS methodology, it was confirmed that evaluating the high-voltage stability of an electrolyte must account for its combination with specific cathode materials. This study clearly demonstrates the superiority of the EC/DMC system and provides crucial experimental evidence and profound insights for understanding and designing safer, high-energy-density lithium-ion batteries.

11.2.2 Mechanism of Cathode Gas Evolution

The development of high-energy-density lithium-ion batteries is heavily reliant on nickel-rich layered cathode materials such as NMC622. However, during synthesis and storage, these materials are prone to the formation of intrinsic defects like Li/Ni cation mixing and oxygen vacancies, as well as the retention or environmental reaction-induced formation of contaminants such as Li_2CO_3 and LiOH. These initial surface states can trigger severe interfacial side reactions under high-voltage conditions (>4.4 V vs. Li/Li^+), leading to electrolyte oxidative decomposition and loss of lattice oxygen from the cathode, accompanied by the generation of significant amounts of CO_2 and O_2 gases. Currently, commercially available high-energy-density lithium-ion batteries primarily utilize nickel-rich layered cathode materials. However, during the synthesis and storage of nickel-rich ternary cathode materials, residual alkali (such as Li_2CO_3 and LiOH) from reaction byproducts and synthesis processes remains on the surface. Additionally, high-nickel layered cathode materials are prone to defects like lithium/nickel cation co-stacking and oxygen vacancies. These residual alkalis on the material surface can trigger severe interfacial side reactions under high-voltage conditions (>4.4 V vs. Li/Li^+). This leads to electrolyte oxidation decomposition and loss of lattice oxygen from the cathode, accompanied by the generation of large amounts of CO_2 and O_2 gas. The substantial gas production from these detrimental side reactions causes battery swelling and poses severe thermal runaway risks. During these side reactions, the depletion of active lithium and electrolyte also contributes to capacity fading, consequently reducing the battery's operational lifespan. While research indicates that simple acid-washing or water-rinsing treatments can enhance battery performance, it is still uncertain whether the fundamental mechanism operates primarily by removing residual alkali or by modifying the material's surface structure. Thus, systematically differentiating the individual roles of surface residual alkali and surface structure in gas evolution and interfacial reconstruction, along with clarifying the intrinsic mechanisms, is essential for guiding the development and production of highly stable and safe cathode materials. McCloskey et al. [9] developed an experimental methodology using DEMS to detect gas evolution at material interfaces. Combining acid-base titration and isotope tracing methods to analyze the sources of various gases, they achieved simultaneous detection of gas evolution behavior during high-voltage cycling of cathode materials as shown in Figure 11.9. Using ^{18}O-labeled $LiNi_{0.6}Mn_{0.2}Co_{0.2}O_2$ (NMC622) as the host material, they employed multiple washing methods (methanol and water rinsing), subsequent heat treatment, and intentional surface residual alkali addition to precisely elucidate the influence of surface state on nickel-enriched cathode interface behavior. First, DEMS results indicate that all surface modifications significantly alter gas evolution behavior. Furthermore, combined with isotope analysis, they conclude that CO_2 during charging primarily originates from high-voltage decomposition of the electrolyte, surface residual lithium carbonate, and C*OO generated by the reaction between the cathode material and electrolyte after high-voltage oxygen evolution. Firstly, all surface modifications significantly altered the gas evolution behavior. Notably, the prolonged water soaking treatment (^{18}O-soak) was most effective, nearly completely suppressing the evolution of both O_2 and CO_2 at the high voltage of 4.8 V, highlighting the considerable potential of simple treatments for enhancing interfacial stability. Secondly, through isotopic analysis, the researchers successfully traced the sources of different CO_2 species: samples with intentionally added surface Li_2CO_3 released large amounts of CO_2, confirming its direct origin from surface carbonate decomposition, while the evolution of C*OO was closely associated with surface defects such as lithium vacancies and hydroxides. Additionally, post-cycling titration analysis revealed that the extent of irreversible peroxo-like surface reconstruction formed after the first cycle correlated positively with the initial coverage

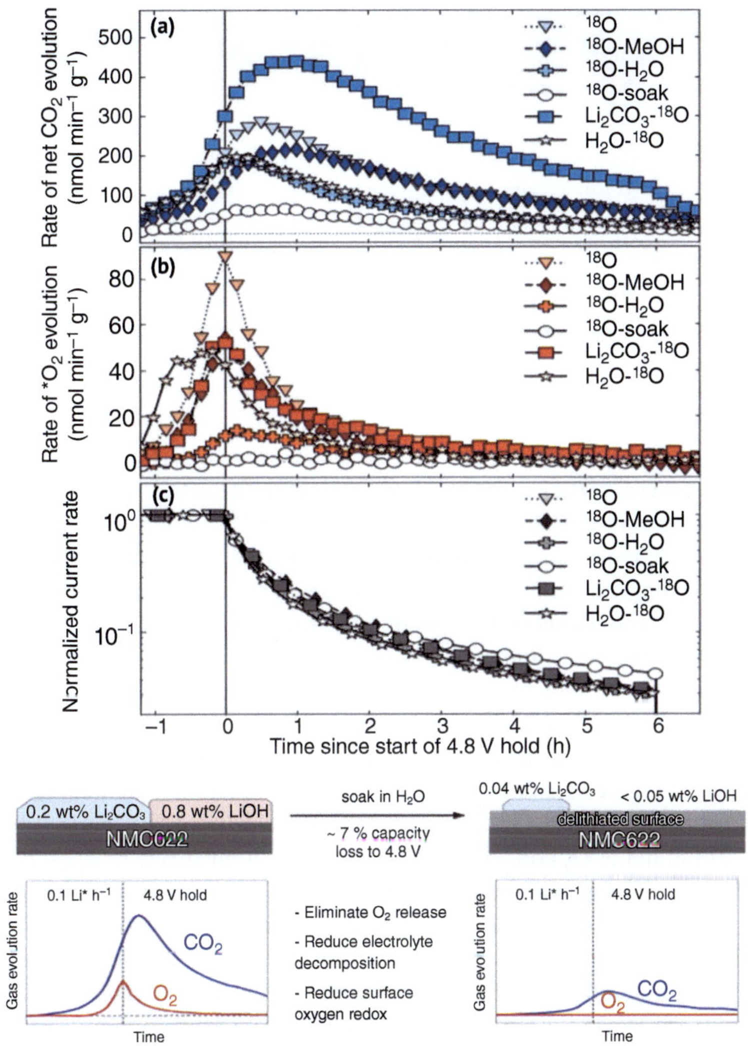

Figure 11.9 Total evolution rates of O_2 and net CO_2 for all NMC622s during the 4.8 V hold. All NMC622s were galvanostatically charged at 0.1 Li^+/h (27.65 mA/g) until 4.8 V, potentiostatically held at 4.8 V for 6 h at 0.1 Li^+/h (27.65 mA/g). (a) shows the net rate of all isotopes of CO_2 evolution and (b) the net O_2 evolution. All NMC622 preparations evolved CO_2 and all but ^{18}O-soak evolved O_2. (c) presents the normalized current (dividing by the galvanostatic rate, 0.1 Li^+/h 27.65 mA/g) on a log scale. Time = 0 h corresponds to the start of the 6 h 4.8 V hold [9] / with permission of American Chemical Society.

of surface contaminants, and all washing treatments reduced this reconstruction extent. These findings collectively indicate that gas evolution and surface reconstruction in nickel-rich cathodes under high voltages are not solely determined by bulk properties but are the result of complex interfacial reactions predominantly governed by the initial surface contaminants (Li_2CO_3) and defects (lithium vacancies).

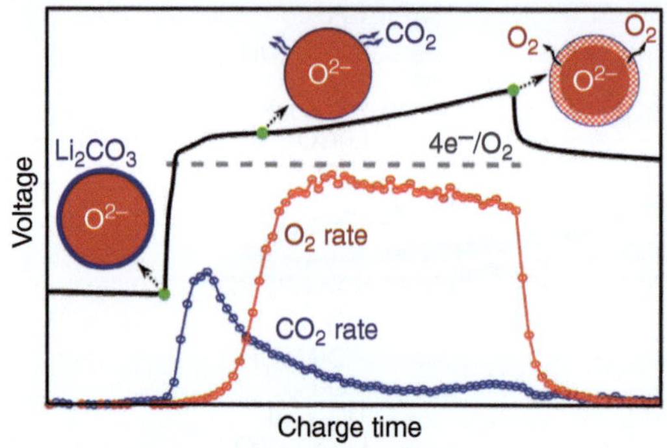

Figure 11.10 Gas evolution and surface studies of Li_2MnO_3 [10] / with permission of American Chemical Society.

In numerous previous studies, it has been observed that Li_2MnO_3 exhibits an abnormally high capacity during its first charge. Rana et al. [10] systematically evaluated all possible charge compensation mechanisms, including bulk oxygen redox, manganese oxidation, and surface degradation, in an attempt to explain the origin of the high capacity. They employed the quantitative capabilities of DEMS in in-situ detection by conducting real time in-situ monitoring of Li_2MnO_3 during the first charging process. This allowed precise tracking and quantification of the evolution kinetics and cumulative amounts of O_2 and CO_2, thereby providing the most direct experimental evidence for the capacity contributions from different reaction pathways (as shown in Figure 11.10). The quantitative DEMS data ultimately revealed the true source of the capacity: the measurements demonstrated that during the first charge, 113 mmol of O_2 and 38 mmol of CO_2 were cumulatively released per mole of active material. Calculations demonstrate that O_2 release alone contributed 125 mAh/g, accounting for 71% of the total capacity, while CO_2 evolution contributed 17 mAh/g, representing approximately 10%. Together, these two gas evolution processes, directly quantified by DEMS, constituted 81% of the total capacity. This conclusive evidence proves that the activation capacity of Li_2MnO_3 primarily originates from irreversible lattice oxygen loss and electrolyte decomposition, rather than the previously hypothesized reversible bulk anionic redox, fundamentally revising the understanding of this material's electrochemical behavior.

Traditionally, it has been widely believed that the CO_2 and CO gases generated during the initial charging process of lithium-ion battery cathode materials are largely due to the oxidative decomposition of carbonate-based electrolytes under high voltage. McCloskey et al. [11] investigated the root cause of CO_2 and CO gas evolution during the first charging cycle of cathode materials. Through precise and meticulous experimental design and rigorous analysis, they found that the residual alkali, namely Li_2CO_3, on the cathode surface is the key factor contributing to this issue. To precisely identify the gas sources, the researchers selected two typical layered transition metal oxide cathodes: lithium-rich (LMR) and Li-stoichiometric (NMC). They innovatively combined isotopic labeling (^{18}O) with a surface carbonate titration method, establishing a robust foundation for subsequent gas source tracing.

During the investigation, DEMS played an indispensable and central role. Leveraging the high sensitivity and real time monitoring capability of DEMS, the researchers were able to dynamically track and quantify the evolution of various gaseous products throughout the charging process. Crucially, the isotopic resolution capability of DEMS served as the definitive tool for identifying gas origins. Experimental data (Figure 11.11) revealed that the oxygen isotope (^{18}O) composition of the CO_2 and CO gases released across the entire charging voltage range (up to 4.8 V vs Li/Li^+) closely matched the isotopic signature of the residual surface Li_2CO_3. In contrast, it differed significantly from the isotopic background of the cathode lattice oxygen or the electrolyte. This direct evidence at the molecular level strongly demonstrates that the CO_2/CO gases evolved during the first cycle originated primarily from the oxidative decomposition of surface Li_2CO_3, rather than from the previously assumed catalytic oxidation of the electrolyte. Further validation was achieved by partially removing the surface Li_2CO_3 via a mild acid wash. DEMS monitoring confirmed a proportional and significant decrease in CO_2 evolution from the treated samples, providing strong corroborating evidence for the conclusion. Additionally, the DEMS data uncovered an unexpected correlation: the presence of surface Li_2CO_3 was also positively correlated with the amount of oxygen (O_2) released from the cathode bulk lattice. This was evidenced by the acid wash treatment simultaneously suppressing both CO_2 and O_2 evolution. This finding challenges the conventional understanding of the oxygen release mechanism from cathodes and implies a complex interaction between surface contaminants and bulk structural stability. In summary, the definitive evidence provided by DEMS technology has successfully shifted the primary responsibility for first-cycle gas evolution from the electrolyte to residual surface Li_2CO_3. This finding not only revises the fundamental understanding of electrochemical reactions at cathode surfaces but also underscores the significant value of DEMS in battery mechanism research, particularly in accurately tracing complex reaction products. These results alert the battery research community to the critical importance of cathode material surface purity and point toward new directions for enhancing battery performance and safety.

McCloskey's team conducted an in-depth investigation into the role of residual surface lithium carbonate (Li_2CO_3) in lithium-excess layered oxide cathode materials [12]. While conventional understanding often attributes gas evolution during the first charge of batteries to the oxidative decomposition of the electrolyte at high voltages, this study, through sophisticated experimental design, revealed that the core function of surface Li_2CO_3 extends far beyond this. To precisely trace the reaction pathways, the researchers first synthesized model cathode materials with "carbonate-free" surfaces in a strictly CO_2 and H_2O free environment. They then controllably exposed these materials to $^{13}CO_2$ or $C^{18}O_2$ gas, depositing Li_2CO_3 with known content and isotopic composition on the surfaces. By combining DEMS with acid titration techniques, the study found that Li_2CO_3 undergoes direct electrochemical oxidation at approximately 4.0 V (vs Li/Li^+), releasing CO_2 without concomitant O_2 evolution. Their experimental results, as shown in Figure 11.12, indicating the generation of other highly reactive oxygen species. More critically, the oxidation of surface Li_2CO_3 triggers the release of reactive oxygen from the bulk

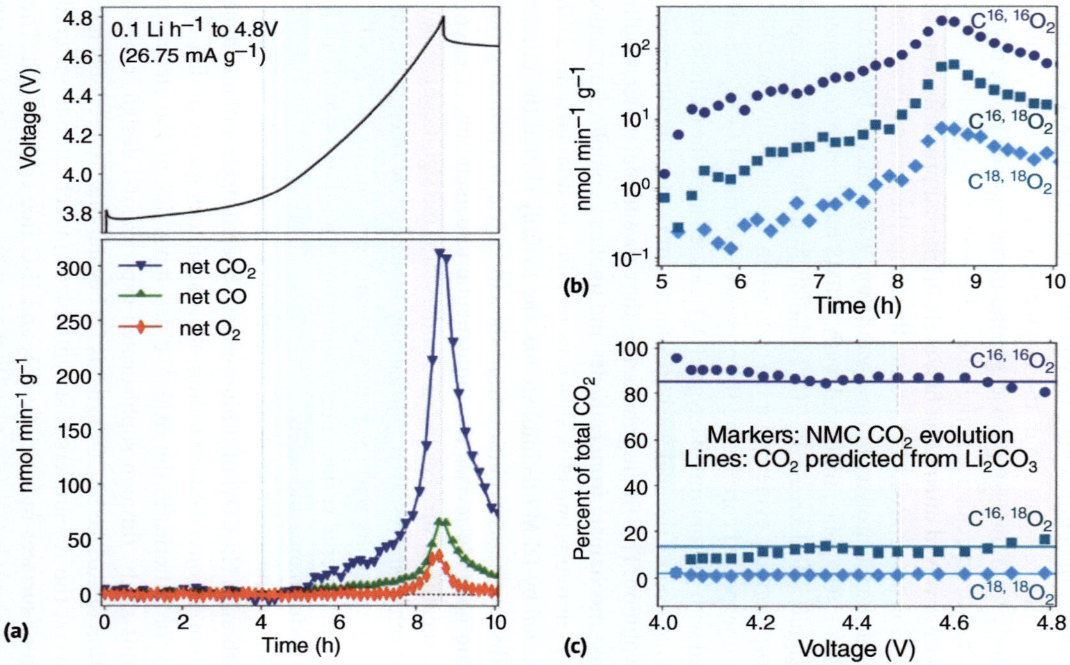

Figure 11.11 Gas evolution for ^{18}O-LMR. (a) Net gas evolution for all CO_2, CO, and O_2 isotopes for the first charge of ^{18}O-NMC followed by a rest period. The lines are to guide the eye. (b) Detail of $C^{16,16}O_2$, $C^{16,18}O_2$, and $C^{18,18}O_2$ evolution plotted on a partial log scale, showing detection of all isotopes in each region. (c) Distribution of the CO_2 evolution versus voltage, with the lines representing the predicted distribution from Li_2CO_3 degradation [11] / with permission of American Chemical Society.

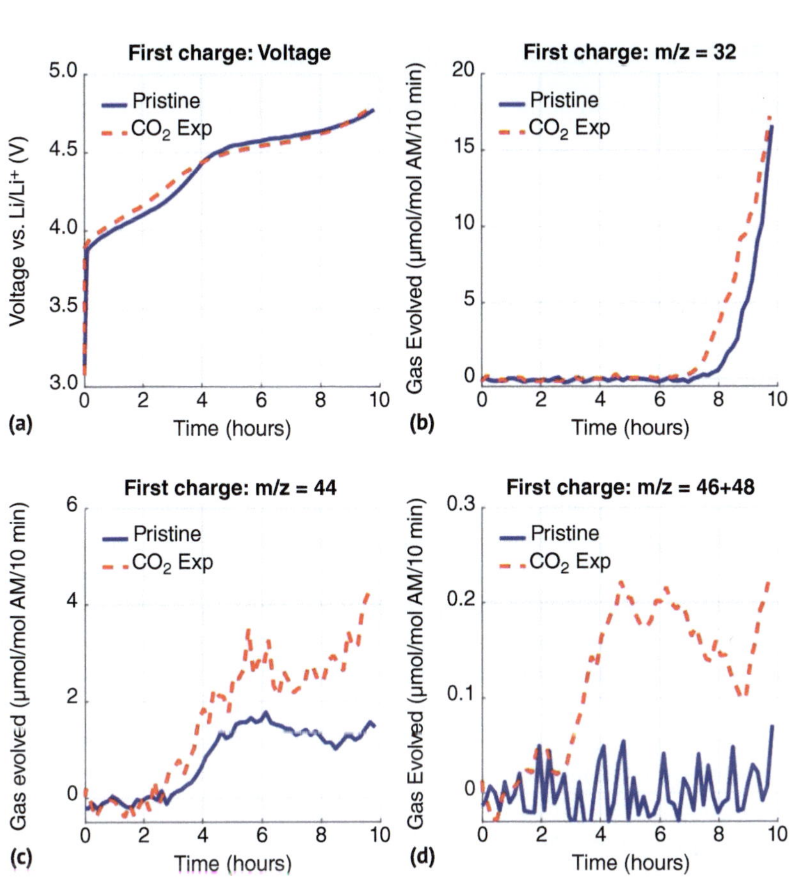

Figure 11.12 Gas evolution during first charge at 0.1 C for pristine (blue) and $C^{18}O_2$ gas-exposed (red) NMC samples. While the overall capacity and voltage profiles are similar between samples (a), there are slight alterations to the gas evolution due to the deposition of $Li_2C^{18}O_3$ on the surface including a slight lowering of the onset potential of oxygen evolution (b), an increase in the amount of $C^{16}O_2$ evolution (c), and the appearance of $C^{18}O_2$ evolution (d), which approximately mirrors the pattern of $C^{16}O_2$ evolution [12] / with permission of American Chemical Society.

cathode lattice and significantly exacerbates the degradation of the carbonate-based electrolyte. Using isotopic labeling, the research obtained direct evidence of reactive oxygen attacking the electrolyte—detecting electrolyte degradation fragments containing diatomic oxygen, where the oxygen atoms primarily originated from the cathode lattice rather than from Li_2CO_3 itself. This indicates that the oxidation process of surface Li_2CO_3 destabilizes the cathode surface structure, subsequently inducing the release of lattice oxygen in highly reactive forms, which then become the primary agents directly attacking and decomposing the electrolyte. This study revises understanding of the electrochemical reactions at the cathode surface and emphasizes the critical importance of strictly controlling and removing surface Li_2CO_3 for the development of high-performance and highly stable lithium-excess cathode materials.

11.2.3 SEI Formation Mechanism at the Anode Interface

Unnecessary gas evolution at the anode directly impairs battery performance and safety: on one hand, the accumulation of flammable products such as hydrogen and ethylene can easily trigger combustion or even thermal runaway; on the other hand, this process continuously depletes active materials, leading to electrode aging, capacity degradation, shortened cycle life, and increased impedance. Therefore, real time monitoring of anode gases is critical for identifying gas-related issues and maintaining high performance. According to conventional understanding, gas evolution at graphite anodes in carbonate-based electrolytes primarily originates from the reductive decomposition of EC, producing ethylene (C_2H_4). However, trace water is inevitably introduced during actual battery manufacturing processes. Its interaction with the highly reactive graphite anode and lithium salts may lead to additional gas evolution (such as H_2 and CO_2), potentially causing safety issues like battery swelling. To accurately quantify the types and amounts of gases generated under different conditions and clarify the underlying reaction mechanisms, Bernhard [13] employed DEMS technology, enabling real time, highly sensitive monitoring of trace gases evolved during battery cycling. Systematic comparisons were conducted for the formation processes of graphite electrodes in both dry electrolyte (H_2O < 20 ppm) and water-containing electrolyte (4000 ppm H_2O). DEMS results clearly demonstrated that in the dry electrolyte, gas evolution primarily consisted of C_2H_4 from EC reduction and minimal H_2 (originating from residual trace water). In contrast, in the water-containing electrolyte, significant H2 and considerable CO_2 evolution were observed in addition to C_2H_4 as shown in Figure 11.13. Notably, the study revealed that a pre-formed, stable SEI layer can substantially suppress the water reduction reaction. By pre-cycling the graphite electrode in an electrolyte containing VC, a robust SEI layer was constructed. DEMS monitoring data showed that for such graphite electrodes with a VC-derived SEI in the water-containing electrolyte, the amount of H_2 evolution was reduced by approximately a factor of 7.5 compared to untreated pristine electrodes. This comparative experiment strongly demonstrates that a complete and compact SEI layer not only prevents continuous electrolyte reduction but also effectively impedes contact between water molecules and the anode surface, thereby kinetically inhibiting H_2 generation.

PC is considered an ideal solvent for constructing low-temperature electrolytes due to its low melting point (−49.2 °C). However, its co-intercalation with lithium ions into the graphite layers leads to graphite exfoliation and battery failure, which limits its application in commercial batteries. To overcome this limitation, this work proposed an innovative two-step strategy aimed at unlocking the potential of PC in low-temperature batteries by pre-forming a stable SEI on the graphite anode to suppress PC co-intercalation. Chen et al. et al. [14] first cycled the graphite electrode (using highly ordered pyrolytic graphite, HOPG, as a model) in an electrolyte of 1M $LiPF_6$ in FEC: DMC (1:1 by volume) to pre-form an initial SEI layer. Subsequently, this pre-formed SEI-coated graphite electrode was placed into the target low-temperature electrolyte—1M LiDFOB in EC: PC: DEC (2:5:3 by volume)—for further cycling. To deeply investigate the mechanism and effectiveness of this

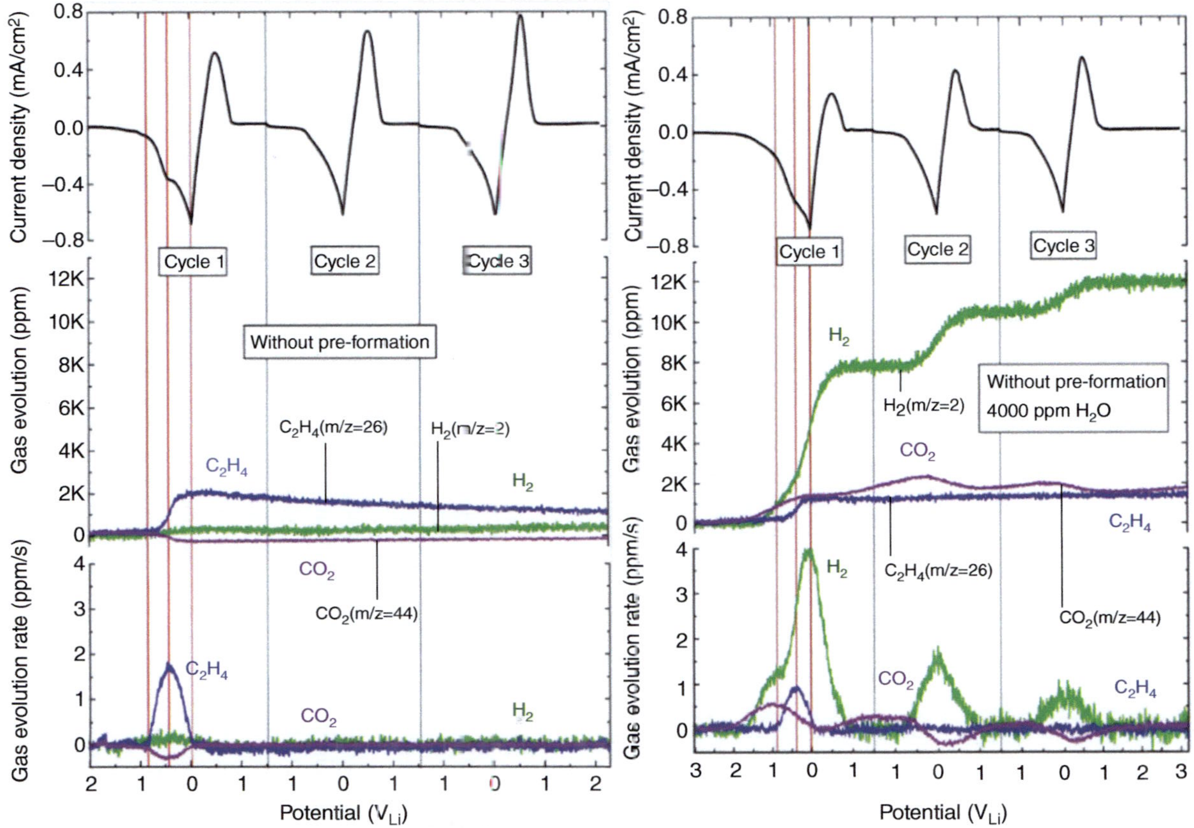

Figure 11.13 Gas evolution during the first three formation cycles of a pristine graphite working electrode vs. a metallic lithium counter electrode in 1 M LiTFSI in EC/EMC (<20 ppm H_2O) and gas evolution during the first three formation cycles of a pristine graphite working electrode vs. a metallic lithium counter electrode in 1 M LiTFSI in EC/EMC containing 4000 ppm H_2O. The upper panel displays the current potential curve corresponding to the CV formation procedure between 0–1.5 V_{Li} at a scan rate of 0.5 mV/s. The middle panel shows the gas concentration in the cell head space during formation in units of ppm for H_2 (m/z = 2), C_2H_4 (m/z = 26), and CO_2 (m/z = 44) [13] / with permission of IOP Publishing Ltd.

strategy, the researchers employed a combination of advanced characterization techniques: in-situ electrochemical atomic force microscopy (EC-AFM) was used to observe the structural evolution and stability of the SEI in real time; DEMS was utilized to precisely monitor gas evolution during battery cycling; and X-ray photoelectron spectroscopy combined with electrochemical impedance spectroscopy were applied to analyze the chemical composition of the SEI and the interfacial impedance. The study found that the SEI film formed directly in the PC-based electrolyte was loose and unstable, failing to prevent PC co-intercalation, which led to surface cracking of the graphite and significant gas generation. However, the situation was fundamentally improved with the pre-formed SEI. EC-AFM results revealed that the pre-formed SEI had a bilayer structure (an inorganic inner layer and an organic outer layer). When exposed to the PC-based electrolyte, the organic top layer partially dissolved, but the robust inorganic bottom layer remained intact and effectively protected the graphite. The DEMS results (Figure 11.14) played a crucial validating role here: for the unprotected graphite, significant amounts of propene (C_3H_6) gas were detected during cycling in the PC electrolyte, providing clear evidence of the PC co-intercalation reaction. In contrast, for the graphite with the pre-formed SEI, DEMS detected no gas signals throughout the entire cycling process, strongly demonstrating that the strategy completely suppressed the PC co-intercalation. Electrochemical tests further confirmed that batteries employing this strategy showed a dramatic improvement in the first-cycle Coulombic efficiency from 56.3% to 97.7%. Moreover, at −20 °C, the discharge capacity could still maintain about 75% of the room temperature capacity, significantly outperforming batteries without the PC electrolyte.

11.3 Applications of TOF-SIMS in Battery Research

11.3.1 Surface and Interface Analysis of Anode Materials

TOF-SIMS analysis, recognized for its capabilities in micro-area in-situ characterization and ultra-high sensitivity detection, has been increasingly applied in secondary battery research. It has demonstrated significant potential for in-depth investigation of electrode surface chemistry, interfacial reaction mechanisms, formation and evolution of the SEI, lithium-ion migration pathways, and battery failure analysis. By precisely analyzing the chemical composition, thickness, depth distribution, and dynamic evolution of electrode samples, TOF-SIMS enables monitoring of electrolyte-electrode interactions and characterization of compositional changes in interfacial layers. This technique provides critical insights into key battery performance aspects such as cycling stability, capacity degradation, and lifetime issues.

The high mass resolution capability of TOF-SIMS enables the accurate identification of specific secondary ion fragments and elements, providing molecular-level evidence for inferring interfacial chemical reaction products and mechanisms. Wu et al. [15] utilized TOF-SIMS to discover that the SEI formed in $LiClO_4$-based electrolytes primarily consists of LiCl, Li_2O, and Li_2CO_3, whereas the $LiPF_6$ system formed an interfacial structure characterized by uniformly distributed LiF coexisting

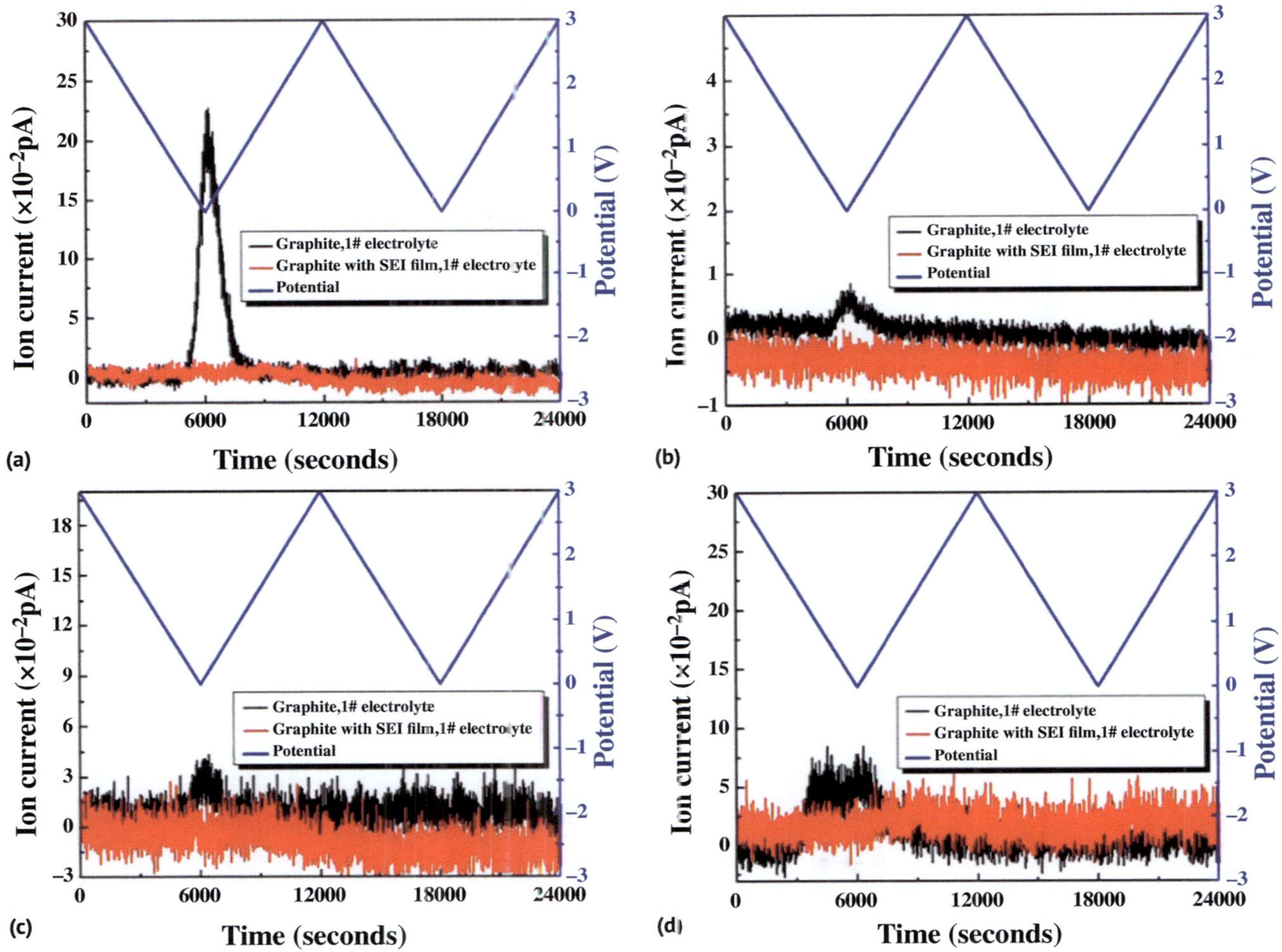

Figure 11.14 Analysis of the gas production, relative molecular mass of gas (a) 42(C_3H_8); (b) 28(C_2H_4); (c) 30(C_2H_6); (d) 44(CO_2) [14] / with permission of Elsevier.

with organic components. Liu et al. [16] further employed TOF-SIMS to reveal that the SEI on a sulfur-modified SiO/C anode exhibited a three-layer gradient distribution in its chemical composition (Figure 11.15a): an outer layer containing sulfur species ($ROSO_2Li$), a middle layer comprising alkyl sulfur compounds and $ROCO_2Li$, and an inner layer of dense LiF inorganic compounds. Similarly, Meng et al. [17] used TOF-SIMS to reveal the unique composition of the SEI on a zinc electrode: organic components (CH^- and CN^-) and inorganic components (ZnS^+ and $CO3^-$)

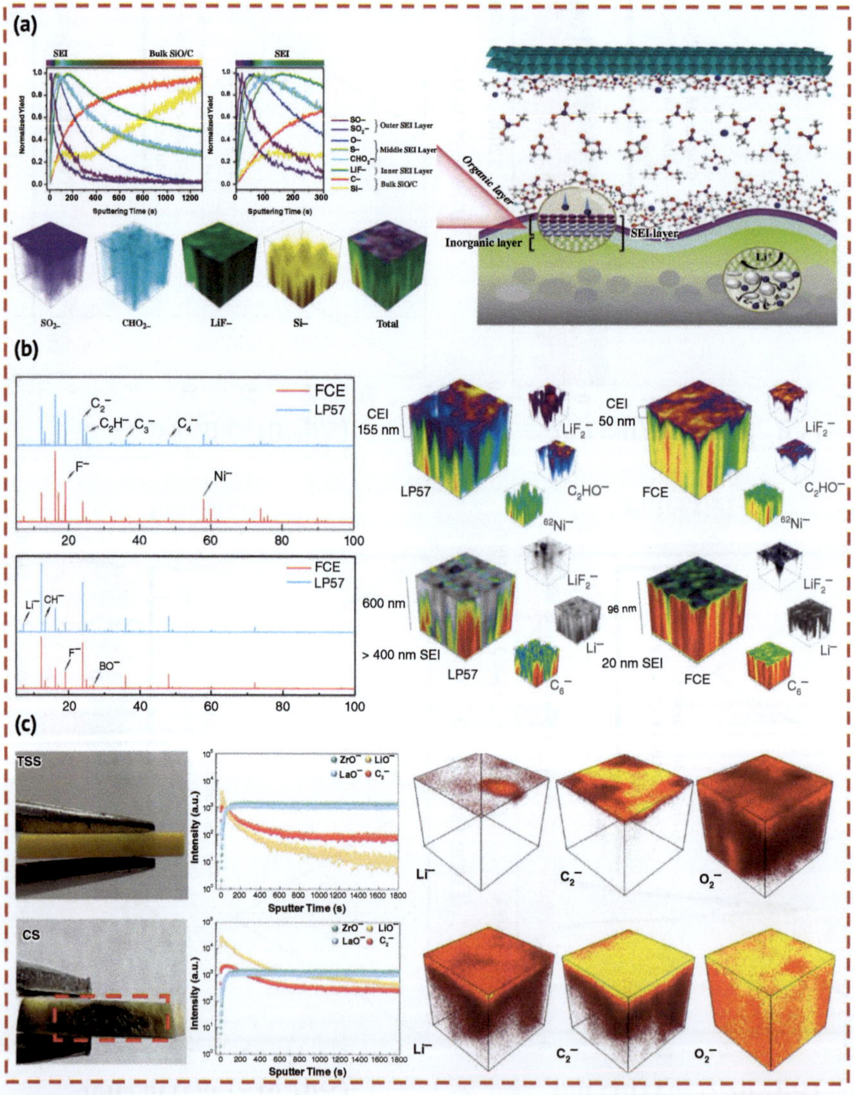

Figure 11.15 (a) TOF-SIMS analysis of the modification layer and schematic diagram of the battery molecular dynamics simulation. Adapted from [2]. Copyright 2022 John Wiley & Sons. (b) TOF-SIMS mass spectra and 3D rendering of the NMC 811 cathode electrode after 1000 cycles. Reproduced from [4] / with permission of John Wiley & Sons. (c) TOF-SIMS analysis of solid electrolytes prepared by two different sintering methods. Adapted from [6].

formed a gradient distribution, constructing a bilayer structure with a crystalline $ZnCO_3$ outer layer and an amorphous ZnS inner layer. Cui and Liu et al. [18] developed a methyl trifluoroacetate (MTFA)-based fast-charging electrolyte, where fluorinated solvent decomposition similarly yielded a uniformly distributed LiF/Li_2CO_3 composite layer. TOF-SIMS imaging confirmed that this electrolyte facilitates the formation of a thinner and more homogeneous SEI (Figure 11.15b), effectively suppressing lithium metal deposition. Li et al. [19] discovered that a high-voltage fluorinated electrolyte (FST) formed an inorganic-rich SEI layer enriched with Li2O and LiF on the silicon-based anode surface through TOF-SIMS analysis, indicating that fluorinated electrolytes significantly mitigate silicon volume expansion effects by enhancing the proportion of the inorganic phase.

Sim et al. [20] utilized TOF-SIMS to compare Ga-LLZO samples prepared by two different sintering methods (Figure 11.15c). Mass spectrometry data revealed the presence of LiO^- and $C2^-$ fragments on the surface of the two-step sintered (TSS) sample, with their concentrations decreasing rapidly in the bulk region, suggesting that the TSS sample effectively suppresses dendrite penetration. Zhang et al. [21] systematically studied the Zr1-LCO-600 electrode modified with surface nanoparticles using TOF-SIMS. They found that key components in its CEI layer, such as LiF_2^-, PO_2F^-, and C_2HO^-, exhibited a gradient distribution from the surface inward, with uniform thickness and complete coverage, indicating that surface engineering can induce the formation of a dense and chemically homogeneous interface protective layer. Chu et al. [22] employed TOF-SIMS to confirm that dual doping with B and W not only effectively suppressed electrolyte decomposition but also promoted the construction of a boron-rich CEI layer, significantly reducing the accumulation of insulating decomposition products.

These studies collectively reflect the unique value of TOF-SIMS mass spectrometry analysis in identifying interfacial chemical compositions. This technique can accurately identify characteristic decomposition products in different electrolyte systems, track the chemical states and distribution patterns of modifiers within the interface, and elucidate the action mechanisms of various interfacial engineering strategies, thereby providing molecular-level evidence for research on interfacial chemical regulation.

11.3.2 Surface and Interface Analysis of Cathode Materials

CEI is formed in lithium-ion batteries during the first charging, which mainly consists of the chemical reaction products between cathode materials and electrolyte, including a complex mixture of inorganic and organic substances. The composition, thickness, stability of CEI film, and its dynamic changes under the battery operating state directly affect the composition, thickness, stability and dynamic changes of the CEI film under the operating condition of the battery directly affect the cycle life, energy efficiency, safety and multiplication performance of the battery, and the analysis of the composition of the CEI film is crucial for understanding the interfacial reaction mechanism inside the battery.

TOF-SIMS provides precise information on interfacial chemical composition and spatial distribution, enabling systematic comparison of interfacial states across different systems to reveal their evolution patterns. Wang et al. [23] investigated the influence of Sc doping on the CEI evolution of ultra-high-nickel cathodes using TOF-SIMS (Figure 11.16a). Their study demonstrated that the Sc-doped Sc-1 sample exhibited a thinner CEI layer compared to the undoped Sc-0 sample. Furthermore, the intersection point of organic species (CH_2O^-) and inorganic species (NiF_3^-) appeared earlier in the Sc-1 sample, confirming significantly suppressed interfacial side reactions. In contrast, the Sc-0 sample displayed a thicker CEI layer containing more CH_2O^- and NiF_3^-, indicating more severe side reactions with the electrolyte. Similarly, Li et al. [24] employed TOF-SIMS to analyze the regulatory effect

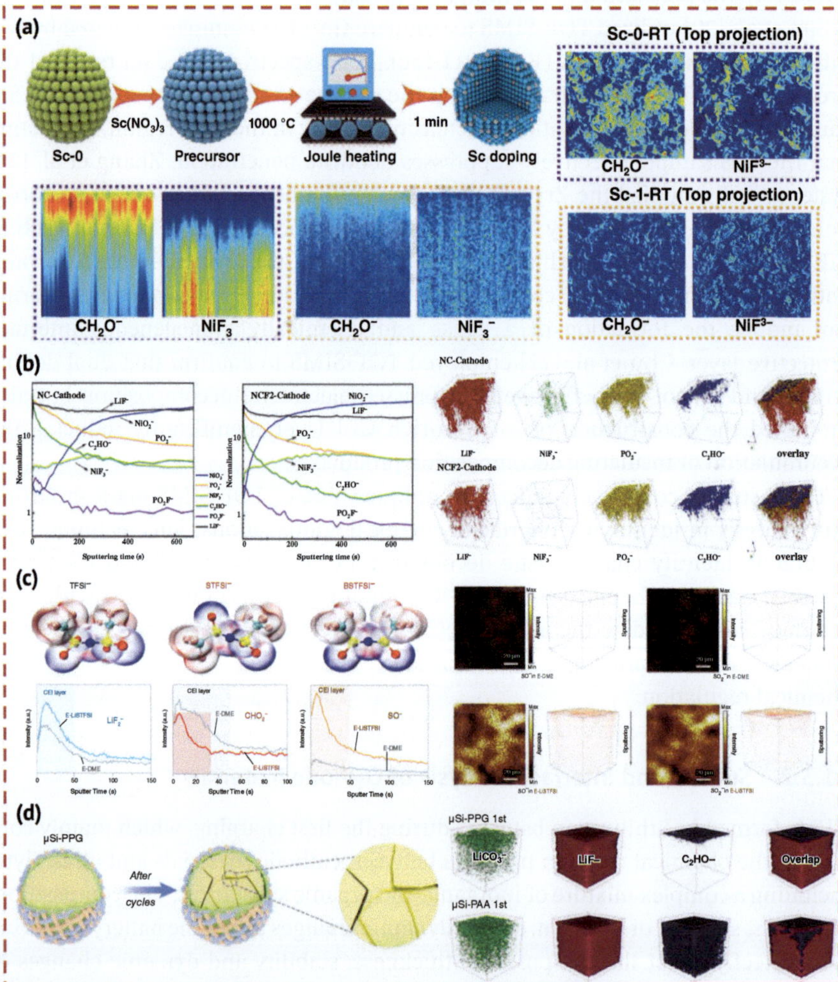

Figure 11.16 (a) Schematic representation of the structural evolution and TOF-SIMS image of CEI on the surface of the Sc-0 and Sc-1. Adapted from [9]. Copyright 2024 Elsevier. (b) Corresponding 3D-reconstruction images of the NC-Li anode and the NCF2-Li anode. Adapted from [10]. (c) TOF-SIMS 3D and 2D distributions of CEIs formed in different electrolytes. Adapted from [11]. (d) μSi electrode with PPG binder and TOF-SIMS 3D image of SEI layer formed after cycling of different electrodes. Adapted from [12].

of fluorination treatment on the interfacial chemistry of $LiNi_{0.9}Co_{0.05}Mn_{0.05}O_2$ material. Their research found that the fluorinated NCF_2 sample exhibited a more uniform and denser distribution of LiF and NiF_3^- within the CEI layer (Figure 11.16b), alongside a thinner interfacial layer, suggesting that moderate fluorination effectively improves CEI quality and delays battery aging.

On the anode side, Lu et al. [25] utilized TOF-SIMS to comparatively study the CEI formed on NMC811 surfaces in LiTFSI versus $LiPF_6$ electrolyte systems (Figure 11.16c). Depth profiling analysis results revealed that the CEI formed in the LiTFSI system possessed a distinct bilayer structure: an inner LiF-rich inorganic layer and an outer layer comprising a negatively charged inorganic polymer containing N-S backbone chains. This unique structure not only provided the CEI with excellent mechanical stability but also significantly accelerated the Li^+ desolvation process, thereby greatly enhancing battery stability and performance under extreme cycling conditions. Yan et al. [26] utilized TOF-SIMS to reveal that μSi-PPG electrodes using PPG binder maintained a stable organic-inorganic bilayer SEI structure (Figure 11.16d), with minimal increase in decomposition products like C_2HO^- upon cycling. In contrast, the SEI in PAA-based electrodes exhibited continuous thickening of the organic layer, leading to structural degradation and performance decay, demonstrating that the PPG binder effectively suppresses repeated SEI formation and growth.

These studies collectively demonstrate that the precise chemical composition and spatial distribution data provided by TOF-SIMS form the cornerstone for systematic comparative analysis, thereby revealing interfacial evolution patterns and stability mechanisms. By systematically comparing the interfacial chemical states under different materials, electrolytes, or cycling conditions, researchers can accurately evaluate the strengths and weaknesses of various interfacial engineering strategies and establish structure-property relationships between interface characteristics and battery macroscopic performance, providing critical experimental evidence for the optimized design of interfaces.

11.3.3 Analysis of Electrode Surface Evolution

Precise control of the chemical composition and elemental distribution of electrode materials is key to improving energy density and cycle stability. In recent years, optimizing elemental distribution through surface engineering strategies (including coating construction, interface control, and bulk doping) has become a hot topic of research [27]. TOF-SIMS, with its unique chemical sensitivity and three-dimensional spatial resolution capabilities, provides a key characterization tool for observing the microstructural evolution of artificially modified interface coatings.

In terms of cathode interface modification, TOF-SIMS provides crucial evidence for elucidating the interface stabilization mechanisms of various functional coatings. Liang et al. [28] used TOF-SIMS to reveal the interfacial stabilization mechanism of the $La_2Mo_2O_7$ modification layer on a nickel-based cathode (L-NCM). The TOF-SIMS depth profiling analysis showed that, compared with unmodified P-NCM, the signal intensities of interfacial byproducts in the L-NCM sample (e.g., PO_2^-, C_2HO^-) with significantly lower signal intensities and thinner distribution gradients of transition metal dissolution-related fragments such as MnF_3^-, confirming that the LMO-modified layer effectively inhibited electrolyte decomposition (Figure 11.17a). 3D chemical imaging further captured the Li^+ near-surface region with stable distribution characteristics, providing direct evidence for the elemental distribution strategy. Du et al. [29] used TOF-SIMS depth profiling and imaging to discover that the $AlPO_4$-Li_3PO_4 composite coating formed an enriched layer (PO_3^-, PO_2^-, Al^-) on the surface of the single-crystal nickel-rich cathode, with trace amounts of $Al3^+$ diffusing into the near-body phase, effectively reducing the concentration of interfacial side reaction products and the thickness of the degradation layer, thereby verifying its ion anchoring and interfacial stabilization effects (Figure 11.17b).

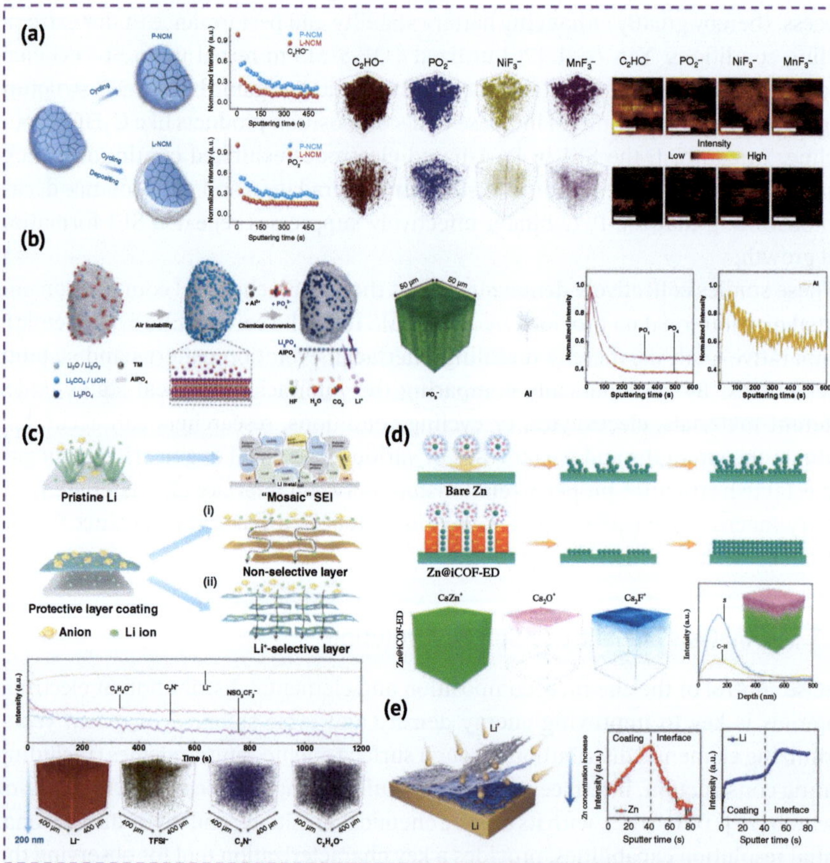

Figure 11.17 (a) TOF-SIMS analysis of the $La_2Mo_2O_7$ modified layer on the surface of a nickel-based cathode. Adapted from [14]. (b) TOF-SIMS analysis of the phospholipid layer on the surface of a single-crystal nickel-rich cathode. Adapted from [15]. (c) TOF-SIMS depth distribution curves and 3D images of Li^+ transport processes with and without protective layers and PCN-PEO protective layers. Adapted from [16]. (d) Zn@iCOF-ED surface deposition process and three-dimensional images of TOF-SIMS of the SEI layer on the anode surface. Adapted from [17]. (e) Schematic of Li deposition on Li metal anode and TOF-SIMS sputtering profile of MLD gradient coating. Adapted from [18].

In the study of anode interface modification, TOF-SIMS similarly demonstrates its precise analytical capability for interfacial chemical processes. Chen et al. [30] focused on the challenge of dendrite suppression in high-energy-density lithium metal anodes, using TOF-SIMS to reveal the Li^+ selective conduction mechanism of the carbon nitride-based (PCN-PEO) protective layer (Figure 11.17c). Depth profiling analysis revealed the $TFSI^-$ anion signal rapidly decays with increasing sputtering depth, while the Li^+ signal continues to strengthen, confirming that the protective layer effectively blocks electrolyte penetration and guides ordered Li^+ transport. 3D imaging further shows that the spatial distribution of C_3N^- and Li^+ highly matches, indicating that the microporous structure of carbon nitride forms low-twist Li^+ channels, promoting uniform lithium deposition. Yang et al. [31] utilized TOF-SIMS in their study of zinc metal anodes to confirm that the iCOF-ED coating uniformly covers the electrode surface (Figure 11.17d), significantly reducing the signal intensity of byproducts in the SEI (such as O, F, ZnO^-, and $Zn(OH)_2^-$), thereby inhibiting hydrogen evolution reactions and the accumulation of corrosion products. Similarly, Sun et al. [32] further highlights the advantage of TOF-SIMS in characterizing the artificial modification layers. Depth profiling analysis clearly revealed the thickness difference between a nanoscale gradient coating (Li@Gr10) and an inverse gradient coating (Li@RGr10) on the lithium metal anode, while also showing bulk phase changes of elemental Zn in both coatings, confirming the capability of molecular layer deposition technology for precise control over coating composition.

Collectively, these studies demonstrate that TOF-SIMS can multi-dimensionally characterize the chemical composition distribution, structural features, and ion transport behavior of various surface modification layers. It provides crucial technical support and a powerful research methodology for understanding interfacial stabilization mechanisms and optimizing electrode interface engineering.

11.4 Summary and Prospects

DEMS has the ability to synchronize electrochemical mass spectrometry signals with high accuracy, making it an indispensable in-situ analysis tool in lithium battery research. By real time quantitative monitoring of the gas release behavior during the electrochemical reaction process, DEMS can observe interface reaction phenomena that are difficult to observe using traditional methods, and can further analyze the reaction mechanism. It becomes an important characterization method for the research of key materials in lithium batteries. Currently, the gas production research of DEMS not only monitors the differences in oxidation stability of various electrolytes at high voltages, the oxygen evolution reaction of the positive electrode under high pressure, and the gas release behavior during the formation of the negative electrode SEI, but also can accurately track the sources of various gases when combined with isotope labeling technology. This further deepens the understanding of the interface reaction mechanism.

Currently, the DEMS technology is developing towards higher spatial and temporal resolution, a wider detection range, and deeper integration with multimodal characterization techniques (such as DEMS and Raman, DEMS and XRD, etc.). It is developing high-throughput gas production detection devices during the activation process of pouch batteries to improve battery performance, safety, and production efficiency. In addition, the key to the development of emerging solid-state batteries, lithium metal anodes, and lithium-sulfur systems is to analyze their main interface failure mechanisms. The use of gas production data can quickly evaluate different improvement strategies. Therefore, building a complete in-situ gas production detection technology for lithium-ion batteries can establish a bridge between macroscopic electrochemical performance and microscopic reaction mechanisms, providing necessary theoretical basis for the development of next-generation energy storage technologies.

As understanding of the complexity and spatial heterogeneity of battery interfacial reactions deepens, relying solely on time-resolved analysis of gaseous products has become increasingly insufficient for systematically resolving the chemical composition, structural distribution, and dynamic evolution of interfaces. Time-of-flight secondary ion mass spectrometry (TOF-SIMS), a surface analysis technique combining high spatial resolution with chemical sensitivity, is increasingly demonstrating its complementary value.

However, TOF-SIMS faces the challenges of matrix effect, lack of dynamic in-situ analysis, sputtering damage and complexity of data resolution, which limit its in-depth application in the study of dynamic evolution of nanoscale heterogeneous interfaces.

The interface of secondary batteries evolves continuously during the charge/discharge process, including lithium dendrite growth, SEI rupture and remodeling, etc. However, traditional TOF-SIMS can only provide a static "snapshot" of the interface after cycling, and is unable to track the electrochemically-driven dynamic processes in real time. Although in-situ TOF-SIMS based on vacuum coupled electrochemical cells has been developed in recent years, its temporal resolution (usually on the order of minutes) is still insufficient to capture millisecond-scale interfacial reactions (e.g., nucleation at the early stage of lithium deposition) due to the compatibility between the battery cycling environment (e.g., liquid electrolyte volatilization, gas by-product release) and the UHV conditions.

SEI and CEI layers usually consist of multilayered heterogeneous structures (e.g., an outer organic phase, an inner inorganic phase) with thicknesses of only a few nanometers to tens of nanometers and a gradient compositional distribution with depth. The detection depth (1–2 nm) of TOF-SIMS, although capable of capturing information on the most superficial layers, requires the revelation of subsurface compositions by argon ion sputtering. However, high-energy sputtering ions (e.g., 1 keV Ar^+) may damage fragile organic components (e.g., polyethylene oxide) or initiate interfacial chemical reactions (e.g., decomposition of $LiPF_3^-$), resulting in analytical results that deviate from the true interfacial structure. In addition, compositional mixing during sputtering (e.g., migration of Li+ from the bottom layer to the surface layer) may blur the original compositional distribution.

After the disassembly of liquid batteries, trace amounts of electrolyte or its decomposition products (e.g., HF, POF_3 generated by the hydrolysis of $LiPF_6$) often remain on the surface of the electrodes, and these contaminants may be volatilized and redeposited on the sample surfaces in the vacuum chamber of the TOF-SIMS, which may interfere with the extraction of the true interfacial signals. To minimize such contaminants, sample preparation is done in an inert atmosphere (e.g., argon glove box) and a vacuum transfer chamber is used or a freeze transfer technique is employed, which is complicated and may introduce other contaminants.

In order to enhance the application value and practical efficacy of TOF-SIMS in battery research, it can be further explored toward the following aspects in the future. Coupling of multimodal characterization techniques (multidimensional monitoring of the dynamic evolution of interfaces): Single-modal characterization techniques are limited by spatial and temporal resolution inconsistencies and lack of information dimensions, and face significant challenges in resolving the structure-performance evolution laws of battery materials under multi-physical field coupling conditions. While traditional TOF-SIMS can only provide static interface snapshots, the dynamic evolution of battery interfaces (e.g., SEI formation, electrolyte decomposition) requires time-resolved in-situ characterization, and the combination of TOF-SIMS and complementary characterization techniques can integrate the multidimensional information of surface chemistry, local structure, and bulk properties to form a complete chain of interfacial analyses, which can realize the relationship among the surface chemical distribution, crystal structure evolution, and interfacial properties of electrodes under dynamic operating. For example, TOF-SIMS is good at providing spatial information on the distribution of elements and molecules on surfaces, but has limited ability to resolve chemical states. By combining chemical state analysis with XPS, the chemical states of interfacial components can be precisely resolved. With the development of technology, in the future it is very promising to develop integrated coupling platforms, such as vacuum interconnection systems, to realize the in-situ sample transfer of TOF-SIMS, XPS, TEM, and other techniques, to avoid the contamination of sensitive interfaces (e.g., lithium metal) by atmospheric exposure, and to enhance the accuracy of data correlation.

References

1 S. Bruckenstein, R.R. Gadde, Use of a porous electrode for in situ mass spectrometric determination of volatile electrode reaction products, Journal of the American Chemical Society 93(3) (1971) 793–794. https://doi.org/10.1021/ja00732a049.

2 O. Wolter, J. Heitbaum, Differential Electrochemical Mass Spectroscopy (DEMS) — a new method for the study of electrode processes, Berichte der Bunsengesellschaft für physikalische Chemie 88(1) (2010) 2–6. https://doi.org/10.1002/bbpc.19840880103.

3 M. Holzapfel, A. Würsig, W. Scheifele, J. Vetter, P. Novák, Oxygen, hydrogen, ethylene and CO_2 development in lithium-ion batteries, Journal of Power Sources 174(2) (2007) 1156–1160. https://doi.org/10.1016/j.jpowsour.2007.06.182.

4 S. Kim, H.S. Kim, B. Kim, Y.J. Kim, J.W. Jung, W.H. Ryu, In situ gas analysis by differential electrochemical mass spectrometry for advanced rechargeable

batteries: A review, Advanced Energy Materials 13(39) (2023) 2301983. https://doi.org/10.1002/aenm.202301983.

5 M. Lanz, P. Novák, DEMS study of gas evolution at thick graphite electrodes for lithium-ion batteries: The effect of γ-butyrolactone, Journal of Power Sources 102 (1–2) (2001) 277–282. https://doi.org/10.1016/s0378-7753(01)00826-6.

6 P.P. Bawol, P. Reinsberg, C.J. Bondue, A.A. Abd-El-Latif, P. Königshoven, H. Baltruschat, A new thin layer cell for battery related DEMS-experiments: The activity of redox mediators in the Li–O_2 cell, Physical Chemistry Chemical Physics 20(33) (2018) 21447–21456. https://doi.org/10.1039/c8cp03592j.

7 Z. Zeng, X. Liu, X. Jiang, Z. Liu, Z. Peng, X. Feng, W. Chen, D. Xia, X. Ai, H. Yang, Y. Cao, Enabling an intrinsically safe and high-energy-density 4.5 V-class Li-ion battery with nonflammable electrolyte, InfoMat 2(5) (2020) 984–992. https://doi.org/10.1002/inf2.12089.

8 H. Wang, E. Rus, T. Sakuraba, J. Kikuchi, Y. Kiya, H.D. Abruña, CO_2 and O_2 evolution at high voltage cathode materials of li-ion batteries: A differential electrochemical mass spectrometry study, Analytical Chemistry 86(13) (2014) 6197–6201. https://doi.org/10.1021/ac403317d.

9 S.E. Renfrew, L.A. Kaufman, B.D. McCloskey, Altering surface contaminants and defects influences the first-cycle outgassing and irreversible transformations of $LiNi_{0.6}Mn_{0.2}Co_{0.2}O_2$, ACS Applied Materials & Interfaces 11(38) (2019) 34913–34921. https://doi.org/10.1021/acsami.9b09992.

10 J. Rana, J.K. Papp, Z. Lebens-Higgins, M. Zuba, L.A. Kaufman, A. Goel, R. Schmuch, M. Winter, M.S. Whittingham, W. Yang, B.D. McCloskey, L.F.J. Piper, Quantifying the capacity contributions during activation of Li_2MnO_3, ACS Energy Letters 5(2) (2020) 634–641. https://doi.org/10.1021/acsenergylett.9b02799.

11 S.E. Renfrew, B.D. McCloskey, Residual lithium carbonate predominantly accounts for first cycle CO_2 and CO outgassing of li-stoichiometric and li-rich layered transition-metal oxides, Journal of the American Chemical Society 139(49) (2017) 17853–17860. https://doi.org/10.1021/jacs.7b08461.

12 L.A. Kaufman, B.D. McCloskey, Surface lithium carbonate influences electrolyte degradation via reactive oxygen attack in lithium-excess cathode materials, Chemistry of Materials 33(11) (2021) 4170–4176. https://doi.org/10.1021/acs.chemmater.1c00935.

13 R. Bernhard, M. Metzger, H.A. Gasteiger, Gas evolution at graphite anodes depending on electrolyte water content and SEI quality studied by on-line electrochemical mass spectrometry, Journal of the Electrochemical Society 162(10) (2015) A1984–A1989. https://doi.org/10.1149/2.0191510jes.

14 L. Chen, J. Shu, Y. Huang, Z. Shi, H. Luo, Z. Liu, C. Shen, Engineering solid electrolyte interphase for the application of propylene carbonate solvent for graphite anode in low temperate battery, Applied Surface Science 598 (2022) 153740. https://doi.org/10.1016/j.apsusc.2022.153740.

15 B. Wu, C. Chen, D.L. Danilov, M. Jiang, L.H.J. Raijmakers, R.-A. Eichel, P.H.L. Notten, Influence of the SEI formation on the stability and lithium diffusion in Si electrodes, ACS Omega 7(36) (2022) 32740–32748. https://doi.org/10.1021/acsomega.2c04415.

16 X. Liu, T. Zhang, X. Shi, Y. Ma, D. Song, H. Zhang, X. Liu, Y. Wang, L. Zhang, Hierarchical sulfide-rich modification layer on SiO/C anode for low-temperature Li-ion batteries, Advanced Science (Weinh) 9(20) (2022) e2104531. https://doi.org/10.1002/advs.202104531.

17 Y. Meng, M. Wang, J. Wang, X. Huang, X. Zhou, M. Sajid, Z. Xie, R. Luo, Z. Zhu, Z. Zhang, N.A. Khan, Y. Wang, Z. Li, W. Chen, Robust bilayer solid electrolyte interphase for Zn electrode with high utilization and efficiency, Nature Communications 15(1) (2024) 8431. https://doi.org/10.1038/s41467-024-52611-z.

18 Z. Cui, C. Liu, A. Manthiram, Enabling stable operation of lithium-ion batteries under fast-operating conditions by tuning the electrolyte chemistry, Advanced Materials 36(41) (2024) 2409272. https://doi.org/10.1002/adma.202409272.

19 A.-M. Li, Z. Wang, T.P. Pollard, W. Zhang, S. Tan, T. Li, C. Jayawardana, S.-C. Liou, J. Rao, B.L. Lucht, E. Hu, X.-Q. Yang, O. Borodin, C. Wang, High voltage electrolytes for lithium-ion batteries with micro-sized silicon anodes, Nature Communications 15(1) (2024) 1206. https://doi.org/10.1038/s41467-024-45374-0.

20 J.-W. Sim, R.-H. Lee, H.-K. Kim, J.K. Lee, J.R. Yoon, S.-H. Lee, Controlling dendrite formation for long-term cyclability and fast Li^+ diffusion kinetics in solid-state electrolyte $Li_7La_3Zr_2O_{12}$, Chemistry of Materials 35(16) (2023) 6538–6548. https://doi.org/10.1021/acs.chemmater.3c01658.

21 W. Zhang, F. Cheng, M. Chang, Y. Xu, Y. Li, S. Sun, L. Wang, L. Xu, Q. Li, C. Fang, M. Wang, Y. Lu, J. Han, Y. Huang, Surface-interspersed nanoparticles induced cathode-electrolyte interphase enabling stable cycling of high-voltage $LiCoO_2$, Nano Energy 119 (2024) 191–204. https://doi.org/10.1016/j.nanoen.2023.109031.

22 B. Chu, R. Xu, G. Li, J. Chen, Z. Xu, T. Huang, B. Wang, A. Yu, Simultaneous B/W dual coating on ultra-high nickel single crystal cathode material for lithium-ion batteries, Journal of Power Sources 577 (2023) 233260. https://doi.org/10.1016/j.jpowsour.2023.233260.

23 H. Wang, J. Dong, M. Wang, Y. Lu, H. Zhang, J. Liu, Y. Liu, N. Liu, N. Li, Q. Huang, F. Wu, Y. Su, L. Chen, Synergistic surface restructuring and cation mixing via ultrafast Joule heating enhancing ultrahigh-nickel cathodes for advanced lithium-ion batteries, Journal of Energy Chemistry 103 (2025) 371–382. https://doi.org/10.1016/j.jechem.2024.11.047.

24 C. Li, X. Zhang, Z. Yang, H. Lv, T. Song, S. Lu, Y. Zhang, T. Yang, F. Xu, F. Wu, D. Mu, Stable cycling of $LiNi_{0.9}Co_{0.05}Mn_{0.05}O_2$/lithium metal batteries enabled by synergistic tuning the surface stability of cathode/anode, Journal of Energy Chemistry 87 (2023) 342–350. https://doi.org/10.1016/j.jechem.2023.08.033.

25 Y. Lu, Q. Cao, W. Zhang, T. Zeng, Y. Ou, S. Yan, H. Liu, X. Song, H. Zhou, W. Hou, P. Zhou, N. Hu, Q. Feng, Y. Li, K. Liu, Breaking the molecular symmetricity of sulfonimide anions for high-performance lithium metal batteries under extreme cycling conditions, Nature Energy (10) (2024) 191–204. https://doi.org/10.1038/s41560-024-01679-4.

26 W. Yan, S. Ma, Y. Su, T. Song, Y. Lu, L. Chen, Q. Huang, Y. Guan, F. Wu, N. Li, "Shooting three birds with one stone": Bi-conductive and robust binder enabling low-cost micro-silicon anodes for high-rate and long-cycling operation, Energy Storage Materials 76 (2025) 2404360. https://doi.org/10.1016/j.ensm.2025.104140.

27 Y. Liu, C. Guo, G. Wu, W. Lv, R. Zhou, W. Qiu, Y. Zhou, Q. Xu, C. Xu, Y. Niu, Uniformly dispersed bismuth metal nano catalyst modified carbon cloth electrode for iron-chromium flow battery, Nano Research Energy 4(1) (2025) e9120135. https://doi.org/10.26599/nre.2024.9120135.

28 J. Liang, Y. Zhu, X. Li, J. Luo, S. Deng, Y. Zhao, Y. Sun, D. Wu, Y. Hu, W. Li, T.-K. Sham, R. Li, M. Gu, X. Sun, A gradient oxy-thiophosphate-coated Ni-rich layered oxide cathode for stable all-solid-state Li-ion batteries, Nature Communications 14(1) (2023) 146. https://doi.org/10.1038/s41467-022-35667-7.

29 Y.-H. Du, H. Sheng, X.-H. Meng, X.-D. Zhang, Y.-G. Zou, J.-Y. Liang, M. Fan, F. Wang, J. Tang, F.-F. Cao, J.-L. Shi, X.-F. Cao, Y.-G. Guo, Chemically converting residual lithium to a composite coating layer to enhance the rate capability and stability of single-crystalline Ni-rich cathodes, Nano Energy 94 (2022) 106901. https://doi.org/10.1016/j.nanoen.2021.106901.

30 Q. Chen, B. Gao, Z. Yang, Y. Li, Q. Zhai, Y. Jia, Q. Zhang, X. Gu, J. Zuo, L. Wang, T. Wang, P. Zhai, C. Yang, Y. Gong, Macroscopically uniform interface layer with Li^+ conductive channels for high-performance Li metal batteries, Nature Communications 15(1) (2024) 10045. https://doi.org/10.1038/s41467-024-54310-1.

31 L. Yang, Q. Ma, Y. Yin, D. Luo, Y. Shen, H. Dou, N. Zhu, R. Feng, Y. Kong, A. Yu, B. Cheng, X. Wang, Z. Chen, Construction of desolvated ionic COF artificial SEI layer stabilized Zn metal anode by in-situ electrophoretic deposition, Nano Energy 117 (2023) 108779. https://doi.org/10.1016/j.nanoen.2023.108799.

32 Y. Sun, C. Zhao, K.R. Adair, Y. Zhao, L.V. Goncharova, J. Liang, C. Wang, J. Li, R. Li, M. Cai, T.-K. Sham, X. Sun, Regulated lithium plating and stripping by a nano-scale gradient inorganic–organic coating for stable lithium metal anodes, Energy & Environmental Science 14(7) (2021) 4085–4094. https://doi.org/10.1039/d1ee01140e.

12

Future Directions in Advanced Characterization of Battery Materials

12.1 In-situ Characterizations

In recent years, remarkable progress has been achieved in the development of ex-situ characterization techniques for battery materials. Although ex-situ characterization can provide valuable information about electrode, electrolyte, and interfacial properties, it requires disassembling the cell after charging, discharging or cycling to a specific state and subsequently extracting the components (such as cathode, anode, or electrolyte) for analysis. Consequently, the continuous evolution of these components from their pristine state to the measured state remains unclear. For example, detailed structural changes that occur during synthesis process or battery cycling cannot be effectively captured by ex-situ characterization. Furthermore, due to the high sensitivity of those components to air and moisture, electrochemical information such as valence-state variation, surface reactions, and interfacial evolution obtained from ex-situ characterization does not fully reflect the true processes occurring inside the working cell. This limitation introduces uncertainties and delays in data interpretation, thereby restricting the ability of such characterization to guide practical battery development. Therefore, acquiring information under real operating conditions is essential for advancing lithium/sodium-ion battery research.

To address the growing demand for more comprehensive characterization approaches, many recent studies have shifted from conventional ex-situ techniques toward high-resolution in-situ characterization methods. In-situ characterization enables simultaneous observation and analysis of materials during reactions or processing, providing real time and dynamic insights into chemical, compositional, structural, and morphological evolution. Depending on the specific technique, in-situ characterization can achieve various levels of temporal, spatial, and energy resolution. By strategically selecting appropriate in-situ characterization, researchers can obtain multidimensional, real time information on the materials evolution mechanisms of lithium/sodium-ion batteries.

Recently, several advanced in-situ characterization techniques have achieved significant progress in the field of lithium/sodium-ion battery. Among them, in-situ X-ray diffraction (XRD) has become a powerful tool for investigating electrochemical processes, widely employed to study structural evolution and the underlying mechanisms of capacity degradation. This technique can provide a wealth of valuable information during electrochemical cycling, such as changes in lattice parameters, unit cell volume variations, and phase transitions. These insights enable the establishment of correlations between electrochemical performance and structural evolution, thereby offering guidance for material optimization and design. As illustrated in Figure 12.1, Li et al. [1] utilized in-situ XRD to analyze the electrochemical and structural evolution of lithium-rich materials during the first cycle, confirming the structural stability of the designed material.

In-situ X-ray absorption spectroscopy (XAS) employs tunable photon energies to monitor X-ray absorption, thereby enabling the determination of the electronic structure of materials at the atomic and molecular levels. This technique is commonly used to investigate the electronic configuration of specific absorbing atoms and their local coordination environments in battery materials. By analyzing in-situ XAS, valuable information can be obtained to elucidate and verify electrochemical reaction mechanisms. For instance, the technique allows direct observation of changes in transition metal oxidation states and local electronic structures during electrochemical cycling. As shown in Figure 12.2, Wang et al. [2] utilized this technique to construct a three-dimensional nanotomography based on variations

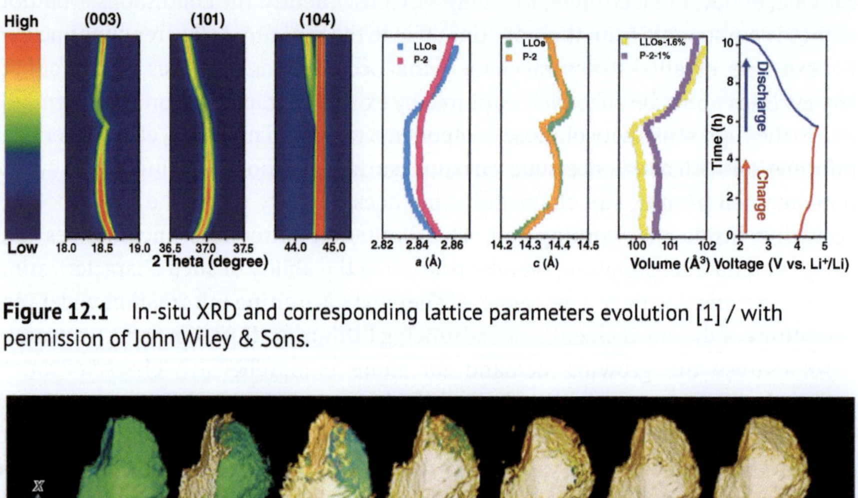

Figure 12.1 In-situ XRD and corresponding lattice parameters evolution [1] / with permission of John Wiley & Sons.

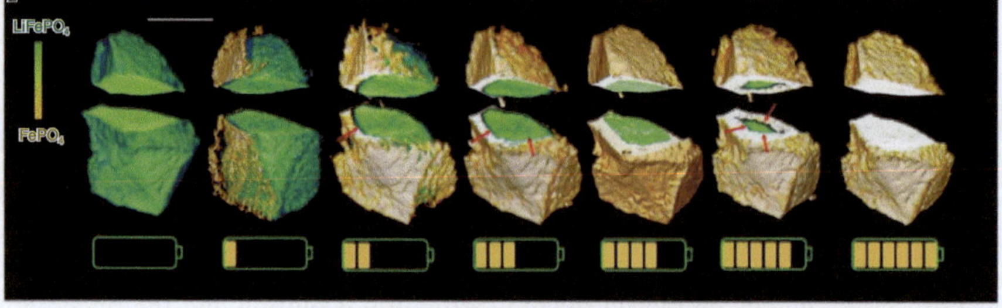

Figure 12.2 Phase transformation behavior of $LiFePO_4$ [2] / Springer Nature / CC BY 4.0.

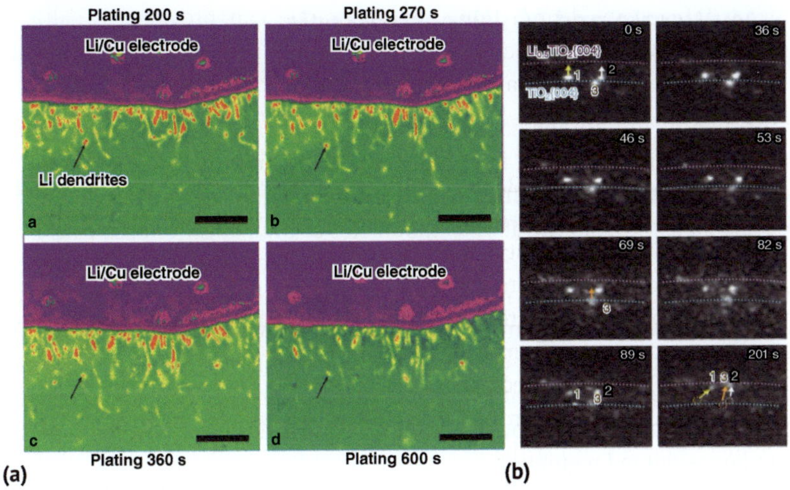

Figure 12.3 (a) A series of SEM images of the lithium plating process [3] / with permission of John Wiley & Sons. (b) Lithiation of TiO_2 particles through single-phase pathway [4] / with permission of John Wiley & Sons.

in valence states of Fe among different chemical phases ($LiFe^{2+}PO_4$ *vs.* $Fe^{3+}PO_4$), directly revealing the phase transformation behavior of $LiFePO_4$ during the charging process.

In-situ electron-based characterization, such as in-situ scanning electron microscopy (SEM), rely on the interaction between electron beams and solid samples to monitor the evolution of morphological and compositional features in real time. Zhang et al. [3] employed in-situ SEM to directly observe the dynamic growth process of lithium dendrites (Figure 12.3a), providing critical insights for the research of lithium metal anodes. Meanwhile, in-situ transmission electron microscopy (TEM) has emerged as one of the most powerful tools for investigating the microstructure and structural evolution of electrode materials. As illustrated in Figure 12.3b, Mao et al. [4] demonstrated that the lithiation process of α-TiO_2 nanoparticles follows a single-phase solid-solution reaction pathway through in-situ TEM.

In addition, several other in-situ techniques, such as atomic force microscopy (AFM), transmission X-ray microscopy, and X-ray photoelectron spectroscopy, enable precise investigation of nucleation and growth processes occurring at electrode interfaces. Meanwhile, in-situ Raman spectroscopy and in-situ nuclear magnetic resonance (NMR) spectroscopy provide direct insights into phase transitions and species evolution during lithiation and delithiation processes. The application of in-situ characterization techniques in lithium/sodium-ion batteries allows the dynamic, real time, and visual observation of chemical reactions and structural evolution within key components such as electrodes and electrolytes under realistic operating conditions. Moreover, these techniques effectively minimize contamination and oxidation that may occur during ex-situ measurements, thereby offering crucial guidance for the continued development of the entire batteries field.

12.2 Multiscale and Multispectroscopic Characterizations

A well-functioning lithium-ion battery with excellent electrochemical performance requires coordination among all its components across multiple length scales. The architecture of a lithium-ion battery is inherently hierarchical, spanning from the macroscopic meter scale to the atomic nanometer or even angstrom scale, as illustrated in Figure 12.4. Taking an electric vehicle power battery pack as an example, the pack consists of thousands of individual lithium-ion batteries (cells). At the meter scale, the battery management system plays a critical role in monitoring and maintaining the stable and safe operation of the entire system. Another essential subsystem, the thermal management system, ensures precise temperature control through techniques such as temperature prediction algorithms, zonal fuzzy control, and boundary-weighted regulation.

At the decimeter scale, each cell comprises four primary components: the cathode, separator, electrolyte, and anode. Among these, the cathode has the greatest variety of materials and the most complex structure, and therefore serves as a representative example here. The cathode consists of active materials, conductive additives, and polymer binders. At the micrometer scale, the active material (such as $LiCoO_2$, $LiFePO_4$, or NMC) is composed of numerous micrometer-sized particles. These particles may exist as single crystals or as polycrystalline secondary particles formed by the aggregation of nanosized primary crystallites. At the nanoscale or even angstrom level, lithium ions and transition metal ions are periodically and orderly arranged within the crystal lattice. For instance, in layered oxide cathodes, lithium and transition metal ions are alternately stacked in well-defined layers, forming the fundamental framework that governs ion transport and redox activity during electrochemical cycling.

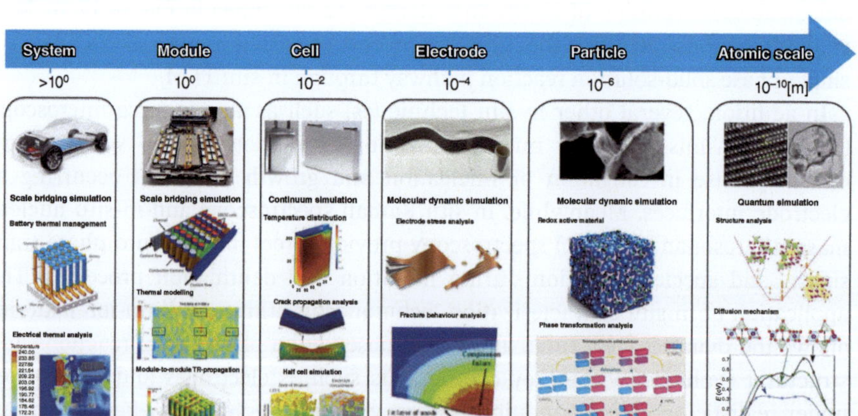

Figure 12.4 Multiscale components from system design to material properties of a battery pack [5] / with permission of John Wiley & Sons.

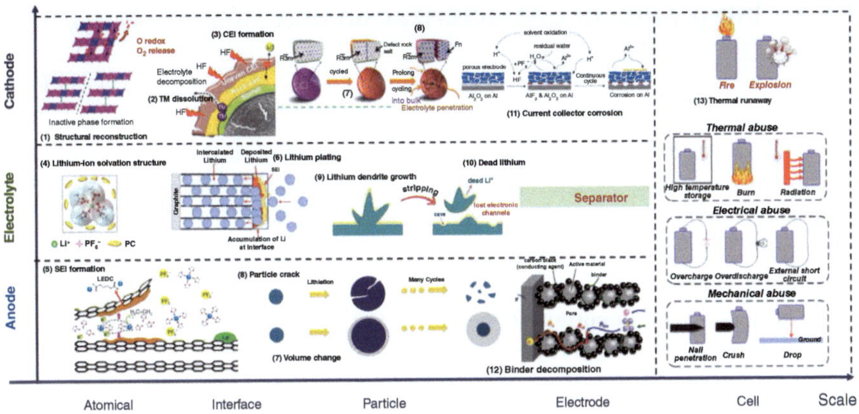

Figure 12.5 Multiscale degradation mechanism [5] / with permission of John Wiley & Sons.

At each of these scales, even subtle variations can significantly influence the overall performance of a battery pack. For instance, at the atomic scale, trace-element doping can markedly alter the electrochemical behavior of active materials, thereby affecting the driving range or fast-charging capability of power batteries. As shown in Figure 12.5, Yang et al. [5] illustrated the degradation mechanisms of lithium-ion batteries across scales ranging from the atomic level to the full cell level. An in-depth understanding of the reaction and degradation mechanisms is essential for optimizing high-performance electrode materials and ensuring the practical implementation of large-scale energy storage systems. Appropriate characterization techniques are thus crucial for effectively identifying the origins of degradation, elucidating underlying degradation mechanism, and guiding targeted material improvements or management system enhancement. Therefore, mastering multiscale and correlative characterization techniques is of paramount importance for advancing the research and development of lithium-ion batteries.

On the other hand, within the same length scale, multispectroscopic characterization techniques can be employed to analyze the same material parameter, each providing complementary insights. For instance, the crystalline structure of cathode materials can be investigated using XRD, Spherical Aberration-Corrected Transmission Electron Microscope (AC-TEM), neutron diffraction (ND), or NMR. The morphological features of materials can be examined through SEM, TEM, or AFM. To probe surface and interfacial chemistry, spectroscopic techniques such as Raman spectroscopy and Fourier transform infrared (FT-IR) spectroscopy are commonly used. Furthermore, the study of ion diffusion behavior within materials can be conducted using ND, NMR, or AFM. These examples demonstrate that advanced characterization techniques have been extensively applied in the field of lithium-ion batteries and can yield significant insights across diverse research topics. Therefore, the integration of multiple spectroscopic and microscopic methods provides a more comprehensive understanding of the intrinsic mechanisms of lithium-ion batteries, ultimately contributing to enhanced material optimization and electrochemical performance.

12.3 Integration of Characterization and Simulation

Research on lithium-ion batteries encompasses diverse topics and involves highly complex scientific and engineering challenges. With the rapid advancement of modern electrochemical analysis and characterization techniques, traditional experimental-based approaches have become increasingly capable of summarizing material properties and elucidating their effects on electrochemical performance. However, these methods still face inherent limitations, such as difficulty in controlling experimental variables and simultaneously accounting for multiple physical fields. Most importantly, they often fail to establish an intuitive link between material design and full-device performance evaluation.

On one hand, comprehensive evaluation of battery materials typically requires a large number of repetitive experiments. The intrinsic diversity of material properties further extends the overall optimization cycle of battery systems. On the other hand, the global performance of a battery and its influencing factors are closely coupled and dynamically interactive. For instance, electrochemical processes are strongly influenced by the structural design of the cell and its thermal environment. Traditional experimental-based techniques, however, struggle to capture such multivariate and hierarchical couplings among interdependent variables. Hence, developing new research paradigms capable of addressing multiphysics interactions has long been a critical pursuit in this field.

Fortunately, with the rapid development of computational technologies, numerical simulation has become an indispensable tool for analyzing and solving complex multiphysics problems, as well as for product prediction, design, and optimization. In the field of lithium-ion batteries, simulation techniques allow researchers to construct physical models that comprehensively capture the coupling mechanisms among various physical fields during operation, thereby providing theoretical guidance for optimizing battery design and performance. Multiscale numerical simulation methods have enabled the development of integrated computational platforms that encompass electrode fabrication processes, structural optimization, lifetime prediction, safety assessment, module construction, and system-level operation and maintenance. Such platforms make it possible to predict electrochemical behavior under the combined influence of multiple coupled physical fields under complex environmental conditions, including electrochemical, thermal, magnetic, fluid, and electric domains.

Taking electrochemical models as an example, since their application in lithium-ion batteries is closely related to the structural architecture of the cell, these models can be categorized, in terms of spatial scale, into three hierarchical levels: one-dimensional models describing individual material particles, two-dimensional models representing the electrode level, and three-dimensional models corresponding to the entire single cell. In addition, when classified by temporal scale, these models also span from short to long time domains, corresponding to the evolution of physical and electrochemical processes across multiple time scales, as illustrated in Figure 12.6.

At present, a wide variety of electrode materials have been developed for lithium-ion batteries, each exhibiting distinct microstructural and morphological features.

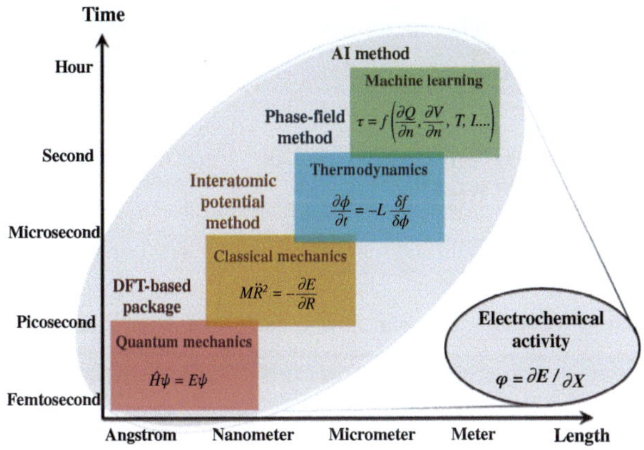

Figure 12.6 Multiple spatial and temporal scales of electrochemical modeling [6] / with permission of John Wiley & Sons.

The microstructure and morphology of electrode materials play a decisive role in determining the electrochemical performance of the entire device. For example, even for the same type of electrode material, variations in morphology often lead to significant differences in electrochemical performance and behavior. Therefore, controlling the morphology and microstructure of electrode materials has become a critical aspect of materials design. However, during the construction of electrochemical models, macroscopic homogeneous theory is generally adopted, whether based on one-dimensional, two-dimensional, or three-dimensional frameworks. This theory neglects the real pore architecture within the electrode, instead of treating the porous electrode as a superposition of two continuous phases including the solid electrodes and the liquid electrolyte. Such models primarily focus on macroscopic phenomena and commonly assume that both the electrode and the separator are homogeneous porous media. To account for lithium-ion diffusion within the active material, the solid phase is often idealized as spherical particles with fixed size. These assumptions deviate significantly from the complex and irregular microporous structures that exist in real electrodes and separators. Consequently, macroscopic models contribute only limited insight into transport mechanisms at the fine scale, as well as the design and optimization of mesoscale electrode microstructures.

To more accurately describe and analyze the influence of mesoscopic electrode and separator structures on multiphysical transport coupled with electrochemical reactions inside lithium-ion batteries, and to identify the key parameters governing performance, the development of advanced mesoscale models has become a major research focus in the numerical simulation of lithium-ion batteries. For example, as shown in Figure 12.7, Chen et al. [7] employed COMSOL Multiphysics simulations to elucidate the origin of the cycling stability improvement in sodium superionic conductor (NASICON)-coated single-crystal nickel-rich (SCNCM-LASTP) cathode materials. The simulation results revealed that, after cycling, the unmodified SCNCM exhibited a highly inhomogeneous lithium-ion concentration distribution and severe stress accumulation within the particles, which are prone to

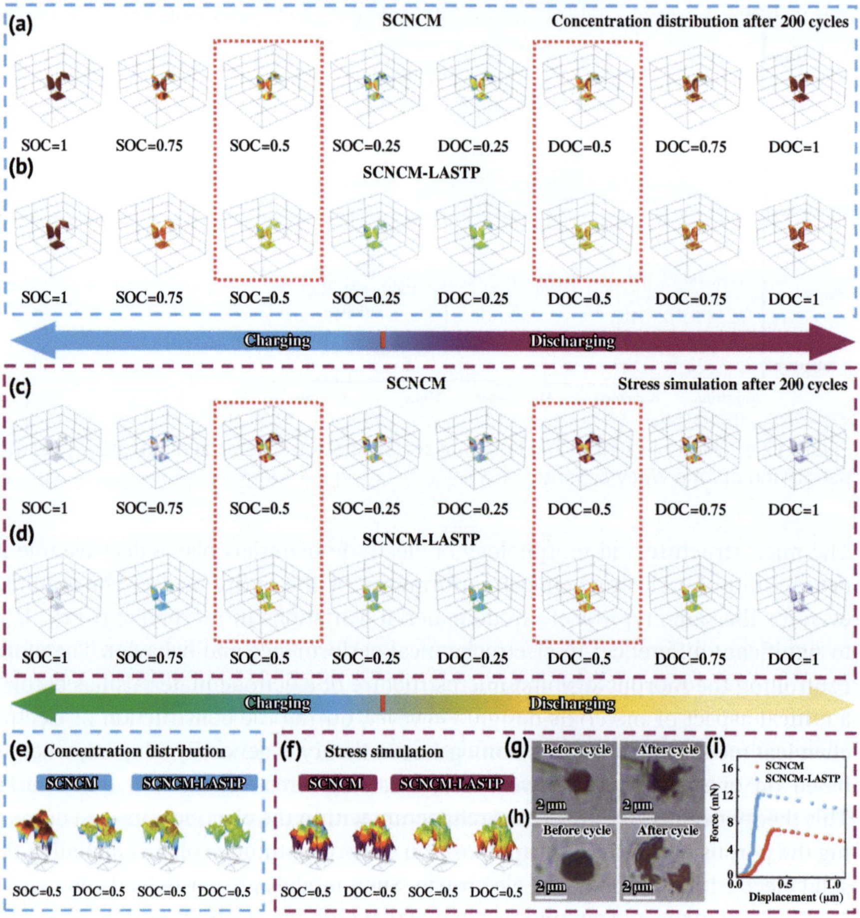

Figure 12.7 The simulation diagram with COMSOL software after 200 cycles [7] / with permission of John Wiley & Sons.

induce cracking. In contrast, SCNCM-LASTP demonstrated a uniform lithium-ion distribution and significantly reduced stress levels, leading to enhanced mechanical robustness. This improvement in mechanical stability effectively suppressed structural degradation, thereby substantially promoting the overall structural integrity and cycling stability of the material.

12.4 Artificial Intelligence in Materials Characterizations

In the context of the global energy transition, the traditional "trial-and-error" approach in battery research has reached a bottleneck due to the complexity of multiple variables and the explosion of data. With the continuous advancement of battery technology, efficiently and rapidly screening materials, optimizing manufacturing processes, and improving diagnostic accuracy have become key challenges. Artificial Intelligence (AI), as a powerful tool, accelerates material discovery, optimizes production workflows, and enhances performance predictions by processing vast amounts of experimental and simulation data, thus offering new solutions to overcome the performance bottlenecks in battery technology. AI-driven materials science heralds the dawn of a new era, particularly in fields such as energy, environment, and biomedicine, where AI can provide environmentally friendly, efficient, and people-beneficial technological solutions. AI will not only act as an accelerator for materials innovation but will also benefit from advancements in materials science, further enhancing AI's predictive capabilities. This feedback mechanism, where material innovation propels AI development and vice versa, will drive the symbiotic growth of both fields, paving the way for a future driven by advanced AI materials.

In materials science, AI extends beyond material discovery and design or performance prediction. AI also plays a critical role in material characterization and battery diagnostics. For example, through machine learning and deep learning algorithms, AI can extract key performance indicators from operational data of batteries, identify potential failure modes, and provide early warnings, thereby improving battery reliability and safety. Furthermore, AI can accelerate microstructural characterization, analyze complex multiphysical issues, and help researchers uncover intricate electrochemical processes, interfacial reactions, and structural changes within batteries. These AI-based characterization methods significantly enhance analytical precision, reduce experimental costs, and promote accurate predictions of battery lifespan and performance.

For instance, Gayon-Lombardo and colleagues [8] recently published an intriguing application based on GANs in the field of tomography imaging. They developed a GAN trained on tomography images (workflow illustrated in Figure 12.8), demonstrating that the GAN can reproduce electrode microstructures comparable to experimental results within seconds, once the model is trained. Given the high cost of obtaining experimentally reliable three-dimensional microstructures both in terms of time and resources, typically through focused ion beam SEM (FIB-SEM) or tomography scanning, this application may represent a significant breakthrough in the characterization and optimization of microstructures. Furthermore, this method allows for the generation of electrode microstructures of arbitrary size, which differ from their experimental counterparts. The authors showed that periodic boundary conditions could be applied, which is particularly relevant when these structures are used as inputs for electrochemical models.

As AI technology continues to mature, particularly in data processing and predictive modeling, AI will play an increasingly pivotal role in future battery development and materials science. Through intelligent characterization and diagnostic approaches, AI is expected to accelerate the innovation process in battery technologies and provide robust technological support for advancements in energy storage. The deep integration of AI and materials science will open up unprecedented research possibilities, driving sustainable global energy systems and ushering in revolutionary breakthroughs in future battery technologies.

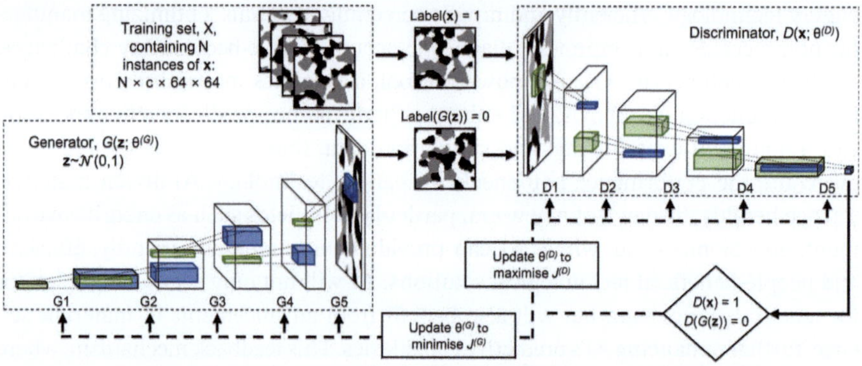

Figure 12.8 The schematic illustrates the architecture of the DC-generative adversarial networks (GAN) designed for processing 2D microstructural data. While extending this approach to 3D samples is conceptually straightforward, it is challenging to represent visually due to the requirement of illustrating 4D tensors. In each layer, the green sub-volume represents the convolutional kernel applied at an arbitrary location within the volume, while the blue sub-volume indicates the result of this convolution. The kernel maintains the same depth as the single nonspatial dimension, c, but it must scan across the two spatial dimensions to construct the image in the subsequent layer [8] / Springer Nature / CC BY 4.0.

References

1 C. Zheng, J. Zhang, H. Mao, Y. Wang, X. Yang, J. Li, D. Zhang, T. Feng, X. Wang, F. Kang, J. Li, Entropic design of anionic site to improve anionic redox stability in lithium-rich cathode, Advanced Materials 36(52) (2024) 2413785. https://doi.org/10.1002/adma.202413785.

2 J. Wang, Y.-c. Karen Chen-Wiegart, C. Eng, Q. Shen, J. Wang, Visualization of anisotropic-isotropic phase transformation dynamics in battery electrode particles, Nature Communications 7(1) (2016) 12372. https://doi.org/10.1038/ncomms12372.

3 G. Rong, X. Zhang, W. Zhao, Y. Qiu, M. Liu, F. Ye, Y. Xu, J. Chen, Y. Hou, W. Li, W. Duan, Y. Zhang, Liquid-phase electrochemical scanning electron microscopy for in situ investigation of lithium dendrite growth and dissolution, Advanced Materials 29(13) (2017) 1606187. https://doi.org/10.1002/adma.201606187.

4 L. Zhong, Y. Liu, W.-Q. Han, J.Y. Huang, S.X. Mao, In situ observation of single-phase lithium intercalation in sub-25-nm nanoparticles, Advanced Materials 29(26) (2017) 1700236. https://doi.org/10.1002/adma.201700236.

5 X. Liu, L. Zhang, H. Yu, J. Wang, J. Li, K. Yang, Y. Zhao, H. Wang, B. Wu, N.P. Brandon, S. Yang, Bridging multiscale characterization technologies and digital modeling to evaluate lithium battery full lifecycle, Advanced Energy Materials 12(33) (2022) 2200889. https://doi.org/10.1002/aenm.202200889.

6 W. Qiu, Y. Wang, J. Liu, Multiscale computations and artificial intelligent models of electrochemical performance in Li-ion battery materials, WIREs Computational Molecular Science 12(4) (2022) e1592. https://doi.org/10.1002/wcms.1592.

7 Y. Liu, J. Dong, Y. Guan, X. Wang, Z. Chen, M. Gao, S. Guo, K. Yan, Y. Lu, M. Wang, N. Li, Y. Su, F. Wu, L. Chen, In situ 3D conductive networks and interfacial bonding to stabilize oxygen vacancies for single-crystal Ni-rich cathodes, Advanced Materials 38(5) (2025) e15106. https://doi.org/10.1002/adma.202515106.

8 A. Gayon-Lombardo, L. Mosser, N.P. Brandon, S.J. Cooper, Pores for thought: Generative adversarial networks for stochastic reconstruction of 3D multi-phase electrode microstructures with periodic boundaries, npj Computational Materials 6(1) (2020) 82. https://doi.org/10.1038/s41524-020-0340-7.

Index

Note: Page numbers in *italics* and **bold** refers to figures and tables respectively.